STUDENT SOLUTIONS MANUAL

for use with

MAHAFFY | BUCAT | TASKER
KOTZ | MCMURRY | TREICHEL | WEAVER

CHEMISTRY

HUMAN ACTIVITY, CHEMICAL REACTIVITY

Prepared by Randy Dumont
McMaster University

NELSON EDUCATION

Student's Solution Manual for use with
CHEMISTRY: Human Activity, Chemical Reactivity
First Canadian Edition
Prepared by Randy Dumont

Vice President and Editorial Director:
Evelyn Vetch

Editor-in-Chief, Higher Education:
Anne Williams

Senior Marketing Manager:
Sean Chamberland

Developmental Editor:
My Editor Inc.

Senior Content Production Manager:
Anne Nellis

Proofreader:
Integra

Manufacturing Coordinator:
Ferial Suleman

Design Director:
Ken Phipps

Managing Designer:
Franca Amore

Printer:
Globus

For more information contact Nelson, 1120 Birchmount Road, Scarborough, Ontario M1K 5G4. Or you can visit our Internet site at www.nelson.com

Table of Contents

To the Student

This *Student's Solutions Manual* was prepared to assist the student in mastering the skills required for an understanding of chemistry. The selected questions have been chosen by the authors of your text to allow you to discover the range and depth of your understanding of chemical concepts. The importance of mastering the "basics" cannot be overemphasized. You will find that the text, *Chemistry: Human Activity, Chemical Reactivity*, First Canadian Edition, has a wealth of study questions to assist you.

Many of the questions contained in your book—and this solutions manual—have multiple parts. In many cases, comments have been added to aid you in the process of gathering available data and applicable conversion factors, and connecting them via fundamental concepts. Working multiple problems is one very good approach to clarifying and solidifying the fundamental concepts. I suggest that your *Student's Solutions Manual* will provide maximum benefit if you consult it *after* you have attempted to solve a problem.

This *Student's Solutions Manual* contains the worked-out solutions to in-chapter exercises, and the odd numbered solutions to the end-of-chapter review questions, summary questions, and conceptual questions that are in the textbook. Answers to the Interactive Exercises are provided, where applicable and where instructors felt necessary, online to students after completing these online questions, and are not included in this Manual.

Randy Dumont
McMaster University

Chapter 1
Human Activity, Chemical Reactivity

REVIEW QUESTIONS

Section 1.2

1.1
PDT or photodynamic therapy requires a photo sensitizer, light and oxygen. The photo sensitizer is a molecule that readily absorbs light, and transfers its extra energy to neighbouring triplet oxygen (the usual form of oxygen found in tissue and elsewhere). The transferred energy takes triplet oxygen to its singlet state. The very reactive singlet oxygen kills rapidly growing cancer cells.

1.3
The tumour must be located in a place that can be subjected to light. For example, Dolphin and collaborators developed porphyrin PDT for the treatment of skin cancers, for which an external light source is sufficient.

1.5
Toxicology is the study of the ill effects (toxicity) of substances on living organisms. Before introducing a porphyrin into the body for PDT, it must be established that the porphyrin, by itself, has little or no significant toxicity.

1.7
Chemotherapy is the use in medicine of substances that are selectively toxic to malignant cells or to a disease-causing virus or bacterium. As such, a vaccine would not be considered chemotherapy. Vaccines expose the body to substances that cause the body to make "antibodies" that fight particular disease-causing viruses or bacteria. The vaccine is administered before exposure to the virus or bacterium, and does not directly attack the disease-causing agent itself. The use of garlic to treat gangrene, on the other hand, is an example of chemotherapy. Garlic is a mild antiseptic which kills bacteria infecting tissue leading to gangrene

1.9
Yes, arsenic is generally considered to be toxic. However, Sec. 6.1 discusses how the toxicity of arsenic varies dramatically depending on the species containing the arsenic atoms. For example, whereas elemental arsenic is toxic, the arsenic containing species in lobster are not. This is why we can eat lobster with no ill effects.

1.11
Both Vitamin B12 and Visudyne are porphyrin-based.

Section 1.3

1.13
A natural product is a compound produced by a living organism.

1.15
(a) The Haber process combines hydrogen and nitrogen to make ammonia. Ammonia is used to make fertilizer.
(b) In the Bohr model, a hydrogen atom consists of an electron in a circular orbit about a proton. The angular momentum of the electron is a multiple of Planck's constant divided by 2π.
(c) A conical flask used in chemistry labs to carry out reactions.
(d) Van der Waals equation is a relation between the pressure, temperature and volume of a gas that accounts for the non-zero size of the gas molecules and the attractive forces between them.
(e) Gibbs free energy, $G = H - TS$, combines enthalpy and entropy to give a quantity which must decrease for any processes that actually happens.
(f) Lewisite is a chlorinated alkyl arsenic compound which was produced as a chemical weapon causing blisters and lung irritation.
(g) A Lewis base has a lone pair of electrons that it can donate to an electron pair acceptor—a Lewis acid.
(h) The Schroedinger equation determines the wave function that describes the state of an atom.

Section 1.4

1.17

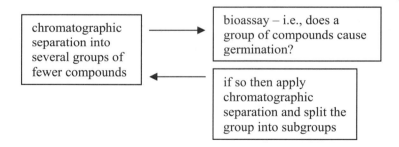

1.19
Organic is commonly used to label food grown without the use of chemical fertilizer, pesticides, hormones, genetic alteration, and related technologies. At a supermarket, one can buy "organic" apples and carrots. These foods contain water and salts which are not considered organic by organic chemists.

1.21
The chemical structure of the synthetic drugs is identical to those extracted from plants. It is the impurities associated with the synthetic procedure, in the case of synthetic drugs, and those associated with the biological source and the means of extraction, in the case of natural products, that are different.

1.23
Mycotoxins are toxins produced by fungi such as poisonous mushrooms. Penicillin is a mycotoxin that is toxic to bacteria, making it the original antibiotic. It is a natural product and an organic compound.

SUMMARY AND CONCEPTUAL QUESTIONS

1.25
Due to their sensitivity, porphyria sufferers would avoid sunlight like the vampires of legend. Other symptoms of porphyria include skin blisters, skin colour change, dark red urine, and mental disorders. Skin blisters on the neck might be interpreted as bite marks. Dark red urine might be thought to be associated with the drinking of blood. Mental disorders are consistent with the aberrant behaviour attributed to vampires.

1.27
Infrared light excites bends and vibrations of molecular bonds. Infrared spectra are used to determine the types of bonds present in a molecule.

Chapter 2
Building Blocks of Materials

IN-CHAPTER EXERCISES

Exercise 2.1—Classification of substances

Pure	Mixture	
Compounds	*Solutions*	
(h) testosterone (f) sodium chloride	(b) air (c) vinegar (g) athlete's urine sample	(a) mud (e) milk
Elements		
(d) gold		
Homogeneous	*Heterogeneous*	

Exercise 2.2—Thinking at different levels

(a) symbolic
(b) observable
(c) molecular

Exercise 2.3—Names and symbols of elements

(a) Na, Cl and Cr are the symbols for sodium, chlorine and chromium, respectively.
(b) Zn, Ni and K are the symbols for zinc, nickel and potassium.

Exercise 2.4—Difference in properties between a compound and its elements

Sucrose is a colourless molecular solid with a moderate melting temperature. Carbon is a black solid (graphite) or a hard crystalline material (diamond). Both are network solids. Hydrogen and oxygen are both colourless diatomic gases.

Exercise 2.5—Meaning of subscripts in a chemical formula

(a) In CO, there is one carbon atom for every oxygen atom (or the ratio of C to O atoms is 1:1).
(b) In CH_4, the C to H atom ratio is 1:4.
(c) In C_2H_2, the C to H atom ratio is 1:1.
(d) In $C_6H_{12}O_6$, C, H and O atoms are in proportion to 1:2:1.

Exercise 2.6—Meaning of chemical equations in words

$$6\,CO_2 + 6\,H_2O \longrightarrow C_6H_{12}O_6 + 6\,O_2$$

For every six carbon dioxide molecules that react, six water molecules react with them to form one glucose and six oxygen molecules.

Exercise 2.7—Meaning of chemical equations in words

$$O_3(g) + NO(g) \longrightarrow O_2(g) + NO_2(g)$$

Ozone and nitric oxide gases combine to form oxygen and nitrogen dioxide gases. The relative numbers of molecules of the four respective species are 1:1:1:1—i.e., the stoichiometric coefficients.

Exercise 2.8—Relative sizes of the nucleus and the atom

The atom is bigger than the nucleus by the factor, 100 pm/0.001 pm = 10^5. If the radius of the atom were 100 m, then the nucleus radius would be 100 m/10^5 = 0.001 m = 1 mm. Very small pebbles and small peas are this size.

Exercise 2.9—Atomic composition

(a) Iron has atomic number 26. Mass number = number of protons + number of neutrons = 26 + 30 = 56.
(b) Zinc has atomic number 30. Therefore, ^{64}Zn has 30 protons.
Number of neutrons = mass number – number of protons = 64 – 30 = 34.
Since it is an atom—rather than an ion—the number of electrons = number of protons = 30.

Exercise 2.12—Isotopes

(a) The three isotopes of argon are $^{36}_{18}Ar$, $^{38}_{18}Ar$ and $^{40}_{18}Ar$.
The percent isotopic abundances must add up to 100%.
∴ % abundance of argon 22 is 100.000 – 0.337 – 0.063% = 99.600%
(b) Gallium is element 31. All gallium isotopes have 31 protons.
^{69}Ga has 69 – 31 = 38 neutrons
^{71}Ga has 71 – 31 = 40 neutrons
The percent abundance of ^{71}Ga is 100.0 – 60.1% = 39.9%

Exercise 2.13 — Mass ratios of isotopes

(a) Table 2.2 shows the mass of ^{24}Mg to be 23.9850. The mass of ^{12}C is, by definition, 12.0000.

$\therefore$ the ratio of ^{24}Mg to ^{12}C is $\dfrac{23.9850}{12.0000} = 1.99875$

Note that the answer has the right number of significant digits.

(b) The mass of ^{26}Mg is 25.9826

$\therefore$ the ratio of ^{26}Mg to ^{12}C is $\dfrac{25.9826}{12.0000} = 2.16522$

(c) The ratio of ^{26}Mg to ^{24}Mg is $\dfrac{25.9826}{23.9850} = 1.08329$

Exercise 2.14 — Calculating atomic weight of an element

Atomic weight of chlorine $= \dfrac{75.77}{100} \times 34.96885 + \dfrac{24.23}{100} \times 36.96590 = 35.45$

Exercise 2.18 – Ratios of amounts and numbers of atoms

It is just the ratio of the numbers of moles, $\dfrac{2 \; \text{mol}}{0.2 \; \text{mol}} = 10$

Exercise 2.19 — Mass-amount conversions

(a) Mass of 1.5 mol of potassium, $m = n \times M = 1.5 \; \text{mol} \times 39.098 \, \text{g mol}^{-1} = 59 \, \text{g}$

(b) Amount, $n(Cu) = \dfrac{m}{M} = \dfrac{200 \, \text{g}}{63.55 \, \text{g mol}^{-1}} = 3.15 \, \text{mol}$

Number of Cu atoms $= 3.15 \; \text{mol} \times 6.022 \times 10^{23} \; \text{atoms mol}^{-1} = 1.90 \times 10^{24} \; \text{atoms}$

Exercise 2.20 — Relative numbers of atoms in samples

Amount, $n(Zn) = \dfrac{m}{M} = \dfrac{1.00 \, \text{g}}{65.39 \, \text{g mol}^{-1}} = 0.0153 \, \text{mol}$

We want the mass of 0.0153 mol of copper—same number of moles means same number of atoms.

Mass of 0.0153 mol of copper, $m = n \times M = 0.0153 \; \text{mol} \times 63.55 \, \text{g mol}^{-1} = 0.97 \, \text{g}$

REVIEW QUESTIONS

Section 2.3 Three levels of operation: observable, molecular, symbolic

2.27
The piece of table salt is the macroscopic view. The representation of its internal structure is the particulate view. The macroscopic view arises when the number of particles is very large—vast.

Section 2.6 Chemical reactions, chemical change

2.29
(a) The orange colour of bromine is a physical property.
(b) The rusting of iron in the presence of air and water is a chemical property.
(c) The exploding of hydrogen when ignited in air is a chemical property.
(d) The density of titanium metal is a physical property.
(e) The melting point of tin is a physical property.
(f) The colour of chlorophyll, a plant pigment, is a physical property.

2.31
(a) "colourless" is a physical property; "burns in air" is a chemical property
(b) "shiny" and "metal" are physical properties of aluminum; "orange" and "liquid" are physical properties of bromine; "aluminum reacts readily with … bromine" is a chemical property of aluminum and bromine

2.33
Mass number = number of protons + number of neutrons
(a) Magnesium has atomic number 12 = number of protons. Therefore, the mass number is 12 + 15 = 27.
(b) Titanium has atomic number 22 = number of protons. Therefore, the mass number is 22 + 26 = 48.
(c) Zinc has atomic number 30 = number of protons. Therefore, the mass number is 30 + 32 = 62.

2.35
mass number = 27 + 30 = 57, 27 + 31 = 58, and 27 + 33 = 60.
The symbols are $^{57}_{27}\text{Co}$, $^{58}_{27}\text{Co}$ and $^{60}_{27}\text{Co}$

2.37
Without using the exact masses of the ^{203}Tl and ^{205}Tl isotopes, we can deduce that ^{205}Tl is more abundant because the atomic weight of thallium, 204.4, is closer to 205 than 203. The atomic weight of thallium is a "weighted" average of 203 and 205 with the abundances providing the "weights."

2.39
Since we are only estimating (we are given values that are easily distinguished without an accurate calculation), we can use the values 106.9 and 108.9 for the two isotopic masses, and 107.9 for the atomic weight. At this level of accuracy, the atomic weight is exactly halfway between the two isotopic masses. The percentage abundance of ^{107}Ag is therefore 50%.

2.41

Symbol	^{65}Cu	^{86}Kr	^{195}Pt	^{81}Br
Number of protons	29	36	78	35
Number of neutrons	36	50	117	46
Number of electrons in the neutral atom	29	36	78	35
Name of element	copper	krypton	platinum	Bromine

Section 2.11 Amount of substance and its unit of measurement, the mole

2.43

Molecule ratio = mole ratio = $\dfrac{2 \text{ mol}}{1 \text{ mol}} = 2$

2.45

Molecule ratio = mole ratio = $\dfrac{0.1 \text{ mol}}{0.1 \text{ mol}} = 1$

2.47

Amount, $n(\text{Ba}) = \dfrac{m}{M} = \dfrac{137.3\,\text{g}}{137.3\,\text{g mol}^{-1}} = 1.000\,\text{mol}$

We want the mass of 5.000 mol of aluminium—five times the number of moles means five times the number of atoms.

Mass of 5.000 mol of aluminium, $m = n \times M = 5.000\,\text{mol} \times 26.98\,\text{g mol}^{-1} = 134.9\,\text{g}$

Section 2.12 The periodic table of elements

2.49
Two periods have 8 elements, two periods have 18 elements, and two periods have 32 elements. Though, the last 6 elements in the second 32 element period have not yet been synthesized.

2.51
(a) Bk is a radioactive element
(b) Br is a liquid at room temperature
(c) B is a metalloid.
(d) Ba is an alkaline earth metal.
(e) Bi is a group 5A element.

2.53
(a) Si is a metalloid, while P is a non-metal
(b) Si has some electrical conductivity, while P does not
(c) Both Si and P are solids at 25ºC

SUMMARY AND CONCEPTUAL QUESTIONS

2.55
Reviewing the periodic table:
(a) Any from beryllium, Be, to radium, Ra
(b) Any from sodium, Na, to argon, Ar
(c) carbon, C
(d) sulfur, S
(e) iodine, I
(f) magnesium, Mg
(g) krypton, Kr
(h) germanium, Ge, or arsenic, As.

2.57
The sample with the lowest atomic weight has the greatest amount (mol) and so has largest number of atoms. It is boron.

2.59

$$\text{Amount, } n(\text{Kr}) = \frac{m}{M} = \frac{0.00789 \text{ g}}{83.80 \text{ g mol}^{-1}} = 9.42 \times 10^{-5} \text{ mol}$$

Number of Kr atoms $= 9.42 \times 10^{-5} \text{ mol} \times 6.022 \times 10^{23} \text{ atoms mol}^{-1} = 5.67 \times 10^{19}$ atoms

2.61
(a) $A_r'(^1\text{H}) = 1$, $A_r'(^{11}\text{B}) = 11.0093/1.0078 = 10.9241$,
$A_r'(^{24}\text{Mg}) = 23.9850/1.0078 = 23.7994$, and $A_r'(^{63}\text{Cu}) = 62.9396/1.0078 = 62.4525$
(b) $A_r'(^1\text{H}) = 63 \times 1.0078/62.9396 = 1.0088$, $A_r'(^{11}\text{B}) = 63 \times 11.0093/62.9396 = 11.0199$,
$A_r'(^{24}\text{Mg}) = 63 \times 23.9850/62.9396 = 24.0080$, and $A_r'(^{63}\text{Cu}) = 63$
(c) $A_r'(^1\text{H}) = 1.0084$, $A_r'(^{11}\text{B}) = 11.0162$, $A_r'(^{24}\text{Mg}) = 24$, and $A_r'(^{63}\text{Cu}) = 62.9790$

2.63
$\delta^{13}\text{C} = 218248/20000002 = 0.0109124$ is consistent with a biogenic sample

2.65
We need the density of iron, (d), in order to determine the mass of 1 cm^3 of iron. Dividing this mass by the molar mass of iron, (b), determines the number of moles in 1 cm^3 of iron. Multiplying that number by Avogadro's constant, (c), provides the final answer—the number of atoms in 1 cm^3 of iron.

2.67
You could count how many jelly beans it takes to fill a smaller container—fewer jelly beans, easier to count—then multiply by the ratio of the volume of jar to that of the smaller container. If you only have the picture, you could count the number of jelly beans half way around the perimeter of the cylindrical jar, and from bottom to top of the jar. You have determined $\pi \times$ radius of jar and the height of the jar in jelly bean length units. Squaring the first quantity, then dividing by π and multiplying by the second quantity provides an estimate of the number of jelly beans in the jar.

Chapter 3
Models of Structure to Explain Properties

IN-CHAPTER EXERCISES

Exercise 3.1—Formulas of covalent network substances

BN is the formula of boron nitride.

Exercise 3.2—Ionic compounds

Sodium chloride is held together in a lattice by strong ionic bonds—attraction between positively and negatively charged ions. Silicon carbide is a network solid. The atoms are similarly held together in a lattice, but in this case it is by covalent bonds. Covalent bonds are generally stronger than ionic bonds. Silicon carbide is expected to have the higher melting point. This is in fact the case—silicon carbide decomposes at 2730 °C, while sodium chloride melts at 800 °C.

Exercise 3.3—Ions

(a) An S atom has 16 protons and 16 electrons. An S^{2-} ion has 16 protons and 18 electrons for a net charge of -2
(b) An Al atom has 13 protons and 13 electrons. An Al^{3+} ion has 13 protons and 10 electrons for a net charge of $+3$
(c) An H atom has 1 proton and 1 electron. An H^+ ion has 1 proton and no electrons for a net charge of $+1$

Exercise 3.4—Ions

N^{3-}, O^{2-}, F^-, Na^+, Mg^{2+}, and Al^{3+} ions have 7, 8, 9, 11, 12 and 13 protons, respectively. All of these species have 10 electrons, the same as the number of electrons in a neutral Ne atom (10 protons).

Exercise 3.6—Formulas of ionic compounds

(a) Equal numbers of Na^+ and F^- ions in NaF(s)
Twice as many NO_3^- ions as Cu^{2+} ions in $Cu(NO_3)_2$(s)
Equal numbers of Na^+ ions and CH_3COO^- in $NaCOOCH_3$(s)
(b) $FeCl_2$(s) and $FeCl_3$(s) are the compounds formed by Fe^{2+} and Fe^{3+}, respectively
(c) Na_2S, Na_3PO_4, BaS and $Ba_3(PO_4)_2$

Exercise 3.10—Molar masses of ionic compounds

(a) $M(\text{KBr}) = M(\text{K}) + M(\text{Br}) = 39.1 \text{ g mol}^{-1} + 79.9 \text{ g mol}^{-1} = 119.0 \text{ g mol}^{-1}$

(b) $M(\text{NH}_4\text{Cl}) = M(\text{N}) + 4 \times M(\text{H}) + M(\text{Cl}) = 14.0 + 4 \times 1.0 + 35.5 \text{ g mol}^{-1} = 53.5 \text{ g mol}^{-1}$

(c) $M(\text{Al(NO}_3)_3) = M(\text{Al}) + 3 \times (M(\text{N}) + 3 \times M(\text{O})) =$
$$27.0 + 3 \times (14.0 + 3 \times 16.0) \text{ g mol}^{-1} = 213 \text{ g mol}^{-1}$$

(d) $M(\text{Mg(HPO}_4)) = M(\text{Mg}) + M(\text{H}) + M(\text{P}) + 4 \times M(\text{O}) =$
$$24.3 + 1.0 + 31.0 + 4 \times 16.0 \text{ g mol}^{-1} = 120.3 \text{ g mol}^{-1}$$

Exercise 3.11—Molar masses of ionic compounds

(a) $M(\text{NaOH}) = M(\text{Na}) + M(\text{O}) + M(\text{H}) = 22.99 + 16.00 + 1.01 \text{ g mol}^{-1} = 40.00 \text{ g mol}^{-1}$

Amount, $n(\text{NaOH}) = \dfrac{m}{M} = \dfrac{1.00 \text{ g}}{40.00 \text{ g mol}^{-1}} = 0.0250 \text{ mol}$

(b) $M(\text{Fe(NO}_3)_2) = M(\text{Fe}) + 2 \times (M(\text{N}) + 3 \times M(\text{O})) = 55.85 + 2 \times (14.01 + 3 \times 16.00) \text{ g mol}^{-1} = 179.87 \text{ g mol}^{-1}$

Amount, $n(\text{Fe}(\text{NO}_3)_2) = \dfrac{m}{M} = \dfrac{0.0125 \text{ g}}{179.87 \text{ g mol}^{-1}} = 6.95 \times 10^{-5} \text{ mol}$

(c) $M(\text{CaCO}_3) = M(\text{Ca}) + M(\text{C}) + 3 \times M(\text{O}) = 40.08 + 12.01 + 3 \times 16.00 \text{ g mol}^{-1} = 100.09 \text{ g mol}^{-1}$
0.500 tonnes = 500 kg = 5×10^5 g

Amount, $n(\text{CaCO}_3) = \dfrac{m}{M} = \dfrac{5 \times 10^5 \text{ g}}{100.09 \text{ g mol}^{-1}} = 5 \times 10^3 \text{ mol}$

Exercise 3.14—Metallic substances

(a) electrons—solid magnesium is a metal
(b) electrons and to a lesser extent Mg^{2+} ions (they are less mobile)
(c) Mg^{2+} and Cl^- ions—ions are mobile in an ionic liquid
(d) Mg^{2+} and Cl^- ions—ions are mobile in an aqueous solution

Exercise 3.15—Molecular substances

(b) nitric oxide, NO(g), (d) ammonia, NH_3(g), and (e) sulfur, S_8(s) are all molecular substances Magnesium oxide and molten sodium chloride consist of ions, while solid magnesium is a metal consisting of ions and electrons bound in a lattice.

Exercise 3.16—Molecular substances

Silicon has a much higher melting point because it is a network solid. It does not consist of discrete molecules. Covalent bonding extends throughout a lattice giving the material exceptional strength. Sulfur consists of discrete S_8 molecules which are, in turn, held in a lattice by relatively weak bonds between the molecules. The strength of the covalent bonds between S atoms in the molecules does not affect the melting point.

Exercise 3.17—Relative molecular mass from high resolution mass spectrometry data

^{12}C, ^{16}O and ^{1}H are the most common isotopes of carbon, oxygen and hydrogen. The molar mass of the most abundant isotopologue, $^{12}C_2{}^1H_6{}^{16}O$, is $2 \times 12.00000 + 6 \times 1.00783 + 15.99491$ g mol^{-1} = 46.04189 g mol^{-1}. This is not in exact agreement with the experimental value. But, it is much closer to the experimental value than to the average value obtained using natural abundances, 46.06844 g mol^{-1}.

Exercise 3.18

(a) $M(CH_4) = M(C) + 4 \times M(H) = 12.011 + 4 \times 1.008$ g mol^{-1} = 16.043 g mol^{-1}

(b) $M(C_3H_6O) = 3 \times M(C) + 6 \times M(H) + M(O) = 3 \times 12.011 + 6 \times 1.008 + 15.999$ g mol^{-1} = 58.080 g mol^{-1}

(c) $M(C_6H_5NH_2) = 6 \times M(C) + 7 \times M(H) + M(N) = 6 \times 12.011 + 7 \times 1.008 + 14.007$ g mol^{-1} = 93.129 g mol^{-1}

Exercise 3.19

(a) $M(C_3H_7OH) = 3 \times M(C) + 8 \times M(H) + M(O) = 3 \times 12.011 + 8 \times 1.008 + 15.999$ g mol^{-1} = 60.096 g mol^{-1}

mass of propanol = $m = n(C_3H_7OH) \times M(C_3H_7OH) = (0.0255$ mol$) \times (60.096$ g mol$^{-1}) = 1.53$ g

(b) $M(C_{11}H_{16}O_2) = 11 \times M(C) + 16 \times M(H) + 2 \times M(O) = 11 \times 12.011 + 16 \times 1.008 + 2 \times 15.999$ g mol^{-1} = 180.247 g mol^{-1}

mass of $C_{11}H_{16}O_2 = m = n(C_{11}H_{16}O) \times M(C_{11}H_{16}O) = (0.0255$ mol$) \times (60.096$ g mol$^{-1}) = 4.60$ g

(c) $M(C_9H_8O_4) = 9 \times M(C) + 8 \times M(H) + 4 \times M(O) = 9 \times 12.011 + 8 \times 1.008 + 4 \times 15.999$ g mol^{-1} = 180.159 g mol^{-1}

mass of aspirin = $m = n(C_9H_8O_4) \times M(C_9H_8O_4) = (0.0255$ mol$) \times (60.096$ g mol$^{-1}) = 4.59$ g

Exercise 3.23—Molecular formula from mass spectrum

There are pairs of peaks separated by 2 in *m/z* value at 112 and 114, and 113 and 115. These pairs are consistent with a single chlorine atom with two isotopes ^{35}Cl and ^{37}Cl in roughly 3:1 abundance (75.8% vs. 24.2%). Subtracting 35 from 112 gives 77, which could correspond to C_6H_5. Therefore a reasonable molecular formula is C_6H_5Cl. Exact mass values corresponding to peaks at 112 and 114 would be given by
$6 \times 12.00000 + 5 \times 1.00783 + 34.968852 = 112.008002$ and
$6 \times 12.00000 + 5 \times 1.00783 + 36.965902 = 114.005052$, respectively.

Exercise 3.24 —Molecular formulas

$C_6H_5CHCH_2$ or C_8H_8
The first formula provides some information about the structure—a C_6H_5 "phenyl" group attached to an ethyl group, $CHCH_2$.

Exercise 3.25—Interpreting IR spectra

(a) C=O stretch—ketone, aldehyde or carboxylic acid
(b) C-N-H bend of an amine
(c) C≡N stretch of a nitrile
(d) C=O and O-H stretches of a carboxylic acid
(e) N-H and C=O stretches of an amide

REVIEW QUESTIONS

Section 3.2   Classifying substances by properties—An overview

3.27
(a) Iodine is molecular
(b) silicon dioxide is covalent network
(c) sulfur dioxide is molecular
(d) aluminium oxide is ionic (though it does have some covalent character)

Section 3.3   Covalent network substances

3.29
(a) Se^{2-} **(c)** Fe^{2+} and Fe^{3+}
(b) F^- **(d)** N^{3-}

3.31
(a) An ammonium ion, NH_4^+, has one more proton than electron—hence the $+1$ charge.
(b) A phosphate ion, PO_4^{3-}, has three more electrons than protons—hence the -3 charge.
(c) A dihydrogenphosphate ion, $H_2PO_4^-$, has one more electron than protons—hence the -1 charge.

3.33
Na_2CO_3 sodium carbonate
$BaCO_3$ barium carbonate
NaI sodium iodide
BaI_2 barium iodide

3.35

(a)	ClF_3 chlorine trifluoride	**(f)**	OF_2 oxygen difluoride
(b)	NCl_3 nitrogen trichloride	**(g)**	KI potassium iodide ionic
(c)	$SrSO_4$ Strontium sulfate ionic	**(h)**	Al_2S_3 aluminium sulfide or dialuminum trisulfide ionic—but with some covalent character because of the large +3 charge on aluminium and the large polarizability of sulfide (See Chapter 26)
(d)	$Ca(NO_3)_2$ calcium nitrate ionic	**(i)**	PCl_3 Phosphorus trichloride
(e)	XeF_4 xenon tetrafluoride	**(j)**	K_3PO_4 potassium phosphate ionic

Section 3.5   Metals and metallic substances

3.37
(iii) electrons are set free within a lattice of positive ions

Section 3.6   Molecular substances

3.39
A chemical reaction of oxygen requires breaking the oxygen-oxygen double bond, and generally forming other bonds. As such, the propensity of oxygen to react depends on the strength of the O=O bond, as well as the strength of bonds formed in the reaction. The boiling point of oxygen depends only on the strength of the weak intermolecular bonds between neighbouring oxygen molecules.

Section 3.8   Composition and formula by mass spectrometry

3.41   Mass spectrum and molecular formula
Only the calculated molecular mass, 28.03132, of C_2H_2 comes really close to this value. If the data were from a low resolution spectrometer, we could not rule out H_2CN (molecular mass = 28.01873), N_2 (molecular mass = 28.00614) and CO (molecular mass = 27.99491).

3.11   Connectivity: Evidence from infrared spectroscopy

3.43 Infrared spectroscopy and structure
(a) absorption frequencies in cm^{-1}:

2850-2980	due to alkyl C-H stretches
3000-3100	due to aromatic C-H stretches
2700-2850	due to aldehyde C-H stretch
1720-1740	due to aldehyde C=O stretch
1700-1725	due to carboxylic C=O stretch
675-900, 1400-1500 and 1585-1600	due to aromatic C-C bends and stretches

(b)

2850-2980	due to alkyl C-H stretches
1730-1750	due to ester C=O stretch
1640-1670	due to C=C stretch
1000-1300	due to ester C-O stretch

(c)

3200-3550	due to alcohol O-H stretch
2850-2980	due to alkyl C-H stretches
1705-1725	due to ketone C=O stretch
1000-1260	due to alcohol C-O stretch

3.45   Infrared spectroscopy and structure
(a) absorption frequencies in cm^{-1}:

2500-3300	broad peak due to carboxylic O-H stretch
3000-3100	due to aromatic C-H stretches
1700-1725	due to carboxylic C=O stretch
675-900, 1400-1500 and 1585-1600	due to aromatic C-C bends and stretches

(b)

3000-3100	due to aromatic C-H stretches
2850-2980	due to alkyl C-H stretches
1700-1725	due to ester C=O stretch
1000-1300	due to ester C-O stretch
675-900, 1400-1500 and 1585-1600	due to aromatic C-C bends and stretches

(c)

3200-3550	broad peak due to phenol O-H stretch
3000-3100	due to aromatic C-H stretches
2210-2260	due to nitrile C≡N stretch—not given in Table 3.5
675-900, 1400-1500 and 1585-1600	due to aromatic C-C bends and stretches

3.47
(a) $M(\text{Fe}_2\text{O}_3) = 2\times M(\text{Fe}) + 3\times M(\text{O}) = 2\times55.845 + 3\times15.9994 \text{ g mol}^{-1} = 159.688 \text{ g mol}^{-1}$

(b) $M(\text{BCl}_3) = M(\text{B}) + 3\times M(\text{Cl}) = 10.811 + 3\times35.453 \text{ g mol}^{-1} = 117.170 \text{ g mol}^{-1}$

(c) $M(\text{C}_6\text{H}_8\text{O}_6) = 6\times M(\text{C}) + 8\times M(\text{H}) + 6\times M(\text{O}) = 6\times12.0107 + 8\times1.00794 + 6\times15.9994 \text{ g mol}^{-1} = 176.1241 \text{ g mol}^{-1}$

3.49

$M((CH_3)_2CO) = 3 \times M(C) + 6 \times M(H) + M(O) = 3 \times 12.0107 + 6 \times 1.00794 + 15.9994$ g mol^{-1}
     $= 58.0791$ g mol^{-1}

1260 million kg $= 1.26 \times 10^{12}$ g

amount of acetone, $n = m((CH_3)_2CO) / M((CH_3)_2CO) = (1.26 \times 10^{12}$ g$) / (58.0791$ g mol$^{-1}) = 2.1 \times 10^{10}$ mol

$= 21$ million kMol

3.51

Each mol of $BeCl_2$ has 3 mol of ions (1 mol of Be^{2+} ions, and 2 mol of Cl^- ions).

$n(BeCl_2) = \dfrac{1.0 \text{ g}}{79.91 \text{ g mol}^{-1}} = 0.0125 \text{ mol}$ and $\Sigma n = 3 \times 0.0125 \text{ mol} = 0.0375 \text{ mol}$

Each mol of $MgCl_2$ also has 3 mol of ions. Since its molar mass is greater than that of $BeCl_2$, the amount (mol) of $MgCl_2$ must be less than that of $BeCl_2$, as must the total number of ions.

$n(MgCl_2) = \dfrac{1.0 \text{ g}}{95.21 \text{ g mol}^{-1}} = 0.0105 \text{ mol}$ and $\Sigma n = 3 \times 0.0105 \text{ mol} = 0.0315 \text{ mol}$

Each mol of CaS has 2 mol of ions: 1 mol of Ca^{2+} ions, and 1 mol of S^{2-} ions:

$n(CaS) = \dfrac{1.0 \text{ g}}{72.14 \text{ g mol}^{-1}} = 0.0139 \text{ mol}$ and $\Sigma n = 2 \times 0.0139 \text{ mol} = 0.0278 \text{ mol}$

Each mol of $SrCO_3$ has 2 mol of ions: 1 mol of Sr^{2+} ions, and 1 mol of CO_3^{2-} ions:

$n(SrCO_3) = \dfrac{1.0 \text{ g}}{147.63 \text{ g mol}^{-1}} = 0.0068 \text{ mol}$ and $\Sigma n = 2 \times 0.0068 \text{ mol} = 0.0136 \text{ mol}$

Each mol of BaO_4 has 2 mol of ions: 1 mol of Ba^{2+} ions, and 1 mol of SO_4^{2-} ions:

$n(SrCO_3) = \dfrac{1.0 \text{ g}}{233.4 \text{ g mol}^{-1}} = 0.0043 \text{ mol}$ and $\Sigma n = 2 \times 0.0043 \text{ mol} = 0.0086 \text{ mol}$

There are more ions in 1 g of $BeCl_2$ than in 1 g of any of the other compounds.

3.53

(a) $M(C_{18}H_{27}NO_3) = 18 \times M(C) + 27 \times M(H) + M(N) + 3 \times M(O)$
        $= 18 \times 12.0107 + 27 \times 1.00794 + 14.00674 + 3 \times 15.9994$ g mol^{-1} $= 305.4119$ g mol^{-1}

   55 mg $= 0.055$ g

(b) Amount of capsaicin $= n(\text{capsaicin}) = $ mass / (molar mass) $= (0.055$ g$) / (305.4119$ g mol$^{-1}) = 1.8 \times 10^{-4}$ mol

(c) 55 mg of capsaicin is 1.8×10^{-4} mol of capsaicin $\rightarrow$ contains $18 \times (1.8 \times 10^{-4}$ mol) carbon which has

Mass, $m = 18 \times (1.8 \times 10^{-4}$ mol$) \times (12.0107$ g mol$^{-1}) = 0.039$ g $= 39$ mg

3.55

(a) Volume of Ni foil = (0.550 mm) × (1.25 cm) × (1.25 cm)

$\qquad\qquad$ = (0.0550 cm) × (1.25 cm) × (1.25 cm) = 0.0859 cm^3

Mass = volume × density = 0.0859 cm^3 × 8.902 g/cm^3 = 0.765 g

Molar mass of nickel = 58.6934 g mol^{-1}

Amount of water = n(water) = mass / (molar mass) = (0.765 g) / (58.6934 g mol^{-1}) = 0.0130 mol

(b) 0.0130 mol of Ni produces 0.0130 mol of NiF$_a$

$\quad$ Ni $\qquad$ + $\qquad$ (a/2) F$_2$ $\quad\rightarrow\qquad$ NiF$_a$

$\quad$ 0.0130 mol $\qquad\qquad\qquad\qquad\qquad$ 0.0130 mol

Assuming 100% yield—that is, all nickel fluoride product formed is isolated—we can determine the molar mass of NiF$_a$

M(NiF$_a$) = mass / amount = (1.261 g) / (0.0130 mol) = 97.0 g mol^{-1}

So, $\quad M$(N) + a × M(F) = 97.0 g mol^{-1}

$\qquad$ 58.69 + a 19.00 = 97.0

$\qquad$ a = (97.0 − 58.69) / 19.00 = 2.02

The formula must be NiF$_2$—that a differs slightly from 2 is likely due to experimental error.

(c) Nickel (II) fluoride

SUMMARY AND CONCEPTUAL QUESTIONS

3.57
Sodium chloride is ionic. Its solid is held together by the strong attraction between oppositely charged ions. Chlorine is a molecular substance—it consists of Cl_2 molecules. In the solid, the molecules are held together by weak forces between the molecules, independently of the strength of the Cl-Cl bonds within the molecules.

3.59
(a) Silicon dioxide is a covalent network solid which melts at very high temperature.
Carbon dioxide is a molecular substance which melts (sublimes at 1 atm pressure) at very low temperature because the molecules re held together by weak forces between the molecules.
(b) Sodium sulfide is an ionic substance melting at high temperature.
Hydrogen sulfide is a molecular substance which melts at very low temperature.

3.61
(a) Calcium and chloride ions are in a 1:2 ratio in $CaCl_2$
(b) Calcium, carbon and oxygen atoms are in a 1:1:3 ratio in $CaCO_3$
(c) Nitrogen and hydrogen atoms are in a 1:3 ratio in NH_3
(d) Silicon and carbon atoms are in a 1:1 ratio in SiC
(e) Hydrogen and chlorine atoms are in a 1:1 ratio in HCl

3.63
Al^{3+} is most attracted to water because it has the largest magnitude charge. (It is also the smallest, so can get closer to the negatively charged parts of water molecules.)

3.64

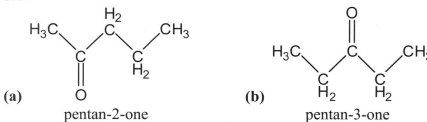

(a) pentan-2-one   **(b)** pentan-3-one
(b) C=O stretch at 1705-1725 cm^{-1}
(c) $^{12}C_5{}^1H_{10}{}^{16}O$ – molar mass = $5 \times 12.00000 + 10 \times 1.00783 + 15.99491$ g mol^{-1} = 86.07321 g mol^{-1}
(d) Pentan-2-one should produce peaks at $86 - 15 = 71$ and $86 - 43 = 43$ when it loses a CH_3 or $CH_2—CH_2—CH_3$ fragment.
Pentan-3-one should produce a peak at $86 - 29 = 57$ when it loses a $CH_2—CH_3$ fragment.
Therefore, the top spectrum corresponds to pentan-3-one—the bottom to pentan-2-one.

3.65
(a)

Clostebol has alkyl groups, an alkene, a ketone, a chloride and an alcohol group—the last four groups are circled.
(b) The testosterone molecular ion peak is at 288, whereas the clostebol molecular ion peak is at 322. There are peaks at both these positions, though the peak at 322 is smaller. It would appear that this mass spectrum does not rule out clostebol. However, chlorinated compounds generally show peaks associated with loss of a chlorine atom from the molecular ion. We therefore would expect to see a peak at 287. No such peak is seen. The peak at 322 must be due to some other compound. The data is consistent with the expert analysis.

3.67
Absorption frequencies in cm^{-1}
2850-2980   due to alkyl C-H stretches
2210-2260   due to nitrile $C\equiv N$ stretch—not given in Table 3.5
1705-1725   due to ketone C=O stretch
1640-1670   due to C=C stretch

3.69
(a) Look for the N-H stretch of $CH_3CH_2NHCH_3$—a peak around 3250-3400 cm^{-1} not seen for $(CH_3)_3N$.
(b) Look for the ketone C=O stretch of CH_3COCH_3—a peak around 1705-1725 cm^{-1}—or the OH stretch—a broad peak around 2500-3300 cm^{-1}.
(c) Look for the aldehyde C-H peaks (2 of them) of CH_3CH_2CHO, around 2700-2850 cm^{-1}.

Chapter 4
Carbon Compounds

IN-CHAPTER EXERCISES

Exercise 4.1—Forces of attraction in methane

Covalent forces hold the carbon and hydrogen atoms together within a methane molecule.

Exercise 4.2

Weak dispersion forces—a type of intermolecular or non-bonding force—hold neighbouring methane molecules together.

Exercise 4.3—Reaction of carbon compounds with hydroxyl radical

The CFCs do not have H atoms bonded to C. Such H atoms readily react with hydroxyl radicals.

Exercise 4.4—Temperature profile of the atmosphere

Most greenhouse gas absorption takes place in troposphere, the atmospheric layer next to the earth. As the visualization shows, the atmosphere is most dense close to the surface of the earth, so this is where most of the IR absorbing compounds are found.

Exercise 4.5—'Greenhouse' gases

Nitrous oxide absorbs strongly in regions of the IR spectrum emitted by the earth that carbon dioxide and water do not absorb. Also, molecules of N_2O have a longer average lifetime in the atmosphere than that of CO_2 molecules.

Exercise 4.6—Concentration and total mass

The amount of carbon = the amount of CO_2. There are 7.0×10^{16} mol of carbon in atmospheric CO_2. Using the molar mass of C, we get total mass of carbon in atmospheric CO_2 to be

$m(\text{C}) = (7.0 \times 10^{16} \,\text{mol}) \, (12.011 \text{g} \,\text{mol}^{-1}) = 8.4 \times 10^{17} \text{g}$

Exercise 4.7—Carbon dioxide storage

Today, I drank from a plastic cup. The plastic material was polymerized from monomers obtained from fractionation and cracking, etc., of petroleum extracted from fossil deposits. I am using a number of plastic products today—for example, the keyboard keys I am typing on, and the computer monitor I am viewing. Another fraction of the same petroleum was used to drive my car. Natural gas—from fossil deposits of mostly methane—is used to heat my house.

Exercise 4.8—Structural representations

Exercise 4.11—Units of unsaturation

There are 4 units of unsaturation: $14 - 6 = 8 = 4 \times 2$. There is unit of unsaturation for each of the double bonds plus one for the ring.

Exercise 4.12—Polymers

Teflon:

REVIEW QUESTIONS

Section 4.1 Ice on fire

4.13
(a) Covalent bonds hold the nitrogen atoms together in the N_2 molecules, inside the clathrate hydrate.
(b) Dispersion forces attract the N_2 and H_2O molecules within the clathrate hydrate cage.
(c) Hydrogen bonding holds neighbouring H_2O molecules together to make the clathrate hydrate cage.
(d) Dispersion forces attract the N_2 and O_2 molecules within the clathrate hydrate cage.

Section 4.3 Climate change and 'greenhouse' gases

4.15
A greenhouse prevents convection of heat—a key component of the heat flow from the earth's surface out into the atmosphere. Greenhouse gases do not prevent convection. However, like the greenhouse walls and roof, they prevent some of the radiation emitted by the earth from being lost to space. As shown in the visualization in "Think about it" e4.4, greenhouse gas molecules absorb some of the infrared radiation emitted from earth, causing the molecules to be vibrationally excited. They transfer their energy to surrounding molecules of substances such as N_2 and O_2 that are unable themselves to absorb infrared radiation. This process is called collisional de-excitation.

4.17
$CH_3CH_3(g)$; ozone, $O_3(g)$; and chloroform, $CHCl_3$
Stretching and/or bending bonds in these molecules changes the distribution of charge allowing the molecules to absorb light—an electromagnetic field.
He has no vibrations—it is a monatomic gas. Chlorine is a homonuclear diatomic gas—its vibration does not change its distribution of charge.

4.19
1750 is the beginning of the industrial revolution. Radiative forcing of our climate was subject only to natural fluctuations before that time.

4.21
Clouds cause cooling by reflecting incoming sunlight back out into space.

4.23
Consulting the ranking of levels of scientific uncertainty in Figure 2.20 of the 2007 IPCC summary report:
tropospheric carbon dioxide > tropospheric nitrous oxide > stratospheric ozone > jet contrails (we understand that its effect is very small) > solar irradiance

4.25

Stratospheric ozone has a negative radiative forcing effect. It absorbs UV light preventing it from reaching the earth's surface. The stratosphere is heated, but the surface cools.

4.27

Halocarbons are chemically most active in the stratosphere where they can photodissociate (absorb light, breaking chemical bonds) into reactive halogen atoms halo-alkyl radicals. It is these species that can deplete ozone in chain reactions.

4.29

The strong absorptions of methane, in comparison with carbon dioxide, in the IR spectrum emitted by the earth give methane a larger global warming potential.

Section 4.4 Carbon capture and storage

4.31

The advantages of using biofuels are as follows:
(1) They are renewable—we get a new crop each year.
(2) There is no net addition to atmospheric carbon dioxide. Atmospheric CO_2 is consumed as the plant grows. It is returned to the atmosphere when the fuel is combusted.
(3) Biofuel can be produced in most places—provided a suitable crop can be found. Fossil fuels are found only in certain places.
The disadvantages of the biofuel strategy, in practice, are as follows:
(1) Energy is needed to produce the crop and manufacture the biofuel. These energy demands must be minimized to make biofuel production worth the effort.
(2) Only a small portion of corn, for example, can be used to produce biofuels. The cellulosic part of the plant (its bulk), is simply ploughed back into the field. This significantly reduces the viability of biofuel as an alternative energy strategy, as currently produced in North America.
(3) Biofuel production takes land away from food production. This has driven up food prices in North America in recent years.

4.33

The principal difference between the data sets is the size of the annual oscillation in CO_2 level. The CO_2 level at Mauna Loa drops in the spring and summer months due to the growth of vegetation on its land mass and rises again when this growth stops. The Antarctic data oscillates less because of the lack of vegetation in the polar region.

Section 4.6 Polymers and unsaturated hydrocarbons

4.35
For example,
(a)

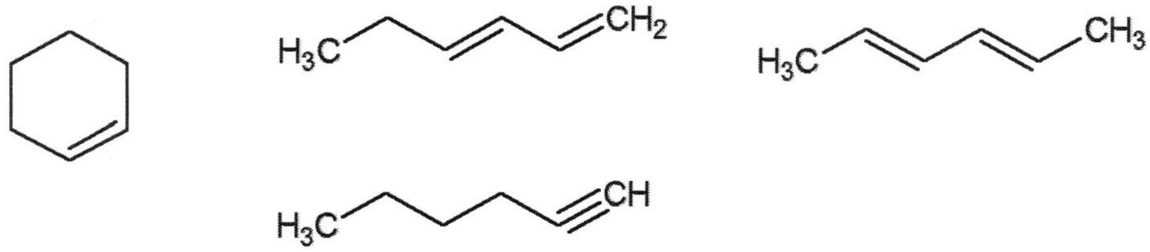

(b)

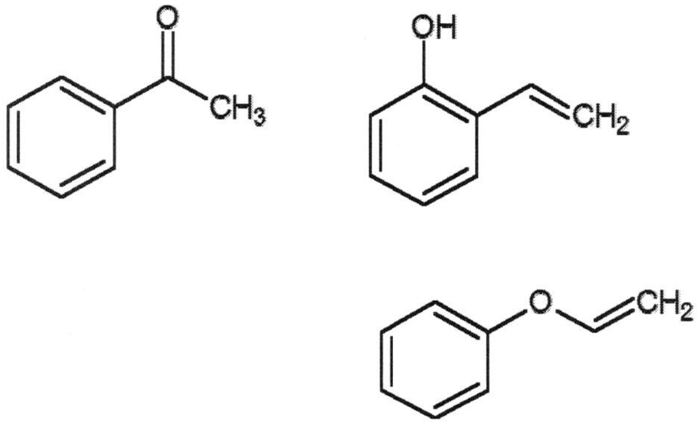

(c)

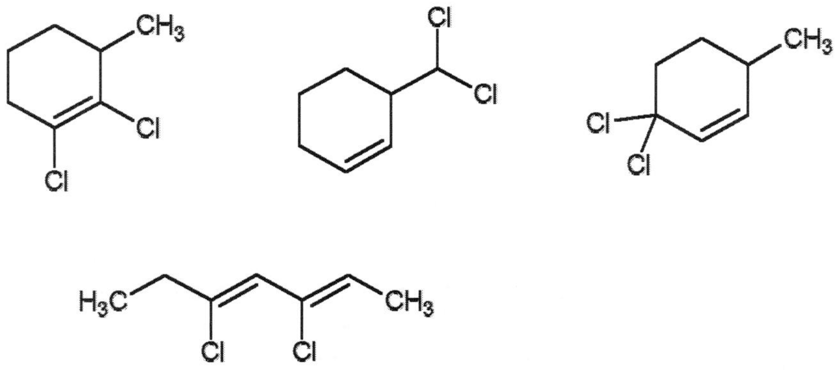

4.37
(a) Hexane has only the alkane functional group
(b) Hex-3-ene has the alkene and alkane functional groups
(c) Cyclohex-2-enol has the alkene, alkane, and alcohol functional groups
(d) Benzaldehyde has alkane, aromatic, and aldehyde functional groups

4.39

(a)

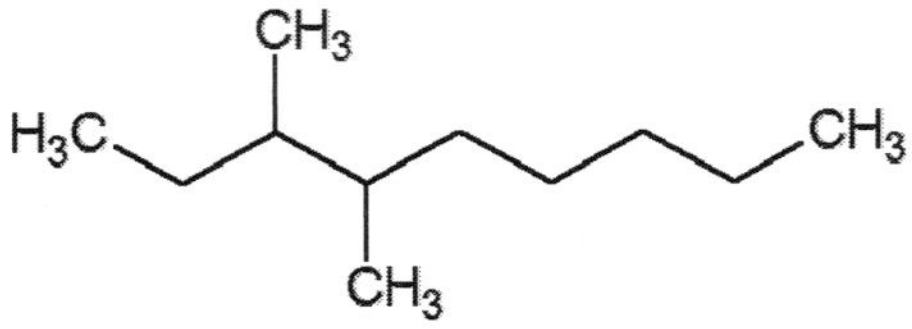

(b)

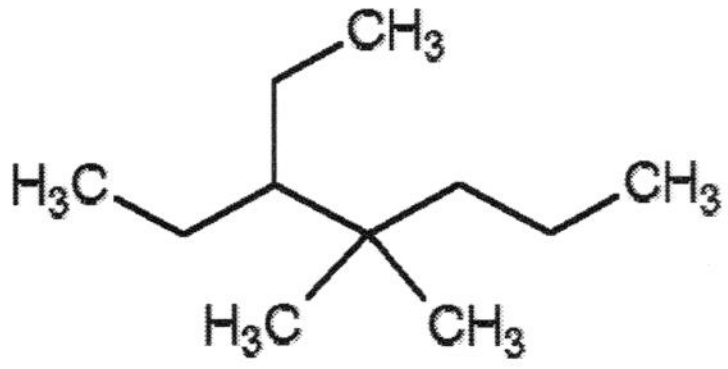

(c)

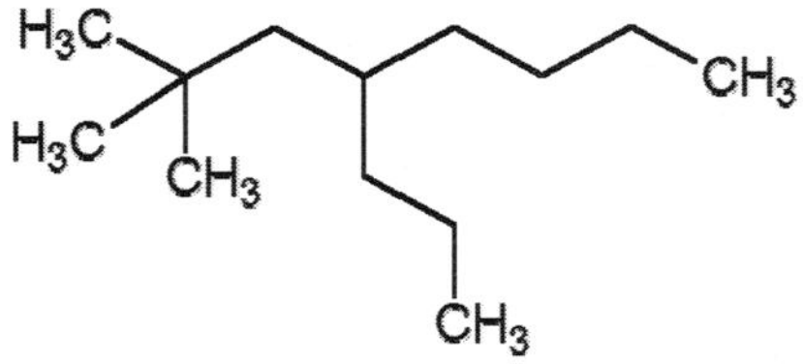

(d)

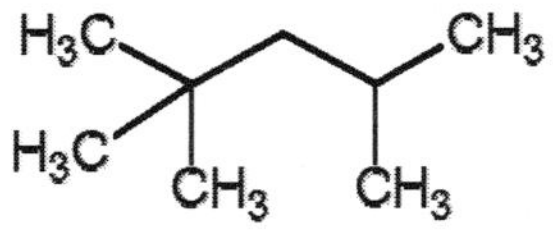

4.41

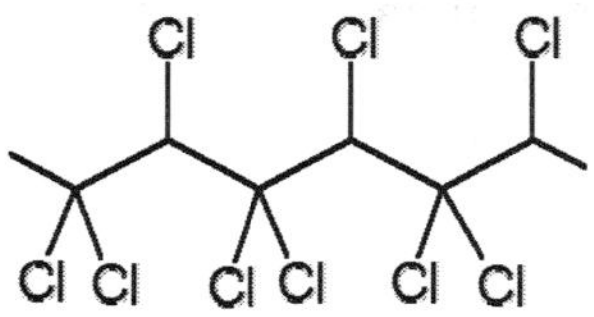

4.43

Presence of a strong oxidizing agent on Mars means the environment is hostile to many fragile molecules found on earth. So the presence of hydrogen peroxide may explain why some expected alkanes besides methane have not been detected. They may have existed, but been rapidly destroyed by oxidation reactions.

SUMMARY AND CONCEPTUAL QUESTIONS

4.45

Table 7.3 shows the bond energies of C-H, C-F and C-Cl bonds to be 413, 485, and 339 kJ mol^{-1}, respectively. When a hydroxyl radical reacts it breaks the C-X bond and forms an O-X bond. The bond energies of the bonds with O are 463 for O-H and 218 for O-Cl. There is no energy given for the O-F bond. Here, we see that the most favourable reaction is that which breaks the C-H bond to form the stronger O-H bond.

4.47

SF_6 has more bonds than CO_2 and consequently more vibrational modes—and more associated IR absorption peaks. This and its long life in the atmosphere increase its global warming potential.

4.49

Perhaps we should look for oxygen, water, and carbon dioxide—at least if the importance of these molecules to life on earth is a guide. Plants use light and carbon dioxide to drive their life processes. Animals react fuels (food) with oxygen to drive their processes. Water is the medium in which the chemistry primarily takes place.

4.51

My car consumes about 10 L of gasoline per 100 km travelled (actually it consumes a little more than this, but let's keep the numbers simple). I travel about 10 000 km per year. This corresponds to consuming 1000 L of gasoline. Looking up the density of gasoline on the Internet yields a value of about 0.7 g mL^{-1}. Therefore, 1000 L of gasoline corresponds to Mass = volume × density = $(10^6$ mL) × $(0.7$ g mL$^{-1}) = 7 \times 10^5$ g of gasoline. If gasoline can be treated as octane, C_8H_{18}, then the fraction $8 M(C) / M(C_8H_{18}) = 96/114 = 0.84$ of the mass of gasoline is carbon atoms. In one year, my car emits about $0.84 \times 7 \times 10^5$ g $= 6 \times 10^5$ g of carbon atoms. For each g of carbon atoms in CO_2 the mass of CO_2 is 1 g × ($M(CO_2) / M(C)$) = 1 g × 44/12 = 3.7 g Therefore, the mass of carbon dioxide emitted by my car in one year is about $3.7 \times 6 \times 10^5$ g $= 2 \times 10^6$ g. Compare this with the total mass of atmospheric carbon dioxide, 3×10^{18} g.

4.53

Fossil fuels are more valuable as feedstocks for the production of polymers. This should be the most important use.

4.55

No, we should not be concerned about human breathing as a source of atmospheric carbon dioxide. On average, each human exhales about 1 kg per day. But this carbon dioxide includes carbon atoms that were originally taken out of the carbon dioxide in the air by plants through photosynthesis—whether you eat the plants directly or animals that eat the plants. Thus, there is a closed loop, with no net addition to the atmosphere. (http://cdiac.ornl.gov/pns/faq.html)

Chapter 5
Chemical Reaction, Chemical Equations

IN-CHAPTER EXERCISES

Exercise 5.2—Balanced chemical equations

(a) $Fe_2O_3(s) + 3\,CO(g) \longrightarrow 2\,Fe(s) + 3\,CO_2(g)$

is balanced for electrical charge—there are no charged species among reactants or products. There are ...

2 mol Fe atoms in 1 mol of Fe_2O_3 on the left, and 2 mol Fe on the right;

3 mol C atoms in 3 mol of CO on the left, and 3 mol C atoms in 3 mol of CO_2 on the right; and

$1 \times 3 + 3 \times 1 = 6$ mol O atoms in 1 mol of Fe_2O_3 and 3 mol of CO on the left, and 6 mol O atoms in 3 mol of CO_2 on the right.

Therefore, the reaction is balanced for all relevant atoms.

(b) $CH_3COOH(aq) + C_6H_5CH_2OH(aq) \longrightarrow CH_3COOCH_2C_6H_5(aq) + H_2O(l)$

is balanced for electrical charge—there are no charged species among reactants or products. There are ...

$4 + 8 = 12$ mol H atoms in 1 mol of CH_3COOH and 1 mol of $C_6H_5CH_2OH$ on the left, and $10 + 2 = 12$ mol H in 1 mol $CH_3COOCH_2C_6H_5$ and 1 mol H_2O on the right;

$2 + 7 = 9$ mol C atoms in 1 mol of CH_3COOH and 1 mol of $C_6H_5CH_2OH$ on the left, and 9 mol C atoms in 1 mol $CH_3COOCH_2C_6H_5$ on the right; and

$2 + 1 = 3$ mol O atoms in 1 mol of CH_3COOH and 1 mol of $C_6H_5CH_2OH$ on the left, and $2 + 1 = 3$ mol O atoms in 1 mol $CH_3COOCH_2C_6H_5$ and 1 mol H_2O on the right.

Therefore, the reaction is balanced for all relevant atoms.

(c) $CaCO_3(s) + H_3O^+(aq) \longrightarrow Ca^{2+}(aq) + CO_2(g) + H_2O(l)$

is NOT balanced for electrical charge—there is 1 mol of a 1+ species on the left and 1 mol of a 2+ species on the right. There are ...

1 mol Ca atoms in 1 mol of $CaCO_3$ on the left, and 1 mol Ca (in the form, Ca^{2+}) on the right;

1 mol C atoms in 1 mol of $CaCO_3$ on the left, and 1 mol C atoms in 1 mol of CO_2 on the right; and

$3 + 1 = 4$ mol O atoms in 1 mol of $CaCO_3$ and 1 mol of H_3O^+ on the left, and $2 + 1 = 3$ mol O atoms in 1 mol of CO_2 and 1 mol of H_2O on the right.

Therefore, the reaction is balanced for Ca and C, but NOT for O atoms.

Exercise 5.4—Relative amounts of reactants and products

Multiplying this reaction by $6/4 = 1.5$ gives the stoichiometry in case of 6.0 mol of Al(s).

$6.0\ Al(s) + 4.5\ O_2(g) \longrightarrow 3.0\ Al_2O_3(s)$

We see that 4.5 mol of $O_2(g)$ is required to completely react with 6.0 mol Al(s). 3.0 mol of $Al_2O_3(s)$ is the product.

Exercise 5.5—Spontaneous reaction direction and relative stability

(a) A small amount of solid sodium chloride in water has a higher chemical potential than a dilute solution of sodium chloride. Chemical potential is lowered—the spontaneous direction—when the sodium chloride dissolves.

(b) A dilute solution of sodium chloride is more stable than a small amount of solid sodium chloride in water. Dissolution of the solid sodium chloride takes a less stable state to a more stable state.

Exercise 5.6—Equilibrium, spontaneity and stability

(a) Neither
$$CaCO_3(s) \longrightarrow \underset{6\times10^{-5}\,mol\,L^{-1}}{Ca^{2+}(aq)} + \underset{6\times10^{-5}\,mol\,L^{-1}}{CO_3^{2-}(aq)}$$

nor $\underset{6\times10^{-5}\,mol\,L^{-1}}{Ca^{2+}(aq)} + \underset{6\times10^{-5}\,mol\,L^{-1}}{CO_3^{2-}(aq)} \longrightarrow CaCO_3(s)$

is the direction of spontaneous reaction. The chemical potentials on both sides of these reactions are equal. The reactants and products are in equilibrium under the stated conditions. 6×10^{-5} mol L^{-1} is the solubility of $CaCO_3(s)$ in water—i.e., it is the Ca^{2+} and CO_3^{2-} concentration of a saturated solution.

(b) A mixture of solid calcium carbonate in pure water has a higher chemical potential than a mixture containing solid calcium carbonate and $Ca^{2+}(aq)$ ions and $CO_3^{2-}(aq)$ both at a concentration of 6×10^{-5} mol L^{-1}.

(c) A mixture containing solid calcium carbonate and $Ca^{2+}(aq)$ ions and $CO_3^{2-}(aq)$ both at a concentration of 6×10^{-5} mol L^{-1} is more stable than a mixture of solid calcium carbonate in pure water.

Exercise 5.7—Relative masses of reactants and products

Always start with the balanced chemical equation. In this case
$$C_6H_{12}O_6(s) + 6\,O_2(g) \longrightarrow 6\,CO_2(g) + 6\,H_2O(l)$$
We are given the mass of glucose. To use the balanced equation we need to work in amounts, in moles.

$$\text{Amount of glucose, } n(C_6H_{12}O_6) = \frac{m}{M} = \frac{25.0\,g}{180.16\,g\,mol^{-1}} = 0.139\,mol$$

The balanced equation tells us how much $O_2(g)$ is consumed, and how much $CO_2(g)$ and $H_2O(l)$ is produced. These amounts must be converted to masses to answer the question. The answer is summarized in the following amounts table:

Equation	$C_6H_{12}O_6(s)$	+ 6 O_2(g)	→ 6 CO_2(g)	+ 6 H_2O(l)
Initial amount (Initial mass)	0.139 mol (25.0 g)	6 × 0.139 = 0.834 mol (mass of oxygen reacted = 0.834 mol × 32.00 g mol^{-1} = 26.7 g)	0 mol (0 g)	0 mol (0 g)
Change in amount (Change in mass)	−0.139 mol (−25.0 g)	−0.834 mol (−26.7 g)	+6 × 0.139 = 0.834 mol (mass of CO_2 produced = 0.834 mol × 44.01 g mol^{-1} = 36.7 g)	+6 × 0.139 = 0.834 mol (mass of H_2O produced = 0.834 mol × 18.02 g mol^{-1} = 15.0 g)
Final amount (Final mass)	0 mol (0 g)	0 mol (0 g)	0.834 mol (36.7 g)	0.834 mol (15.0 g)

Check that the total mass of products formed is the same as the total mass of reactants that react.

Exercise 5.11 — Limiting reactant situations

(a) The balanced equation is

$$CO(g) + 2 H_2(g) \longrightarrow CH_3OH(\ell)$$

Initial amount of CO, $n(CO) = \dfrac{m}{M} = \dfrac{356\ \cancel{g}}{28.01\ \cancel{g}\ mol^{-1}} = 12.7\ mol$

Initial amount of H_2, $n(H_2) = \dfrac{m}{M} = \dfrac{65.0\ \cancel{g}}{2.02\ \cancel{g}\ mol^{-1}} = 32.2\ mol$

To react all the initial CO requires 2×12.7 mol = 25.4 mol of H_2. Note that the 2× factor is the ratio of the stoichiometric coefficients of H_2 and CO. Clearly, there is more than enough initial H_2. So, CO is the **limiting reactant**—i.e., all of the initial CO can react and consume 25.4 mol of H_2. This leaves 32.2 – 25.4 = 6.8 mol of unreacted H_2.

(b) The amount of methanol that can be produced is 12.7 mol (CO and methanol have the same stoichiometric coefficient in the balanced equation). The associated mass of methanol is
$m(CH_3OH) = (12.7\ mol) \times (32.04\ g\ mol^{-1}) = 407\ g$.

(c) We have already established that 6.8 mol of H_2 is unreacted. This corresponds to the mass,
$m(H_2) = (6.8\ mol) \times (2.02\ g\ mol^{-1}) = 14\ g$.

The results are summarized in the following amounts table:

Equation	CO(g)	+ 2 H₂ (g)	→ CH₃OH(l)
Initial amount (Initial mass)	12.7 mol (356 g)	32.2 mol (65 g)	0 mol (0 g)
Change in amount (Change in mass)	−12.7 mol (−356 g)	−2× 12.7 = −25.4 mol (65 − 14 = 51 g) ← not asked for	+12.7 mol (+407 g)
Final amount (Final mass)	0 mol (0 g)	6.8 mol (14 g)	12.7 mol (407 g)

Exercise 5.12—Limiting reactant situations

(a) The balanced equation is

$$Fe_2O_3(s) + 2\ Al(s) \longrightarrow 2\ Fe(s) + Al_2O_3(s)$$

Initial amount of Fe_2O_3, $n(Fe_2O_3) = \dfrac{m}{M} = \dfrac{50.0\ g}{159.69\ g\ mol^{-1}} = 0.313\,mol$

Initial amount of Al, $n(Al) = \dfrac{m}{M} = \dfrac{50.0\ g}{26.98\ g\ mol^{-1}} = 1.85\,mol$

To react all the initial Fe_2O_3 requires 2×0.313 mol = 0.626 mol of Al. There is more than enough initial Al. So, Fe_2O_3 is the **limiting reactant**—i.e., all the initial Fe_2O_3 can react and consume 0.626 mol of Al. This leaves 1.85 − 0.626 = 1.22 mol of unreacted Al.
(b) The amount of iron that can be produced is 2×0.313 mol = 0.626 mol. The associated mass of iron is
$m(\text{Fe}) = (0.626\ mol) \times (55.847\text{g mol}^{-1}) = 35.0\ g.$

These results are summarized in the following amounts table:

Equation	Fe₂O₃(s)	+ 2 Al (s)	→ 2 Fe(s)	+ Al₂O₃(s)
Initial amount (Initial mass)	0.313 mol (50.0 g)	1.85 mol (50.0 g)	0 mol (0 g)	0 mol (0 g)
Change in amount (Change in mass)	−0.313 mol (−50.0 g)	−2×0.313 mol = −0.626 mol	+2×0.313 mol = 0.626 mol (+35.0 g)	+0.313 mol
Final amount (Final mass)	0 mol (0 g)	1.22 mol	0.626 mol (35.0 g)	0.313 mol

Exercise 5.17—Percent yield and theoretical yield

First, write the balanced chemical equation,
$CH_3OH(l) \longrightarrow 2\ H_2(g) + CO(g)$
The amount of methanol reacted is (125 g) / (32.04 g mol⁻¹) = 3.90 mol
Theoretical yield of hydrogen = 2×3.90 mol × 2.016 g mol⁻¹ = 15.7 g
The actual yield is 13.6 g of hydrogen. Thus,

$$\text{Percent yield of hydrogen} = \frac{13.6 \text{ g}}{15.7 \text{ g}} \times 100\% = 86.6\%$$

Exercise 5.19—Quantitative analysis based on stoichiometry

The key to answering this question is to note that the Ni atoms in the precipitate all came from NiS in the original sample. Since we know the mass of precipitate, we can get the amount of precipitate which equals the amount of Ni^{2+} which is, in turn, equal to the amount of NiS. There are no stoichiometric ratio factors here because Ni^{2+} and $Ni(C_4H_7N_2O_2)_2$ have equal stoichiometric coefficients in the second equation, while NiS and Ni^{2+} have equal coefficients in the first equation. First, we determine the amount of $Ni(C_4H_7N_2O_2)_2$.

Amount of $Ni(C_4H_7N_2O_2)_2 = m(Ni(C_4H_7N_2O_2)_2) / M(Ni(C_4H_7N_2O_2)_2) = (0.206 \text{ g}) /(288.91 \text{ g mol}^{-1}) = 7.13 \times 10^{-4}$ mol

Next, we note that this is the amount of NiS in the sample of millerite. Therefore, the mass of NiS in the sample was

mass $= n(NiS) \times M(NiS) = (7.13 \times 10^{-4}$ mol$) \times (90.76 \text{ g mol}^{-1}) = 0.0647$ g

Finally, we compare the mass of NiS with the mass of the sample which clearly contains additional non-nickel compounds.

$$\text{Mass percent of NiS} = \frac{0.0647 \text{ g}}{0.468 \text{ g}} \times 100\% = 13.8\%$$

Exercise 5.20—Atom economy

(a)

For the production of maleic anhydride by oxidation of benzene,
(i) The C atom efficiency $= 100\% \times$ (number of C atoms in product)/(number of C atoms in all reactancts)
$= 100\% \times 4/6 = 66.7\%$
The O atom efficiency $= 100\% \times$ (number of O atoms in product)/(number of O atoms in all reactancts)
$= 100\% \times 3/(4.5 \times 2) = 33.3\%$
The H atom efficiency $= 100\% \times$ (number of H atoms in product)/(number of H atoms in all reactancts)
$= 100\% \times 2/6 = 33.3\%$
(ii) The overall atom efficiency, OAE, is the percent ratio of the mass of desired product to the total mass of reactants.
OAE $= 100\% \times M$(maleic anhydride) / $(M$(benzene $) + 4.5\ M(O_2))$
$= 44.1\%$
(iii) The E-factor is the ratio of waste products to desired product(s). It is given by
$(2\ M(CO_2) + 2\ M(H_2O)) / M$(maleic anhydride),
but can also be obtained from OAE. If 44.1% of the mass goes into desired product, then 55.9% of the mass goes into waste products. Therefore, E-factor $= 55.9\% / 44.1\% = 1.27$

(b)

For the production of maleic anhydride by oxidation of butene,
(i) The C atom efficiency = 100% × (number of C atoms in product)/(number of C atoms in all reactancts)
    = 100% × 4/4 = 100.0%
The O atom efficiency = 100% × (number of O atoms in product)/(number of O atoms in all reactancts)
    = 100% × 3/(3×2) = 50.0%
The H atom efficiency = 100% × (number of H atoms in product)/(number of H atoms in all reactancts)
    = 100% × 2/8 = 25.0%
(ii) The overall atom efficiency, OAE, is the percent ratio of the mass of desired product to the total mass of reactants.
OAE = 100% × M(maleic anhydride) / (M(butene) + 3 M(O_2))
    = 64.5%
(iii) The E-factor is the ratio of waste products to desired product(s). It is given by
(3 M(H_2O)) / M(maleic anhydride),
but can also be obtained from OAE. If 64.5% of the mass goes into desired product, then 35.5% of the mass goes into waste products. Therefore,
E-factor = 35.5% / 64.5% = 0.55

REVIEW QUESTIONS

Section 5.3 Chemical equations: chemical accounting

5.21

(a) $C_6H_6(l) + 3H_2(g) \longrightarrow C_6H_{12}(l)$

is balanced for electrical charge—there are no charged species among reactants or products. There are ...

6 mol C atoms in 1 mol of C_6H_6 on the left, and 6 mol C atoms in 1 mol of C_6H_{12} on the right; and $1\times6 + 3\times2 = 12$ mol H atoms in 1 mol of C_6H_6 and 3 mol of H_2 on the left, and 12 mol H atoms in 1 mol of C_6H_{12} on the right.

Therefore, the reaction is balanced for all relevant atoms.

(b) $MnO_4^-(aq) + 5Fe^{2+}(aq) + 8H^+(aq) \longrightarrow Mn^{2+}(aq) + 5Fe^{3+}(aq) + 4H_2O(l)$

has charges, $-1 + 5\times(+2) + 8\times(+1) = +17$ on the left and $+2 + 5\times(+3) = 17$ on the right. It is balanced for electrical charge.

There are ...

1 mol Mn atoms on the left and right

5 mol Fe atoms on the left and right

8 mol H atoms on the left, and 4×2 H atoms on the right; and

4 mol O atoms on the left, and 4 mol O atoms on the right.

Therefore, the reaction is balanced for all relevant atoms.

(c) $Cu(OH_2)_6^{2+}(aq) + 4NH_3(aq) \longrightarrow Cu(NH_3)_4^{2+}(aq) + 6H_2O(l)$

has a charge $+2$ on both the left and right. It is balanced for electrical charge.

There are ...

1 mol Cu atoms on the left and right

4 mol N atoms on the left and right

$6\times2 + 4\times3 = 24$ mol H atoms on the left, and $4\times3 + 6\times2 = 24$ mol H atoms on the right; and

6 mol O atoms on the left and right.

Therefore, the reaction is balanced for all relevant atoms.

Section 5.4 Spontaneous direction of reaction

5.23

(a) If the pressure of $H_2O(g)$ is 1 bar and temperature 25°C, then

$H_2O(g) \longrightarrow H_2O(l)$ is the spontaneous direction of reaction

There is net condensation of water under these conditions.

(b) Water vapour at 1 bar pressure has a higher chemical potential than liquid water at 25°C.

Section 5.5 The condition of dynamic chemical equilibrium

5.25

(a) A mixture containing a piece of copper in a 1 mol L^{-1} aqueous silver nitrate solution has a higher chemical potential than a mixture containing some solid silver in an aqueous solution in which the concentration of Cu^{2+}(aq) ions is about 0.5 mol L^{-1} and the concentration of Ag^+(aq) ions is about 1 x 10^{-10} mol L^{-1}.

(b) A mixture containing some solid silver in an aqueous solution in which the concentration of Cu^{2+}(aq) ions is about 0.5 mol L^{-1} and the concentration of Ag^+(aq) ions is about 1 x 10^{-10} mol L^{-1} is more stable than a mixture containing a piece of copper in a 1 mol L^{-1} aqueous silver nitrate solution.

Section 5.6 Masses of reactants and products: stoichiometry

5.27

In this case, the theoretical yield of CO_2 is the total mass of reactants because there are no other products—an exact stoichiometric amount of oxygen was used. See the balanced equation,
$C(s) + O_2(g) \rightarrow CO_2(g)$
theoretical yield of CO_2 = 10.0 + 26.6 g = 36.6 g

5.29

$$C_6H_{14}N_4O_2 + H_2O \longrightarrow NH_2CONH_2 + C_5H_{12}N_2O_2$$
arginine, urea, ornithine

m(urea) = 95 mg = 0.095 g
Amount of urea (NH_2CONH_2) produced = m(urea) / M(urea)
  = (0.095 g) / (60.06 g mol^{-1}) = 0.00158 mol
Amount of arginine ($C_6H_{14}N_4O_2$) required to produce this much urea = amount of urea = 0.00158 mol
Mass of arginine required = n(arginine) × M(arginine) = (0.00158 mol) ×(174.20 g mol^{-1}) = 0.275 g
  = 275 mg
Amount of ornithine ($C_5H_{12}N_2O_2$) produced = amount of urea = 0.00158 mol
Mass of ornithine produced = n(ornithine) × M(ornithine) = (0.00158 mol) ×(132.16 g mol^{-1}) = 0.209 g
  = 210 mg
In this last step, we revert to correct number of significant digits—we kept an extra one along the way to avoid round-off errors.

5.31

The total mass of the beakers and solutions after reaction is the same as it was before reaction. Chemical reactions conserve mass. Atoms are rearranged, but the total number of each type of atom and the total mass are conserved.
Total mass after reaction = 167.170 g

5.33

(a) $4\,NH_3(g) + 5\,O_2(g) \longrightarrow 4\,NO(g) + 6\,H_2O(g)$

Amount of NH_3 reacted, $n = m(NH_3) / M(NH_3) = (750\ g) / (17.03\ g\ mol^{-1}) = 44.0\ mol$

Amount of oxygen required to react with all the ammonia, $n = \dfrac{5}{4} \times$ amount of NH_3 reacted $= 1.25$

$\times\ 44.0\ mol = 55.0\ mol$

Mass of oxygen required, $m = n(O_2) \times M(O_2) = (55.0\ mol) \times (32.00\ g\ mol^{-1}) = 1760\ g = 1.76\ kg$

(b) Amount of water produced, $n = \dfrac{6}{4} \times$ amount of NH_3 reacted $= 1.50 \times 44.0\ mol = 66.0\ mol$

Mass of water produced, $m = n(H_2O) \times M(H_2O) = (66.0\ mol) \times (18.02\ g\ mol^{-1}) = 1190\ g = 1.19\ kg$

Section 5.9 Stoichiometry and chemical analysis

5.35

Mass of $Cu_2S(s)$, $m = 1.00$ tonne $= 1000\ kg = 1.00 \times 10^6\ g$

Amount of Cu_2S, $n = m(Cu_2S) / M(Cu_2S) = (1.00 \times 10^6\ g) / (159.16\ g\ mol^{-1}) = 6.29 \times 10^3\ mol$

Amount of H_2SO_4 that can be produced, $n =$ amount of $Cu_2S = 6.29 \times 10^3\ mol$

Each S atom in $Cu_2S(s)$ leads to one molecule of H_2SO_4

Mass of H_2SO_4 that can be produced, $m = n(H_2SO_4) \times M(H_2SO_4) = (6.29 \times 10^3\ mol) \times (98.08\ g\ mol^{-1}) = 6.17 \times 10^5\ g$

$\quad = 617\ kg$

5.37

Here is the balanced reaction:

$x\,TiO_2(s) + H_2(g) \rightarrow Ti_xO_y(s) + (y - 2x)\ H_2O(l)$

Amount of TiO_2, $n = m(TiO_2) / M(TiO_2) = (1.598\ g) / (79.866\ g\ mol^{-1}) = 0.02001\ mol$

Amount of Ti_xO_y, $n = \dfrac{1}{x} \times (1.598\ g) / (79.866\ g\ mol^{-1}) = \dfrac{1}{x} \times (0.02001\ mol)$

$\qquad = m(Ti_xO_y) / M(Ti_xO_y) = (1.438\ g) / M(Ti_xO_y)$

Therefore, $\dfrac{M\left(Ti_xO_y\right)}{x} = (1.438\ g) / (0.02001\ mol) = 71.86\ g\ mol^{-1}$

But, we also have $\dfrac{M\left(Ti_xO_y\right)}{x} = M(Ti) + \dfrac{y}{x} \times M(O) = 47.867 + \dfrac{y}{x} \times (15.999)$.

So $\dfrac{y}{x} \times (15.999\ g\ mol^{-1}) = 71.86 - 47.867\ g\ mol^{-1} = 23.993\ g\ mol^{-1}$

and $\dfrac{y}{x} = \dfrac{23.993}{15.999} = 1.500$

Therefore, $Ti_xO_y = TiO_{1.5}$ or Ti_2O_3 .

5.39

Mass of AgCl, $m = 0.1804$ g

Amount of AgCl, $n = m(AgCl) / M(AgCl) = (0.1804$ g$) / (143.32$ g mol$^{-1}) = 0.001259$ mol

2 mol of Cl atoms in 1 mol 2,4-D (2,4-dichlorophenoxyacetic acid) leads to 2 mol of AgCl (one AgCl for each Cl)

Amount of 2,4-D (2,4-dichlorophenoxyacetic acid), $C_8H_6Cl_2O_3 = \dfrac{1}{2}$ (amount of AgCl) $=$

6.30×10^{-4} mol

Mass of 2,4-D (2,4-dichlorophenoxyacetic acid) in sample, $m = n(C_8H_6Cl_2O_3) \times M(C_8H_6Cl_2O_3)$
 $= (6.30\times10^{-4}$ mol$) \times (221.04$ g mol$^{-1}) = 0.139$ g

Mass percent of 2,4-D in sample $= 100\% \times m(2,4$-D$) / m($sample$) = 100\% \times (0.139$ g$) / (1.236$ g$)$
$= 11.2\%$

5.41

$$CaCO_3(s) \longrightarrow CaO(s) + CO_2(g)$$

Mass $CaCO_3$ in limestone, $m = \dfrac{95\%}{100\%} \times 125$ kg $= 119$ kg $= 1.19\times10^5$ g

Amount of $CaCO_3$, $n = m(CaCO_3) / M(CaCO_3) = (1.19\times10^5$ g$) / (100.09$ g mol$^{-1}) = 1.19\times10^3$ mol

1 mol of lime, CaO(s), is produced for every 1 mol of $CaCO_3$

Amount of lime produced, $n = $ amount of $CaCO_3 = 1.19\times10^3$ mol

Mass of CaO, $m = n(CaO) \times M(CaO) = (1.19\times10^3$ mol$) \times (56.08$ g mol$^{-1}) = 6.67\times10^4$ g $= 66.7$ kg

5.43

Mass of copper in product metal, $m = \dfrac{89.5\%}{100\%} \times 75.4$ g $= 67.5$ g

Amount of copper, $n = m(Cu) / M(Cu) = (67.5$ g$) / (63.55$ g mol$^{-1}) = 1.06$ mol

This copper came from 100.0 g of ore, 89.0 g of which consists of CuS and Cu_2S (because there is an 11% impurity). 67.5 g of the 89.0 g come from copper atoms (the copper recovered). Therefore, $89.0 - 67.5$ g $= 21.5$ g of the 89.0 g come from sulfur atoms. This corresponds to ...

Amount of sulfur, $n = m(S) / M(S) = (21.5$ g$) / (32.07$ g mol$^{-1}) = 0.670$ mol

If we let the amount of CuS be a and the amount of Cu_2S be b, then we have

Amount of Cu $= a + 2b = 1.06$ mol   (1)

Amount of S $= a + b = 0.670$ mol   (2)

Therefore – subtract (2) from (1),

$b = 1.06 - 0.670$ mol $= 0.39$ mol

and – substitute back into (2)

$a = 0.670 - 0.39$ mol $= 0.28$ mol

Mass of CuS in ore $= a \times M(CuS) = 0.28$ mol $\times 95.62.17$ g mol$^{-1} = 27$ g

Mass of Cu_2S in ore $= b \times M(Cu_2S) = 0.39$ mol $\times 159.17$ g mol$^{-1} = 62$ g

Mass percent CuS in ore $= 100\% \times 27 / 89 = 30\%$

Mass percent Cu_2S in ore $= 100\% \times 62 / 89 = 70\%$

SUMMARY AND CONCEPTUAL QUESTIONS

5.45
(a) From the graph,

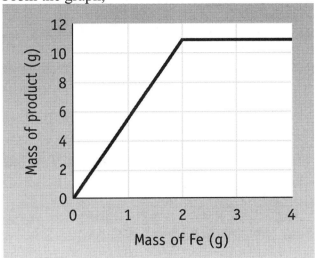

when the reaction consumes 2.0 g of Fe(s), about 10.8 g of iron bromide is produced.
2.0 g of the 10.8 g product is due to Fe atoms
The remaining 10.8 − 2.0 = 8.8 g of the iron bromide comes from bromine atoms.
The same mass, 8.8 g, of bromine liquid (the sole source of bromine atoms in the product) is used to make the product.
(b) Amount of Fe in product, $n = m,(\text{Fe}) / M(\text{Fe}) = (2.0 \text{ g}) / (55.85 \text{ g mol}^{-1}) = 3.6 \times 10^{-2}$ mol
Amount of Br_2 needed to make product, $n = m(\text{bromine}) / M(Br_2) = (8.8 \text{ g}) / (159.80 \text{ g mol}^{-1}) = 5.5 \times 10^{-2}$ mol
Mole ratio of bromine to iron $= (5.5 \times 10^{-2} \text{ mol}) / (3.6 \times 10^{-2} \text{ mol}) = 1.5$
(c) The formula of the product is $FeBr_3$
(d) Balanced chemical equation for the reaction of iron and bromine:
$$Fe(s) + 1.5 \ Br_2(l) \rightarrow FeBr_3$$
or $\quad 2 \ Fe(s) + 3 \ Br_2(l) \rightarrow 2 \ FeBr_3 \quad$ in terms of smallest integer coefficients
(e) The product is iron (III) bromide
(f)

(i) When 1.00 g of Fe(s) is added to the $Br_2(l)$, Fe(s) is the limiting reagent. **TRUE**
(ii) When 3.50 g of Fe(s) is added to the $Br_2(l)$, there is an excess of $Br_2(l)$. **FALSE**
The amount of product stops increasing above 2.0 g of Fe. Any Fe above 2.0 g is excess and does not react.
Under these conditions bromine is the limiting reactant—it reacts completely.
(iii) When 2.50 g of Fe(s) is added to the $Br_2(l)$, both reactants are used up completely. **FALSE**
2.0 g is the mass of Fe that reacts completely with all the bromine.
(iv) When 2.00 g of Fe(s) is added to the $Br_2(l)$, 10.0 g of product is formed.
The percent yield must therefore be 20.0%. **FALSE**
Percent yield = 100% × (actual yield) / (theoretical yield)
$= 100\% \times (10.0 \text{ g}) / (10.8 \text{ g}) = 93\%$

5.47

(a) cisplatin, $Pt(NH_3)_2Cl_2$, is

$100\% \times M(Pt) / M(Pt(NH_3)_2Cl_2) = 100\% \times (195.08 \text{ g mol}^{-1}) / (300.05 \text{ g mol}^{-1}) = 65.0\%$ Pt by mass

$100\% \times 2 \times M(N) / M(Pt(NH_3)_2Cl_2) = 100\% \times 2 \times (14.01 \text{ g mol}^{-1}) / (300.05 \text{ g mol}^{-1}) = 9.3\%$ N by mass

$100\% \times 2 \times M(Cl) / M(Pt(NH_3)_2Cl_2) = 100\% \times 2 \times (35.45 \text{ g mol}^{-1}) / (300.05 \text{ g mol}^{-1}) = 23.6\%$ Cl by mass

(b) $K_2PtCl_4(aq) + 2 NH_3(aq) \longrightarrow Pt(NH_3)_2Cl_2(aq) + 2 KCl(aq)$

Starting mass of K_2PtCl_4, $m = 16.0$ g

Amount of K_2PtCl_4, $n = m(K_2PtCl_4) / M(K_2PtCl_4) = (16.0 \text{ g}) / (415.09 \text{ g mol}^{-1}) = 0.0385$ mol

To completely consume this reactant requires amount of ammonia, $n = 2 \times 0.0385 \text{ mol} = 0.0770$ mol

Mass of ammonia required, $m = 0.0770 \text{ mol} \times 17.03 \text{ g mol}^{-1} = 1.31$ g

Amount of $Pt(NH_3)_2Cl_2$ produced, $n =$ amount of $K_2PtCl_4 = 0.0385$ mol

Mass of $Pt(NH_3)_2Cl_2$ produced, $m = n(Pt(NH_3)_2Cl_2) \times M(Pt(NH_3)_2Cl_2) = (0.0385 \text{ mol}) \times (300.05 \text{ g mol}^{-1}) = 11.6$ g

5.49

$Zn(s) + 2 H^+(aq) \longrightarrow Zn^{2+}(aq) + H_2(g)$

Amount of HCl $= 0.100$ mol

(1) Mass of zinc, $m = 7.00$ g

Amount of zinc, $n = m(Zn) / M(Zn) = (7.00 \text{ g}) / (65.39 \text{ g mol}^{-1}) = 0.107$ mol

Amount of zinc needed to react with all of the HCl, $n = \dfrac{1}{2} \times$ amount of HCl $= 0.050$ mol

There is more than enough zinc. In this case, HCl is the limiting reactant—all of it reacts, producing enough H_2 to inflate the balloon. $0.107 - 0.050 \text{ mol} = 0.057$ mol of zinc remains.

(2) Mass of zinc, $m = 3.27$ g

Amount of zinc, n $= m(Zn) / M(Zn) = (3.27 \text{ g}) / (65.39 \text{ g mol}^{-1}) = 0.0500$ mol

In this case, there is just enough zinc to consume all of the HCl. The same amount of H_2 is produced as in part (1)—i.e., enough to inflate the balloon. No zinc remains.

(3) Mass of zinc, $m = 1.31$ g

Amount of zinc, n $= m(Zn) / M(Zn) = (1.31 \text{ g}) / (65.39 \text{ g mol}^{-1}) = 0.0200$ mol

In this case, there is not enough zinc to consume all of the HCl. Zinc is the limiting reactant. Only 40% as much H_2 is produced as in part (1)—i.e., not enough to inflate the balloon. No zinc remains.

Chapter 6
Chemistry of Water, Chemistry in Water

IN-CHAPTER EXERCISES

Exercise 6.1—Temperature dependence of water density

According to Figure 6.3 (from the text), the densities of ice and liquid water (presumably at 1 bar) at the temperatures indicated are tabulated as follows:

Temperature / °C	Density of Ice / g mL^{-1}	Density of Liquid Water / g mL^{-1}
−5	0.9175	
0	0.9169	0.99987
4		1.00000
10		0.99974

The volume of 100.0 g of ice at −5 °C is
    100.0 g / 0.9175 g mL^{-1} = 109.0 mL
The other volumes are similarly computed—results are tabulated as follows:

Temperature / °C	Volume of 100.0 g of Ice / mL	Volume of 100.0 g of Liquid Water / mL
−5	108.99	
0	109.06	100.01
4		100.00
10		100.03

These data are displayed graphically as follows:

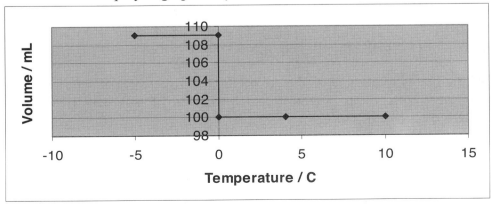

Exercise 6.2—Specific heat capacities of liquids

The specific heat capacities of water and ethanol are 4.18 and 2.44 J K^{-1} g^{-1}, respectively.
The amount of heat required to raise the temperature of water by $100 - 25\,°C = 75\,°C = 75$ K, is

$$q_{water} = m(water) \times c(water) \times \Delta T = (1000\text{ g}) \times (4.18\text{ J K}^{-1}\text{ g}^{-1}) \times (75\text{ K}) = 3.14 \times 10^5\text{ J}$$

Note that ΔT in °C = ΔT in K.
The amount of heat required to raise the temperature of ethanol by 75 K, is

$$q_{ethanol} = m(ethanol) \times c(ethanol) \times \Delta T = (1000\text{ g}) \times (2.44\text{ J K}^{-1}\text{ g}^{-1}) \times (75\text{ K}) = 1.83 \times 10^5\text{ J}$$

If the heat is delivered to water at a uniform rate in 60 s, then the time needed to heat the ethanol (note that it would actual boil before getting to 100 °C at 1 atm pressure) is

$$\frac{q_{ethanol}}{q_{water}} \times (60\text{ s}) = \frac{1.83 \times 10^5\text{ J}}{3.14 \times 10^5\text{ J}} \times (60\text{ s}) = 35\text{ s}$$

Note that the above fraction of 60 s is just the ratio of the specific heat capacities. The calculation is simplified to just

$$\frac{c_{ethanol}}{c_{water}} \times (60\text{ s}) = \frac{2.44\text{ J g}^{-1}\text{ K}^{-1}}{4.18\text{ J g}^{-1}\text{ K}^{-1}} \times (60\text{ s}) = 35\text{ s}$$

the same as above because the masses and temperature changes are the same for ethanol and water

Exercise 6.3—Enthalpy change of vaporization

(a) Amount of methanol, $n(CH_3OH) = \dfrac{m}{M} = \dfrac{1.00 \times 10^3\text{ g}}{32.04\text{ g mol}^{-1}} = 31.2\text{ mol}$

Energy absorbed $= n \times \Delta_{vap}H = 31.2\text{ mol} \times 35.2\text{ kJ mol}^{-1} = 1.10 \times 10^3\text{ kJ}$

(b) To vaporize the same amount of ammonia requires

Energy absorbed $= n \times \Delta_{vap}H = 31.2\text{ mol} \times 23.3\text{ kJ mol}^{-1} = 727\text{ kJ}$

Exercise 6.4—Equilibrium vapour pressure and volatility

Decreasing volatility corresponds to decreasing vapour pressure at the same temperature.
In order of decreasing volatility, we have
diethyl ether > carbon disulfide > acetone > bromine > hexane > ethanol > water

Exercise 6.5—Boiling point and molar enthalpy change of vaporization

The order of normal boiling points is the same as the order of enthalpy changes of vaporization. This is what we expect because both boiling point and enthalpy change of vaporization increase with increasing strength of intermolecular forces.

Exercise 6.6—Surface tension of liquids

The surface tension of diethyl ether is 1.7×10^{-2} J m^{-2}.
It is much smaller than the value for water, 7.3×10^{-2} J m^{-2}.
The surface tension is that which pulls the drop into a spherical shape—minimizing surface area.
Since this force is smaller for diethyl ether we expect its drops to be even more distorted by gravity than those of water.

Exercise 6.7—Intermolecular forces and intramolecular forces

(a) Water molecules in the cold morning air settle on even colder objects on the ground such as blades of grass. Here, the intermolecular force of attraction between water molecules and other water molecules or the molecules on the surface of the blades of grass is sufficient to allow water molecules to stick together on the blades of grass.
(b) When a piece of paper burns, carbon-carbon and carbon-hydrogen bonds are broken while new bonds with oxygen are formed. Intramolecular forces are the losing (in the breaking reactant bonds) and winning forces (in the making of bonds) at work when paper burns.
(c) Intermolecular forces holding water molecules together, and next to clothes fibre molecules, are overcome when wet clothes are hung out to dry.
(d) Intermolecular forces cause water molecules to arrange themselves into a lattice of lowest potential energy. The freezer takes away the kinetic energy thereby liberated, leaving water in its crystalline, solid form—ice.

Exercise 6.9—Bond polarities

(a) H→F is more polar than H→I
because $\chi(F) > \chi(I)$ while both are $> \chi(H)$
Note the dipole arrows on the bonds.
(b) B→C is less polar than B→F
because $\chi(F) > \chi(C)$ while both are $> \chi(B)$
(c) C←Si is more polar than C→S
In this case, $|\chi(C) - \chi(Si)| = |2.5 - 1.9| = 0.6$ while
$|\chi(C) - \chi(S)| = |2.5 - 2.6| = 0.1$.

Exercise 6.10—Bond polarities

(a) $\chi(C) = 2.5$, $\chi(H) = 2.2$, and $\chi(O) = 3.5$
The C=O and C-H bonds are polar, while C-C and C=C are non-polar.
(b) The C=O bond is the most polar, with O at the negative end.

Exercise 6.12—Bond polarity and molecular polarity

CS_2 and CO_2 are non-polar because of their linear shape. The two bond dipoles in these molecules exactly cancel out, leaving NO net dipole moment.
SO_2 and H_2O are polar because of their bent shape. The two bond dipoles vectors only partially cancel out, leaving a NET dipole moment.

Exercise 6.13—Molecular polarity

(a) $BFCl_2$ is trigonal planar. It has a NET dipole because one of the three bonds is different—more polar—than the other two. If the three substituents on B were the same, as in BCl_3, then the bond dipoles would cancel.
(b) NH_2Cl is trigonal pyramidal. It has a NET dipole because it is asymmetrical and has one bond different from the other two—N-Cl vs. N-H.
(c) SCl_2 is bent like water. It has a NET dipole because the S-Cl bond dipoles only partially cancel out.

Exercise 6.15—Dipole-dipole forces

(a) Bent SO_2 has a net dipole. So, there are dipole-dipole forces between neighbouring molecules.
(b) Linear CO_2 has no net dipole. There are no dipole-dipole forces between these molecules—only dispersion forces.
(c) HCl has just one polar bond, and so has a dipole moment. There are dipole-dipole forces between neighbouring molecules.

Exercise 6.17—Molecular structure and hydrogen bonding

$$CH_3\text{—O—H} \cdots\cdots \text{O—H}$$
$$\overset{|}{CH_3}$$

shows the hydrogen bonding between neighbouring methanol molecules as they move past each other.

Exercise 6.19—Dispersion forces

(a) Fluorine has lower melting and boiling points than bromine. Clearly, bromine has the stronger dispersion forces. This is because bromine atoms are bigger with more electrons. The bromine molecule charge distribution has larger fluctuations, and its neighbours are more easily polarized.
(b) Ethane has lower melting and boiling points than butane. Butane has the stronger dispersion forces. It is simply a bigger molecule—more C and H atoms—with more electrons. The butane molecule charge distribution has larger fluctuations, and its neighbours are more easily polarized.

Exercise 6.23—Intermolecular forces and properties
(a) ICl has a higher boiling point than Br$_2$ because it has dipole-dipole intermolecular forces, whereas bromine only has dispersion forces.
(b) Krypton has a higher boiling point than neon because Kr atoms are bigger—more electrons—than Ne atoms.
(c) Ethanol has a higher boiling point than ethylene oxide. Ethanol has hydrogen-bonding, whereas ethylene oxide has only non-H-bonding dipole-dipole forces.

Exercise 6.24—Intermolecular forces and properties

(a) In hexane, dispersion forces are the only intermolecular forces affecting vapour pressure.
(b) In water, hydrogen bonds (special dipole-dipole bonds) and dispersion forces are at work.
(c) In carbon tetrachloride, dispersion forces are the only intermolecular forces at work.

Exercise 6.27—Aquated ions

Mg^{2+}(aq) and Br$^-$(aq) are the main species present in solution when some magnesium bromide, MgBr$_2$(s), is dissolved in water. These species can be represented in a similar way to the un-named aquated cation and anion in Figure 6.26. A more schematic representation is provided as follows:

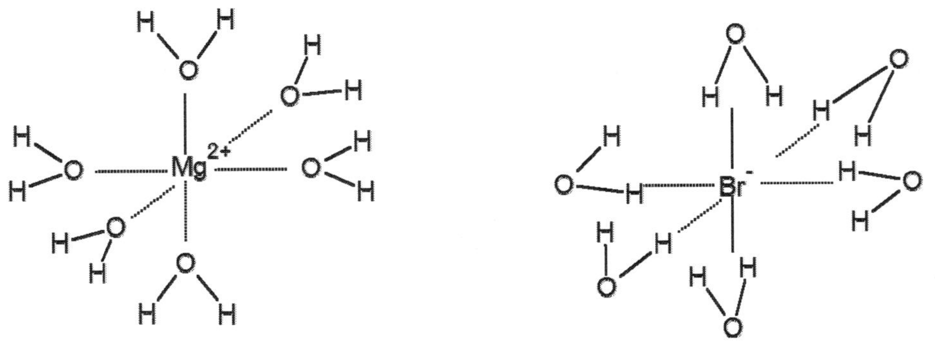

Exercise 6.28—Solubility of ionic compounds

(a) soluble
$$LiNO_3(s) \xrightarrow{H_2O} Li^+(aq) + NO_3^-(aq)$$
(b) soluble
$$CaCl_2(s) \xrightarrow{H_2O} Ca^{2+}(aq) + 2\,Cl^-(aq)$$
(c) insoluble, CuO(s)
(d) soluble
$$NaCH_3COO(s) \xrightarrow{H_2O} Na^+(aq) + CH_3COO^-(aq)$$

Exercise 6.29—Solubilities of molecular substances

(a) Ammonia, $NH_3(g)$, is soluble in water because water molecules can make strong hydrogen bonds with ammonia molecules –O-H and H-N.

(b) Hydrogen chloride, $HCl(g)$, is soluble in water because water molecules can make strong hydrogen bonds with HCl molecules –O-H only. Because of strong interaction between water molecules and the charged ends of the polar HCl molecules, ionization to form $H^+(aq)$ and $Cl^-(aq)$ is highly favourable.

(c) Iodine, $I_2(s)$, is non-polar and insoluble in water. Only dispersion-like forces can operate between water and I_2. This interaction is weaker than the dispersion force between I_2 molecules and the hydrogen bonds in water.

(d) Octane, $C_8H_{18}(l)$ is non-polar and insoluble in water. Only dispersion-like forces can be made between water and octane. This interaction is weaker than the dispersion force between octane molecules and the hydrogen bonds between water molecules.

Exercise 6.30—Solubilities of molecular substances

(a) Benzene, $C_6H_6(l)$, is non-polar and soluble in non-polar octane. Dispersion forces are all that are at play in the pure substances and mixtures. The entropy of mixing is sufficient to drive dissolution.

(b) Water is polar and insoluble in non-polar octane. The energy of interaction between water molecules and octane molecules (dispersion forces only) is insufficient to compensate for the energy required to pull apart water molecules and allow the natural tendency for mixing of molecules.

(c) Iodine, $I_2(s)$, is non-polar and soluble in non-polar octane. Dispersion forces are all that are at play in the pure substances and mixtures. The entropy of mixing is sufficient to drive dissolution.

Exercise 6.31—Polar and non-polar parts of molecules

Butan-1-ol, $CH_3CH_2CH_2CH_2OH(l)$, should be more soluble in hexane than butan -2,4-diol, $HOCH_2CH_2CH_2CH_2OH(l)$. The latter compound has two hydroxyl (-OH) groups vs. only one for the former. Like the –O-H of water, these hydroxyls do not spontaneously mix with hexane—that is, not without the help of the favourable dispersion interactions of the alkyl chains. The energy of interaction between butan-1-ol molecules is too great to allow interpenetration by hexane molecules (in the absence of strong forces of attraction between butan-1-ol molecules and hexane molecules).

Exercise 6.32—Ionization of molecular solutes

(a) $HBr(g) \xrightarrow{\ H_2O\ } H^+(aq) + Br^-(aq)$

or $\quad HBr(g) + H_2O(l) \longrightarrow H_3O^+(aq) + Br^-(aq)$

(b) $HF(l) \xrightarrow{\ H_2O\ } HF(aq)$

$HF(aq) + H_2O(l) \underset{}{\overset{\text{only some ionised}}{\rightleftharpoons}} H_3O^+(aq) + F^-(aq)$

(c) $HCOOH(l) \xrightarrow{\ H_2O\ } HCOOH(aq)$

$HCOOH(aq) + H_2O(l) \underset{}{\overset{\text{only some ionised}}{\rightleftharpoons}} HCOO^-(aq) + H_3O^+(aq)$

(d) $C_{12}H_{22}O_{11}(s) \xrightarrow{\ H_2O\ } C_{12}H_{22}O_{11}(aq)$

Sucrose is soluble in water, but a non-electrolyte

Exercise 6.36—Precipitation reactions

(a) The precipitate, $CuCO_3(s)$, is formed via

$$Cu^{2+}(aq) + CO_3^{2-}(aq) \longrightarrow CuCO_3(s)$$

$Na^+(aq)$ and $Cl^-(aq)$ are spectator ions

(b) No precipitation occurs when a potassium carbonate solution is mixed with a sodium nitrate solution. Potassium nitrate and sodium carbonate are soluble salts.

(c) The precipitate, $Ni(OH)_2(s)$, is formed via

$$Ni^{2+}(aq) + 2\,OH^-(aq) \longrightarrow Ni(OH)_2(s)$$

$K^+(aq)$ and $Cl^-(aq)$ are spectator ions

Exercise 6.37—Oxidation-reduction reactions

$Fe^{2+}(aq)$ ions are oxidized to $Fe^{3+}(aq)$ ions. The oxidizing agent is $MnO_4^-(aq)$ ions. $MnO_4^-(aq)$ ions are reduced to $Mn^{2+}(aq)$ ions. $Fe^{2+}(aq)$ ions are the reducing agent.

Exercise 6.38—Acids and bases

(a) $HCOOH(aq)$ is the acid reactant, while $OH^-(aq)$ ions are the base reactant.

(b) $H_2CO_3(aq)$ is the acid reactant, while $NH_3(aq)$ is the base reactant.

(c) $H_2C_4H_4O_6(aq)$ is the acid reactant, while $HCO_3^-(aq)$ is the base reactant.

(d) $H_3O^+(aq)$ is the acid reactant, while $CH_3NH_2(aq)$ is the base reactant.

Exercise 6.39—Acids and bases

(a) $H_3O^+(aq) + OH^-(aq) \longrightarrow 2\,H_2O(l)$

$Li^+(aq)$ and $Br^-(aq)$ ions are spectator ions—they do not appear in the net reaction.

(b) $H_2CO_3(aq) + OH^-(aq) \longrightarrow HCO_3^-(aq) + H_2O(l)$

and $HCO_3^-(aq) + OH^-(aq) \longrightarrow CO_3^{2-}(aq) + H_2O(l)$

$K^+(aq)$ ion is a spectator ion.

(c) $H_3O^+(aq) + CH_3NH_2(aq) \longrightarrow H_2O(l) + CH_3NH_3^+(aq)$

$NO_3^-(aq)$ is a spectator ion

(d) $H_3Citrate(aq) + OH^-(aq) \longrightarrow H_2Citrate^-(aq) + H_2O(l)$

$H_2Citrate^-(aq) + OH^-(aq) \longrightarrow HCitrate^{2-}(aq) + H_2O(l)$

and $HCitrate^{2-}(aq) + OH^-(aq) \longrightarrow Citrate^{3-}(aq) + H_2O(l)$

$Na^+(aq)$ ion is a spectator ion.

Exercise 6.40—Complexation reactions

(a) Co^{2+} is the Lewis acid, while Cl^- is the Lewis base.
(b) Fe^{2+} is the Lewis acid, while CN^- is the Lewis base.
(c) Ni^{2+} is the Lewis acid, while CH_3NH_2 is the Lewis base.
(d) Cu^{2+} is the Lewis acid, while Cl^- and NH_3 are the Lewis bases.

Exercise 6.41—Aquation as complexation

(a) $[Fe(OH_2)_6]^{2+}$

(b) $[Mn(OH_2)_6]^{2+}$

(c) $[Al(OH_2)_6]^{3+}$

(d) $[Cr(OH_2)_6]^{3+}$

These complex ions can be pictorially represented as follows.

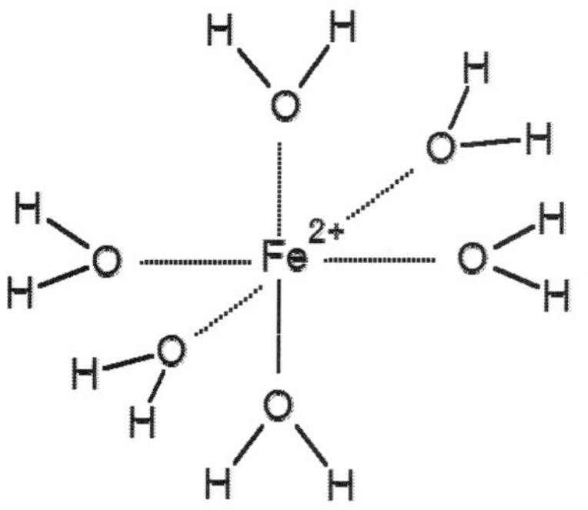

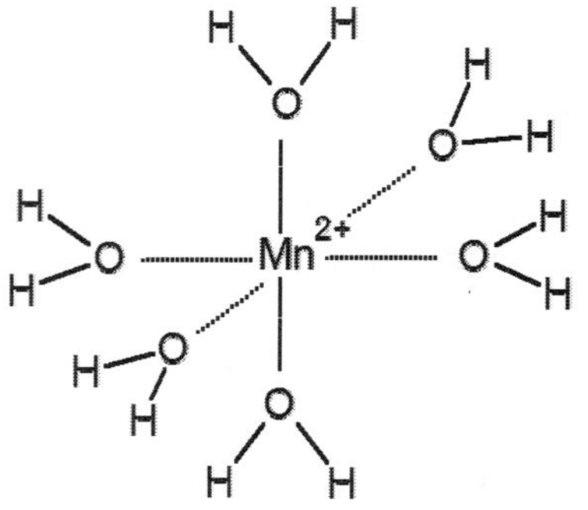

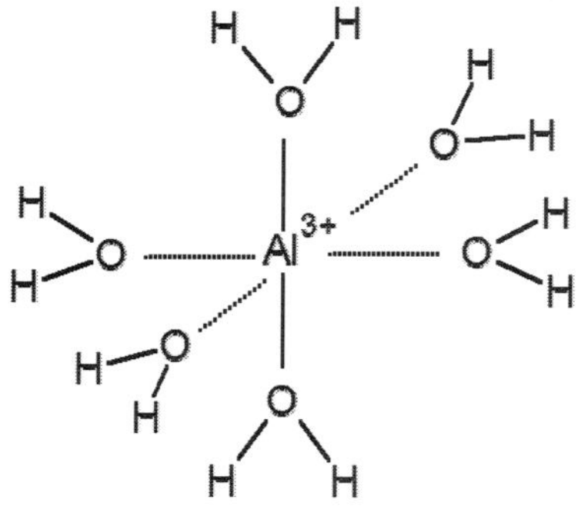

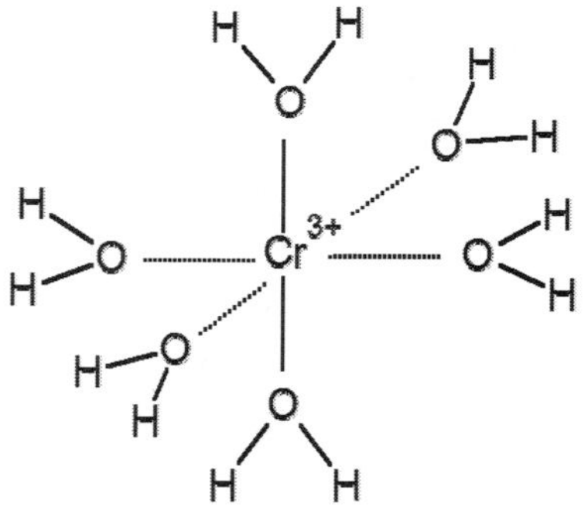

Exercise 6.42—Aquation as complexation

(a) $[Co(OH_2)_6]^{2+}(aq) + 4\,Cl^-(aq) \rightarrow [CoCl_4]^{2-}(aq) + 6\,H_2O(l)$
Lewis base, Cl^-, competes successfully with the Lewis base, H_2O, to bond to Co^{2+} ions.
(b) $[Fe(OH_2)_6]^{2+}(aq) + 6\,CN^-(aq) \rightarrow [Fe(CN)_6]^{4-}(aq) + 6\,H_2O(l)$
Lewis base, CN^-, competes successfully with the Lewis base, H_2O, to bond to Fe^{2+} ions.
(c) $[Ni(OH_2)_6]^{2+}(aq) + 2\,CH_3NH_2(aq) \rightarrow [Ni(CH_3NH_2)_2]^{2+}(aq) + 6\,H_2O(l)$
Lewis base, CH_3NH_2, competes successfully with the Lewis base, H_2O, to bond to Ni^{2+} ions.
(d) $[Cu(OH_2)_6]^{2+}(aq) + 4\,Cl^-(aq) + 2\,NH_3(aq) \rightarrow [Cu(NH_3)_2Cl_4]^{2-}(aq) + 6\,H_2O(l)$
Lewis bases, Cl^- and NH_3, compete successfully with the Lewis base, H_2O, to bond to Cu^{2+} ions.

Exercise 6.43—Solution concentration

$$\text{Amount of } Na_2CO_3,\ n = \frac{m}{M} = \frac{25.3\ \text{g}}{105.99\ \text{g mol}^{-1}} = 0.239\ \text{mol}$$

Molarity of sodium carbonate solution

$$= c = \frac{\text{amount of solute (mol)}}{\text{volume (L)}} = \frac{0.239\ \text{mol}}{0.500\ \text{L}} = 0.478\ \text{mol L}^{-1}$$

Exercise 6.47—Solution concentration

(a) (A) $(0.20\ \text{mol}) / (0.200\ \text{L}) = 1.0\ \text{mol L}^{-1}$ is more concentrated than (B) $(0.05\ \text{mol}) / (0.200\ \text{L}) = 0.25\ \text{mol L}^{-1}$
(b) (A) $(0.20\ \text{mol}) / (0.200\ \text{L}) = 1.0\ \text{mol L}^{-1}$ is more concentrated than (B) $(0.50\ \text{mol}) / (1.00\ \text{L}) = 0.50\ \text{mol L}^{-1}$
(c) (B) $(0.20\ \text{mol}) / (0.100\ \text{L}) = 2.0\ \text{mol L}^{-1}$ is more concentrated than (A) $(0.20\ \text{mol}) / (0.200\ \text{L}) = 1.0\ \text{mol L}^{-1}$
(d) (A) $(0.020\ \text{mol}) / (0.020\ \text{L}) = 1.0\ \text{mol L}^{-1}$ is more concentrated than (B) $(0.20\ \text{mol}) / (0.500\ \text{L}) = 0.40\ \text{mol L}^{-1}$
(e) (A) 0.20 mol sucrose in 200 mL solution and (B) a 10 mL sample of solution A have the same concentration—(B) is just a sample of (A).

Exercise 6.49—Concentrations of species

(a) $2 \times 0.25\ \text{mol L}^{-1} = 0.50\ \text{mol L}^{-1}$ $NH_4^+(aq)$ ions and $\times 0.25\ \text{mol L}^{-1}$ $SO_4^{2-}(aq)$ ions
$NH_4^+(aq)$ ions are a weak acid which do dissociate to a small extent—we neglect this here
(b) $2 \times 0.123\ \text{mol L}^{-1} = 0.246\ \text{mol L}^{-1}$ $Na^+(aq)$ ions and $0.123\ \text{mol L}^{-1}$ $CO_3^{2-}(aq)$ ions
(c) $0.056\ \text{mol L}^{-1}$ $H_3O^+(aq)$ ions and $NO_3^-(aq)$ ions

REVIEW QUESTIONS

Section 6.2 The remarkable properties of water

6.51
Water has a lower vapour pressure than hexane, 3.17 vs. 20.2 kPa at 25 °C. It takes a higher temperature for the vapour pressure of water to reach 1 atm, than it does for hexane—i.e., water has a higher boiling point.

Section 6.3 Intermolecular forces

6.53
(a) Bonds are broken and formed when iron rusts—these are intramolecular forces.
(b) Long-stranded polymer molecules are pulled together when a rubber band is stretched—chemical bonds are not made or broken. Intermolecular molecular forces are at play here.
(c) Ultraviolet light, in the stratosphere, breaks O-O bonds to form O atoms. Intramolecular forces are overcome.
(d) Mothballs gradually 'disappear' by sublimation wherein a solid material passes directly into the gas phase. Here, intermolecular forces holding the molecules in the solid phase are overcome. There is a steady outflux of naphthalene—the mothball material—until the mothball has disappeared.

6.55
(a) $C\rightarrow O > C\rightarrow N$
    i.e., $C\rightarrow O$ is more polar than $C\rightarrow N$
(b) $P\rightarrow Cl > P\rightarrow Br$
(c) $B\rightarrow O > B\rightarrow S$
(d) $B\rightarrow F > B\rightarrow I$

6.57
The dipole moments are ordered as HF > HCl > HBr > HI. In each case, the dipole points toward the halogen. While the dipole moment depends on the charge separation which is indicative of difference in electronegativity, it also depends on the bond distance. Note that even though the bond distance increases in this series, the dipole moments decrease. This indicates a decreasing sequence of differences in electronegativity which corresponds to a decreasing sequence of halogen electronegativities—F > Cl > Br > I.

6.59
(a) Linear $BeCl_2$ is non-polar because of symmetry—the bond dipoles cancel.
(b) Trigonal planar HBF_2 is polar because one of the three bonds is different—the bond dipoles do not cancel exactly. The negative end is at the two F atoms ($\chi = 4.0$).
(c) Tetrahedral CH_3Cl, is polar because one of the four bonds is different—the bond dipoles do not cancel. The negative end is at the Cl atom. Note that in this case the three H-C bond dipoles reinforce the C-Cl bond dipole.
(d) Trigonal pyramidal SO_3 is non-polar because of symmetry—the bond dipoles cancel.

6.61
(a) There are NO dipole-dipole forces between hexane molecules.
(b) There are dipole-dipole forces between H_2S molecules.
(c) There are NO dipole-dipole forces between methane molecules.

6.63
(b)
i.e., hydrogen bonding operates in only formic acid, HCOOH. In H_2Se, HI, CH_3COCH_3, hydrogen is bonded to a more electronegative atom, but not one of N, O or F. Hydrogen bonds form only between these small (second period), electronegative atoms—the concentration of charge is important for the effect.

6.65
Kr atoms are more polarizable than Ne atoms because Kr atoms are bigger, with more electrons.

6.67
(a) When O_2 vaporizes, dispersion forces are overcome.
(b) When mercury vaporizes, metallic bonds are overcome.
(c) When CH_3I vaporizes, dipole-dipole and dispersion forces are overcome.
(d) When CH_3CH_2OH vaporizes, H-bonds (special dipole-dipole interactions) and dispersion forces are overcome.

6.69
In order of increasing force of attraction, we have
(a) < (b) < (c) < (d)
i.e., Ne < CH_4 < CO < CCl_4
The series simply reflects the size of dispersion forces—they increase with increasing number of electrons. The dipole-dipole forces in CO are not as big as the dispersion forces in CCl_4.

Section 6.4 Explaining the properties of water

6.71
(a) The stronger are the intermolecular forces in a liquid, its normal boiling point is *higher*.
(b) The weaker are the intermolecular forces in a liquid, its equilibrium vapour pressure at a specified temperature is *higher*.
(c) The smaller the volume of water in a sealed flask, its equilibrium vapour pressure is *unchanged*.
(d) The higher the temperature of a liquid, its equilibrium vapour pressure is *higher*.
(e) The more volatile a liquid, its boiling point is *smaller*.
(f) The higher the boiling point of a liquid, its enthalpy change of vaporization is *higher*.

6.73
Propan-1-ol ($CH_3CH_2CH_2OH$) has a higher boiling point than methyl ethyl ether ($CH_3CH_2OCH_3$), a compound with the same empirical formula, because propan-1-ol molecules can form hydrogen bonds, whereas methyl ethyl ether cannot. Both molecules have dipole-dipole forces, but they are the especially strong hydrogen bonds in the case of propan-1-ol.

6.75

With respect to boiling point, we have

(a) $O_2 > N_2$ because the oxygen molecules are bigger—bigger dispersion forces

(b) $SO_2 > CO_2$ because SO_2 is bent—it has a net dipole moment, and dipole-dipole intermolecular forces

(c) HF > HI because of hydrogen bonding in the case of HF

(d) $SiH_4 < GeH_4$ because GeH_4 is bigger

6.77

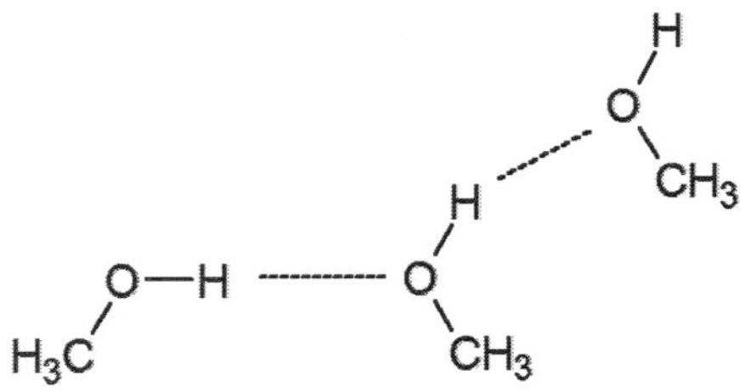

6.79

(a) Water has a higher viscosity than hexane, in spite of its smaller dispersion forces, because of its strong hydrogen bonds.

(b) Glycerol (propan-1,2,3-triol, $HOCH_2CHOHCH_2OH$) is even more viscous than water because it has three O-H groups and can form more hydrogen bonds than water.

6.81

The melting point of fumaric acid (287 °C) is much higher than that of maleic acid (131 °C) even though these substances are just *cis* and *trans* isomers.

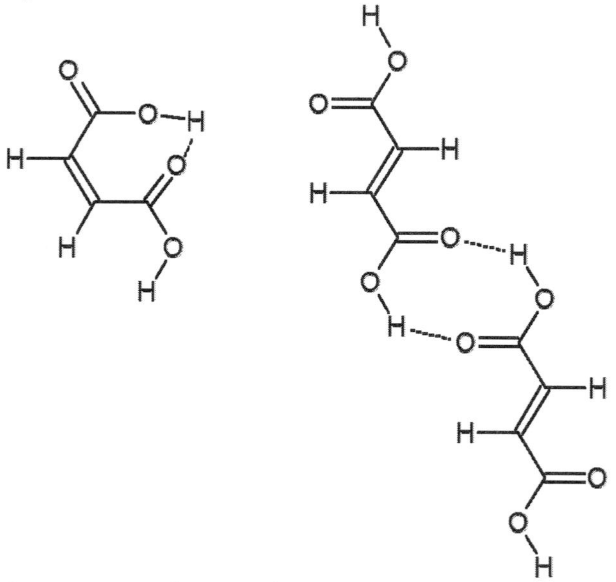

Maleic acid, on the left, makes a strong intramolecular hydrogen bond—this reduces opportunities for intermolecular hydrogen bonds, as an O and H are already hydrogen bonding. Strong intermolecular pairs of hydrogen bonds are formed between adjacent fumaric acid molecules.

Section 6.5 Water as a solvent

6.83
(a) The O end (the negative end) of water points to Ca^{2+}(aq).
(b) The H end (the positive end) of water points to Br^-(aq).
(c) The H end (the positive end) of water points to $Cr_2O_7^{2-}$(aq).
(d) The O end (the negative end) of water points to NH_4^+(aq).

6.85
When silver nitrate, $AgNO_3$(s) and potassium chloride, KCl(s) are dissolved in some water, the possible reactant species are H_2O(l), Ag^+(aq) ions, NO_3^-(aq) ions, K^+(aq) ions and Cl^-(aq) ions. In fact, Ag^+(aq) ions and Cl^-(aq) ions are the reactants in a precipitation reaction. The other two solvated ions are spectators – they do not participate in the reaction.

6.87
Cooking oil is not miscible with water because its molecules are non-polar (or very weakly polar). They do not interact with water strongly enough for water to solvate them—water molecules prefer to interact with other water molecules. Cooking oil is soluble in hexane—a non-polar solvent.

Section 6.6 Self-ionization of water

6.89
The concentrations of hydronium ions and hydroxide ions are equal in pure water because of the stoichiometry of the self-ionization reaction,
$$2\,H_2O(l) \rightarrow H_3O^+(aq) + OH^-(aq)$$
i.e., we get 1 hydronium ion for each hydroxide ion formed.

Section 6.7 Categories of chemical reaction in water

6.91
Since barium sulfate, $BaSO_4$(s), precipitates from water, whereas iron (II) sulfate does not, we conclude that $BaSO_4$(s) is insoluble in water (it has a very low solubility) whereas $FeSO_4$(s) is soluble.

6.93
$$2\,Cr_2O_7^{2-}(aq) + 28\,H^+(aq) + 12\,e^- \longrightarrow 4\,Cr^{3+}(aq) + 14\,H_2O(l)$$
$$3\,C_2H_5OH(aq) + 3\,H_2O(l) \longrightarrow 3\,CH_3COOH(aq) + 12\,H^+(aq) + 12\,e^-$$
Here, ethanol, C_2H_5OH(aq), is oxidized to acetic acid, CH_3COOH(aq). Aquated dichromate ions, $Cr_2O_7^{2-}$(aq), are the oxidizing agent. $Cr_2O_7^{2-}$(aq) ions are reduced to Cr^{3+}(aq) ions. Ethanol is the reducing agent.

6.95

When nitric acid dissolves in water, nitrate and hydronium ions are produced. Because nitric acid is a strong acid, there is a negligible concentration of undissociated aquated HNO_3.
When barium hydroxide dissolves in water, $Ba^{2+}(aq)$ ions and $OH^-(aq)$ ions are produced.

6.97

(a) An iron(II) ion is bound to six ammonia molecules:

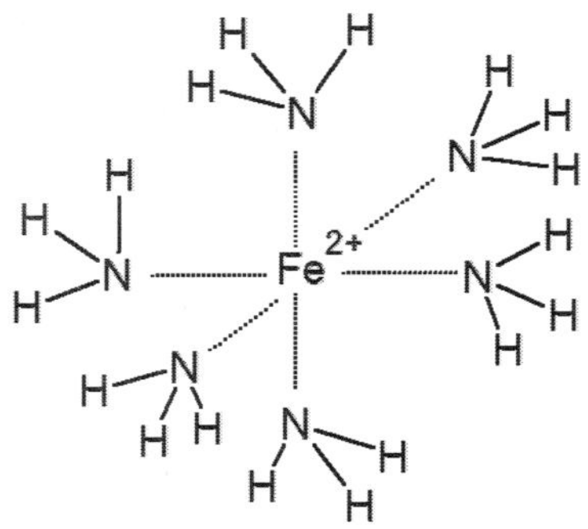

The net charge on this complex is +2.

(b) A zinc ion is bound to four cyanide ions:

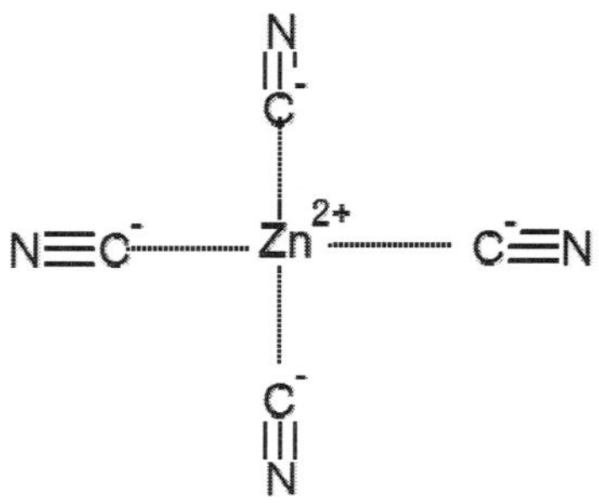

The net charge is $-4 + 2 = -2$.

(c) A manganese(II) ion is bound to six fluoride ions:

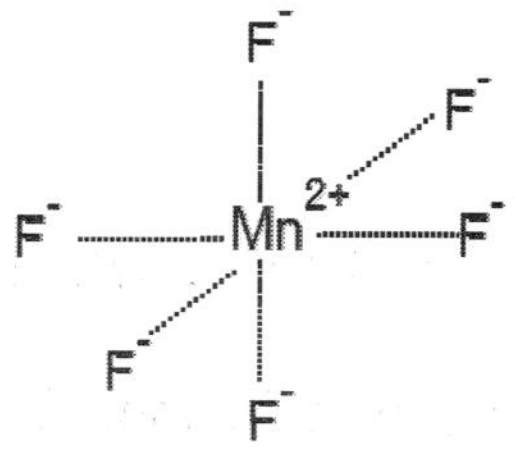

The net charge is $-6 + 2 = -4$.

(d) An iron(III) ion is bound to six cyanide ions:

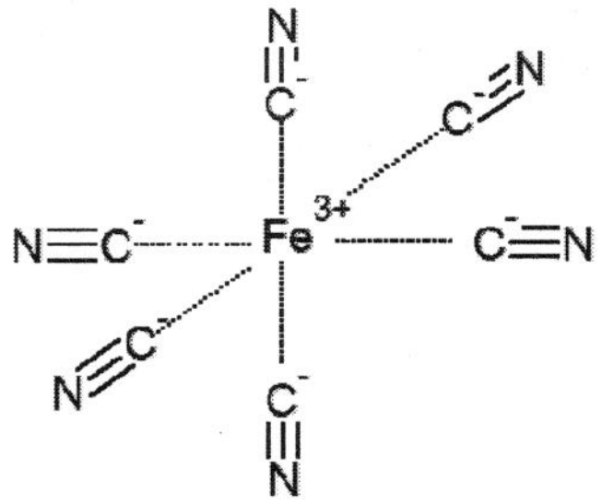

The net charge is $-6 + 3 = -3$.

(e) A cobalt(II) ion is bound to four chloride ions:

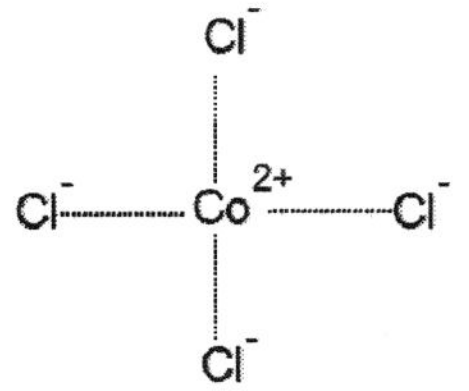

The net charge is $-4 + 2 = -2$.

(f) A nickel(II) ion is bound to four ammonia molecules and two water molecules:

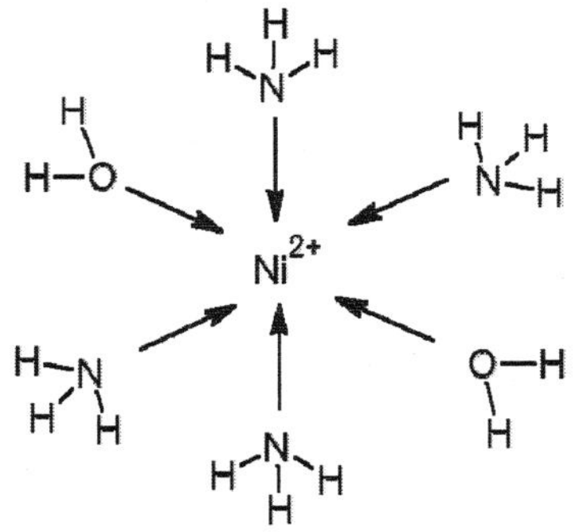

This is the *trans* isomer. The net charge is $+2$

Section 6.8 Solution concentration

6.99

Volume = 250 mL = 0.250 L

Amount of $AgNO_3$ required = volume × (concentration of $AgNO_3$) = (0.250 L) × (0.0200 mol L^{-1}) = 0.00500 mol

Mass of $AgNO_3$ required = $n(AgNO_3)$ × $M(AgNO_3)$ = (0.00500 mol) × (169.87 g mol^{-1}) = 0.849 g

To make the desired solution, carefully weigh 0.849 g of $AgNO_3(s)$, add it to the volumetric flask, add ~150 mL of de-ionized water, stopper the flask and shake to dissolve the $AgNO_3$ and ensure a homogenenous solution. After the $AgNO_3$ has dissolved, top up the volume to the 250 mL mark with de-ionized water—add the additional de-ionized water in steps, swirling between each step to ensure a homogeneous solution.

6.101
Amount of NaCl, n = (volume of NaCl solution) × (concentration of NaCl) = (1.0 L) × (0.1 mol L^{-1}) = 0.1 mol
Mass of NaCl, m = n(NaCl) × M(NaCl) = (0.1 mol) × (58.44 g mol^{-1}) = 6 g
Volume of Na_2CO_3 solution = 1250 mL = 1.250 L
Amount of Na_2CO_3, n = (volume of Na_2CO_3 solution) × (concentration of Na_2CO_3) = (1.250 L) × (0.060 mol L^{-1})
= 0.075 mol
Mass of Na_2CO_3, m = n(Na_2CO_3) × M(Na_2CO_3) = (0.075 mol) × (105.99 g mol^{-1}) = 7.9 g > 6 g
1250 mL of 0.060 mol L^{-1} Na_2CO_3 solution has the greater mass of solute

6.103
Volume of solution = 250 mL = 0.250 L
Amount of $KMnO_4$, n = (volume of solution) × (concentration of $KMnO_4$) = (0.250 L) × (0.0125 mol L^{-1})
= 0.003125 mol
Mass of $KMnO_4$, m = n($KMnO_4$) × M($KMnO_4$) = (0.003125 mol) × (158.03 g mol^{-1}) = 0.494 g

6.105
Amount of NaOH, n = m(NaOH) / M(NaOH) = (25.0 g) / (40.00 g mol^{-1}) = 0.625 mol
Volume of solution = (amount of NaOH) / (concentration of NaOH) = (0.625 mol) / (0.123 mol L^{-1}) = 5.08 L

6.107
Volume of solution = 250 mL = 0.250 L
Amount of $H_2C_2O_4$, n = Volume × concentration of $H_2C_2O_4$ = (0.250 L) × (0.15 mol L^{-1}) = 0.0375 mol
Mass of $H_2C_2O_4$, m = n($H_2C_2O_4$) × M($H_2C_2O_4$) = (0.0375 mol) × (90.04 g mol^{-1}) = 3.38 g

6.109
Amount of Na_2CO_3, n = m(Na_2CO_3) / M(Na_2CO_3) = (6.73 g) / (105.99 g mol^{-1}) = 0.0635 mol
Volume of solution = 250 mL = 0.250 L
Concentration of Na_2CO_3, c = n(Na_2CO_3) / Volume = (0.0635 mol) / (0.250 L) = 0.254 mol L^{-1}
$[Na^+]$ = 2 × c(Na_2CO_3) = 2 × 0.254 mol L^{-1} = 0.508 mol L^{-1}
$[CO_3^{2-}]$ = c(Na_2CO_3) = 0.254 mol L^{-1}

SUMMARY AND CONCEPTUAL QUESTIONS

6.111 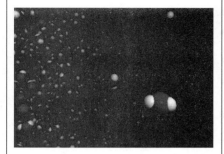	The figure on the left is a frame from the animation in E6.4 Molecular-level modelling activity depicting the inside of a bubble in boiling water. **(a)** Kinetic energy of the water molecules within and on the surface of the bubble prevents its collapse. These "hot" molecules smash into their neighbours pushing open the bubble. **(b)** Bubbles form when the vapour pressure within the heated liquid slightly exceeds atmospheric pressure—enough to push open the bubble.
6.113 	The figure on the left is a frame from the animation in E6.21 Molecular-level modelling activity depicting a hydronium ion transferring a proton to another water molecule. **(a)** Hydronium ions appear to diffuse very quickly through an aqueous solution—faster than they should be able to via diffusion—because the transfer of hydronium across a cluster of water molecules is achieved through successive H^+ ion transfers between adjacent water molecules. **(b)** Each proton transfer involves one O atom of a hydronium ion taking the pair of electrons from an H atom bonded to it, and another O atom (on a water molecule on the other side of the H) donating a lone pair of electrons to the released proton—i.e., accepting the proton.
6.115 	The figure on the left is a frame from the animation in E6.22 Molecular-level modelling activity depicting the formation of a AgCl precipitate. **(a)** Here, the electrostatic attraction of the Ag^+ and Cl^- ions overcomes the dipole-ion forces between solvating water molecules and the ions to cause growth of the ionic lattice. **(b)** The lattice is an array of Ag^+ and Cl^- ions. It does not consist of distinct AgCl molecules.

6.117 	The figure on the left is a frame from the animation in E6.25 Molecular-level modelling activity depicting proton transfer from an acetic acid molecule to a water molecule. **(a)** The electrons in the O–H bond of the CH_3COOH molecule go entirely to the O atom releasing a proton which attaches to a lone pair of electrons on a water molecule. **(b)** A covalent bond is broken and a new one made.

6.119 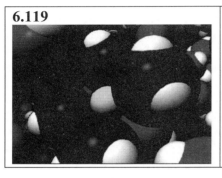	The figure on the left is a frame from the animation in E6.27 Molecular-level modelling activity depicting the formation of a tetraamminecopper(II) complex ion. **(a)** Ammonia is the Lewis base, donating lone pairs of electrons to the copper atom, in the form of a coordinate bond. Cu^{2+} is the Lewis acid. **(b)** Ammonia molecules replace water molecules because the N in ammonia forms a stronger coordinate bond with copper than the O in water—ammonia is a better ligand.

6.121

(a) $Ca(OH)_2(s) + 2H_3O^+(aq) \longrightarrow Ca^{2+}(aq) + 4H_2O(l)$

is an acid-base reaction, (iii).

(b) $CH_3COOH(aq) + Ag^+(aq) + H_2O(l) \longrightarrow AgCH_3COO(s) + H_3O^+(aq)$

is a precipitation reaction, (i).

(c) $Fe(OH)_3(s) + 3H_2C_2O_4(aq) \longrightarrow [Fe(C_2O_4)_3]^{3-}(aq) + 3H_2O(l)$

is an acid-base, (iii), and complexation, (iv), reaction. The oxalic acid neutralizes the hydroxyls initially coordinated to Fe^{2+} (an acid-base reaction), then forms a bond to the iron ion itself—it is a bidentate ligand.

(d) $5Fe^{2+}(aq) + MnO_4^-(aq) + 8H^+(aq) \longrightarrow 5Fe^{3+}(aq) + Mn^{2+}(aq) + 4H_2O(l)$

is an oxidation-reduction reaction, (ii).

Chapter 7
Chemical Reactions and Energy Flows

EXERCISES FROM CHAPTER

Exercise 7.1—Exothermic and endothermic processes

(a) Since $H^+(aq) + OH^-(aq) \longrightarrow H_2O(l)$ is an exothermic process (the change in energy is negative),
1 mol each of $H^+(aq)$ ions and $OH^-(aq)$ ions, that have not reacted, have more energy than 1 mol of water. The difference is released as heat.
(b) If this reaction takes place in a test tube, where you are holding your fingers would feel hot. This is because the decrease of energy of the system requires removal of excess energy—it leaves in the form of heat that your fingers sense.

Exercise 7.3—Energy conversion

The chemical potential energy stored in a battery can be converted to the mechanical energy of sound waves—from your mp3 player—the electrical and magnetic energy of an image taken by your digital camera, or the light energy emitted by a flashlight—to name a few possibilities.

Exercise 7.4—Temperature and direction of heat flow

Although the energy of the air in a classroom at 25 °C is much greater than the energy in a 50 mL beaker of water at 35 °C, heat will flow from the 50 mL beaker of water at 35 °C to the air at 25 °C. The energy is more concentrated in the beaker of water, as indicated by its higher temperature. Heat always flows from higher temperature to lower temperature.

Exercise 7.5—System and surroundings

What constitutes the system can be defined at will. The following are sensible suggestions:
(a) The system is the contents of the combustion chamber of the gas furnace—a mixture of air and methane. The surroundings are the furnace and everything around it.
(b) The system is the water drops plus the air around you. The surroundings consist of your body and the sun—they provide the heat that evaporates the water drops.
(c) The water, initially at 25 °C, is the system. The container and the rest of the freezer contents—including the air—are the surroundings.
(d) The aluminum and Fe2O3(s) mixture is the system (initially—later it consists of Al2O3(s) and iron). The flask and the laboratory bench are the surroundings.

Exercise 7.6—State functions

(a) The volume of a balloon is a state function. Note that the volume might change due to leakage from the balloon, but at each time it depends only on the thermodynamic state of the balloon plus air inside.

(b) The time it takes to drive from your home to your college or university is NOT a state function. It characterizes a process (travelling between places) rather than a state (such as where you are when you are at home).

(c) The temperature of the water in a coffee cup is a state function.

(d) The potential energy of a ball held in your hand is a state function—it is a function of the height of the ball – the ball's "vertical" state.

Exercise 7.7—Enthalpy change and change of state

When 1.0 L of water at 0 °C freezes to form ice at 0 °C,

$$\text{heat evolved} = -(\text{change in enthalpy})$$
$$= (\text{amount of water - in mol}) \times (\text{molar enthalpy of fusion} - \text{in kJ mol}^{-1})$$
$$\text{mass of water} = \text{volume} \times \text{density} = (1.0 \text{ L}) \times (1.0 \text{ g mL}^{-1}) = (1.0 \times 10^3 \text{ mL}) \times (1.0 \text{ g mL}^{-1})$$
$$= 1.0 \times 10^3 \text{ g}$$

$$\text{Heat evolved, } q = 6.00 \text{ kJ mol}^{-1} \times \frac{1.0 \times 10^3 \text{ g}}{18.02 \text{ g mol}^{-1}} = 333 \text{ kJ}$$

Exercise 7.10—Enthalpy change and change of state

Step 1: To warm 25.0 g of liquid methanol from 25.0 °C to its boiling point, 64.6 °C:
Heat required for step 1 = (specific heat capacity of methanol) × (mass of methanol) × (temperature change)

$$= (2.53 \text{ J K}^{-1} \text{ g}^{-1}) \times (25.0 \text{ g}) \times (64.4 - 25.0 \text{ K}) = 2490 \text{ J} = 2.49 \text{ kJ}$$

Step 2: To completely evaporate the methanol at its boiling point:
Heat required for step 2 = (molar enthalpy change of vaporization of methanol) × (amount of methanol)

$$= q = 37.4 \text{ kJ mol}^{-1} \times \frac{25.0 \text{ g}}{32.04 \text{ g mol}^{-1}} = 29.2 \text{ kJ}$$

Altogether, the total heat required for both steps is
$$2.9 \text{ kJ} + 29.2 \text{ kJ} = 32.1 \text{ kJ}$$

Exercise 7.11—Enthalpy change and change of state

Mass of mercury = volume × density = (1.00 mL) × (13.6 g mL^{-1}) = 13.6 g
Heat released upon freezing = (molar enthalpy change of fusion of mercury) × (amount of mercury)

$$= q = 2.3 \text{ kJ mol}^{-1} \times \frac{13.6 \text{ g}}{200.59 \text{ g mol}^{-1}} = 0.16 \text{ kJ}$$

Exercise 7.12—Enthalpy change of reaction

$$2\,NO(g) + O_2(g) \longrightarrow 2\,NO_2(g) \qquad \Delta_r H = -114.1\,kJ$$

is exothermic – the enthalpy change is negative. 114.1 kJ of heat is evolved for every 2 mol of NO(g) that reacts

If 1.25 g of NO is converted completely to NO_2, then

heat evolved by reaction $= (-$ enthalpy change of reaction$) \times 1/2($amount of NO reacted$)$

$$\text{Heat evolved} = q = 114.1\,kJ\ \text{mol}^{-1} \times \frac{1}{2} \times \frac{1.25\ \text{g}}{30.01\ \text{g mol}^{-1}} = 2.38\ kJ$$

Exercise 7.15—Using a calorimeter to measure $\Delta_r H$

First, we assume that all of the heat evolved stays within the calorimeter. This is the purpose of the styrofoam insulation—the coffee cup. The specific heat capacity of the reaction solution is 4.20 J K^{-1} g^{-1}, while the density of the solution is 1.00 g mL^{-1}.

Mass inside calorimeter $=$ (total volume) $\times$ (density of solution) $=$ $(200 + 200\ mL) \times (1.00\ g\ mL^{-1})$ $= 400\ g$

The heat evolved when 200 mL of 0.400 mol L^{-1} HCl solution is mixed with 200 mL of 0.400 mol L^{-1} NaOH solution in a coffee-cup calorimeter is given by

Heat evolved $=$ (specific heat capacity) $\times$ (mass inside calorimeter) $\times$ (temperature change)

$= (4.20\ J\ K^{-1}\ g^{-1}) \times (400\ g) \times (27.78 - 25.10\ K) = 4502.4\ J$

The reaction in the calorimeter can be written

$$H_3O^+(aq) + OH^-(aq) \rightarrow 2\,H_2O(l)$$

so the $H_3O^+(aq)$ and $OH^-(aq)$ ions react in a 1:1 ratio of amounts. HCl and NaOH are strong electrolytes in water, so:

In 200 mL of 0.400 mol L^{-1} HCl solution, $n(H^+) = (0.200\ L) \times (0.400\ mol\ L^{-1}) = 0.0800\ mol$

In 200 mL of 0.400 mol L^{-1} NaOH solution, $n(OH^-) = (0.200\ L) \times (0.400\ mol\ L^{-1}) = 0.0800$ mol

The reacting species are present in stoichiometric amounts: neither is a limiting reagent, so 0.0800 mol of each ion reacts.

Molar enthalpy change of neutralization of $H^+(aq)$ ions $=$ heat liberated when 1 mol of $H^+(aq)$ ions react

$= -$ (heat evolved when 0.0800 mol of $H^+(aq)$ ions react) / (0.0800 mol) $= -$ (4502.4 J) / (0.0800 mol)

$= -56300\ J = -56.3\ kJ$

The minus sign denotes an exothermic reaction. Note that the reaction has enthalpy change -56.3 kJ at 25.10 °C. In the experiment the temperature increased to 27.78 °C. Imagine removing the heat evolved from the calorimeter, returning its temperature to 25.10 °C.

Exercise 7.16—Standard states of substances

(a) The standard state of bromine at 25 °C is molecular liquid $Br_2(l)$.
(b) The standard state of mercury at 25 °C is metallic liquid, $Hg(l)$.
(c) The standard state of sodium sulfate at 25 °C is ionic solid $Na_2SO_4(s)$.
(d) The standard state of ethanol at 25 °C is molecular liquid $CH_3CH_2OH(l)$.
(e) The standard state of aquated chloride ions in aqueous sodium chloride solution, at 25 °C, is 1 mol L^{-1} concentration $Cl^-(aq)$.

Exercise 7.17—Standard enthalpy change of reaction

(a) $CO(g) + \frac{1}{2} O_2(g) \longrightarrow CO_2(g)$

The standard enthalpy change of this reaction, $\Delta_r H°$, is the heat absorbed (hence a negative number when is evolved) per mole of $CO(g)$ that reacts at a constant temperature of 25 °C, in vessel in which each of $CO(g)$, $O_2(g)$ and $CO_2(g)$ have a partial pressure of 1 bar.

(b) $Mg(s) + 2 H^+(aq) \longrightarrow Mg^{2+}(aq) + H_2(g)$

The standard enthalpy change of this reaction, $\Delta_r H°$, is the heat absorbed per mole of $Mg(s)$ that reacts in a reaction mixture at 25 °C in which simultaneously there is pure $Mg(s)$, in solution $[H^+]$ = $[Mg^{2+}] = 1$ mol L^{-1}, and $H_2(g)$ is present at 1 bar pressure.

(c) $H^+(aq) + OH^-(aq) \longrightarrow H_2O(l)$

The standard enthalpy change of this reaction, $\Delta_r H°$, is the heat absorbed per mole of $H^+(aq)$ ions that react, under constant pressure conditions of 1 bar, and constant temperature at 25 °C, if in solution $[H^+] = [OH^-] = 1$ mol L^{-1}.

Exercise 7.19—Using Hess's law

(a) The desired reaction,
C(graphite) → C(diamond)
can be written as the sum of the combustion reaction for graphite and the reverse of the combustion reaction for diamond,

Equation 1: C(graphite) + O₂(g) ⟶ CO₂(g) $\Delta_r H° = \Delta_r H°_1 = -393.5$ kJ
Equation 2: CO₂(g) ⟶ C(diamond) + O₂(g) $\Delta_r H° = \Delta_r H°_2 = -(-395.4$ kJ$)$
Net Equation: C(graphite) ⟶ C(diamond) $\Delta_r H°_{net} = \Delta_r H°_1 + \Delta_r H°_2 = -393.5 + 395.4$ kJ $= 1.9$ kJ

The standard enthalpy change of this reaction, $\Delta_r H°$, is 1.9 kJ.
(b)

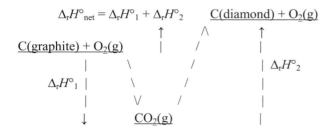

Exercise 7.20—Using Hess's law

$$C(s) + O_2(g) \longrightarrow CO_2(g) \qquad \Delta_r H°_1 = -393.5 \text{ kJ}$$
$$2\,S(s) + 2\,O_2(g) \longrightarrow 2\,SO_2(g) \qquad \Delta_r H°_2 = 2 \times (-296.8 \text{ kJ})$$
$$\underline{CO_2(g) + 2\,SO_2(g) \longrightarrow CS_2(l) + 3\,O_2(g) \qquad \Delta_r H°_3 = -(-1103.9 \text{ kJ})}$$
$$C(s) + 2\,S(s) \longrightarrow CS_2 \qquad \Delta_r H° = 116.8 \text{ kJ}$$

Exercise 7.21—Standard molar enthalpy changes of formation

(a) The standard molar enthalpy of formation of bromine, $Br_2(l)$, is the standard enthalpy change of the reaction:

$Br_2(l) \rightarrow Br_2(l)$ at 25 °C

(b) The standard molar enthalpy of formation of solid iron (III) chloride, $FeCl_3(s)$, is the standard enthalpy change of the reaction:

$Fe(s) + Cl_2(g) \rightarrow FeCl_3(s)$ at 25 °C

(c) The standard molar enthalpy of formation of solid sucrose, $C_{12}H_{22}O_{11}(s)$, is the standard enthalpy change of the reaction:

$12\ C(graphite) + 11\ H_2(g) + \dfrac{11}{2}O_2(g) \rightarrow C_{12}H_{22}O_{11}(s)$ at 25 °C

Exercise 7.22—Standard molar enthalpy changes of formation

The standard molar enthalpy of formation of liquid methanol, $CH_3OH(l)$, is the standard enthalpy change of the following reaction:

$C(graphite) + 2\ H_2(g) + \dfrac{1}{2}O_2(g) \rightarrow CH_3OH(l)$ at 25 °C

Its value is $\Delta_f H°[CH_3OH(l)] = -238.4 \text{ kJ mol}^{-1}$. Note that enthalpies of "named" reactions (such as formation reaction) are always given in kJ mol^{-1}.

Exercise 7.26—Calculation of $\Delta_r H°$ from $\Delta_f H°$ values

$$C_6H_6(l) + 7.5\,O_2(g) \longrightarrow 6\,CO_2(g) + 3\,H_2O(l)$$

$\Delta_r H°$ at 25 °C $= 6 \times \Delta_f H°[CO_2(g)] + 3 \times \Delta_f H°[H_2O(l)] - \Delta_f H°[C_6H_6(l)] - 7.5 \times \Delta_f H°[O_2(g)]$

$\qquad = 6 \times (-393.5 \text{ kJ mol}^{-1}) + 3 \times (-285.8 \text{ kJ mol}^{-1}) - (49.0 \text{ kJ mol}^{-1}) - 7.5 \times (O \text{ kJ mol}^{-1})$

$\qquad = -3267.4 \text{ kJ mol}^{-1}$

Exercise 7.27—Using bond energies to estimate enthalpy change of reaction

acetone   isopropanol

$$\Delta_r H = \sum D(\text{bonds broken}) - \sum D(\text{bonds formed})$$
$$= D(\text{C=O}) + D(\text{H-H}) - (D(\text{C-O}) + D(\text{C-H}) + D(\text{O-H}))$$
$$= 745 \text{ kJ mol}^{-1} + 436 \text{ kJ mol}^{-1} - (358 \text{ kJ mol}^{-1} + 413 \text{ kJ mol}^{-1} + 463 \text{ kJ mol}^{-1})$$
$$= -53 \text{ kJ mol}^{-1}$$

Exercise 7.28—Using bond energies to estimate enthalpy change of reaction

$$CH_4(g) + 2 O_2(g) \rightarrow CO_2(g) + 2 H_2O(g)$$
$$\Delta_r H = \sum D(\text{bonds broken}) - \sum D(\text{bonds formed})$$
$$= 4 \times D(\text{C-H}) + 2 \times D(\text{O=O}) - (2 \times D(\text{C=O}) + 2 \times 2 \times D(\text{O-H}))$$
$$= 4 \times 413 \text{ kJ mol}^{-1} + 2 \times 498 \text{ kJ mol}^{-1} - (2 \times 745 \text{ kJ mol}^{-1} + 2 \times 2 \times 463 \text{ kJ mol}^{-1})$$
$$= -694 \text{ kJ mol}^{-1}$$

Exercise 7.31—Energy from food

(a) Breaking the P—O bond in ATP is an exothermic process. **FALSE**
Breaking bonds is always endothermic—it requires energy input.
(b) Making a new bond between the phosphorus atom in the phosphate group being cleaved off ATP and the OH group of water is an exothermic process. **TRUE**
Making a bond (in the absence of other concurrent processes) is always exothermic—it releases energy.
(c) Breaking bonds is an endothermic process. **TRUE**
(d) The energy released in the hydrolysis of ATP may be used to run endothermic reactions in a cell. **TRUE.**
The cell uses ATP as an energy source—it couples other reactions to the hydrolysis of ATP to harness the chemical energy released in the latter process—processes achieved this way are mostly endothermic—they need the energy input.

REVIEW QUESTIONS

7.2—Chemical changes and energy redistribution

7.33
The liquid water has more energy. Melting is an endothermic process—energy is put into the ice to convert it to liquid water.

7.5—Enthalpy changes accompanying changes of state

7.35
Amount of benzene, $n = (125 \text{ g}) / (78.11 \text{ g mol}^{-1}) = 1.60 \text{ mol}$
Heat required to vaporize 125 g of benzene, $q = (1.60 \text{ mol}) \times (30.8 \text{ kJ mol}^{-1}) = 49.3 \text{ kJ}$

7.37
Heat required to convert 60.1 g of ice to liquid water at 0.0 °C
$\quad = (333 \text{ J g}^{-1}) \times (60.1 \text{ g}) = 2.00 \times 10^4 \text{ J} = 20.0 \text{ kJ}$
Heat required to raise temperature of liquid water from 0.0 °C to 100.0 °C
$\quad = (4.18 \text{ J K}^{-1} \text{g}^{-1}) \times (60.1 \text{ g}) \times (100 \text{ K}) = 2.51 \times 10^4 \text{ J} = 25.1 \text{ kJ}$
Heat required to convert liquid water to vapour at 100.0 °C
$\quad = (2260 \text{ J g}^{-1}) \times (60.1 \text{ g}) = 1.36 \times 10^5 \text{ J} = 136 \text{ kJ}$
Total heat required for the three steps $= (20.0 + 25.1 + 136) \text{ kJ} = 181 \text{ kJ}$

7.6—Enthalpy change of reaction, $\Delta_r H$

7.39
$$H_2(g) + \frac{1}{2}O_2(g) \longrightarrow H_2O(l) \qquad \Delta_r H = -285.8 \text{ kJ}$$
Decomposition of liquid water to hydrogen and oxygen gases is the reverse of this reaction. Formation of water is exothermic, so its decomposition is endothermic (heat is absorbed)
$\quad$ Heat absorbed, $q = (285.8 \text{ kJ mol}^{-1}) \times \{(12.6 \text{ g}) / (18.02 \text{ g mol}^{-1})\} = 200 \text{ kJ}$

7.41
$$CH_3OH(l) + CO(g) \longrightarrow CH_3COOH(l) \qquad \Delta_r H = -355.9 \text{ kJ}$$
The reaction is exothermic. Producing 1.00 L of acetic acid in this way evolves
$q = (355.9 \text{ kJ mol}^{-1}) \times \{(1.00 \times 10^3 \text{ mL}) \times (1.044 \text{ g mL}^{-1}) / (60.05 \text{ g mol}^{-1})\}$
$\quad = 6190 \text{ kJ}$

7.43

Heat absorbed by dissolution of 5.44 g of $NH_4NO_3(s)$

$= (4.20\ J\ K^{-1}g^{-1}) \times (155.4\ g) \times (18.6 - 16.2\ K) = 1.57 \times 10^3\ J = 1.57\ kJ$

Note the reversal of initial and final temperatures because we compute heat "absorbed"—rather than heat evolved.

Amount of $NH_4NO_3(s) = (5.44\ g) / (80.04\ g\ mol^{-1}) = 0.0680\ mol$

Per mole of $NH_4NO_3(s)$ that dissolves,

heat evolved $= (1.57\ kJ) / (0.0680\ mol) = 23\ kJ$

7.45

(a) Pure $H_2O(l)$ is the standard state of water at 25 °C.

(b) Pure $NaCl(s)$ is the standard state of sodium chloride at 25 °C.

(c) Pure $Hg(l)$ is the standard state of mercury at 25 °C.

(d) $CH_4(g)$ at 1 bar pressure is the standard state of methane at 25 °C.

(e) The standard state for $Na^+(aq)$ ions in solution, regardless of their origin, is $[Na^+] = 1.0\ mol\ L^{-1}$.

7.7—Hess's law

7.47

$\frac{1}{2}\{N_2(g) + 3H_2(g) \longrightarrow 2NH_3(g)\}$ $\qquad \Delta_r H° = \frac{1}{2} \times (-91.8\ kJ\ mol^{-1})$

$\frac{1}{4}\{4NH_3(g) + 5O_2(g) \longrightarrow 4NO(g) + 6H_2O(g)\}$ $\qquad \Delta_r H° = \frac{1}{4} \times (-906.2\ kJ\ mol^{-1})$

$\frac{3}{2}\{H_2O(g) \longrightarrow H_2(g) + \frac{1}{2}O_2(g)\}$ $\qquad \Delta_r H° = -\frac{3}{2} \times (-241.8\ kJ\ mol^{-1})$

_____

$\frac{1}{2}N_2(g) + \frac{1}{2}O_2(g) \longrightarrow NO(g)$ $\qquad \Delta_r H° = 90.3\ kJ\ mol^{-1}$

7.8—Standard molar enthalpy change of formation

7.51

$4NH_3(g) + 5O_2(g) \longrightarrow 4NO(g) + 6H_2O(g)$

$\Delta_r H° = 4 \times \Delta_r H°[NO(g)] + 6 \times \Delta_r H°[H_2O(g)] - 4 \times \Delta_r H°[NH_3(g)] - 5 \times \Delta_r H°[O_2(g)]$

$\quad = (4 \times 90.25\ kJ\ mol^{-1}) + (6 \times -241.818\ kJ\ mol^{-1}) - (4 \times -46.11\ kJ\ mol^{-1}) - (5 \times 0)$

$\quad = -905.47\ kJ\ mol^{-1}$

```
                         2 N₂(g) + 6 H₂(g) + 5 O₂(g)
                        ↑ −4 Δ_fH°[NH₃(g)]          |
  4 NH₃(g) + O₂(g)     |                            |
                        |                            | 4×Δ_rH°[NO(g)] + 6×Δ_rH°[H₂O(g)]
  4×Δ_rH°[NO(g)] + 6×Δ_rH°[H₂O(g)] |                |
    − 4×Δ_rH°[NH₃(g)]   |                            |
                        ↓             ↓ 4 NO(g) + 6 H₂O(g)
```

7.53

$$C_{10}H_8(s) + 12\,O_2(g) \longrightarrow 10\,CO_2(g) + 4\,H_2O(l) \qquad \Delta_rH° = -5156.1 \text{ kJ mol}^{-1}$$

$\Delta_rH° = 10\;\Delta_fH°[CO_2(g)] + 4\;\Delta_fH°[H_2O(l)] - \Delta_fH°[C_{10}H_8(s)] - \Delta_fH°[O_2(g)] = -5156.1 \text{ kJ}$

Note that $\Delta_fH°[O_2(g)] = 0$

We can solve the above equation for the unknown, $\Delta_fH°[C_{10}H_8(s)]$.

$\Delta_fH°[C_{10}H_8(s)] = -5156.1 \text{ kJ} - (\,10\;\Delta_fH°[CO_2(g)] + 4\;\Delta_fH°[H_2O(l)]\,)$

$\qquad\qquad = \{-5156.1 - (\,10 \times (-393.509) + 4 \times (-285.83)\,)\} \text{ kJ mol}^{-1} = -77.69 \text{ kJ mol}^{-1}$

7.55

$$Mg(s) + 2\,H_2O(l) \longrightarrow Mg(OH)_2(s) + H_2(g)$$

$\Delta_rH° = \Delta_fH°[Mg(OH)_2(s)] + \Delta_fH°[(H)_2(g)] - \Delta_fH°[Mg(s)] - 2 \times \Delta_fH°[H_2O(l)] = \{-924.54 + 0 - 0 - 2 \times (-285.830)\} \text{ kJ} = -352.9 \text{ kJ mol}^{-1}$

That is, when 1.000 mol of Mg(s) reacts 352.9 kJ of energy is released.

To warm 25 mL of water from 25 °C to 85 °C requires $(4.20 \text{ J g}^{-1} \text{ K}^{-1}) \times ((25 \text{ mL}) \times (1.00 \text{ g mL}^{-1})) = 105 \text{ J}$

Amount of Mg(s) that reacts to evolve this much heat, $n = (105 \text{ J}) / (352.9 \text{ kJ mol}^{-1}) = (105 \text{ J}) / (352.9 \times 10^3 \text{ J mol}^{-1})$

$\qquad = 2.98 \times 10^{-4} \text{ mol}$

Mass of Mg(s) required, $m = (2.98 \times 10^{-4} \text{ mol}) \times (24.31 \text{ g mol}^{-1}) = 7.24 \times 10^{-3} \text{ g} = 7.24 \text{ mg}$

7.57

$$N_2H_4(l) + O_2(g) \longrightarrow N_2(g) + 2\,H_2O(g)$$

hydrazine

$$N_2H_2(CH_3)_2(l) + 4\,O_2(g) \longrightarrow 2\,CO_2(g) + 4\,H_2O(g) + N_2(g)$$

1,1-dimethylhydrazine

For hydrazine combustion,

$\Delta_rH° = 2\;\Delta_fH°[H_2O(g)] + \Delta_fH°[N_2(g)] - \Delta_fH°[N_2H_4(l)] - \Delta_fH°[O_2(g)] = \{2 \times (-241.826) + 0 - (50.6) - 0)\} \text{ kJ mol}^{-1}$

$\qquad = -534.3 \text{ kJ mol}^{-1}$ (per mol of hydrazine)

$\qquad$ Enthalpy change per g of hydrazine $= (-534.3 \text{ kJ mol}^{-1}) / (32.05 \text{ g mol}^{-1}) = -16.67 \text{ kJ g}^{-1}$

For 1,1-dimethylhydrazine combustion,

$\Delta_rH° = 2\;\Delta_fH°[CO_2(g)] + 4\;\Delta_fH°[H_2O(g)] + \Delta_fH°[N_2(g)] - \Delta_fH°[N_2H_2(CH_3)_2(l)] - 4\;\Delta_fH°[O_2(g)]$

$= \{2 \times (-393.509) + 4 \times (-241.826) + 0 - (48.9) - 4 \times (0)\} \text{ kJ mol}^{-1}$

$\qquad = -1803.2 \text{ kJ mol}^{-1}$ (per mol of 1,1-dimethylhydrazine)

$\qquad$ Enthalpy change per g of 1,1-dimethylhydrazine $= (-1803.2 \text{ kJ mol}^{-1}) / (60.10 \text{ g mol}^{-1}) = -30.00 \text{ kJ g}^{-1}$

1,1-dimethylhydrazine evolves more heat on a per gram basis.

7.9—Enthalpy change of reaction from bond energies

7.59

$$O_3 + O \longrightarrow 2 O_2 \qquad \Delta_r H° = \text{-394 kJ mol}^{-1}$$

$\Delta_r H° = 2 D[\text{O-O in } O_3] - 2 D[\text{O=O in } O_2] = -394 \text{ kJ mol}^{-1}$

So, $D[\text{O-O in } O_3] = \frac{1}{2} \times (-394 \text{ kJ mol}^{-1} + 2 D[\text{O=O in } O_2]) = \frac{1}{2} \times (-394 + 2 \times 498) \text{ kJ mol}^{-1} = 301 \text{ kJ mol}^{-1}$

This value is between the oxygen single and double bond energies, 146 and 498 kJ mol^{-1}, respectively.

7.61

$$2 CH_3OH(g) + 3 O_2(g) \longrightarrow 2 CO_2(g) + 4 H_2O(g)$$

(a) From bond energies,

$\Delta_r H° = 2 \times (3 D[\text{C-H}] + D[\text{C-O}] + D[\text{O-H}]) + 3 D[\text{O=O}] - (2 \times 2 D[\text{C=O}] + 4 \times 2 D[\text{O-H}])$

$\qquad = 6 D[\text{C-H}] + 2 D[\text{C-O}] + 3 D[\text{O=O}] - (4 D[\text{C=O}] + 6 D[\text{O-H}])$

$\qquad = \{6 \times 413 + 2 \times 358 + 3 \times 498 - (4 \times 745 + 6 \times 463)\} \text{ kJ mol}^{-1} = -1070 \text{ kJ mol}^{-1}$

(b) From enthalpies of formation,

$\Delta_r H° = 2 \Delta_f H°[CO_2(g)] + 4 \Delta_f H°[H_2O(g)] - 2 \Delta_f H°[CH_3OH(g)] - 3 \Delta_f H°[O_2(g)]$

$= \{2 \times (-393.509) + 4 \times (-241.818) - 2 \times (-200.66) - 3 \times (0)\} \text{ kJ mol}^{-1}$

$\qquad = -1352.97 \text{ kJ mol}^{-1}$

This is the more correct enthalpy change for this reaction. The value computed in part (a) from average bond energies is only an estimate—about 20% off in this case.

7.10—Energy from food

7.63

$$C_6H_{12}O_6(s) + 6 O_2(g) \longrightarrow 6 CO_2(g) + 6 H_2O(l) \qquad \Delta_r H° = \text{-2803 kJ}$$

$\Delta_r H° = 6 \Delta_f H°[CO_2(g)] + 6 \Delta_f H°[H_2O(l)] - \Delta_f H°[C_6H_{12}O_6(s)] - 6 \Delta_f H°[O_2(g)] = -2803 \text{ kJ}$

$\qquad = \{6 \times (-393.509) + 6 \times (-285.830) - \Delta_f H°[C_6H_{12}O_6(s)] - 6 \times (0)\} \text{ kJ mol}^{-1}$

Therefore,

$\Delta_r H°[C_6H_{12}O_6(s)] = \{6 \times (-393.509) + 6 \times (-285.830) - (-2803)\} \text{ kJ mol}^{-1} = -1273 \text{ kJ mol}^{-1}$

SUMMARY AND CONCEPTUAL QUESTIONS

7.65

(a) An exothermic reaction releases energy which must be removed to return the system to its original temperature—heat leaves the system. An endothermic reaction absorbs energy which must be supplied for the system to stay at its original temperature—heat enters the system.

(b) The system is the set of all substances of interest—e.g., the reactants and products of a reaction. The surroundings consist of everything else.

(c) The specific heat capacity of substance is the amount of heat (usually expressed in J) required to raise the temperature of exactly 1 g of the substance 1 C°—assuming no phase transitions occur during heating.

(d) A state function is anything that depends only on the "state" of a system. A state function is any property of the system, such as enthalpy or entropy, which has a particular value at a specified set of conditions, regardless of its history (how those conditions were obtained).

(e) The standard state of a substance is the stable form of the substance at 1 bar pressure and— unless specified otherwise – 25 °C.

(f) The enthalpy change of reaction, $\Delta_r H$, is the change in enthalpy when the extent of reaction is 1 mol (i.e., reactants form products with numbers of moles given by the stoichiometric coefficients), and the temperature of the products is returned to the initial temperature of reactants. It is the difference between the total enthalpy of all the products and the total enthalpy of all the reactants.

(g) The standard enthalpy change of reaction, $\Delta_r H°$, is the enthalpy change of reaction under standard conditions—all reactants and products in their standard states at the specified temperature.

(h) The standard molar enthalpy change of formation, $\Delta_f H°$, is the standard enthalpy change of a formation reaction wherein 1 mol of a substance is formed from its elemental substances in their standard states.

7.67

A perpetual motion machine is impossible as soon as there is friction or other forms of energy dissipation. Because energy is constantly lost to friction, there must be a constant supply of incoming useful energy—here we invoke conservation of energy. For the machine to run forever, it must have an infinite supply of energy—impossible in a finite machine.

7.69

We can combust sulfur to $SO_3(g)$ and Ca(s) to CaO(s), separately in calorimeters, measuring the enthalpy change for both processes. We can also combine $SO_3(g)$ and CaO(s) in a calorimeter, and measure the enthalpy change for this reaction. The enthalpy change for the desired reaction is the sum of the three enthalpy changes just described—just as the desired reaction itself is the sum of the three reactions.

$$\tfrac{1}{8}S_8(s) + \tfrac{3}{2}O_2(g) \longrightarrow SO_3(g)$$
$$Ca(s) + \tfrac{1}{2}O_2(g) \longrightarrow CaO(s)$$
$$\underline{CaO(s) + SO_3(g) \longrightarrow CaSO_4(s)}$$
$$Ca(s) + \tfrac{1}{8}S_8(s) + 2O_2(g) \longrightarrow CaSO_4(s)$$

Calorimetric measurements involve reacting known amounts of reactants in a calorimeter and measuring the temperature change.

$$\Delta_r H° = \Delta_r H°_1 + \Delta_r H°_2 + \Delta_r H°_3 = \{\Delta_f H°[SO_3(g)] + \Delta_f H°[CaO(s)] + (-402.7)\} \text{ kJ mol}^{-1}$$
$$= \{(-395.72) + (-635.09) + (-402.7) \text{ kJ mol}^{-1} = -1433.5 \text{ kJ mol}^{-1}$$

7.70

(a) Simply sum the reactions—all species cancel (they appear on reactant and product sides with the same stoichiometric coefficients)—except water as a reactant and hydrogen and oxygen as products

(b) Starting with 1000 kg of water, the mass of $H_2(g)$ that can be produced is easily determined as the mass of hydrogen atoms within the 1000 kg of water. Note that both hydrogen atoms in water end up as $H_2(g)$ is the net reaction,

$$H_2O(g) \rightarrow H_2(g) + \tfrac{1}{2} O_2(g)$$

$$\text{Mass of hydrogen} = \left(\frac{2.016 \text{ g mol}^{-1}}{18.015 \text{ g mol}^{-1}} \right) \times 1000 \text{ kg} = 111.9 \text{ kg}$$

(c)

$$CaBr_2(s) + H_2O(g) \longrightarrow CaO(s) + 2\,HBr(g)$$
$$Hg(l) + 2\,HBr(g) \longrightarrow HgBr_2(s) + H_2(g)$$
$$HgBr_2(s) + CaO(s) \longrightarrow HgO(s) + CaBr_2(s)$$
$$HgO(s) \longrightarrow Hg(l) + \tfrac{1}{2} O_2(g)$$

$\Delta_r H^\circ_1 = \Delta_f H^\circ[\,CaO(s)\,] + 2\,\Delta_r H^\circ[\,HBr(g)\,] - (\,\Delta_f H^\circ[\,CaBr_2(s)\,] + \Delta_r H^\circ[\,H_2O(g)\,]\,)$
$\quad = \{-635.09 + (-36.40) - ((-683.2) + (-241.818))\}\text{ kJ mol}^{-1} = 253.5 \text{ kJ mol}^{-1}$

$\Delta_r H^\circ_2 = \Delta_f H^\circ[\,HgBr_2(s)\,] - (\,2\,\Delta_r H^\circ[\,HBr(g)\,]\,) = \{-169.5 - (\,2 \times (-36.40))\}\text{ kJ mol}^{-1} = -96.7 \text{ kJ mol}^{-1}$

$\Delta_r H^\circ_3 = \Delta_f H^\circ[\,HgO(s)\,] + \Delta_r H^\circ[\,CaBr_2(s)\,] - (\,\Delta_f H^\circ[\,HgBr_2(s)\,] + \Delta_r H^\circ[\,CaO(s)\,]\,)$
$\quad = \{-90.83 + (-683.2) - ((-169.5) + (-635.09))\}\text{ kJ mol}^{-1} = 30.6 \text{ kJ mol}^{-1}$

$\Delta_r H^\circ_4 = -\Delta_f H^\circ[\,HgO(s)\,] = 90.83 \text{ kJ mol}^{-1}$

Reactions 1, 3 and 4 are endothermic. Reaction 2 is exothermic.

(d) It is not clear whether this set of reactions is a commercially feasible means of producing hydrogen. Three of the steps are endothermic and will require energy input. The amount of energy input required will be the same as any other means of producing hydrogen from water—by energy conservation. The above approach has the disadvantage of involving mercury compounds which can be very toxic. There will be serious handling issues concerning the last three reactions on this account.

7.71

A bond dissociation energy is the enthalpy change associated with breaking a single bond. All bond breaking processes are endothermic—the enthalpy change is positive. It always takes energy input to break a bond.

7.73

Volume of rain water $= \text{thickness} \times \text{area} = (25 \text{ mm}) \times (1 \text{ km}^2) = (2.5 \text{ cm}) \times (10^5 \text{ cm})^2 = 2.5 \times 10^{10} \text{ cm}^3$

Mass of rain water $= \text{volume} \times \text{density} = (2.5 \times 10^{10} \text{ cm}^3) \times (1.0 \text{ g cm}^{-3}) = 2.5 \times 10^{10} \text{ g}$

Amount of rain water $= (2.5 \times 10^{10} \text{ g}) / (18.02 \text{ g mol}^{-1}) = 1.4 \times 10^9 \text{ mol}$

Heat released upon condensation of this much water vapour $= (44.0 \text{ kJ mol}^{-1}) \times (1.4 \times 10^9 \text{ mol}) = 6.2 \times 10^{10} \text{ kJ}$

This is more than 10,000 times the energy released when a tonne of dynamite explodes.

Chapter 8
Modelling Atoms and Their Electrons

EXERCISES FROM CHAPTER

Exercise 8.1—Chemical periodicity

(a) Element with atomic number 8 greater than F—i.e., atomic number = 17 – is Cl.
Element with atomic number 18 greater than Cl—i.e., atomic number = 35 – is Br.
Element with atomic number 18 greater than Br—i.e., atomic number = 53 – is I.
Element with atomic number 32 greater than I—i.e., atomic number = 85 – is At.
These are all halogens.
(b) The atomic numbers of the group 15 elements, N, P, As, Sb & Bi, are
7, 15, 33, 51, and 83.
Differences between successive atomic numbers = 8, 18, 18, and 32.

Exercise 8.2—Periodic variation of metallic/non-metallic character

(a) Strontium, Sr(s), is more metallic than magnesium, Mg(s), because it is lower in the same group in the periodic table.
(b) Calcium, Ca(s), is more metallic than selenium, Se(s), because it is to the left in the same period of the periodic table.
(c) Rubidium, Rb(s), is more metallic than arsenic, As(s), because it is to the left and lower in the periodic table.

Exercise 8.3—Periodic trends of oxidizing and reducing abilities

(a) Bromine, $Br_2(g)$, to the right of arsenic, As(s), is the more powerful oxidizing agent.
(b) Sodium, Na(s), to the left of silicon, Si(s), is the more powerful reducing agent.
(c) Chlorine, $Cl_2(g)$, well to the right of aluminium, Al(s), is most likely to use to bring about oxidation of a substance that is difficult to oxidize.
(d) The elements with atomic numbers 34, 35, 36, 37, and 38 are Se, Br, Kr, and Rb. Rubidium, Rb, is the lowest and furthest to the left of these. It is the most powerful reducing agent.

Exercise 8.5—Periodic trends of atomic size

In order of increasing atomic radius, we have C < Si < Al.

Exercise 8.7—Trends of first ionization energies

(c) Li < Si < C < Ne
is the correct ordering. Note that although Si is a period below, it is far enough to the right of Li that its ionization energy is larger.
(a) can be eliminated because Si (being lower is group 4A) must have a lower ionization energy than C.
(b) and **(d)** can be eliminated because Ne must have a higher ionization energy than Li & C—to the left in the same period (row).

Exercise 8.8—Successive ionization energies of the elements

(a) The fourth ionization energy of Al is much larger than the third—more so than successive ionization energies usually increase. After removing three electrons from an aluminum atom—the result of the first three ionization steps – the atom (now an ion) is left with the electron configuration of Ne, a noble gas. Al^{3+} has filled n=1 and n=2 shells. The next electron to be removed is a tightly held member of the second shell. The first three electrons came from the third shell.
(b) The successive ionization energies,

$$IE_1 = \quad 738 \text{ kJ mol}^{-1}$$
$$IE_2 = \quad 1451 \text{ kJ mol}^{-1}$$
$$IE_3 = \quad 7733 \text{ kJ mol}^{-1}$$
$$IE_4 = \quad 10543 \text{ kJ mol}^{-1}$$

are those of magnesium. Here, we see the big jump in ionization energy between the second and third. Ionizing Mg2+, via the third ionization energy, requires removing a tightly held electron of the second shell. The first two ionizations removed third shell electrons.

Exercise 8.9—Charges on monatomic ions of the elements

(a) The ions of barium, a group 2 element, are normally Ba^{2+}.
(b) The ions of selenium, a group 16 element, are normally Se^{2-}.

Exercise 8.13—Relative sizes of atoms and ions

Select the atom or ion in each pair that has the larger radius.

(a) Cl^- ions are larger than Cl atoms
Anions are larger than the parent atom

(b) Ba^{2+} ions are larger that Mg^{2+} ions
Ba^{2+} is further down the same group

(c) K atoms are larger than K^+ ions
Cations are smaller than the parent atom

(d) Se^{2-} ions are larger than O^{2-} ions
Se^{2-} is further down the same group

(e) Cl^- ions are larger than K^+ ions
These are isoelectronic. K^+ ions have larger nuclear charge

(f) Pb^{2+} ions are larger than Pb^{4+} ions
Pb^{4+} ions have fewer electrons, but the same nuclear charge

Exercise 8.14—Periodic variation of electron affinities

(a) It will take more energy to remove an electron from an O^- ion than an N^- ion. See Table 8.7.
(b) It will take more energy to remove an electron from a Cl^- ion than a Br^- ion. See Table 8.7.

Exercise 8.15—Periodic trends of properties

(a) In order of increasing atomic radius, we have
$O < C < Si$
Here, we move to the left and then down within the periodic table.
(b) The first ionization energy of O is larger than that of C and Si.
In order of increasing ionization energy, we have
$Si < C < O$
Here, we move up and then to the right within the periodic table.
(c) Which has the greatest electron affinity: O or C?
The electron affinity of O is larger than that of C and Si.
In order of increasing electron affinity, we have
$Si < C < O$
Here, we move up and then to the right within the periodic table.

Exercise 8.16—Energies of electrons and transitions

(a)

$$\text{Energy (in } n = 3 \text{ state of an H atom)} = -2.18 \times 10^{-18} \left(\frac{1^2}{3^2} \right) J$$

$$= -2.42 \times 10^{-19} \, J$$

$$\text{Energy per mole of H atoms} = (-2.42 \times 10^{-19} \, J) \times (6.022 \times 10^{23} \, mol^{-1})$$

$$= -1.46 \times 10^5 \, J \, mol^{-1} = -146 \, kJ \, mol^{-1}$$

(b)
$$\text{Energy per mole of H atoms in } n = 2 \text{ state} = (-5.45 \times 10^{-19} \, J) \times (6.022 \times 10^{23} \, mol^{-1})$$

$$= -3.28 \times 10^5 \, J \, mol^{-1} = -328 \, kJ \, mol^{-1}$$

Energy emitted per mole of electrons changing from $n = 3$ to $n = 2$ state

$$= -146 \, kJ \, mol^{-1} - (-328 \, kJ \, mol^{-1})$$

$$= 182 \, kJ \, mol^{-1}$$

Exercise 8.18—Energy of a spectral line of atomic hydrogen

The energy of the photon emitted when an H atom in the $n = 4$ state changes to the $n = 1$ state is

$$E_{photon} = E_4 - E_1 = \frac{-2.179 \times 10^{-18}\,J}{4^2} - \frac{-2.179 \times 10^{-18}\,J}{1^2}$$

$$= \left(-\frac{1}{16} + \frac{1}{1}\right)(2.179 \times 10^{-18}\,J) = 2.043 \times 10^{-18}\,J$$

The frequency and the wavelength of the third line of the Lyman series are given by

$$\nu = \frac{E_{photon}}{h} = \frac{2.043 \times 10^{-18}\,J}{6.626 \times 10^{-34}\,J\,s} = 3.083 \times 10^{15}\,s^{-1} = 308.3\ THz$$

$$\lambda = \frac{c}{\nu} = \frac{(2.998 \times 10^8\,m\,s^{-1})}{3.083 \times 10^{15}\,s^{-1}} = 9.724 \times 10^{-8}\,m = 97.24\,nm$$

Exercise 8.19—Using De Broglie's equation

We need the mass of the golf ball in kg
$1.0 \times 10^2\ g = 1.0 \times 10^{-1}\ kg$
Also, note $1\ J\,s = 1\ (kg\,m^2\,s^{-2}) \times s = 1\ kg\,m^2\,s^{-1}$

$$\lambda = \frac{h}{m\nu} = \frac{6.626 \times 10^{-34}\,kg\,m^2\,s^{-1}}{(1.0 \times 10^{-1}\,kg) \times (30\,m\,s^{-1})} = 2.2 \times 10^{-34}\,m$$

$$= 2.2 \times 10^{-25}\,nm$$

This de Broglie wavelength (of a macroscopic object—a golf ball) is even very small on the scale of an atom.

To have a wavelength of 5.6×10^{-3} nm, the ball must travel

$$\nu = \frac{h}{m\lambda} = \frac{6.626 \times 10^{-34}\,kg\,m^2\,s^{-1}}{(1.0 \times 10^{-1}\,kg) \times (5.6 \times 10^{-12}\,m)} = 1.2 \times 10^{-21}\ m\,s^{-1}$$

Exercise 8.20—Standing waves

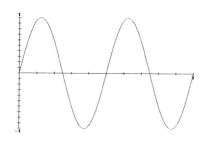 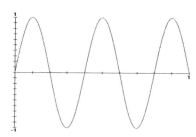

Exercise 8.22—Atomic orbitals

(a) $n = 4$, $l = 2$, $m_l = -2$ is a valid set of H atom quantum numbers—it determines an electron standing wave
(b) $n = 1$, $l = 2$, $m_l = -2$ is NOT a valid set of H atom quantum numbers—l must be less than n
(c) $n = 3$, $l = 1$, $m_l = -2$ is NOT a valid set of H atom quantum numbers—$|m_l|$ must be less than l

Exercise 8.24—Standing waveforms from the wave equation

(a) When $n = 2$, the value of l must be **0 or 1**.
(b) When $l = 1$, the value of m_l must be **−1, 0 or 1**, and the subshell is labelled the **p subshell**.
(c) When $l = 2$, the subshell is called a **d subshell**.
(d) When a subshell is labelled s, the value of **l is 0** and **m_l has the value 0**.
(e) When a subshell is labelled p, it has **3 orbitals**.
(f) When a subshell is labelled f, there are **7** values of m_l and it has **7 orbitals**.

Exercise 8.25—Quantum numbers of electrons in atoms

(a) $n = 4$, $l = 2$, $m_l = 0$, $m_s = 0$ is not valid because m_s is always ± ½ because $s = ½$ for an electron. m_s is never equal to zero.
(b) $n = 3$, $l = 1$, $m_l = -3$, $m_s = -½$ is not valid because $m_l = -3$ does not go with $l = 1$. $m_l = -1, 0$, or 1 are the "allowed" values.
(c) $n = 3$, $l = 3$, $m_l \square = -1$, $m_s = +½$ is not valid because l must be less than or equal to $n - 1$.

Exercise 8.27—Orbital Shapes

(a)

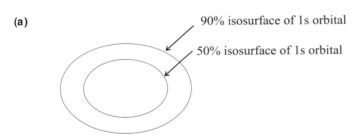

90% isosurface of 1s orbital
50% isosurface of 1s orbital

(b)

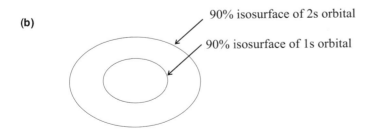

90% isosurface of 2s orbital
90% isosurface of 1s orbital

Exercise 8.28—Orbital Shapes

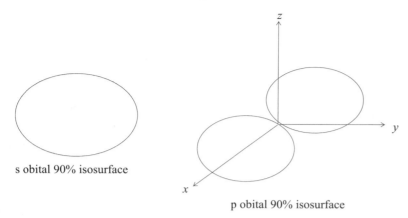

s obital 90% isosurface

p obital 90% isosurface

Exercise 8.30—Electrons, orbitals, subshells and shells

(a) $n = 3$ corresponds to
(number of spin states) × (orbitals in s subshell + orbitals in p subshell + orbitals in d subshell)
$= 2 \times (1 + 3 + 5) = 18$ possible electron states. These are:

$n = 3, l = 0, m_l = 0, m_s = -\frac{1}{2}$ and $+\frac{1}{2}$
$n = 3, l = 1, m_l = -1, m_s = -\frac{1}{2}$ and $+\frac{1}{2}$
$n = 3, l = 1, m_l = 0, m_s = -\frac{1}{2}$ and $+\frac{1}{2}$
$n = 3, l = 1, m_l = 1, m_s = -\frac{1}{2}$ and $+\frac{1}{2}$
$n = 3, l = 2, m_l = -2, m_s = -\frac{1}{2}$ and $+\frac{1}{2}$
$n = 3, l = 2, m_l = -1, m_s = -\frac{1}{2}$ and $+\frac{1}{2}$
$n = 3, l = 2, m_l = 0, m_s = -\frac{1}{2}$ and $+\frac{1}{2}$
$n = 3, l = 2, m_l = 1, m_s = -\frac{1}{2}$ and $+\frac{1}{2}$
$n = 3, l = 2, m_l = 2, m_s = -\frac{1}{2}$ and $+\frac{1}{2}$

(b) $n = 3, l = 2$ corresponds to
(number of spin states) × (orbitals in d subshell) $= 2 \times 5 = 10$ possible electron states. See part (a) for specification of the states.

(c) $n = 4, l = 1, m_l = -1, m_s = -\frac{1}{2}$ corresponds to just one electron state—all the quantum numbers are specified.

(d) $n = 5, l = 0, m_l = +1$ corresponds to no electron state—m_l must equal 0 if $l = 0$.

Exercise 8.32— Electron configurations of ground-state atoms

$$[\uparrow\downarrow] \quad [\uparrow\downarrow] \quad [\uparrow\downarrow][\uparrow\downarrow][\uparrow\downarrow] \quad [\uparrow\downarrow] \quad [\uparrow\,][\uparrow\,][\uparrow\,]$$
$$1s \qquad 2s \qquad\quad 2p \qquad\qquad 3s \qquad\quad 3p$$

can be written as

$$[Ne] \quad [\uparrow\downarrow] \quad [\uparrow\,][\uparrow\,][\uparrow\,]$$
$$3s \qquad\quad 3p$$

The last five electrons correspond to the quantum numbers,

$n = 3$,

$l = 0$ and $m_l = 0$, and $m_s = -\frac{1}{2}$ or $+\frac{1}{2}$ 2 electron states

$l = 1$ and $m_l = -1$, 0 or 1 and $m_s = +\frac{1}{2}$ 3 electron states

Exercise 8.34—Effective nuclear charge

$$Z^*\left(_{12}Mg\right) = +12 - [(1 \times 0.35) + (8 \times 0.85) + (2 \times 1.0)] = +12 - 9.15 = +2.85$$

$$Z^*\left(_{15}P\right) = +15 - [(4 \times 0.35) + (8 \times 0.85) + (2 \times 1.0)] = +15 - 10.2 = +4.8$$

$$Z^*\left(_{18}Ar\right) = +18 - [(7 \times 0.35) + (8 \times 0.85) + (2 \times 1.0)] = +18 - 11.25 = +6.75$$

These estimates show that valence electrons are held increasingly tightly by these elements in this order. This is consistent with the increasing first ionization energies of the elements.

Exercise 8.35—Effective nuclear charge in ions

$$Z^*\left(_{8}O\right) = +8 - [(5 \times 0.35) + (2 \times 0.85)] = +8 - 3.45 = +4.55$$

$$Z^*\left(_{8}O^{2-}\right) = +8 - [(7 \times 0.35) + (2 \times 0.85)] = +8 - 4.15 = +3.85$$

The less tightly held valence electrons of O^{2-} make this ion much larger than an O atom – 2 extra valence electrons, and they are less tightly held.

Exercise 8.36—Effective nuclear charge and atom sizes

Although $_{84}Po$ has a lot more electrons than $_{11}Na$, the electrons of Po are held so much more tightly that it is smaller. The effective nuclear charges, +5.45 and +2.2, show that even the valence electrons of Po are held more tightly than the valence electrons of Na. The valence electrons determine the size of the atom.

Exercise 8.39—Effective nuclear charge and ionization energies

$$Z^*\left(Na\right) = +11 - [(0 \times 0.35) + (8 \times 0.85) + (2 \times 1.0)] = +11 - 8.8 = +2.2$$

$$Z^*\left(Si\right) = +14 - [(3 \times 0.35) + (8 \times 0.85) + (2 \times 1.0)] = +14 - 9.85 = +4.15$$

$$Z^*\left(Ar\right) = +18 - [(7 \times 0.35) + (8 \times 0.85) + (2 \times 1.0)] = +18 - 11.25 = +6.75$$

The effective nuclear charge increases here. As the valence electrons are held more tightly, the first ionization energy increases—it gets harder to remove an electron.

Exercise 8.40—Effective nuclear charge and charges on simple ions

(a) N^{3-} ion configuration: $1s^2 \, 2s^2 \, 2p^6$

$$Z^*\left(N^{3-}\right) = +7 - [(7 \times 0.35) + (2 \times 0.85)] = +7 - 4.15 = +2.85$$

Compare this with the N^{4-} ion: $1s^2 \, 2s^2 \, 2p^6 \, 3s^1$

$$Z^*\left(N^{4-}\right) = +7 - [(0 \times 0.35) + (8 \times 0.85) + (2 \times 1.0)] = +7 - 8.8 = -1.8$$

N can form a 3− ion because the valence electrons are still held by an effective nuclear charge of +2.85. However, a fourth additional electron, in the $3s$ orbital, would experience no attraction to the nuclear charge because of the degree of repulsion (shielding) by other electrons.

(b) S^{2-} ion configuration: $1s^2 \, 2s^2 \, 2p^6 \, 3s^2 \, 3p^6$

$$Z^*\left(S^{2-}\right) = +16 - [(7 \times 0.35) + (8 \times 0.85) + (2 \times 1.0)] = +16 - 11.3 = +4.7$$

The S^{3-} ion has an electron in the next shell: $1s^2 \, 2s^2 \, 2p^6 \, 3s^2 \, 3p^6 \, 4s^1$

$$Z^*\left(S^{3-}\right) = +16 - [(0 \times 0.35) + (8 \times 0.85) + (10 \times 1.0)] = +16 - 16.8 = -0.8$$

S cannot form such an ion because its valence electron would be repelled by a net negative effective nuclear charge. The valence electron in the $4s$ orbital is well-shielded by the filled shells.

Exercise 8.41—Effective nuclear charge and ion sizes

$$Z^*\left(Na^+\right) = +11 - [(7 \times 0.35) + (2 \times 0.85)] = +11 - 4.15 = +6.85$$

$$Z^*\left(Mg^{2+}\right) = +12 - [(7 \times 0.35) + (2 \times 0.85)] = +12 - 4.15 = +7.85$$

$$Z^*\left(Al^{3+}\right) = +13 - [(7 \times 0.35) + (2 \times 0.85)] = +13 - 4.15 = +8.85$$

The valence electrons of these three ions are increasingly tightly held. This is consistent with the trend in sizes—the ions decrease in size.

Exercise 8.42—Effective nuclear charge and electronegativities

Electron configuration of $_{12}Mg$: $1s^2 \, 2s^2 \, 2p^6 \, 3s^2$

$$Z^*\left(Mg\right) = +12 - [(1 \times 0.35) + (8 \times 0.85) - 2 \times 1.0] = +12 - 9.15 = +2.85$$

This value is a measure of the net force of attraction between the nucleus of an Mg atom and electrons in its valence shell. If an Mg atom attracted another electron into its valence shell, we would have

$$Z^*\left(Mg^-\right) = +12 - [(2 \times 0.35) + (8 \times 0.85) + (2 \times 1.0)] = +12 - 9.5 = +2.5$$

Electron configuration of $_{17}Cl$: $1s^2 \, 2s^2 \, 2p^6 \, 3s^2 \, 3p^5$

$$Z^*\left(Cl\right) = +17 - [(6 \times 0.35) + (8 \times 0.85) + (2 \times 1.0)] = +17 - 10.9 = +6.1$$

This value illustrates that the force of attraction between the nucleus of a Cl atom and electrons in its valence shell is much greater than is the case for Mg atoms, and so the electronegativity of Cl is greater than that of Mg. Even if another electron is attracted into the valence shell of a Cl atom, it experiences strong attraction:

$$Z^*\left(Cl^-\right) = +17 - [(7 \times 0.35) + (8 \times 0.85) + (2 \times 1.0)] = +17 - 11.25 = +5.75$$

Exercise 8.43—Effective nuclear charge and electron affinities

Consideration of electron affinities requires that we consider the effective nuclear charge of the anion formed by capture of an electron by atoms—in this case of Mg^-, Cl^- and Ar^- ions.

Electron configuration of $_{12}Mg^-$ ion: $1s^2\,2s^2\,2p^6\,3s^2\,3p^1$

$$Z*(Mg^-) = +12 - [(2 \times 0.35) + (8 \times 0.85) + (2 \times 1.0)] = +12 - 9.5 = +2.5$$

Electron configuration of $_{17}Cl^-$ ion: $1s^2\,2s^2\,2p^6\,3s^2\,3p^6$

$$Z*(Cl^-) = +17 - [(7 \times 0.35) + (8 \times 0.85) + (2 \times 1.0)] = +17 - 11.25 = +5.75$$

Electron configuration of $_{18}Ar^-$ ion: $1s^2\,2s^2\,2p^6\,3s^2\,3p^6\,4s^1$

$$Z*(Ar^-) = +18 - [(0 \times 0.35) + (8 \times 0.85) + (10 \times 1.0)] = +18 - 16.8 = +1.2$$

We see from these values that an extra electron is attracted much more strongly in a Cl^- ion, than in an Mg^- ion, and that it is very weakly attracted in an Ar^- ions. This accounts for the relative values of the electron affinities of Cl, Mg and Ar.

REVIEW QUESTIONS

8.2—Periodic variation of properties of the elements

8.45
$MgCl_2$ is an ionic material—a compound of a metal and non-metal. It has a high melting point—i.e., 714 °C—and it conducts electricity in its molten state but not its solid state.
PCl_3 is a molecular substance—a compound of two non-metals. Its solid consists of molecules held together by intermolecular forces. It has a low melting point—i.e., 112 °C—and it does not conduct electricity.

8.47
In order of increasing size of atoms, we have
C < B < Al < Na < K
Each step in this sequence corresponds to progressing to the left or down within the periodic table.

8.49
In order of increasing ionization energy, we have
K < Ca < Si < P
Here, we are moving to the right in the third period (row) of the periodic table.

8.51
(a) It takes more energy to remove an electron from an S^- ion than a Te^- ion (that is, the electron affinity of S is larger than that of Te) because S is higher in the same group as Te. The valence electrons of S^- are held more closely to the nucleus—they are in the third shell versus the fifth shell for Te.
(b) It takes more energy to remove an electron from a Cl^- ion than a P^- ion (that is, the electron affinity of Cl is larger than that of P) because Cl is to the right of P in the same period of the periodic table.

8.53
(a) In order of increasing atomic radii, we have
C < B < Al
Here we move to the left, then down within the periodic table.
(b) In order of increasing ionization energy, we have
Al < B < C
Here we move up, then to the right within the periodic table.
(c) Carbon has the highest electron affinity of these elements. It is highest and furthest to right within the periodic table.

8.55

(a) In order of increasing ionization energy, we have

S < O < F

Here, we move up and to the right within the periodic table.

(b) Among the group 16 elements, O, S and Se, O has the largest first ionization energy. O is highest in the group.

(c) Among Se, Cl and Br, Cl has the greatest electron affinity. Cl is highest and furthest to the right within the periodic table.

(d) Among O^{2-}, F^-, or F, O^{2-} has the largest radius. O^{2-} is larger than F^-—it has the same number of electrons with one less nuclear charge. F^- is larger than F—one more electron, same nuclear charge.

8.3—Experimental evidence about electrons in atoms

8.59

$$E_\infty - E_1 = \frac{-2.179\times10^{-18}\text{ J}}{\infty^2} - \frac{-2.179\times10^{-18}\text{ J}}{1^2}$$

$$= \left(0 + \frac{1}{1}\right)(2.179\times10^{-18}\text{ J}) = 2.179\times10^{-18}\text{ J}$$

is the ionization energy of an H atom in its ground state.

8.61

The 410.2 nm line in the Balmer series of excited H atom emission lines is blue. The energy of photons in this line is

$$E_{\text{photon}} = \frac{hc}{\lambda} = \frac{(6.626\times10^{-34}\text{ J s})\times(2.998\times10^8\text{ m s}^{-1})}{410.2\times10^{-9}\text{ m}} = 4.8426\times10^{-19}\text{ J}$$

and this energy =

$$(-2.179\times10^{-18}\text{ J})\times\left(\frac{1}{n_{\text{initial}}^2} - \frac{1}{n_{\text{final}}^2}\right)$$

So,

$$\left(\frac{1}{n_{\text{final}}^2} - \frac{1}{n_{\text{initial}}^2}\right) = \left(\frac{1}{2^2} - \frac{1}{n_{\text{initial}}^2}\right) = \frac{4.8426\times10^{-19}\text{ J}}{2.179\times10^{-18}\text{ J}} = 0.22224$$

and

$$n_{\text{initial}}^2 = \frac{1}{0.2500 - 0.22224} = \frac{1}{0.0277} = 36.02 \cong 36$$

This line results from the H atom in its $n_{\text{initial}} = 6$ excited state.

8.63
(a) There are 4 (from $n = 5$) + 3 (from $n = 4$) + 2 (from $n = 3$) + 1 (from $n = 2$) = 10 possible emission lines.
(b) Photons of the highest frequency (highest energy difference) are emitted in a transition from the level with $n = 5$ to a level with $n = 1$.
(c) The emission line having the longest wavelength (lowest energy difference) corresponds to a transition from the level with $n = 5$ to the level with $n = 4$.

8.65
(a) An electron energy transition from $n = 4$ to $n = 2$ emits less energy than an electron transition from $n = 3$ to $n = 2$. The level spacings increase as we descend the ladder of energy levels.
(b) An electron energy transition from $n = 4$ to $n = 1$ emits more energy than an electron transition from $n = 5$ to $n = 2$. The level spacings decrease as we ascend the ladder of energy levels.

8.67
First, we must compute the velocity of the neutron. Note that we need the mass of the neutron in kg.
1.675×10^{-24} g $= 1.675 \times 10^{-27}$ kg

$$v = \left(\frac{2E_{kinetic}}{m}\right)^{1/2} = \left[\frac{2 \times (6.21 \times 10^{-21} \text{ kg m}^2\text{s}^{-2})}{1.675 \times 10^{-27} \text{ kg}}\right]^{1/2}$$

$$= [7.41 \times 10^6 \text{ m}^2\text{s}^{-2}]^{1/2} = 2.72 \times 10^3 \text{ m s}^{-1}$$

$$\lambda = \frac{h}{mv} = \frac{6.626 \times 10^{-34} \text{ kg m}^2 \text{ s}^{-1}}{(1.675 \times 10^{-27} \text{ kg}) \times (2.72 \times 10^3 \text{ m s}^{-1})} = 1.45 \times 10^{-10} \text{ m}$$

$$= 0.145 \text{ nm}$$

8.4—The quantum mechanical model of electrons in atoms

8.69
(a)

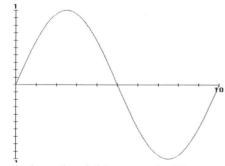

The wavelength of this wave is 10 cm.

(b)

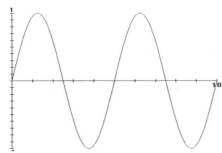

The wavelength of this wave is 5 cm.

(c) Four waves fit in the interval if the wavelength of the standing wave is 2.5 cm. The number of nodes between the ends is $2 \times 4 - 1 = 7$.

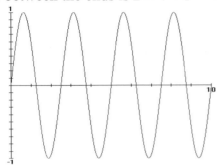

8.71
For a $4p$ orbital, there are three sets of allowed values of n, l, and m_l (that is, three sets of these quantum numbers which give rise to solutions for standing waves from the Schrodinger equation).

$n = 4$, $l = 1$ and $m_l = -1$
$n = 4$, $l = 1$ and $m_l = 0$
$n = 4$, $l = 1$ and $m_l = +1$

8.73
(a) $n = 2, l = 2, m_l = 0$ is NOT allowed because l must be no greater than 1 when $n = 2$.
(b) $n = 3, l = 0, m_l = -2$ is NOT allowed because m_l must equal 0 when $l = 0$.
(c) $n = 6, l = 0, m_l = 1$ is NOT allowed because m_l must equal 0 when $l = 0$.

8.77
There are 4 subshells in the shell defined by principal quantum number $n = 4$. (These comprise the orbitals with $l = 3$, those with $l = 2$, those with $l = 1$, and those with $l = 0$.)

8.81

(a) The n and l values for $6s$, $4p$, $5d$, and $4f$ are as follows:

$6s$ means $n = 6$, $l = 0$
$4p$ means $n = 4$, $l = 1$
$5d$ means $n = 5$, $l = 2$
$4f$ means $n = 4$, $l = 3$

(b) For a $4p$ orbital there are $n - 1 = 3$ radial nodes (spherical in shape) and $l = 1$ nodal planes.

$3 + 1 = 4$ altogether

For a $6d$ orbital there are $n - 1 = 5$ radial nodes (spherical in shape) and $l = 2$ nodal planes (or cones).

$5 + 2 = 7$ altogether

8.83

(a) The quantum number n describes the **energy** of an atomic orbital.
(b) The shape of an atomic orbital is given by the **quantum number l**.
(c) A photon of green light has **more** energy than a photon of orange light.
(d) The maximum number of orbitals that may be associated with the set of quantum numbers $n = 4$ and $l = 3$ is **7**.
(e) The maximum number of orbitals that may be associated with the quantum number set $n = 3$, $l = 2$, and $m_l = -2$ is 1.
(f) When $n = 5$, the allowed values of l are **0, 1, 2, 3 and 4**.
(g) The number of orbitals in the $n = 4$ shell is **$1 + 3 + 5 + 7 = 16$** (expressed here in terms of contributions from each subshell).

8.5—Electron configurations in atoms

8.87

(a) The ground state electron configuration of sulfur is $1s^2 2s^2 2p^6 3s^2 3p^4$
(b) The ground state electron configuration of aluminum is $1s^2 2s^2 2p^6 3s^2 3p^1$

8.89

(a) The ground state electron configuration of arsenic, As, is $1s^2 2s^2 2p^6 3s^2 3p^6 4s^2 3d^{10} 4p^3$ or $[Ar]4s^2 3d^{10} 4p^3$
(b) The ground state electron configuration of krypton, Kr, is $1s^2 2s^2 2p^6 3s^2 3p^6 4s^2 3d^{10} 4p^6$ or $[Ar]4s^2 3d^{10} 4p^6 = [Kr]$

8.7—Rationalizing the periodic variation of properties

8.93

$Z*(\text{Mg}) = +12 - [(1 \times 0.35) + (8 \times 0.85) + (2 \times 1.0)] = +12 - 9.15 = +2.85$

$Z*(\text{Mg}^+) = +12 - [(0 \times 0.35) + (8 \times 0.85) + (2 \times 1.0)] = +12 - 8.8 = +3.2$

$Z*(\text{Mg}^{2+}) = +12 - [(7 \times 0.35) + (2 \times 0.85)] = +12 - 4.15 = +7.85$

The increasing effective nuclear charge for these three species correlates with the magnitudes of their successive ionization energies: $IE_1 < IE_2 \ll IE_3$ (page 241). Note, especially the large jump from IE_2 to IE_3

8.95

$Z*(\text{N}^{3-}) = +7 - [(7 \times 0.35) + (2 \times 0.85)] = +7 - 4.15 = +2.85$

$Z*(\text{O}^{2-}) = +8 - [(7 \times 0.35) + (2 \times 0.85)] = +8 - 4.15 = +3.85$

$Z*(\text{F}^-) = +9 - [(7 \times 0.35) + (2 \times 0.85)] = +9 - 4.15 = +4.85$

These effective nuclear charges correlate with the relative sizes of the ions: $\text{F}^- < \text{O}^{2-} < \text{N}^{3-}$

SUMMARY AND CONCEPTUAL QUESTIONS

8.97

(c) Electrons are moving from a given energy level to one of lower n. The energy difference emerges as an emitted photon.

8.99

Electrons and other subatomic "particles" are found to exhibit properties of both waves and particles. We can measure the position of the "particles"—a particle property. However, the results of such measurements can show interference patterns characteristic of a wave.

8.101

$Li^{\square+}$ ions are so much smaller than Li atoms because the effective nuclear charge 'felt' by the valence electron (in the $n = 2$ shell) of a Li atom is less than that 'felt' by the $n = 1$ electrons in a Li^+ ion. Li has a lone valence electron in the $n = 2$ shell, whereas $Li^{\square+}$ just has the two $n = 1$ electrons. In any case, with a net positive charge, cations hold on to their electrons more tightly because there are less repulsions (shielding) of the valence electrons.

F^- ions are so much larger than F atoms, because the effective nuclear charge experienced by the valence electrons in F^- ions is less than is the case with F atoms. The additional electron in F^- ions causes more repulsion (shielding) of the valence electrons. Anions, with a net negative charge hold on to their electrons more loosely, and are much larger than the 'parent' neutral atoms.

8.103

The ionization energy of atoms increases from left to right across the periodic table, and decreases going down a group. The decrease going down a group is attributable to the valence electrons being further from the nucleus. The effective nuclear charge is the same (or similar) because the elements are in the same group. What changes is the distance from the nucleus. Electrons further from the nucleus are more easily removed.

Chapter 9
Molecular Shapes and Structures

IN-CHAPTER EXERCISES

Exercise 9.1

$CH_3CH_2CH_2CH_2OH$ $CH_3CH_2OCH_2CH_3$
butan-1-ol   diethyl ether
Diethyl ether has only two distinct C atoms, because of symmetry. Butan-1-ol has four.

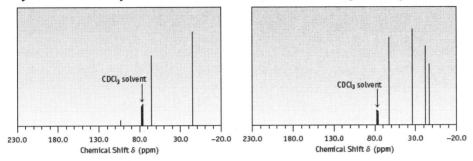

The spectrum on the left shows two peaks, while that on the right shows four—beyond the solvent and reference peaks. The spectra on the left and right are those of diethyl ether and butan-1-ol, respectively.

Exercise 9.2

One could distinguish butan-1-ol and diethyl ether by looking for the distinctive O-H stretch peak (broad peak around 3200-3550 cm^{-1}) of butan-1-ol in the IR spectra.

Exercise 9.3

$CH_3CH_2CH_2NH_2$ CH_3CHCH_3
                                           |
                                          NH_2
propan-1-amine   propan-2-amine
Propan-2-amine has only two distinct C atoms, because of symmetry. Propan-1-amine has three.

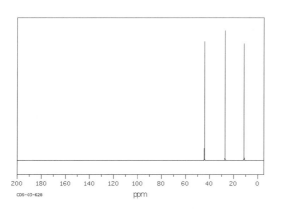 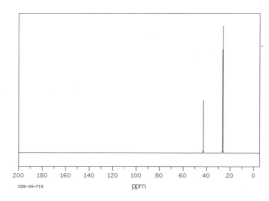

The spectrum on the right shows two peaks, while that on the left shows three. The spectra on the right and left are those of propan-2-amine and propan-1-amine, respectively.

Exercise 9.4

(a) For cyclopentane, we expect to see 1 peak in the ^{13}C NMR spectrum. All C atoms are equivalent.

(b) For 1,3-dimethylbenzene, we expect to see 5 peaks in the ^{13}C NMR spectrum—the two equivalent methyl C atom peaks; the equivalent benzene ring C1 and C3 peaks; the C2 position, the C4 and 6 positions; and the C5 position.

(c) For 1,2-dimethylbenzene, we expect to see 4 peaks in the ^{13}C NMR spectrum—the two methyl C atom peaks; the benzene ring C1 and C2 positions; the C3 and C6 positions; and the C4 and C5 positions.

(d) For 1-methylcyclohexene, we expect to see 7 peaks in the ^{13}C NMR spectrum. There is no symmetry—all C atoms are inequivalent.

Exercise 9.5

(a) CH_4 methane
1 peak in both carbon-13 and proton spectra—the 4 H atoms are equivalent
(b) CH_3CH_3 ethane
1 peak in both carbon-13 and proton spectra—the 2 C atoms are equivalent, as are the 6 H atoms
(c) $CH_3CH_2CH_3$ propane
2 peaks in both carbon-13 and proton spectra—the 2 methyl C atoms are equivalent, while the 6 methyl H atoms and 2 methylene H atoms are equivalent
(d) cyclohexane

All C atoms are equivalent by symmetry—1 peak in carbon-13 spectrum.

There are two kinds of H atoms—axial and equatorial. However, at ordinary temperatures, the ring flips—chair to boat to chair (facing opposite direction) – back and forth so fast that the NMR experiment "sees" the average environment—i.e., we see one peak in the proton spectrum. If the sample were cooled sufficiently, it may be possible to see separate axial and equatorial peaks.

(e) CH_3OCH_3 dimethyl ether
1 peak in both carbon-13 and proton spectra

(f) benzene

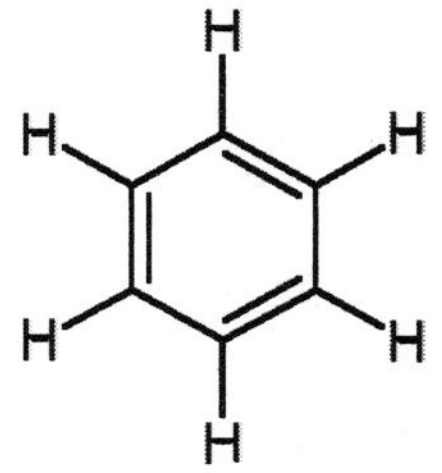

1 peak in both carbon-13 and proton spectra

(g) $(CH_3)_3COH$ 1,1-dimethylethanol
2 peaks in both carbon-13 and proton spectra

(h) CH_3CH_2Cl chloroethane
2 peaks in both carbon-13 and proton spectra

(i) $(CH_3)_2C=C(CH_3)_2$ 2,3-dimethylbut-2-ene
2 peaks in carbon-13 spectrum and 1 peak in proton spectrum

Exercise 9.6

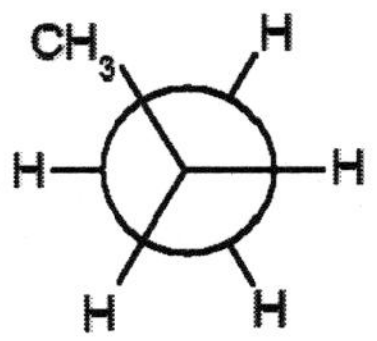

This is the staggered conformation—it is the most stable.

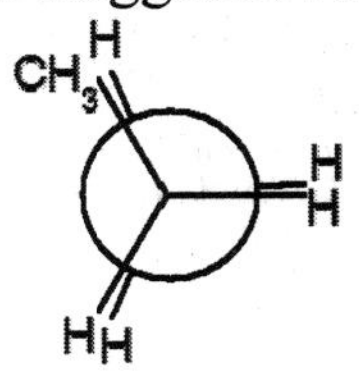

This is the eclipsed conformation—it is the least stable.

Exercise 9.7

These are the Newman projections of the three staggered conformers. The conformer on the left has the lowest energy. The other two have the same energy.

These are the Newman projections of the three eclipsed conformations. The conformation on the left has the highest energy. The other two have the same energy.

Exercise 9.8

There are three equivalent staggered conformers and three equivalent eclipsed conformations of propane.

For propane, the energy as a function of angle of rotation about either of the C-C bonds looks like

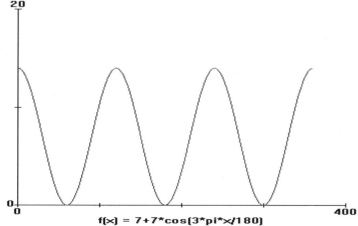

f(x) = 7+7*cos(3*pi*x/180)

Energy (in kJ) vs. angle (in °) for propane

Exercise 9.10

(a) pyridine C_5H_5N one H atom is bonded to each C atom
(b) cyclohexanone $C_6H_{10}O$ two H atoms are bonded to each C atom, except the C=O carbon
(c) indole C_8H_6NH one H atom is bonded to each C atom, except the two C atoms

belonging to both rings

Exercise 9.14

chair conformations of methylcyclohexane—equatorial methyl conformation (on left) and axial methyl conformation (on right)

Exercise 9.15

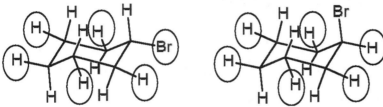

Here are the equatorial (on the left) bromine and axial bromine conformers of bromocyclohexane. The equatorial positions are circled.

Exercise 9.17

Since $[\square]_D = -16°$ is positive for cocaine, cocaine is termed levorotatory.

Exercise 9.18

Specific rotation $[\square]_D$ equals observed rotation, $\square$, divided by the product of sample concentration c and path length l, measured in g mL^{-1} and dm, respectively.

$[\square]_D = \dfrac{\alpha}{c \times l}$, where $\alpha = 1.21°$, $l = 5.00$ cm $= 0.500$ dm, and $c = 1.50$ g in 10.00 mL $= 0.150$ g mL^{-1}.

$[\square]_D = \dfrac{1.21°}{(0.150 \text{ g mL}^{-1}) \times (0.500 \text{ dm})} = 16.1°$

Exercise 9.21

From highest to lowest priority of substituents, we have
(a) —Br > —CH₂CH₂OH > —CH₂CH₃ > —H
(b) —OH > —CO₂CH₃ > —CO₂H > —CH₂OH
(c) —Br > —Cl > —CH₂Br > —CH₂Cl

Exercise 9.22

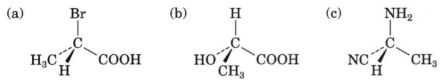

At the carbon stereocentre, the configurations are
(a) S, **(b)** S and **(c)** R

Exercise 9.23

(a)   (b)   (c)

At the top and bottom (respectively) carbon stereocentre, the configurations are
(a) R and R,
(b) S and R, and
(c) R and S

Exercise 9.24

Chloramphenicol
$[\alpha]_D = +18.6°$

At the top and bottom (respectively) carbon stereocentre, the configurations are
R and R.

Exercise 9.25

(a) 2,3-Dibromobutane

```
    CH₃              CH₃
     ⋮                ⋮
Br—C—H           H—C—Br
    |      =       |            is a meso compound—it is achiral.  The molecule on the right is
identical to its
Br—C—H           H—C—Br
     ⋮                ⋮         mirror image on the left.
    CH₃              CH₃
```

(c) 2,4-Dibromopentane

```
    CH₃              CH₃
     ⋮                ⋮
Br—C—H           H—C—Br
    |      =       |            is a meso compound—it is achiral.  The molecule on the right is
identical
H—C—H            H—C—H
    |                |          to its mirror image on the left.
Br—C—H           H—C—Br
     ⋮                ⋮
    CH₃              CH₃
```

(b) 2,3-Dibromopentane

```
    CH₃              CH₃
     |                |
H—C—H            H—C—H
    |      ≠       |            are enantiomers.
Br—C—H           H—C—Br
    |                |
Br—C—H           H—C—Br
    |                |
    CH₃              CH₃
```

Exercise 9.26

(a)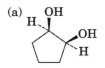

(b)

(c)

```
        H
   Br     CH₃
      \ | /
        C
        |
        C
   H₃C / \ H
        Br
```

(a) and **(c)** are meso molecules—they are *R, S*. **(b)** is an *R, R* stereoisomer.

Exercise 9.27

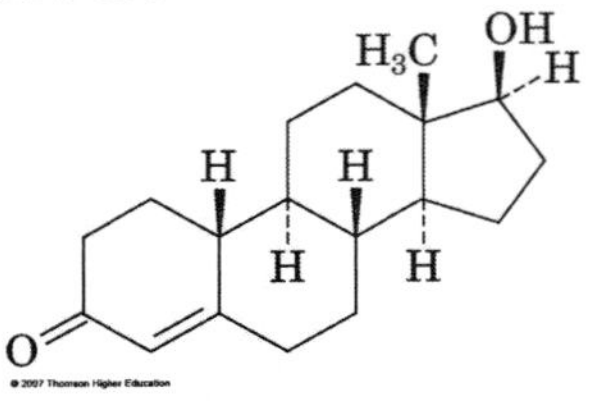

Nandrolone

Nandrolone has 6 stereocentres. There could in principle be as many as $2^6 = 64$ different stereoisomers.

Exercise 9.28

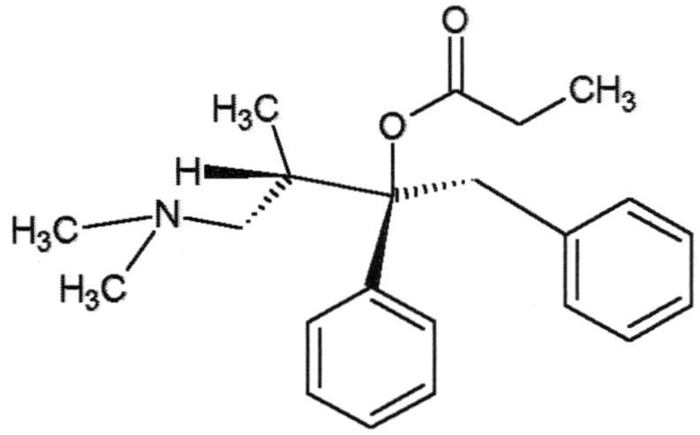

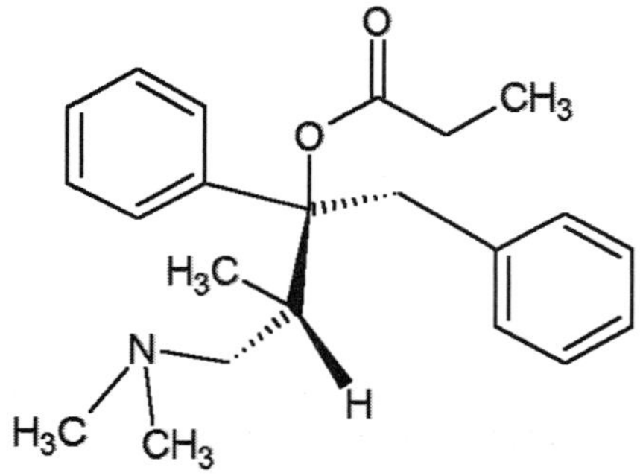

Darvon is the 2*R*,3*S* (+) enantiomer of propoxyphene.
Novrad is the 2*S*,3*R* (-) enantiomer.

REVIEW QUESTIONS

Section 9.4 ^{13}**C NMR—Mapping the carbon framework**

9.29
(a)

CH$_3$—CH$_2$—C=C—CH$_2$—CH$_2$—CH$_3$

with H above and H below the C=C

is a hydrocarbon with seven peaks in its ^{13}C NMR spectrum.

(b)

CH$_3$—CH—CH$_2$— CH$_2$— CH$_3$
   |
   CH$_3$

is a six-carbon compound with only five peaks in its ^{13}C NMR spectrum.

(c)

O (double bond)

CH$_3$—CH—C—H
   |
   CH$_3$

is a four-carbon compound with three peaks in its ^{13}C NMR spectrum and a carbonyl functional group.

9.31
NMR uses photons with lower energy than those used in IR spectroscopy. 400 MHz is a much lower frequency than the 10-100 THz range of IR spectroscopy. NMR causes transitions between spin energy levels that are very close. IR causes vibrational energy transitions that are much higher in energy.

9.33

Reasonable values for the carbon atom chemical shifts are as follows:
—carbonyl carbon atom at 200 ppm
—two methyl carbon atoms at around 20 ppm—two distinct peaks
—C-O methylene carbon atom at 70 ppm
—C-Cl methine carbon atom at 50 ppm
The carbon-13 spectrum might look something like:

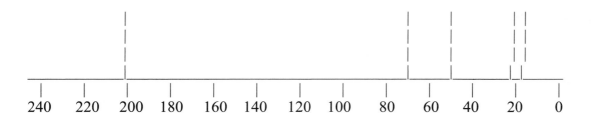

240   220   200   180   160   140   120   100   80   60   40   20   0

9.35
(a) Chloroform absorbs 77.23 ppm downfield from TMS.
(b) A chemical shift of 77.23 δ corresponds to 77.23 × 600 Hz = 46338 Hz downfield from TMS on a 600 MHz spectrometer.
(c) In δ units, the chemical shift of chloroform is 77.23 δ on a 600 MHz spectrometer.

9.37
In order of increasing ^{13}C chemical shift, the one peak exhibited by these compounds are expected to be ordered as follows:
CH_4 < CH_2Cl_2 < HC≡CH < benzene < HC(O)C(O)H (two carbonyl carbon atoms bonded together)

Section 9.5 Conformations of alkanes—Rotation about single bonds

9.39
Possible skeletal structures:
(a) C_4H_8

H_2C=

—CH_3

(b) C_3H_6O

H_2C=

O—CH_3

(c) C_4H_9Cl

H_3C—

Cl

9.41

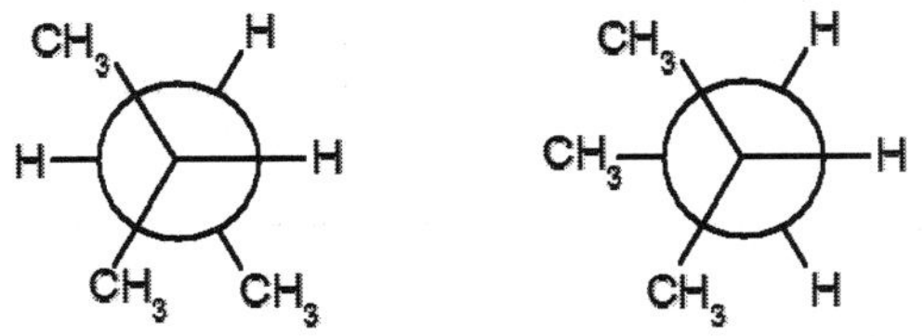

The staggered conformer on the left is expected to have the lower energy because the methyl substituent on the back carbon atom is close to both of the methyl substituents on the front carbon atom in the right conformer (two gauche interactions), but only one in the left conformer (one gauche interaction).

9.43

In bromoethane, there are three atom-atom interactions in the eclipsed conformation giving rise to the rotation barrier of 15.0 kJ mol^{-1} about the C—C bond—two H-H interactions and one H-Br interaction. The two H-H interactions account for 2×3.8 kJ mol^{-1} = 7.6 kJ mol^{-1} of the barrier. The H-Br interaction accounts for
$15.0 - 7.6$ kJ mol^{-1} = 7.4 kJ mol^{-1}.

Section 9.6 Restricted rotation about bonds

9.45

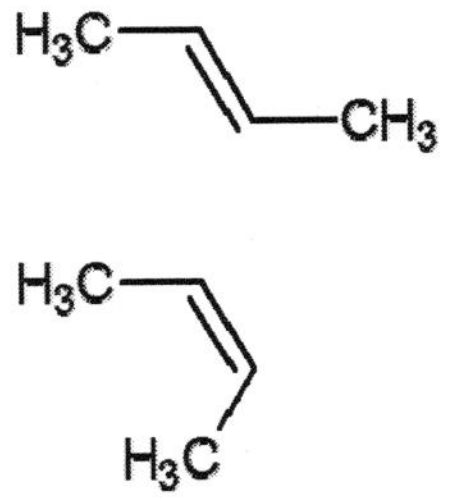

cis and *trans* 2-butene
The other structural isomers of C_4H_8 shown above do not have *cis* and *trans* isomers.

Section 9.7 Cyclic molecules

9.47

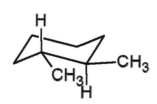

trans-1,2-dichlorocyclohexane in chair conformer. Either both Cl atoms are equatorial (top conformer) or axial (bottom conformer) because on adjacent carbon atoms the two axial positions are on opposite sides of the ring—the two equatorial are necessarily also on opposite sides of the ring. The *trans* isomer has the two Cl atoms on opposite sides of the ring. The diequatorial form is lower in energy.

9.49

ring flip

9.51

trans-1,2-dimethylcyclohexane in its more stable chair conformer. The two methyl groups (larger than the H atom substituents) are in more accommodating equatorial positions. Because this is *trans* methyl groups on adjacent carbon atoms must be the same position—i.e., either both axial or both equatorial.

9.53

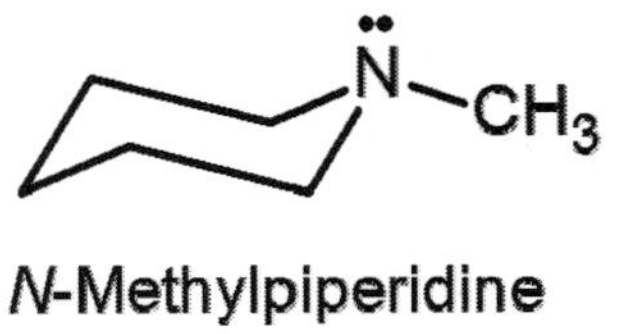

N-Methylpiperidine

The most stable conformer of *N*-methylpiperidine. Clearly the methyl substituent has greater steric requirements than the electron lone pair—the methyl group is in the more accommodating equatorial position.

9.55

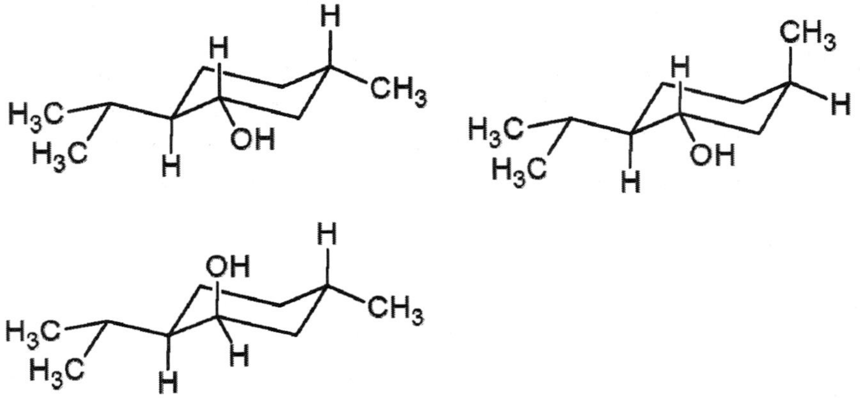

Three of the *cis–trans* isomers of menthol are shown. In the first and second, the hydroxyl group is *trans* to the iospropyl group—it is *cis* in the third isomer. In the first and third, the methyl group is *trans* to the iospropyl group—it is *cis* in the second isomer.

Section 9.8 Stereochemistry

9.57

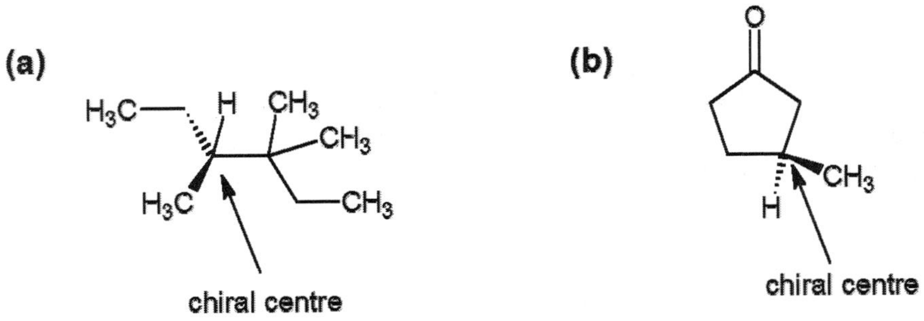

(a)

chiral centre

(b)

chiral centre

(c)

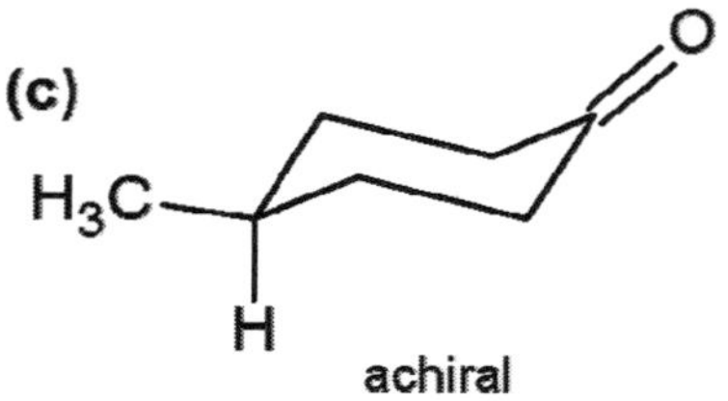

achiral

the ring has a plane of
symmetry through C4
and the carbonyl bond

(d)

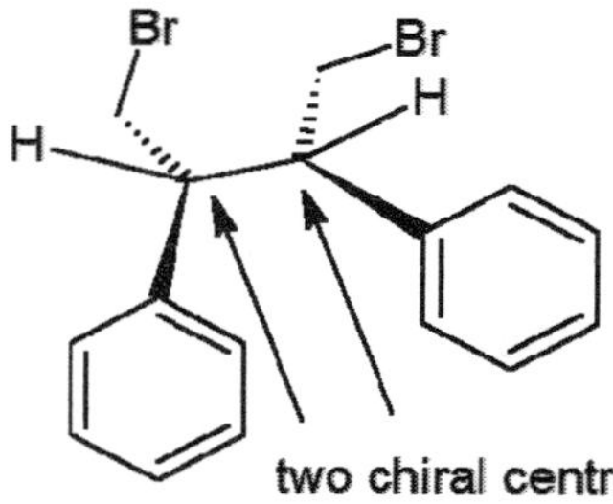

two chiral centres

Depending on whether the stereocenters
are configured S,S or R,R, or S,R (R,S is
the same by symmetry), the molecule is
either chiral (former case - S,S and R,R
are enantiomers) or achiral (S,R = R,S is
called a meso compound)

(e)

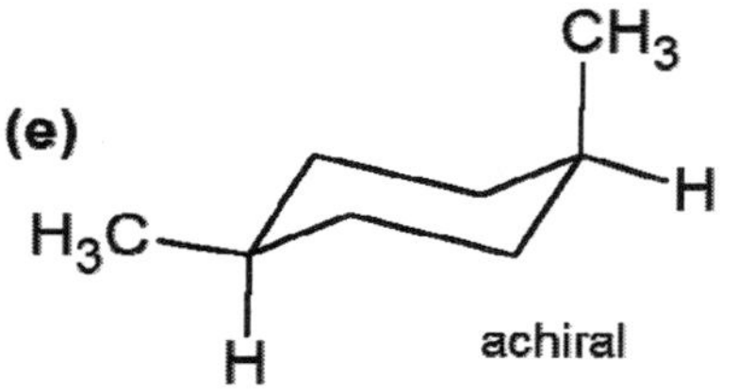

achiral

the ring has a plane of
symmetry through C4
and the carbonyl bond

Section 9.9 Optical activity

9.59

Specific rotation $[\square]_D$ equals observed rotation, $\square$, divided by the product of sample concentration c and path length l, measured in g mL^{-1} and dm, respectively.

$$[\square]_D = \frac{\alpha}{c \times l}, \text{ where } \alpha = 2.22°, l = 1.00 \text{ cm} = 0.100 \text{ dm, and } c = 3.00 \text{ g in } 5.00 \text{ mL} = 0.600 \text{ g mL}^{-1}.$$

$$[\square]_D = \frac{2.22°}{(0.600 \text{ g mL}^{-1}) \times (0.100 \text{ dm})} = 37.0°$$

Section 9.10 Sequence rules for specifying configuration

9.61

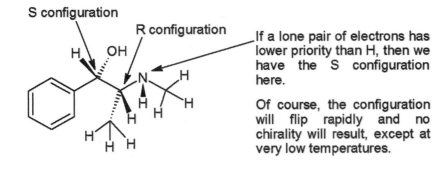

S configuration

R configuration

If a lone pair of electrons has lower priority than H, then we have the S configuration here.

Of course, the configuration will flip rapidly and no chirality will result, except at very low temperatures.

9.63
Sets of substituents are assigned priorities as follows:

(a)

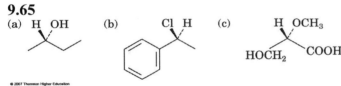

1   2   3   4

(b)
1 —SO$_3$H   2 —SH   3 —OCH$_2$CH$_2$OH   4 —NH$_2$

9.65

(a) H OH   (b) Cl H   (c) H OCH$_3$

HOCH$_2$ COOH

© 2007 Thomson Higher Education

The configurations at the chiral centres are
(a) *S* **(b)** *S* **(c)** *S*

Section 9.11 Enantiomers, diastereomers, and meso stereoisomers

9.67
Substances comprised of (2R,3R)-dihydroxypentane and (2S,3S)-dihydroxypentane will have the same (not equal to zero) numerical value for their optical rotation except for opposite signs. These stereoisomers are enantiomers. (2R,3S)-dihydroxypentane and (2R,3R)-dihyroxypentane are diastereomers of each other, so there will be no predictable relationship between the optical rotation values for substances comprised of these two stereoisomers.

9.69
The stereochemical configurations of the two diastereomers of (2S,4R)-dibromooctane are (2R,4R) and (2S,4S).

Section 9.12 Molecules with more than two stereocentres

9.71

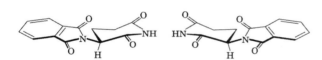

Ribose—three stereocentres indicated with circles. There are 2^3 = 8 stereoisomers of ribose.

9.73

Diastereomer of ribose

Section 9.13 Chiral environments in laboratories and living systems

9.75

Thalidomide

(a) The lone stereocentre is the C atom bonded to the N atom of the 5-membered ring.

(b) The structure on the right is the R enantiomer—the S enantiomer is the structure on the left.

SUMMARY AND CONCEPTUAL QUESTIONS

9.77

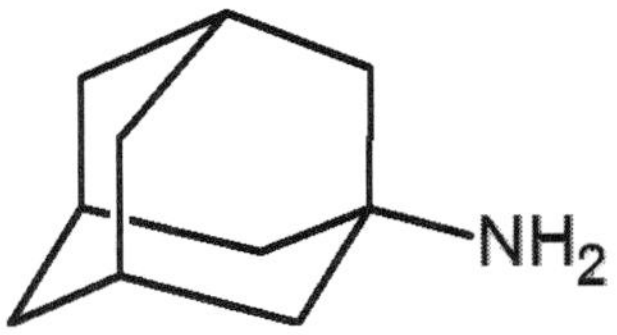

9.79

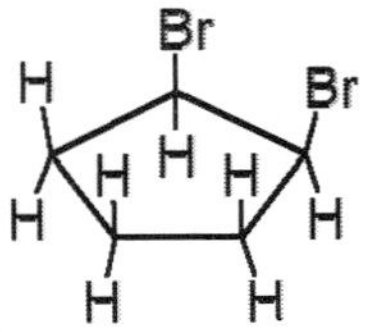

cis-1,2-dibromocyclopentane

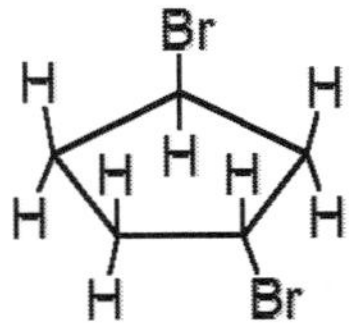

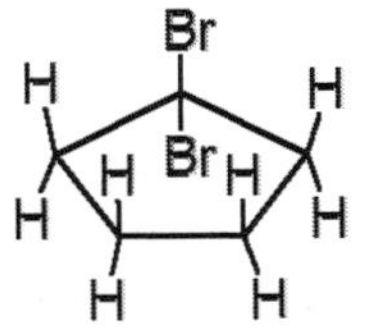

Two constitutional isomers of *cis*-1,2-dibromocyclopentane—with different positions of the Br functional groups (positional isomers).
Any dibromo pentene would have a different carbon skeleton than *cis*-1,2-dibromocyclopentane (skeletal isomers).

9.81

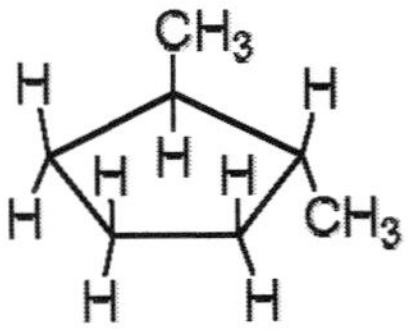

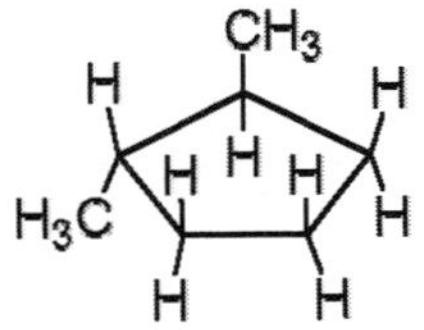

The two enantiomers of *trans*-1,2-dimethylcyclopentane—they are non-superimposable mirror images of each other.

9.83

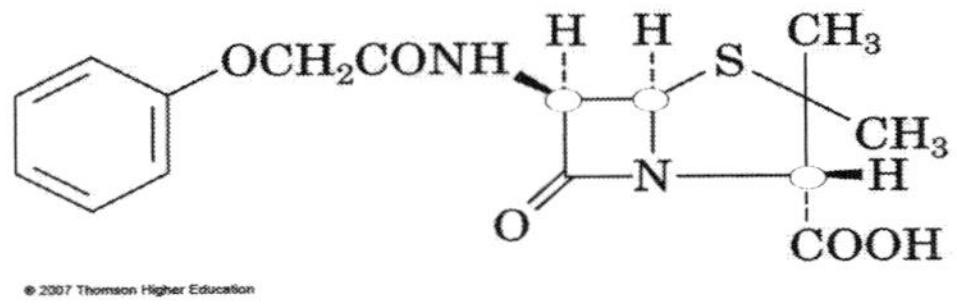

**Penicillin V
(antibiotic)**

© 2007 Thomson Higher Education

The three stereocentres are denoted by solid white circles.

9.85

The specific rotation for (*R*)-serine is $[\square]_D$ = +6.83°.

9.87

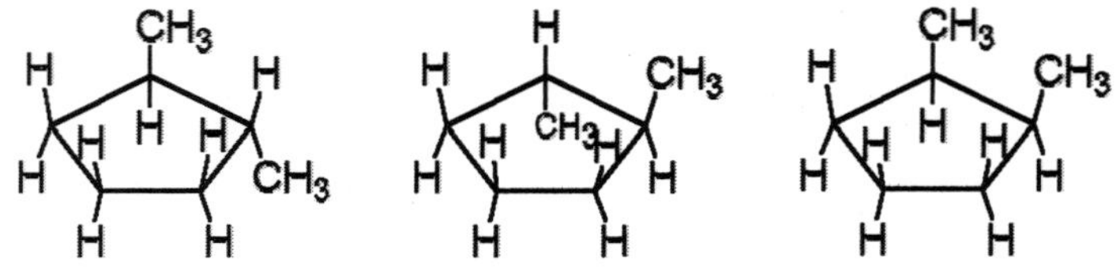

The stereoisomers of 1,2-dimethylcyclopentane. From left to right, the stereocentres are configured as
(*R,R*), (*S,S*), and (*R,S*).
The two on the left (the trans isomers) are a pair of enantiomers. The structure on the right (the *cis* isomer) is a meso molecule.

9.89
(a) The chiral carbon atom here has the *S* configuration.
(b) Both of the chiral carbons here have the *R* configuration.

9.91
(a) (*S*)-5-Chloro-2-hexene and chlorocyclohexane are structural or constitutional isomers. Specifically, they are skeletal isomers.
(b) (2*R*,3*R*)-Dibromopentane and (2*S*,3*R*)-dibromopentane are diastereomers.

9.93

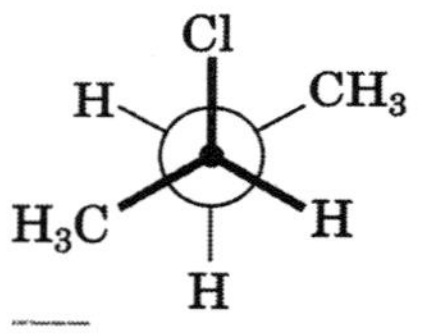

The stereocentre (the top carbon atom in the projection) is in the *S* configuration.

Chapter 10
Modelling Bonding in Molecules

IN-CHAPTER EXERCISES

Exercise 10.3—Drawing Lewis structures

Ammonium ion, NH_4^+
Number of valence electrons = 5 (from N) + 4×1 (from H) − 1 (net +1 charge)
= 8 electrons (4 electron pairs)
Use all four electron pairs to form bonds between N and each of the four H's.

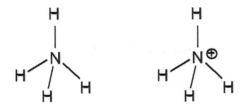

We are finished with the structure on the left. Here, all atoms have a filled valence shell—the N has 8 valence electrons.
Later in the chapter (Section 10.6), you will learn how to attribute the net +1 charge to the N atom, as shown on the right. This is not necessarily a completely accurate description of the distribution of charge. However, it is easy to do—as you will see later—and it gives a first guess at the charge distribution.

CO molecule
Number of valence electrons = 4 (from C) + 6 (from O)
= 10 electrons (5 electron pairs)
Use one electron pair to form a bond between C and O.
    C—O
Place three pairs of electrons on the O atom— as lone pairs— to give O its octet.
There is one pair left. It goes to the C atom. We now have

:C—O̤:

Here, the C atom only has four electrons in its valence shell. Shift two of the O atom lone pairs into the bonding region of CO to give the C atom its octet—a triple bond between C and O is formed in the process. The result is:

:C≡O: :C̠≡Ȯ:

The structure on the right has "formal charges" included—see later in the chapter.

Sulfate ion, SO₄²⁻

Number of valence electrons = 6 (from S) + 4×6 (from the 4 O's) + 2 (net − 2 charge)
= 32 electrons (16 electron pairs)

Use four electron pairs to form bonds between the S atom and each of the four O atoms.

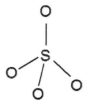

16 − 4 = 12 electron pairs remain. Place three pairs of electrons on each O atom—as lone pairs—to give each O atom its octet.
12 − 4 × 3 = 0 electron pairs remain. We now have:

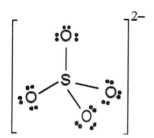

Here, each atom has its octet. Later you will learn how S atoms, being in the third period, are capable of expanding their valence shells beyond the octet. For now, we stop here.

Exercise 10.4—Predicting molecular structures

Methanol, CH₃OH

Number of valence electrons = 4 (from C) + 4×1 (from H) + 1×6 (from O)
= 14 electrons (7 electron pairs)

There are two central atoms here, C and O. The C atom is attached to three H atoms and an O atom, while the O atom is attached to a C atom and another H atom. Here is the skeletal structure.

Five electron pairs are used as single bonds here. Use the remaining two electron pairs to give the O atom an octet.

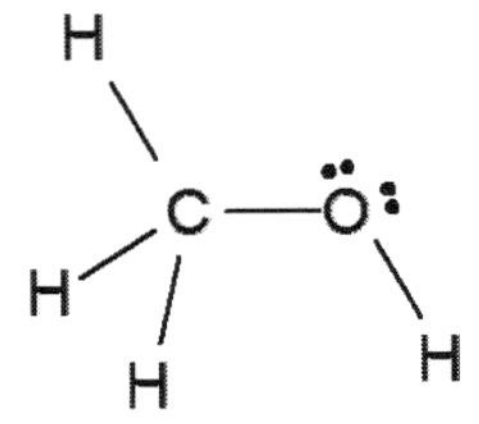

Here, all atoms have filled valence shells.

Hydroxylamine, NH_2OH

Number of valence electrons = 5 (from N) + 3×1 (from H) + 1×6 (from O)
= 14 electrons (7 electron pairs)

Here is the skeletal structure.

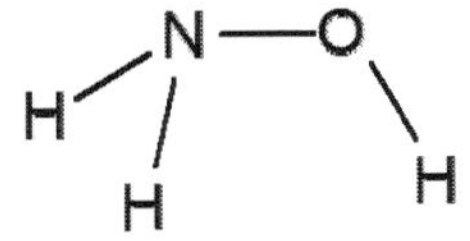

Four electron pairs are used as single bonds here. Use the remaining three electron pairs to give the O atom (2 are needed) and the N atom (one is needed) octets

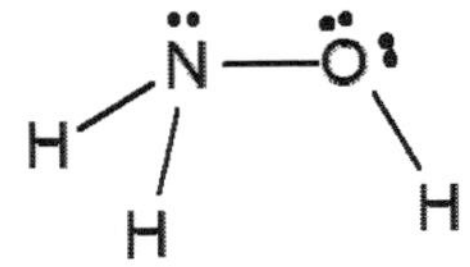

Exercise 10.5—Lewis structures of oxoacids and their anions

Number of valence electrons = 5 (from P) + 2×1 (from H) + 4×6 (from O) + 1 (net −1 charge)
= 32 electrons (16 electron pairs)

Skeletal structure:

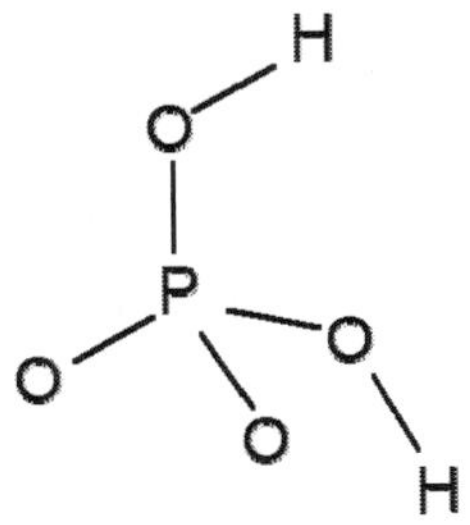

Six electron pairs are used as single bonds here.
$16 - 6 = 10$ electrons pairs remain. Two of the O atoms need three pairs to complete their octet, while the other two only need two pairs. Altogether, $2 \times 3 + 2 \times 2 = 10$ electron pairs are needed to give every atom a filled valence shell—the P atom already has an octet.

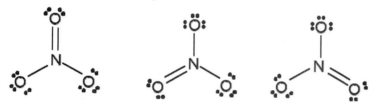

Exercise 10.6—Identifying isoelectronic species

(a) Acetylide ion, C_2^{2-}, is isoelectronic with N_2.
Number of valence electrons in C_2^{2-} = 2 × 4 (from C) + 2 (net −2 charge) = 10
Number of valence electrons in N_2 = 2 × 5 (from N) = 10
(b) SO_2 is isoelectronic with nitrite ion, NO_2^-.
Note that a group 15 to 16 substitution (N → S) together with an increase in charge by 1 (to account for the increased core charge – nucleus + core electrons) produces an isoelectronic species.
(c) Hydroxyl ion, OH^-, is isoelectronic with HF.
A group 17 to 16 substitution, together with a decrease in charge by 1, produces an isoelectronic species.

Exercise 10.9—Drawing resonance structures

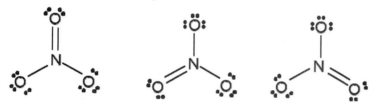

Three resonance structures of a nitrate ion, NO_3^-. Each of these species has a negative charge not indicated.

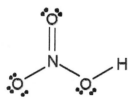

A structure for a nitric acid molecule, HNO_3.

Exercise 10.11—Deciding which resonance structures are most important

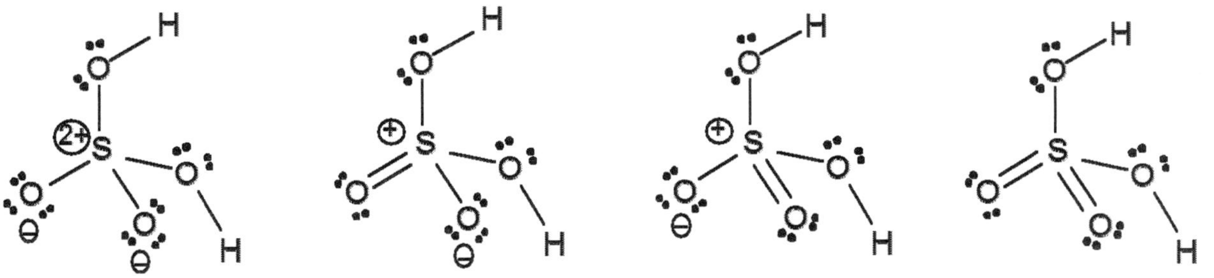

Resonance structures of a sulfuric acid molecule.

Formal charge on an atom =

(# of valence electrons on neutral atom) − (# of non-bonding electrons) − ½ × (# of bonding electrons)

(# of valence electrons on neutral atom) − (# of non-bonding electrons) − (# of BONDS)

Structure on left: formal charge on S atom = $6 - 0 - 4 = +2$

Middle two structures: formal charge on S atom = $6 - 0 - 5 = +1$

Structure on right: formal charge on S atom = $6 - 0 - 6 = 0$

Singly bonded O atom with three lone pairs has formal charge = $6 - 6 - 1 = -1$.

Doubly bonded O atom with two lone pairs has a formal charge = $6 - 4 - 2 = 0$.

In the structure on the right all atoms have zero formal charges—that is, the distribution of valence electrons is closest to that in the neutral atoms. This structure is considered to best describe the actual charge distribution in sulfuric acid molecules.

Exercise 10.12—Deciding which resonance structures are most important

These are the only electron distributions in which the valence shell of each O atom is full. These distributions of electrons in an ozone molecule are equivalent—by symmetry. They make equal contributions to the actual distribution, which is neither of these, but with π electrons delocalized over the two O-O bonds.

Exercise 10.13—Predicting the shapes of molecules

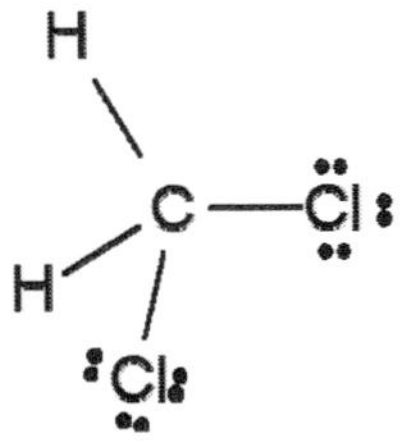

Lewis structure of dichloromethane. It has four electron regions about the central carbon atom—all single bonds. These bonds are tetrahedrally oriented. If the molecule was a regular tetrahedron, the Cl—C—Cl bond angle would be 109.5°. Because the Cl atoms are bigger than the H atoms, we might predict the bond angle to be a little bigger than 109.5°—and experimental measurements tell us that it is.

Exercise 10.16—Predicting the shapes of molecules and ions

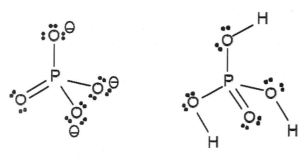

BF$_3$ and BF$_4^{\ominus}$

These species have 3 and 4 electron regions about the central B atom—all single bonds. The bonds have trigonal planar and tetrahedral orientations, respectively. In BF$_3$ molecules, the F-B-F angle is 120°, while in BF$_4^-$ ions, it is 109.5°.

Exercise 10.17—Predicting the shapes of molecules and ions

(a) Phosphate ion, PO$_4^{3-}$, is tetrahedral. It has four electron regions—3 single bonds and a double bond.

(b) Phosphoric acid molecule, H3PO4, is also tetrahedral around the P atom. It has four electron regions—3 single bonds and a double bond.

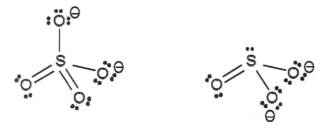

(c) Sulfate ion, SO42$\square$, is tetrahedral. It has four electron regions around the S atom—2 single bonds and 2 double bonds

(d) A sulfite ion, SO32$\square$, is trigonal pyramidal. It has four electron regions around the S atom—2 single bonds, a double bond, and a lone pair—which are oriented tetrahedrally.

(e) Ethanol is tetrahedral at the carbon atoms and bent at the oxygen atom (4 electron regions, two of which are non-bonding).

(f) Methyl ethyl ketone, CH3C(O)CH3 , is trigonal planar at the carbonyl carbon atom (three electron regions, a double bond and two single bonds), and tetrahedral at the methyl carbon atoms.

Exercise 10.18—Valence bond description of bonding

Lewis structures for (a) a hydronium ion and (b) a methylamine molecule

(a) The O atom of a hydronium ion has four electron regions—distributed tetrahedrally. So we invoke sp^3 hybridization on the O atom. Three of the sp^3 orbitals overlap with s orbitals on H atoms to form the O-H bonds, while the fourth accommodates the lone pair on the O atom.

(b) The C atom has four electron regions—distributed tetrahedrally. We invoke sp^3 hybridization on the C atom. Three of the sp^3 orbitals overlap with s orbitals on H atoms to form the C-H bonds, while the fourth makes a bond with the N atom.

The N atom has four electron regions—distributed tetrahedrally. Consequently, we presume sp^3 hybridization on the N atom. Two of the sp^3 orbitals overlap with s orbitals on H atoms to form the N-H bonds, one of them forms the bond with the C atom, while the fourth accommodates the lone pair on the N atom.

Exercise 10.19—Bonding in a molecule of acetone

Lewis structure of acetone

The methyl C atoms have the tetrahedral distribution of electron regions. We presume sp^3 hybridization:Three of the sp^3 orbitals overlap with s orbitals on H atoms to form the C-H bonds, while the fourth makes a bond with the carbonyl C atom.

The carbonyl C atom has a trigonal-planar arrangement of electron regions—consequently, we use sp^2 hybridization here. The C-C bonds are formed from carbonyl-carbon atom sp^2 orbitals, and a methyl-carbon sp^3 orbital. The remaining sp^2 orbital forms the σ bond to oxygen.

The oxygen atom has three electron regions—the trigonal-planar arrangement—as depicted above. This O atom uses one sp^2 orbital to make the σ bond to C. The remaining two sp^2 orbitals accommodate lone pairs.

The C=O double bond is formed from the unhybridized p orbitals on each atom—both atoms are sp^2 hybridized with one unhybridized p orbital. This p-p bond is the C-O π bond.

Since the O atom is terminal, it could also be described in terms of sp hybridization. In this case, one sp orbital forms the σ bond to C, while the other accommodates a lone pair. One of the unhybridized p orbitals on O (the p orbital perpendicular to the plane of the molecule) forms the π bond to carbon, while the other accommodates the remaining lone pair of electrons.

Exercise 10.20—Molecules with triple bonds

:N≡N:

Lewis structure of dinitrogen, N_2

The N atoms have a linear arrangement of two electron regions. This means sp hybridization. An sp orbital from each N atom is used to make the N-N σ bond. The other sp orbital on each atom accommodates a lone pair.

The two remaining p orbitals on one atom overlap sideways with the two p orbitals on the other atom to form the two N-N π bonds.

Exercise 10.21—Bonding and hybridization

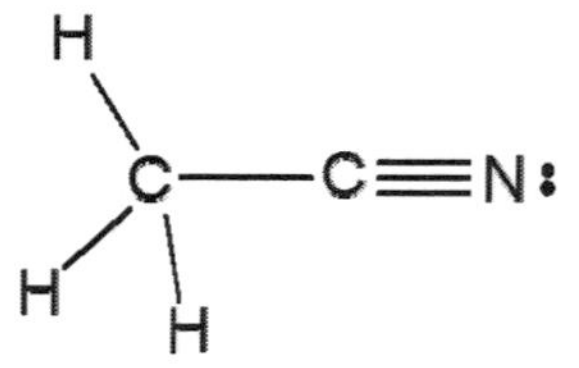

Lewis structure of acetonitrile

The H-C-H, H-C-C, and C-C-N bond angles are predicted to be 109.5°, 109.5°, and 180°, respectively.

The methyl C atom has the tetrahedral distribution of electron regions—presumed sp^3 hybridization. Three of the sp^3 orbitals overlap with s orbitals on H atoms to form the C-H bonds, while the fourth makes a bond with the nitrile C atom.

The nitrile C and N atoms have a linear arrangement of two electron regions. We presume sp hybridization. One sp orbital from each atom is used to make the C-N σ bond. The remaining sp orbital on the nitrile C atom forms the bond with the methyl C atom, while the remaining sp orbital on the N atom accommodates the lone pair.

The two remaining p orbitals on the nitrile C and N atoms overlap in pairs to form the two C-N π bonds.

Exercise 10.22—Valence bond model and electron delocalization

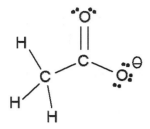

One of the two Lewis resonance structures of acetate.

The methyl C atom has the tetrahedral distribution of electron regions—with presumed sp^3 hybridization. Three of the sp^3 orbitals overlap with s orbitals on H atoms to form the C-H bonds, while the fourth makes a bond with the carboxylate C atom.

The carboxylate C atom has a trigonal-planar arrangement of electron regions—presumed sp^2 hybridization here. The C-C bond is formed from one of these sp^2 orbitals, and a methyl-carbon sp^3 orbital. The remaining two sp^2 orbitals form σ bonds to oxygen atoms.

The oxygen atoms can be described as sp or sp^2 hybridized. Because of symmetry, we should use the same description for both atoms—in contrast to the Lewis structure which implies that one of the O atoms is sp^3 hybridized. We will assume sp hybridized O atoms. Each O atom uses one sp orbital to make a σ bond to C. The other sp orbital on each atom accommodates a lone pair. The other lone pair, on each O atom, occupies one of the two unhybridized p orbitals.

What remains are a p orbital on each of the O atoms and the unhybridized p orbital on the carboxylate carbon atom—three orbitals in all. Two electron pairs also remain. In the Lewis structure above, a π bond is formed with sideways overlap of the p orbital on C and one of the O atom p orbitals. The other O-atom p orbital accommodates the remaining lone pair of electrons. This can be written in one of two equivalent ways—the resonance structures of carboxylate anion.

To accommodate the experimental evidence that the two C-O bonds are equivalent, another approach is to consider that there are two electrons distributed over the O-C-O π-bonding system, so that each C-O bond has three electrons—two in the σ bond, and one in a π-bond. It is difficult to draw a Lewis structure of an acetate ion showing an equal number of lone-pair electrons on each O atom.

Exercise 10.25—Deducing electron configuration and bond order

Electron configurations

$H_2^+ : (\sigma_{1s})^1$

$He_2^{\square}: (\sigma_{1s})^2(\sigma_{1s}^*)^1$

$H_2^{\square}: (\sigma_{1s})^2(\sigma_{1s}^*)^1$

Bond order $= \frac{1}{2} \times$ (# of electrons in bonding MOs $-$ # of electrons in antibonding MOs)

$\qquad = \frac{1}{2} \times (1 - 0) = \frac{1}{2}$ \qquad for H_2^+

$\qquad = \frac{1}{2} \times (2 - 1) = \frac{1}{2}$ \qquad for $He_2^{\square}$

$\qquad = \frac{1}{2} \times (2 - 1) = \frac{1}{2}$ \qquad for $H_2^{\square}$

H_2^+ has a bond order of ½. It should exist. But, it is weakly bound. $He_2^{\square}$ and $H_2^{\square}$ have the same bond order: although they have more electrons in bonding orbitals, but also have an electron in an antibonding orbital.

Exercise 10.26—Electron configuration, bond order and stability of molecules

The electron configuration of $Li_2^{\square}$ ion (7 electrons) is $(\sigma_{1s})^2(\sigma_{1s}^*)^2(\sigma_{2s})^2(\sigma_{2s}^*)^1$.
Bond order $= \frac{1}{2} \times (2 + 2 - 2 - 1) = \frac{1}{2}$
$Li_2^{\square}$ anion should exist. It is weakly bound with bond order, $\frac{1}{2}$.

Exercise 10.27—MO electron configuration for homonuclear diatomic ions

Electron configuration of ground state $O_2^{\square}$ ion (11 **valence** electrons):
    $[\text{core electrons}](\sigma_{2s})^2(\sigma_{2s}^*)^2(\pi_{2p_y})^2(\pi_{2p_z})^2(\sigma_{2p_x})^2(\pi_{2p_y}^*)^1$
Bond order $= \frac{1}{2} \times (2 + 2 + 2 + 2 - 2 - 1) = 2\frac{1}{2}$
$O_2^{\square}$ ion: it has an unpaired electron.

REVIEW QUESTIONS

10.3 Lewis structures

10.31
Elements in group 13 have 3 valence electrons. To obey the octet rule, such elements must form four bonds. Three of these bonds are formed by sharing each of the valence electrons with another atom. This gives the atom 6 valence electrons. To get 8, the atom must form a fourth "coordinate bond" wherein both electrons come from a different atom. B and Al can form four bonds in this way. These species typically have a net negative charge. They are the products of a Lewis acid—the electron-deficient 6 valence electron B or Al—and a (typically negatively charged) Lewis base (e.g., F^- or Cl^-). Elements in group 17 have 7 valence electrons. They can form an octet by forming a single bond.

10.33
(a) NF_3 molecules
Number of valence electrons = 5 (from N) + 3×7 (from F)
= 26 electrons (13 electron pairs)
Use three electron pairs to form bonds between N and each of the three F's.
$13 - 3 = 10$ electron pairs remain.
Use $3 \times 3 = 9$ electron pairs to complete the octet about each of the three F's. The remaining pair of electrons is given to N to complete its octet.

(b) ClO_3^- ions
Number of valence electrons = 7 (from Cl) + 3×6 (from the 3 O's) + 1 (net -1 charge)
= 26 electrons (13 electron pairs)
Use three electron pairs to form bonds between Cl and each of the three O's.
$13 - 3 = 10$ electron pairs remain.
Use $3 \times 3 = 9$ electron pairs to complete the octet about each of the three O's. The remaining pair of electrons is given to Cl to complete its octet.

If you have studied formal charge (see Section 10.6), then you know that this structure can be taken further to get:

The structure with single bonds has a formal charge of +2 on Cl and −1 on each O atom. Forming a double bond by moving a lone pair on O into a bonding pair between O and Cl raises the formal charge on the O to zero and lowers the formal charge on Cl. Doing this twice reduces the formal charge on Cl to zero and minimizes formal charges overall. This is the preferred structure.

(c) HOBr molecules

Number of valence electrons = 7 (from Br) + 6 (from O) + 1 (from H)
= 14 electrons (7 electron pairs)

Use two electron pairs to form the H-O and O-Br bonds.
7 − 2 = 5 electron pairs remain.

Use 3 electron pairs to complete the octet about Br, and 2 electron pairs to complete the octet about O.

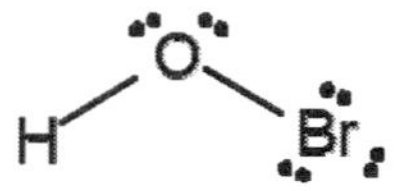

(d) SO_3^{2-} ions

Number of valence electrons = 6 (from S) + 3×6 (from the 3 O's) + 2 (net − 2 charge)
= 26 electrons (13 electron pairs)

Use three electron pairs to form bonds between S and each of the three O's.
13 − 3 = 10 electron pairs remain.

Use 3 × 3 = 9 electron pairs to complete the octet about each of the three O's. The remaining pair of electrons is given to S to complete its octet.

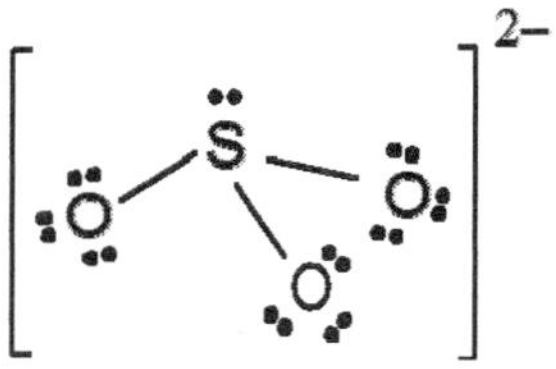

If you have studied formal charge (see Section 10.6), then you know that this structure can be taken further to get:

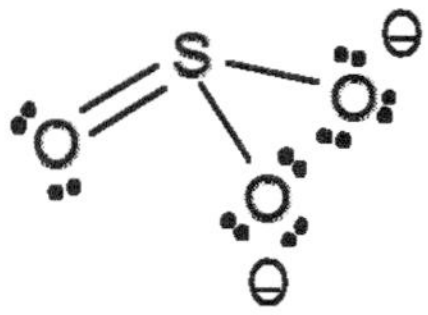

The structure with single bonds has a formal charge of +1 on S and −1 on each O atom. Forming a double bond by moving a lone pair on O into a bonding pair between O and S raises the formal charge on the O to zero and lowers the formal charge on S. Doing this once reduces the formal charge on S to zero and minimizes formal charges overall. This is the preferred structure.

10.4 Exceptions to the octet rule

10.35

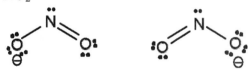

10.5 Resonance and delocalized electron models

10.37
(a) SO_2

(b) NO_2

Note that NO2 cannot have octets on all atoms—it has an odd number of valence electrons (= $5 + 2 \times 6 = 17$).
We choose N to be missing the one electron because it is less electronegative than O.

(c) SCN^-

10.39

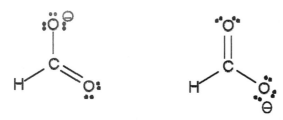

Although in each of these Lewis structures, one C-O bond is a single bond, and the other is a double bond, they are equivalent, with delodcalization of the π electrons, so that bond has three electrons.

Average C-O bond order $= \dfrac{(2+1)}{2} = \dfrac{3}{2}$

10.6 Which resonance structures are most important?

10.41
(a) $NO_2^{\square}$

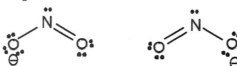

Formal charges are zero on O atoms and +1 on N atom.
Check:
    For the equivalent O atoms, formal charge $= 6 - 4 - 2 = 0$
    For N, formal charge $= 5 - 0 - 4 = +1$
(b) $NO_2^{\square}$

The N has a formal charge of zero—group 15, valence of 3. The formal charges on the two equivalent (although not shown as equivalent in these Lewis structures) O atoms are zero and −1.
Check:
    formal charge (double-bonded O atom) $= 6 - 4 - 2 = 0$
    formal charge (single-bonded O atom) $= 6 - 6 - 1 = -1$

Average formal charge on O atoms $= \dfrac{(0-1)}{2} = -\dfrac{1}{2}$

(c) NF_3

All atoms have zero formal charge. A group 17 element with a valence of 1 and a group 15 element with a valence of 3.

(d) HNO_3

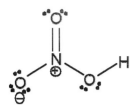

Formal charges are as shown. Note the +1 on the tetravalent N atom (group 15) and −1 on the univalent O (group 16).

10.43

If an $H^{\square}$ ion attaches to $NO_2^{\square}$ ion (to form the acid HNO_2), it attaches to an O atom—not N. This is consistent with the negative charge localized on the O atoms—as indicated in these resonance structures.

10.7 Spatial arrangement of atoms within molecules

10.45
(a) CO_2

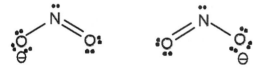

There are two electron regions about the carbon atom. The electron geometry and the molecular geometry (shape of the molecule) are both linear since there are no lone pairs on the C atom.
(b) NO_2^-

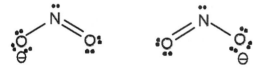

There are three electron regions about the nitrogen atom. The electron geometry is trigonal planar, while the molecular geometry (shape of the ion) is bent—there is one lone pair on the central atom.

(c) O_3

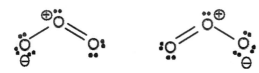

There are three electron regions about the oxygen atom. The electron geometry is trigonal planar, while the molecular geometry (shape of the molecule) is bent—there is one lone pair on the central atom. This molecule is isoelectronic with NO_2^-. They have the same shape and Lewis structure—aside from the formal charge on the central atom.

(d) ClO_2^-

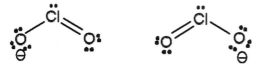

There are four electron regions about the oxygen atom. The electron geometry is tetrahedral, while the molecular geometry (shape of the ion) is bent—there are two lone pairs on the central atom. This ion is probably more bent than O_3 molecules or NO_2^- ions, because there are two lone pairs on the central atom in this case.

10.47

Phenylalanine bond angles:
(1) 120°   (2) 109.5°   (3) 120°
(4) 109.5°   (5) 109.5°

10.8 The valence bond model of covalent bonding

10.49

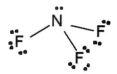

Lewis structure of NF_3

The molecular geometry (shape of the molecule) is trigonal pyramidal—there is a lone pair—while the electron geometry about the N atom is tetrahedral.

A tetrahedral distribution of electron regions implies sp^3 hybridization at nitrogen. Three of the sp^3 orbitals overlap with sp orbitals on F atoms—we assume the F atoms to be sp hybridized. The remaining nitrogen sp^3 orbital accommodates the lone pair on the N atom. The remaining one sp and two p orbitals on each F atom accommodate the lone pairs—three on each F atom. We can equally well invoke sp^3 hybridization on the F atoms (with pairs of electrons in three of the hybrid orbitals, and a single electron in the other). Overlap of the unfilled sp^3 hybrid orbital on each F atom with an sp^3 hybrid orbital on the N atom gives rise to a single bond, and there are three residual electron pairs in sp^3 hybrid orbitals.

10.51

(a) The carbon atoms in dimethyl ether, H_3COCH_3

Use sp^3 hybridized orbitals on the C atoms to form σ bonds with H atoms, and with the O atom.

The oxygen atom in dimethyl ether, H_3COCH_3 uses two sp^3 hybridized orbitals to form σ bonds with C atoms.

The remaining two sp^3 orbitals on the O atom accommodate the two lone pairs.

(b) The carbon atoms in propene, $H_3CCH=CH_2$

Use sp^3 hybridized orbitals on the methyl C atoms to form σ bonds with H atoms, and with the middle C atom.

The vinyl C atoms (i.e., the doubly bonded C atoms) use sp^2 orbitals to form σ bonds with each other, with H atoms and with the methyl C atom. Use the remaining p orbital on each of the two vinyl C atoms to form a π bond.

(c) The alkyl C atom (i.e., the C atom bonded to N) and the N atom in glycine use sp^3 hybridized orbitals to form σ bonds with H atoms, and with each other. The N atom has a lone pair in the remaining sp^3 orbital. The carboxylic C atom uses sp^2 orbitals to form σ bonds with the other C atom and the two O atoms. Use the remaining p orbital on the carboxylic C and O atoms to form a π bond.

10.53

(a) SO_2

The O-S-O bond angle is about 120°. The sulfur atom uses sp^2 orbitals to form two σ bonds—one with each O atom. The other sp^2 orbital accommodates a lone pair. We can describe the Lewis structure with only one double S-O bond with the remaining unhybridized p orbital. It forms a π bond with just one of the O atoms.

(c) $SO_3{}^{2-}$

The O-S-O bond angle is 109.5°. The sulfur atom uses sp^3 orbitals to form three σ bonds—one with each O atom. The other sp^2 orbital accommodates a lone pair on S. In accord with the formal charge minimized structure, we can imagine simultaneous sideways overlap of the p orbital on the S atom with a p orbital on

However, the structure with two S-O double bonds minimizes formal charges and, as such, is preferred. To describe the bonding in this case requires the use of *d* orbitals on the S atom—since in this structure, an S atom accommodates 10 electrons in its valence shell.

(b) SO_3
The O-S-O bond angle is 120°. The sulfur atom uses sp^2 orbitals to form three σ bonds—one with each O atom. In accord with the formal charge minimized structure, we can imagine simultaneous sideways overlap of the *p* orbital on the S atom with a *p* orbital on each of the O atoms.

each of two of the O atoms.

(d) SO_4^{2-}
The O-S-O bond angle is 109.5°. The sulfur atom uses sp^3 orbitals to form four σ bonds—one with each O atom. In accord with the formal charge minimized structure, we can imagine simultaneous sideways overlap of the *p* orbital on the S atom with a *p* orbital on each of two of the O atoms.

10.55
In linear CO_2 the carbon atom uses two *sp* orbitals to form σ bonds with each O atom. Each of the two unhydridized *p* orbitals forms a π bond with an O atom. In CO_3^{2-}, there are two single C-O bonds and one double C-O bond, giving rise to three equivalent resonance structures. The carbon atom uses three sp^2 orbitals to form σ bonds with each of the three O atoms. The unhydridized *p* orbital forms a π bond with an O atom. Since it can be any one of the three, we get three equivalent resonance structures. This differs from the delocalized electron picture in which the unhydridized *p* orbital forms a degree of π bond with all three O atoms.

10.57

(a) The angles *A, B, C,* and *D* are about 120°, 109.5°, 109.5°, and 120°, respectively.
(b) According to the valence bond model, carbon atoms 1, 2, and 3 are sp^2, sp^2, and sp^3 hybridized, respectively.

10.9 Molecular orbital theory of bonding in molecules

10.59
(a)

The bond order in O_2^{2-} is 1.
(b) The molecular orbital theory electron configuration for O_2^{2-} is (using unhybridized orbitals)
[core electrons]$(\sigma_{2s})^2(\sigma_{2s}^*)^2(\pi_{2p_y})^2(\pi_{2p_z})^2(\sigma_{2p_x})^2(\pi_{2p_y}^*)^2(\pi_{2p_z}^*)^2$
bond order $= \frac{1}{2} \times (\, 2+2+2+2-2-2-2\,) = 1$
(c) For O_2^{2-}, the valence bond description and the molecular orbital description predict the same bond order and the same magnetic behaviour (i.e., diamagnetic—NOT paramagnetic—there are no unpaired electrons in either description).

10.61
Molecular orbital; theory electron configurations (core electrons omitted):
(a) NO
$(\sigma_{2s})^2(\sigma_{2s}^*)^2(\pi_{2p_y})^2(\pi_{2p_z})^2(\sigma_{2p_x})^2(\pi_{2p_y}^*)^1$
an unpaired electron $\rightarrow$ paramagnetic
HOMO $= \pi_{2p_y}^*$ or $\pi_{2p_z}^*$ $\leftarrow$ same energy
(b) OF$^\square$
$(\sigma_{2s})^2(\sigma_{2s}^*)^2(\pi_{2p_y})^2(\pi_{2p_z})^2(\sigma_{2p_x})^2(\pi_{2p_y}^*)^2(\pi_{2p_z}^*)^2$
no unpaired electrons $\rightarrow$ NOT paramagnetic—i.e., diamagnetic
HOMO $= \pi_{2p_y}^*$ or $\pi_{2p_z}^*$ $\leftarrow$ same energy
(c) $O_2^{2\square}$
$(\sigma_{2s})^2(\sigma_{2s}^*)^2(\pi_{2p_y})^2(\pi_{2p_z})^2(\sigma_{2p_x})^2(\pi_{2p_y}^*)^2(\pi_{2p_z}^*)^2$
no unpaired electrons $\rightarrow$ NOT paramagnetic—i.e., diamagnetic
HOMO $= \pi_{2p_y}^*$ or $\pi_{2p_z}^*$ $\leftarrow$ same energy
(d) $Ne_2^{\square}$
$(\sigma_{2s})^2(\sigma_{2s}^*)^2(\pi_{2p_y})^2(\pi_{2p_z})^2(\sigma_{2p_x})^2(\pi_{2p_y}^*)^2(\pi_{2p_z}^*)^2(\sigma_{2p_x}^*)^1$
an unpaired electron $\rightarrow$ paramagnetic
HOMO $= \sigma_{2p_x}^*$
(e) CN
$(\sigma_{2s})^2(\sigma_{2s}^*)^2(\pi_{2p_y})^2(\pi_{2p_z})^2(\sigma_{2p_x})^1$
an unpaired electron $\rightarrow$ paramagnetic
HOMO $= \sigma_{2p_x}$

SUMMARY AND CONCEPTUAL QUESTIONS

10.63

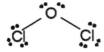

NF_3 has four electron pairs about the central atom, and a trigonal pyramidal shape.

OCl_3 has four electron pairs about the central atom, and a bent shape.

In each case, the bond angles are near 109.5° (the angle between tetrahedrally distributed electron regions).

10.65

The average C-O bond order in the formate ion ($HCOO^-$ is 1.5)—there are two equivalent C-O bonds represented as a single and a double bond in the two resonance structures.

In a methanol molecule (CH_3OH), the C-O bond is just a single bond—there is just one structure.

In a carbonate ion (CO_3^{2-}), there are three equivalent C-O bonds—represented as a double bond and two single bonds in the resonance structures. The average bond order is 4/3 = 1.3333.

The formate ion has the strongest C-O bond, which is expected to be the shortest.

Methanol molecules have the weakest C-O bond, which is expected to be the longest.

10.67

The NO_2^+ ion is linear. Its O-N-O bond angle is 180°.

The NO_2^- ion. Its O-N-O bond angle is about 120°.

The NO_2^+ ion has the larger O-N-O bond angle.

10.69

$$H-C=C-C\equiv N:$$

Acrylonitrile

(a) Angles 1, 2, and 3 equal about 120 °, 180°, and 120°, respectively.

(b) The carbon-carbon double bond is shorter.

(c) The carbon-carbon double bond is stronger.

(d) The C-N triple bond is the most polar—largest difference in electronegativity.

10.71

(a) Bond angles 1, 2 and 3 are about 120°, 109.5°, and 120°, respectively.
(b) The shortest carbon-oxygen bond is the carbonyl C=O double bond.
(c) The most polar bond in the molecule is the O-H bond.

10.73
(a) The geometry about the boron atom in $H_3N \rightarrow BF_3$ is tetrahedral.
(b) In BF_3, the valence orbitals of boron are sp^2 hybridized.
In $H_3N \rightarrow BF_3$, the valence orbitals of boron are sp^3 hybridized.
(c) We expect the hybridization of boron to change when this "coordinate" bond forms.

Chapter 11
States of Matter

IN-CHAPTER EXERCISES

Exercise 11.4—Ideal gas equation

To use the ideal gas equation with R having units of (L kPa K^{-1} mol^{-1}), the pressure must be measured in kPa and the temperature in K. Therefore,

$$p = 100 \text{ kPa}$$
$$T = 23 + 273 = 296 \text{ K}$$

Now substitute the values of p, T, n, and R into the ideal gas equation and solve for the volume of gas, V:

$$V = \frac{nRT}{p} = \frac{(1.3 \times 10^3 \text{ mol})(8.314 \text{ L kPa K}^{-1} \text{mol}^{-1})(296 \text{ K})}{(100 \text{ kPa})} = 32 \times 10^3 \text{ L}$$

Notice that the units kPa, mol, and K cancel to leave the answer in L.

Exercise 11.5—Ideal gas equation

Initial Conditions	Final Conditions
V_1 ☐☐ 22 L	V_2 ☐ ? L
p_1 ☐☐☐152 bar = 1.52×10^4 kPa	p_2 ☐ 1.00 bar = 100 kPa
T_1 ☐ 31°C (304 K)	T_2 ☐☐ 22 °C (295 K)

$$V_2 = \left(\frac{T_2}{p_2}\right) \times \left(\frac{p_1 V_1}{T_1}\right) = V_1 \times \frac{p_1}{p_2} \times \frac{T_2}{T_1}$$

$$= 22 \text{ L} \times \frac{1.52 \times 10^4 \text{ kPa}}{100 \text{ kPa}} \times \frac{295 \text{ K}}{304 \text{ K}}$$

$$= 3250 \text{ L}$$

This is the total volume of gas at the final temperature and pressure—with one additional digit (beyond significant) carried. This gas fills

$$\frac{V_2}{\text{volume of 1 balloon}} = \frac{3250 \text{ L}}{5.0 \text{ L balloon}^{-1}} = 650 \text{ balloons}$$

Exercise 11.6—Ideal gas equation

First, we need to determine the amount of H_2 (the limiting reactant) that reacts with nitrogen—this gives the extent of reaction (×3 because of the stoichiometric coefficient on H_2).

$$n\,(\mathrm{H_2}) = \frac{pV}{RT} = \frac{(72.2 \text{ kPa})(355 \text{ L})}{(8.314 \text{ L kPa K}^{-1} \text{ mol}^{-1})(298.2 \text{ K})} = 10.3 \text{ mol}$$

$$\mathrm{N_2(g) + 3\,H_2(g) \xrightarrow[500\ ^\circ C]{\text{iron catalyst}} 2\,NH_3(g)}$$

For every 3 mol of H_2 reacted, we get 2 mol of NH_3 formed. Therefore,

$$n\,(\mathrm{NH_3}) = \frac{2}{3} \times n\,(\mathrm{H_2}) = \frac{2}{3} \times 10.3 \text{ mol} = 6.87 \text{ mol}$$

If this amount of NH_3 gas is stored in a 125 L tank at 25.0 °C, the pressure would be

$$p = \frac{nRT}{V} = \frac{(6.87 \text{ mol})(8.314 \text{ L kPa K}^{-1} \text{ mol}^{-1})(298.15 \text{ K})}{(125 \text{ L})} = 136 \text{ kPa}$$

Exercise 11.10—Molar mass from density

The molar mass of the unknown gas is

$$M = \frac{\rho RT}{p} = \frac{(5.02 \text{ g L}^{-1})(8.314 \text{ L kPa K}^{-1} \text{mol}^{-1})(288.15 \text{ K})}{(99.3 \text{ kPa})} = 121 \text{ g mol}^{-1}$$

Exercise 11.13——Partial pressures

We have 15.0 g of halothane ($C_2HBrClF_3$) vapour and 23.5 g of oxygen gas. This corresponds to

$$n\,(\mathrm{C_2HBrClF_3}) = \frac{m}{M} = \frac{15.0 \text{ g}}{197.4 \text{ g mol}^{-1}} = 0.0760 \text{ mol}$$

$$n\,(\mathrm{O_2}) = \frac{m}{M} = \frac{23.5 \text{ g}}{32.00 \text{ g mol}^{-1}} = 0.734 \text{ mol}$$

The total amount of gases (number of moles) is

$n\,(\text{total}) = n\,(\mathrm{C_2HBrClF_3}) + n\,(\mathrm{O_2}) = 0.810 \text{ mol}$

The gas mixture is in a 5.00 L tank at 25.0 °C (298.15 K).
The ideal gas law gives the total pressure if we substitute $n\,(\text{total})$,
the partial pressure of halothane if we substitute $n\,(\mathrm{C_2HBrClF_3})$, and
the partial pressure of oxygen if we substitute $n\,(\mathrm{O_2})$.

$$p = \frac{nRT}{V} = \frac{(0.810 \text{ mol})(8.314 \text{ L kPa K}^{-1} \text{ mol}^{-1})(298.15 \text{ K})}{(5.00 \text{ L})} = 402 \text{ kPa}$$

The partial pressures of halothane and oxygen are determined more easily using mole fractions,

$$x(\mathrm{C_2HBrClF_3}) = \frac{n(\mathrm{C_2HBrClF_3})}{n(\text{total})} = \frac{0.0760 \text{ mol}}{0.810 \text{ mol}} = 0.0938$$

$$x(\mathrm{O_2}) = \frac{n(\mathrm{O_2})}{n(\text{total})} = \frac{0.734 \text{ mol}}{0.810 \text{ mol}} = 0.906 = 1.000 - x(\mathrm{C_2HBrClF_3})$$

From which we get

$p(\mathrm{C_2HBrClF_3}) = x(\mathrm{C_2HBrClF_3}) \times p(\text{total}) = 0.0938 \times 402 \text{ kPa} = 37.7 \text{ kPa}$
and
$p(\mathrm{O_2}) = x(\mathrm{O_2}) \times p(\text{total}) = 0.906 \times 402 \text{ kPa} = 364 \text{ kPa}$

Exercise 11.16—RMS speed

The rms speed of CO_2 is

$$\sqrt{\overline{u_{CO_2}^2}} = \sqrt{\frac{3RT}{M_{CO_2}}}$$

and similarly for N_2. Taking the ratio of the two equations gives

$$\frac{\sqrt{\overline{u_{CO_2}^2}}}{\sqrt{\overline{u_{N_2}^2}}} = \sqrt{\frac{3RT}{M_{CO_2}}} \times \sqrt{\frac{M_{N_2}}{3RT}} = \sqrt{\frac{M_{N_2}}{M_{CO_2}}} = \sqrt{\frac{28.01 \text{ g mol}^{-1}}{44.01 \text{ g mol}^{-1}}} = \frac{0.798}{1}$$ The rms speed of N_2 is larger than that of CO_2.

Exercise 11.17—Graham's law of effusion

Taking He as the reference, we have

$$\frac{\text{Rate of effusion of SF}_6}{\text{Rate of effusion of He}} = \sqrt{\frac{M(\text{He})}{M(\text{SF}_6)}} = \sqrt{\frac{4.00 \text{ g mol}^{-1}}{146 \text{ g mol}^{-1}}} = 0.166$$

and

$$\frac{\text{Rate of effusion of N}_2}{\text{Rate of effusion of He}} = \sqrt{\frac{M(\text{He})}{M(\text{N}_2)}} = \sqrt{\frac{4.00 \text{ g mol}^{-1}}{28.01 \text{ g mol}^{-1}}} = 0.378$$

He has the highest rate of effusion, with N_2 at 37.8% the rate of He, and the much bigger SF_6 molecules at 16.6% that rate.

Exercise 11.19—Molar mass using Graham's law of effusion

$$\frac{\text{Rate of effusion of unknown}}{\text{Rate of effusion of CH}_4} = \frac{(n \text{ mol}) / (4.73 \text{ min})}{(n \text{ mol}) / (1.50 \text{ min})} = \frac{1.50 \text{ min}}{4.73 \text{ min}} = 0.317$$

n is the same for both gases. We also have

$$\frac{\text{Rate of effusion of unknown}}{\text{Rate of effusion of CH}_4} = \sqrt{\frac{M(\text{CH}_4)}{M(\text{unknown})}} = \sqrt{\frac{16.04 \text{ g mol}^{-1}}{M(\text{unknown})}} = 0.317$$

Solving for $M(\text{unknown})$ gives

$$M(\text{unknown}) = \frac{16.04 \text{ g mol}^{-1}}{(0.317)^2} = 160 \text{ g mol}^{-1}$$

Exercise 11.23—Relative strength of intermolecular forces

In order of increasing strength, we have
(a) the dispersion forces in liquid O_2 < **(c)** the dipole-induced dipole interactions of O_2 dissolved in H_2O< **(b)** the hydrogen bonding forces in liquid CH_3OH

Exercise 11.25—Vapour pressure curves

(a) From the figure in the textbook (Figure 11.20) we estimate the equilibrium vapour pressure of ethanol at 40 °C to be 14 kPa.
(b) At 60 °C the equilibrium vapour pressure of ethanol is about 44 kPa. If we have a closed container with both liquid and vapour ethanol at 60 °C, and the partial pressure of ethanol is 80.0 kPa, then the vapour and liquid are NOT at equilibrium—the vapour pressure is higher than the equilibrium vapour pressure at that temperature, so some vapour condenses to form more liquid.

Exercise 11.26—Vapour pressure curves

The vapour pressure of water at 60 °C (using Figure 11.20 in the textbook) is about 20 kPa.
To achieve this pressure of water in a 5.0 L flask at 60 °C (333 K) requires

$$n(H_2O) = \frac{pV}{RT} = \frac{(20 \text{ kPa})(5.0 \text{ L})}{(8.314 \text{ L kPa K}^{-1}\text{mol}^{-1})(333 \text{ K})} = 0.036 \text{ mol}$$

or

$$m(H_2O) = n(H_2O) \times M(H_2O) = (0.036 \text{ mol})(18.02 \text{ g mol}^{-1}) = 0.65 \text{ g}$$

0.50 g of water is not enough to achieve this partial pressure. Instead we achieve only

$$20 \times \frac{0.50 \text{ g}}{0.65 \text{ g}} = 15 \text{ kPa}$$

If we started with 2.0 g of water, 0.65 g of it would evaporate to give a vapour in equilibrium with the remaining 2.0 − 0.65 g = 1.35 g of liquid water. The partial pressure of water would be 20 kPa.

Exercise 11.27—Phase diagrams

(a) See Figure 11.23 in the textbook.
Because the slope of the liquid-solid equilibrium curve is positive, we conclude that the density of liquid CO_2 is less than that of solid CO_2. If we increase the pressure of a system containing liquid and solid at equilibrium, some of the less dense phase will change to the more dense phase.
(b) At 500 kPa and 0 °C, the most stable phase of carbon dioxide is the vapour.
(c) CO_2 cannot be liquefied at 45 °C, because this temperature is above the critical temperature. Compressing CO_2 at this temperature produces a dense supercritical fluid.

REVIEW QUESTIONS

11.1 Understanding gases: Understanding our world

11.29
In order of increasing temperature, altitudes in the atmosphere are ranked as follows:
$$T(80 \text{ km}) < T(25 \text{ km}) < T(110 \text{ km}) < T(5 \text{ km})$$

11.31
The temperature stops decreasing, and starts increasing, as we move from the troposphere into the stratosphere. Absorption of UV light by oxygen produces oxygen atoms which, in turn, produce ozone (together with other oxygen molecules). The ozone produce absorbs additional UV light. The energy absorbed increases the local temperature. The density of the atmosphere at the top of the stratosphere and bottom of mesosphere is sufficient to absorb enough UV light to increase the temperature. As we descend into the stratosphere and below, the intensity of incoming UV radiation is diminished due to absorption at higher altitudes.   At 1 km above earth's surface in the troposphere there is too little UV radiation to cause significant absorption by oxygen to produce O atoms, and ozone etc.

11.33
(a) The earth's atmospheric concentration of CO_2 has varied in the following fashion:
$c(CO_2)$ was lower at 30,000 years ago compared to 125,000 years ago, and then was lower again at 260,000 years ago.
(b) The earth's atmospheric temperature has varied in the following fashion:
Temperature was lower at 30,000 years ago compared to 125,000 years ago, and then was lower again at 260,000 years ago.

11.2   Relationships among gas properties

11.35

$$p = 210 \text{ mm Hg} = \frac{210 \ \cancel{\text{mm Hg}}}{760 \ \cancel{\text{mm Hg}} \ \text{atm}^{-1}} = 0.276 \text{ atm}$$

$$= 0.276 \text{ atm} \times 101.3 \text{ kPa atm}^{-1} = 30.0 \text{ kPa} = 0.300 \text{ bar}$$

11.37

Initial Conditions	Final Conditions
V_1 □□ 25.0 mL □□ 0.0250 L	V_2 □ ?
p_1 □□□58.2 kPa	p_2 □ 12.6 kPa
T_1 □ 20.5°C (293.65 K)	T_2 □□ 24.5°C (297.65 K)

$$V_2 = \left(\frac{T_2}{p_2}\right) \times \left(\frac{p_1 V_1}{T_1}\right) = V_1 \times \frac{p_1}{p_2} \times \frac{T_2}{T_1}$$

$$= 0.0250 \text{ L} \times \frac{58.2 \ \cancel{\text{kPa}}}{12.6 \ \cancel{\text{kPa}}} \times \frac{297.65 \ \cancel{\text{K}}}{293.65 \ \cancel{\text{K}}}$$

$$= 0.117 \text{ L}$$

11.39

Initial Conditions	Final Conditions
$V_1 = 0.40$ L	$V_2 = 0.050$ L
$p_1 = 1.00$ atm	$p_2 = $?
$T_1 = 15°C$ (288 K)	$T_2 = 77°C$ (350 K)

$$p_2 = \left(\frac{T_2}{V_2}\right) \times \left(\frac{p_1 V_1}{T_1}\right) = p_1 \times \frac{V_1}{V_2} \times \frac{T_2}{T_1}$$

$$= 1.00 \text{ atm} \times \frac{0.40 \text{ L}}{0.050 \text{ L}} \times \frac{350 \text{ K}}{288 \text{ K}}$$

$$= 9.7 \text{ atm}$$

11.3   Are any properties of different gases the same?

11.41

Gas	Molar Volume at 298.15 K (L mol^{-1})	Molar Volume at 448.15 K (L mol^{-1})
H_2	24.804	37.277
CO_2	24.666	37.218
SF_6	24.512	37.157

The molar volumes of the three gases are more similar at 448.15 K than 298.15 K.

11.4   The ideal gas equation

11.43

$$n(C_2H_5OH) = m(C_2H_5OH) / M(C_2H_5OH) = (1.50 \text{ g}) / (46.07 \text{ g mol}^{-1}) = 0.0326 \text{ mol}$$

$$p = \frac{nRT}{V} = \frac{(0.0326 \text{ mol})(8.314 \text{ L kPa K}^{-1} \text{mol}^{-1})(523.15 \text{ K})}{(0.251 \text{ L})} = 565 \text{ kPa}$$

11.45

$$n = \frac{pV}{RT} = \frac{(95.3 \text{ kPa})(0.452 \text{ L})}{(8.314 \text{ L kPa K}^{-1}\text{mol}^{-1})(296 \text{ K})} = 0.0175 \text{ mol}$$

$$M \text{ (unknown)} = m \text{ (unknown)} / n \text{ (unknown)} = (1.007 \text{ g}) / (0.0175 \text{ mol}) = 57.5 \text{ g mol}^{-1}$$

11.47

Initial Conditions	Final Conditions
$V_1 = 1.2 \times 10^4 \text{ m}^3 = 1.2 \times 10^7 \text{ L}$	$V_2 \; \square \; ?$
$1 \text{ L} = 1 \text{ dm}^3 = 10^{-3} \text{ m}^3$	
$p_1 \; \square\square\square 98.3 \text{ kPa}$	$p_2 \; \square \; 80 \text{ kPa}$
$T_1 \; \square \; 16°\text{C} \; (289 \text{ K})$	$T_2 \; \square\square \; -33°\text{C} \; (240 \text{ K})$

$$V_2 = \left(\frac{T_2}{p_2}\right) \times \left(\frac{p_1 V_1}{T_1}\right) = V_1 \times \frac{p_1}{p_2} \times \frac{T_2}{T_1}$$

$$= 1.2 \times 10^7 \text{ L} \times \frac{98.3 \text{ kPa}}{80 \text{ kPa}} \times \frac{240 \text{ K}}{289 \text{ K}}$$

$$= 1.22 \times 10^7 \text{ L}$$

Though the pressure dropped by almost 20%, the decrease in temperature compensates and the change in volume is not very big. Nevertheless, there would have to be allowance for expansion of the balloon at higher altitudes.

11.5 The density of gases

11.49

$$\frac{n}{V} = \frac{p}{RT} = \frac{25.2 \text{ kPa}}{(8.314 \text{ L kPa K}^{-1}\text{mol}^{-1})(290 \text{ K})} = 1.05 \times 10^{-2} \text{ mol L}^{-1}$$

$$M(\text{organofluorine compound}) = \text{density} \times \left(\frac{n}{V}\right)^{-1} = (0.355 \text{ g L}^{-1}) / (1.05 \times 10^{-2} \text{ mol L}^{-1})$$

$$= 33.97 \text{ g mol}^{-1}$$

11.51

Amount of N_2 required =

$$= n(N_2) = \frac{pV}{RT} = \frac{(132 \text{ kPa})(75.0 \text{ L})}{(8.314 \text{ L kPa K}^{-1}\text{mol}^{-1})(298.15 \text{ K})} = 3.99 \text{ mol}$$

$$2 \text{ NaN}_3(s) \longrightarrow 2 \text{ Na}(s) + 3 \text{ N}_2(g)$$

Amount of NaN_3 required, $n(NaN_3) = \frac{2}{3} \times (\text{amount of } N_2) = 2.66 \text{ mol}$

Mass of NaN_3 required, $m = (2.66 \text{ mol}) \times (65.01 \text{ g mol}^{-1}) = 173 \text{ g}$

11.53

Amount of N_2H_4 to be consumed, $n(N_2H_4) = (1.00 \times 10^3 \text{ g}) / (32.05 \text{ g mol}^{-1}) = 31.2 \text{ mol}$

$$N_2H_4(g) + O_2(g) \longrightarrow N_2(g) + 2 \text{ H}_2O(\ell)$$

Amount of O_2 required, $n(O_2) = \text{amount of } N_2H_4 = 31.2 \text{ mol}$

$$p = \frac{n(O_2)RT}{V} = \frac{(31.2 \text{ mol})(8.314 \text{ L kPa K}^{-1}\text{mol}^{-1})(296 \text{ K})}{(450 \text{ L})} = 171 \text{ kPa}$$

11.6   Gas mixtures and partial pressures

11.55
We have 1.0 g of H_2 and 8.0 g of Ar. This corresponds to

$$n\,(H_2) = \frac{m}{M} = \frac{1.0\ \text{g}}{2.016\ \text{g mol}^{-1}} = 0.50\ \text{mol}$$

$$n\,(Ar) = \frac{m}{M} = \frac{8.0\ \text{g}}{39.95\ \text{g mol}^{-1}} = 0.20\ \text{mol}$$

The total number of moles is
$$n\,(\text{total}) = n\,(H_2) + n\,(Ar) = 0.70\ \text{mol}$$
The gas mixture is in a 3.0 L tank at 27 °C (300 K).
The ideal gas law gives the total pressure if we substitute $n\,(\text{total})$,

$$p = \frac{nRT}{V} = \frac{(0.70\ \text{mol})(8.314\ \text{L kPa K}^{-1}\text{mol}^{-1})(300\ \text{K})}{(3.0\ \text{L})} = 580\ \text{kPa}$$

The partial pressures of H_2 and Ar are determined using mole fractions,

$$x(H_2) = \frac{n(H_2)}{n(\text{total})} = \frac{0.50\ \text{mol}}{0.70\ \text{mol}} = 0.71$$

$$x(Ar) = \frac{n(Ar)}{n(\text{total})} = \frac{0.20\ \text{mol}}{0.810\ \text{mol}} = 0.29$$

From which we get
$$p(H_2) = x(H_2) \times p(\text{total}) = 0.71 \times 580\ \text{kPa} = 412\ \text{kPa}$$

$$p(Ar) = x(Ar) \times p(\text{total}) = 0.29 \times 580\ \text{kPa} = 168\ \text{kPa}$$

11.7   The kinetic-molecular model of gases

11.61

Average speed of a CO_2 molecule $= \sqrt{u_{CO_2}^2} = \sqrt{\frac{3RT}{M_{CO_2}}} = \sqrt{\frac{M_{O_2}}{M_{CO_2}}}\sqrt{\frac{3RT}{M_{O_2}}} = \sqrt{\frac{M_{O_2}}{M_{CO_2}}}\sqrt{u_{O_2}^2}$

$$= \sqrt{\frac{31.9988}{44.0098}} \times 4.28104\ \text{cm s}^{-1} = 3.65041\ \text{cm s}^{-1}$$

11.63
Refer to 11.61.

Average speed of CO molecules $= = \sqrt{\frac{32.00}{28.01}} \times 4.28\ \text{cm s}^{-1} = 4.58\ \text{cm s}^{-1}$

Ratio of speed of Co molecules to speed of Ar atoms (at the same temperature)
$$= \sqrt{\frac{M_{Ar}}{M_{CO}}} = \sqrt{\frac{39.95}{28.01}} = 1.194$$

11.8 Diffusion and effusion

11.67

$$\frac{\text{Rate of effusion of unknown}}{\text{Rate of effusion of He}} = \sqrt{\frac{M(\text{He})}{M(\text{unknown})}} = \frac{1}{3}$$

So,

$$M(\text{unknown}) = 9 \times M(\text{He}) = 36 \text{ g mol}^{-1}$$

11.9 The ideal gas model and real gas behaviour

11.69

According to the ideal gas law,

$$p = \frac{nRT}{V} = \frac{(8.00 \text{ mol})(8.314 \text{ L kPa K}^{-1} \text{mol}^{-1})(300.15 \text{ K})}{(4.00 \text{ L})} = 4990 \text{ kPa} = 49.90 \text{ bar}$$

The van der Waals equation gives a better description of chlorine at such a high pressure.

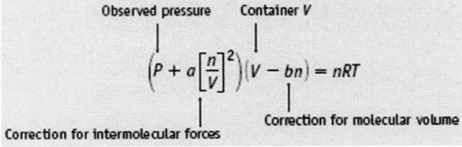

From Table 11.4, the van der Waals' constants for chlorine are

$$a = = 658 \text{ kPa L}^2 \text{ mol}^{-2} \qquad \text{and} \qquad b = 0.0562 \text{ L mol}^{-1}$$

Using these constants and solving for pressure, we get

$$p = \frac{nRT}{V - nb} - a\left(\frac{n}{V}\right)^2$$

$$= \frac{(8.00 \text{ mol})(8.314 \text{ L kPa K}^{-1} \text{mol}^{-1})(300.15 \text{ K})}{(4.00 \text{ L} - (8.00 \text{ mol})(0.0562 \text{ L mol}^{-1}))} - (658 \text{ L}^2 \text{kPa mol}^{-2})\left(\frac{8.00 \text{ mol}}{4.00 \text{ L}}\right)^2$$

$$= 5623 - 2632 \text{ kPa} = 2991 \text{ kPa}$$

11.10 Liquid and solid states: Stronger intermolecular forces

11.71

When solid I_2 dissolves in methanol, CH_3OH, dispersion forces holding I_2 molecules in their lattice positions must be overcome. Hydrogen bonding forces between methanol molecules are disrupted when methanol solvates iodine. The solvation forces between methanol and iodine are dipole-induced dipole interactions.

11.12 Liquids: Properties and phase changes

11.75
(a) The equilibrium vapour pressure of water at 60 °C is about 19 kPa
(b) Water has an equilibrium vapour pressure of 80 kPa at about 92 °C.
(c) The equilibrium vapour pressure of ethanol is higher than that of water at 70 °C (and at all temperatures up to their boiling points).

11.77
The vapour pressure of diethyl ether at 30 °C is (from Figure 11.19) about 75 kPa
To achieve this pressure of diethyl ether in a 0.10 L flask at 30 °C (303 K) requires

$$n((CH_3CH_2)_2O) = \frac{pV}{RT} = \frac{(75 \text{ kPa})(0.10 \text{ L})}{(8.314 \text{ L kPa K}^{-1}\text{mol}^{-1})(303 \text{ K})} = 0.0030 \text{ mol}$$

or

$$m((CH_3CH_2)_2O) = n((CH_3CH_2)_2O) \, M((CH_3CH_2)_2O) = (0.0030 \text{ mol})(74.12 \text{ g mol}^{-1}) = 0.22 \text{ g}$$

1.0 g of diethyl ether is more than enough to achieve this partial pressure. 0.22 g of diethyl ether evaporates, while 0.78 g remains in the liquid phase—the two phases are in equilibrium.
If the flask is placed in an ice bath, the temperature lowers causing the vapour pressure to lower. The gas becomes supersaturated and liquid diethyl ether condenses out of the vapour.

11.81

Oxygen *T-p* phase diagram

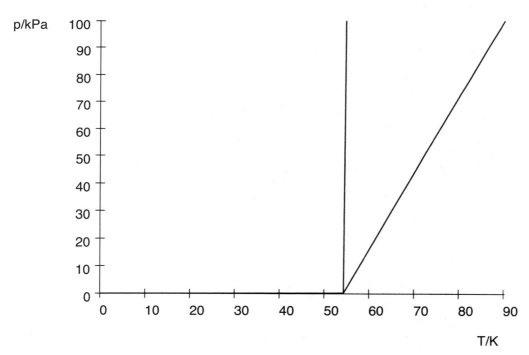

This phase diagram is constructed by (1) connecting ($T = 0$ K, $p = 0$ kPa) to the triple point, ($T = 54.34$ K, $p = 0.267$ kPa), to get the solid-vapour equilibrium curve, (2) connecting the triple point to the normal melting point, ($T = 54.8$ K, $p = 100$ kPa), to get the solid-liquid equilibrium curve, and

(3) connecting the triple point to the normal boiling point, $(T = 90.18 \text{ K}, p = 101.3 \text{ kPa})$, to get part of the liquid-gas equilibrium curve. This curve could be taken further if data for the critical point of oxygen were given. The degree of curvature of both the solid-vapour and liquid vapour curves cannot be qualitatively judged. We will see (Chapter 17) that the curves are such that the data fit a straight-line plot of $\ln p$ vs. $1/T$.

At $T = -196 \text{ °C}$ (i.e., $T = 77 \text{ K}$), $p \approx 50 \text{ kPa}$

The solid-liquid equilibrium line has a positive slope (54.34 K at 0.267 kPa, to 54.8 K at 101.3 kPa). If we increase the pressure on a solid-liquid equilibrium system, we would be taken to the more dense phase—the solid.

11.13 Solids: Properties and phase changes

11.83

To heat 5.00 g of solid silver from 25 °C to 962 °C requires
$$q_1 = (0.235 \text{ J g}^{-1}\text{K}^{-1}) \times (5.00 \text{ g}) \times (962 \text{ K} - 25 \text{ K}) = 1100 \text{ J} = 1.10 \text{ kJ}$$
Amount of silver $= (5.00 \text{ g}) / (107.87 \text{ g mol}^{-1}) = 0.0464 \text{ mol}$

To melt 5.00 g of solid silver at 962 °C requires
$$q_2 = (0.0464 \text{ mol}) \times (11.3 \text{ kJ mol}^{-1}) = 0.524 \text{ kJ}$$
Total heat required $= q_1 + q_2 = 1.10 + 0.524 \text{ kJ} = 1.624 \text{ kJ}$

11.14 Phase diagrams

11.87

CH_3Cl can be liquefied at or above room temperature, up to 416 K (143 °C) the critical temperature. Above the critical temperature, chloromethane cannot be liquefied—it can only be compressed into a supercritical fluid. Below the triple point temperature, 175.4 K (–97.8 °C), chloromethane cannot be liquefied—it solidifies upon compression. However, room temperature is between these limits on liquefication.

11.15 Polymorphic forms of solids

11.89

Ice V only exists over the range of temperatures and pressures shown in the phase diagram. The lowest pressure at which Ice V is stable is about 3 Mbar, i.e., about 3 million atmospheres.

SUMMARY AND CONCEPTUAL QUESTIONS

11.91

Initial Conditions	Final Conditions
$V_1 = 0.0255$ L	$V_2 = 0.0215$ L
p_1 is not given—do not worry	$p_2 = p_1$ ← this is all you need to know
$T_1 \square 90°C$ (363 K)	$T_2 \square\square\square\square$

$$T_2 = p_2 V_2 \times \left(\frac{T_1}{p_1 V_1}\right) = T_1 \times \frac{p_2}{p_1} \times \frac{V_2}{V_1}$$

$$= 363 \text{ K} \times \frac{\not{p_1}}{\not{p_1}} \times \frac{0.0215 \not{L}}{0.0255 \not{L}}$$

$$= 306 \text{ K} \quad \text{or} \quad 33\ °C$$

11.93

(a) A 1.0-L flask containing 10.0 g each of O_2 and CO_2 at 25 °C has greater partial pressure of O_2 than CO_2. The partial pressures depend on the numbers of moles—10.0 g of the lower molar mass O_2 has more moles than 10.0 g of CO_2.
(b) The lighter O_2 molecules have greater average speed—average speed is proportional to $M^{-1/2}$.
(c) At a given temperature, molecules of all substances have the same average kinetic energy.

11.95

(a) A material is in a steel tank at 100 bar pressure. When the tank is opened to the atmosphere, the material suddenly expands, increasing its volume by 10%. This is NOT a gas. An ideal gas would expand 100 fold if its pressure were changed from 100 bar to 1 bar (atmospheric pressure). Only a liquid or solid is so incompressible as to expand only 10% when relieved of such a large pressure.
(b) A 1.0 mL sample of material weighs 8.2 g. Assuming the temperature is about 25 °C, and pressure is about 1 bar, we can say that this is NOT a gas. A density of 8.2 g mL^{-1} = 8200 g L^{-1} is way too high for a gas. An ideal gas at 25 °C and 1 bar pressure has the following molar density.

$$\frac{n}{V} = \frac{p}{RT} = \frac{100\ \not{kPa}}{(8.314\ \text{L}\not{kPa}\ \not{K}^{-1}\text{mol}^{-1})(298\ \not{K})} = 0.0404\ \text{mol L}^{-1}$$

The ratio of mass density to molar density is the molar mass of the substance.
$$M\text{(substance)} = (8200\ \text{g L}^{-1}) / (0.0404\ \text{mol L}^{-1}) = 203000\ \text{g mol}^{-1}$$
There are no substances with such a huge molar mass that are not solids under ordinary conditions.
(c) There is insufficient information. Chlorine is a transparent and pale green gas. But, there are also transparent and pale green liquids and solids.
(d) A material that contains as many molecules in 1.0 m^3 as the same volume of air, at the same temperature and pressure, is definitely a gas (assuming the temperature and pressure are not unusual). Air is approximately an ideal gas under ordinary conditions. Anything else with the same density under the same conditions is behaving like an ideal gas.

11.97

$$3 N_2O(g) + 4 Na(s) + NH_3(\ell) \longrightarrow NaN_3(s) + 3 NaOH(s) + 2 N_2(g)$$

(a)

Amount of sodium, $n(Na) = (65.0 \text{ g}) / (22.99 \text{ g mol}^{-1}) = 2.83$ mol

Amount of N_2O, $n(N_2O) = \dfrac{pV}{RT} = \dfrac{(215 \text{ kPa})(35.0 \text{ L})}{(8.314 \text{ L kPa K}^{-1}\text{mol}^{-1})(296 \text{ K})} = 3.06$ mol

To consume 3.06 mol of N_2O requires

$$\frac{4}{3} \times n(N_2O) = \frac{4}{3} \times 3.06 \text{ mol} = 4.08 \text{ mol of Na}$$

Since we only have 2.83 mol of Na, Na is the limiting reactant.
The amount of NaN_3 that can be produced is

$$\frac{1}{4} \times n(Na) = \frac{1}{4} \times 2.83 \text{ mol} = 0.708 \text{ mol of NaN}_3$$

Mass of NaN_3 that can be produced is

$$m(NaN_3) = (0.708 \text{ mol}) \times (65.01 \text{ g mol}^{-1}) = 46.0 \text{ g}$$

(b)

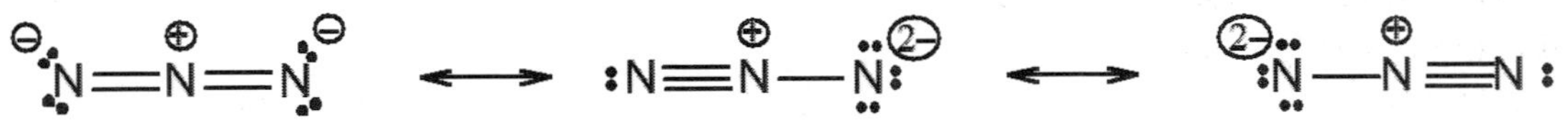

The structure on the left is the most reasonable, as it has the lowest formal charges—the electron distributions are closest to those of the isolated atoms.

(c) The azide ion is linear.

11.99

For the phase diagram of CO_2, see the solution to Exercise 11.27 above.
0.61 kPa corresponds to almost zero pressure in the phase diagram. To solidify carbon dioxide requires a temperature no greater than -90 °C.

11.101

$$B_2H_6(g) + 3 O_2(g) \longrightarrow B_2O_3(s) + 3 H_2O(g)$$

(a) According to increasing average molecular speeds, the gases in this reaction are ordered as follows:

$O_2 \ < \ B_2H_6 \ < \ H_2O$ —i.e., in descending order according to molar mass: the heavier the molecules, the less is their average speed.

(b) We do not need to compute the amount of $B_2H_6(g)$. Under the conditions of this question, the partial pressures of $B_2H_6(g)$ and $O_2(g)$ are in the correct stoichiometric ratio—partial pressures are proportional to amounts.
Therefore,

$$\text{Partial pressure of O}_2 = \frac{3}{1} \text{ partial pressure of B}_2\text{H}_6 = 102.3 \text{ kPa}$$

This presumes that no reaction between $B_2H_6(g)$ and $O_2(g)$ has taken place.

11.103

Amount of air, $n = \dfrac{pV}{RT} = \dfrac{(324 \text{ kPa})(17 \text{ L})}{(8.314 \text{ L kPa K}^{-1}\text{mol}^{-1})(298 \text{ K})} = 2.22 \text{ mol}$

Mass of air, $m = 2.22 \text{ mol} \times 28.96 \text{ g mol}^{-1} = 64.3 \text{ g} = 64 \text{ g}$ (to correct number of significant figures)

11.105

The density of air 20 km above the earth's surface is $92 \text{ g m}^{-3} = 9.2 \times 10^{-2} \text{ g L}^{-1}$. The pressure and temperature are 5.6 kPa and $-63\,°C$, respectively.

(a)

$$\frac{n}{V} = \frac{p}{RT} = \frac{5.6 \text{ kPa}}{(8.314 \text{ L kPa K}^{-1}\text{mol}^{-1})(210 \text{ K})} = 3.21 \times 10^{-3} \text{ mol L}^{-1}$$

$M(\text{air at 20 km altitude}) = \text{density} \times \left(\dfrac{n}{V}\right)^{-1} = (9.2 \times 10^{-2} \text{ g L}^{-1}) / (3.21 \times 10^{-3} \text{ mol L}^{-1})$

$= 28.7 \text{ g mol}^{-1}$

Note that we have used molar mass of air $\Box$ 28.96 g mol^{-1} to solve previous problems. How can the average molar mass of air be different at higher altitude?

The answer is that lighter molecules are distributed to higher altitudes in comparison with heavier molecules. Lighter and heavier molecules (most notably nitrogen and oxygen in the case of air) both have decreasing density with altitude. However, the density of the heavier molecule falls off faster. The result is an average molar mass that decreases slightly with altitude.

(b)

The average molar mass is the average of the O_2 and N_2 molar masses—weighted by their mole fractions.

Average molar mass $= x(N_2) \times M(N_2) + x(O_2) \times M(O_2)$

i.e.,

$28.7 \text{ g mol}^{-1} = x(N_2) \times (28.01 \text{ g mol}^{-1}) + x(O_2) \times (32.00 \text{ g mol}^{-1})$

$= (1 - x(O_2)) \times (28.01 \text{ g mol}^{-1}) + x(O_2) \times (32.00 \text{ g mol}^{-1})$

$28.7 = x(O_2) \times (32.00 - 28.01) + 28.01$

$x(O_2) = (28.7 - 28.01) / (32.00 - 28.01) = 0.17 \quad \leftarrow$ a lower value than on the surface of Earth

11.107

$p(N_2) = p(\text{total}) - p(O_2) - p(CO_2) - p(H_2O)$

$= 33.7 - 4.7 - 1.0 - 6.3 \text{ kPa} = 21.7 \text{ kPa}$

11.109

(a)

At 20 °C, the vapour pressure of water is 23.38 mbar = 2.338 kPa.

At 45% relative humidity, the partial pressure of water is 0.45 × 2.338 kPa = 1.05 kPa

$$\frac{n}{V} = \frac{p}{RT} = \frac{1.05 \; \cancel{kPa}}{(8.314 \; L \, \cancel{kPa} \; \cancel{K}^{-1} mol^{-1})(293 \; \cancel{K})} = 4.31 \times 10^{-4} \; mol \; L^{-1}$$

Density of water = $(18.02 \; g \; mol^{-1}) \times (4.31 \times 10^{-4} \; mol \; L^{-1})$ = 0.00776 g L^{-1} = 7.76 mg L^{-1}

(b)

At 0 °C, the vapour pressure of water is 6.11 mbar = 0.611 kPa.

At 95% relative humidity, the partial pressure of water is 0.95 × 0.611 kPa = 0.580 kPa

$$\frac{n}{V} = \frac{p}{RT} = \frac{0.580 \; \cancel{kPa}}{(8.314 \; L \, \cancel{kPa} \; \cancel{K}^{-1} mol^{-1})(273 \; \cancel{K})} = 2.56 \times 10^{-4} \; mol \; L^{-1}$$

Density of water = $(18.02 \; g \; mol^{-1}) \times (2.56 \times 10^{-4} \; mol \; L^{-1})$ = 0.00461 g L^{-1} = 4.61 mg L^{-1}

Therefore, the density of water vapour is higher at 20 °C with 45% humidity than at 0 °C with 95% humidity.

11.111

Amount density of mercury

$$= \frac{n}{V} = \frac{p}{RT} = \frac{0.225 \times 10^{-3} \; \cancel{kPa}}{(8.314 \; L \, \cancel{kPa} \; \cancel{K}^{-1} mol^{-1})(297 \; \cancel{K})} = 9.11 \times 10^{-8} \; mol \; L^{-1}$$

Atoms in 1 m^3 = $(1.00 \times 10^3 \; L) \times (6.022 \times 10^{23} \; atoms \; mol^{-1}) \times (9.11 \times 10^{-8} \; mol \; L^{-1})$

= 5.49×10^{19} atoms

11.113

The molecules of cooking oil do not form hydrogen bonds with water, and do not have strong dipoles to form significant dipole-dipole interactions.

11.115

(a) The normal boiling point of CCl_2F_2 is the temperature at which the vapour pressure = 1 atm (which is very close to 1 bar, or 100 kPa)

i.e., normal boiling point of CCl_2F_2 = −27 °C

(b) At 25 °C, there is net evaporation of liquid CCl_2F_2 until the dichlorodifluoromethane partial pressure inside the steel cylinder equals about 6.5 bar, the vapour pressure of CCl_2F_2 at 25 °C. Here we assume that the cylinder is not so big that 25 kg of CCl_2F_2 is not enough to fill the cylinder to this pressure, i.e., no bigger than about 780 L.

(c) The CCl_2F_2 vapour rushes out of the cylinder quickly at first because of the large pressure imbalance.

The flow slows as the inside pressure approaches 1 atm.

The outside of the cylinder becomes icy because when the gas expands, intermolecular forces are overcome—the energy required is drawn from the thermal energy of the gas and cylinder.

Expansion of a gas can cause significant cooling. This expansion is driven by the increase in entropy, and the resulting cooling is inexorable.

(d)
(1) Turning the cylinder upside down, and opening the valve, would produce a dangerous situation. The cylinder would behave like a rocket. It would empty quickly though.
(2) Cooling the cylinder to −78 °C in dry ice, then opening the valve would allow the cylinder to be emptied safely. It would not happen quickly though—the vapour pressure of CCl_2F_2 at −78 °C (not shown in the figure) is quite small.
(3) Knocking the top off the cylinder, valve and all, with a sledge hammer would provide rapid and relatively safe discharge of the cylinder. The flow velocity would be smaller because of the large cross-sectional area of the open top of the cylinder.

11.117
Mass of ethanol, $m = (0.125 \text{ L}) \times (784.9 \text{ g L}^{-1}) = 98.1 \text{ g}$
Amount of ethanol, $n = (98.1 \text{ g}) / (46.07 \text{ g mol}^{-1}) = 2.13 \text{ mol}$
Heat required to evaporate 2.13 mol of ethanol at 25 °C is
$$q = (42.32 \text{ kJ mol}^{-1}) \times (2.13 \text{ mol}) = 90.1 \text{ kJ}$$

11.119
(a)   The structure of aspartame.

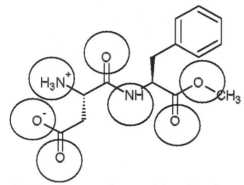

(b) Aspartame is capable of hydrogen bonding. Sites of hydrogen bonding are circled on the structure.

Chapter 12
Solutions and Their Behaviour

IN-CHAPTER EXERCISES

Exercise 12.2—Enthalpy change of solution

(a) The lattice enthalpy of LiCl(s) is larger than that of KCl(s) because the Li^+ ions are smaller than the K^+ ions and so they get closer to the Cl^- ions in the lattice. The electrostatic potential energy is lower—it takes more energy to break the ionic bonds.
(b) The enthalpy of aquation of Li^+ ions is larger than that of K^+ ions (they are both negative enthalpies) because the Li^+ ion is smaller than K^+. The interaction of water molecules with Li^+ is stronger than with K^+ because they can get closer to the smaller lithium ions. Cl^- ions are common to both salts.
We cannot predict the relative magnitudes of enthalpy of solution of LiCl and KCl from the information in parts (a) and (b). The dissolution of an ionic salt in water can be thought of as comprising two components: an endothermic separation of the ions, and an exothermic aquation process. Qualitatively, we can predict that the enthalpies of both of these steps are larger for LiCl than for KCl. Quantitatively, we cannot say without more information how the net enthalpies of solution compare—nor even if each is on balance endothermic or exothermic.

Exercise 12.4—Using Henry's law

Henry's law constant for CO_2 in water is 3.36 x 10^{-4} mol L^{-1} kPa^{-1} (see Table 12.1).
The solubility of CO_2 in water is
$$s = k_H \times p = (3.36 \times 10^{-4} \, mol \, L^{-1} \, kPa^{-1})(33.4 \, kPa) = 1.12 \times 10^{-2} \, mol \, L^{-1}$$
This is the equilibrium concentration of CO_2 in water at 25 °C when the pressure of CO_2(g) above the aqueous phase is 33.4 kPa.

Exercise 12.5—Temperature dependence of solubility

The solubility of Li_2SO_4 in water decreases slightly as we increase the temperature from 10 °C. The amount of solid Li_2SO_4 in the second beaker will increase a little (presuming that the solution was a saturated solution at 10 °C).
The solubility of LiCl in water increases slightly as we increase the temperature from 10 °C. The amount of solid LiCl in the first beaker will decrease.

Exercise 12.7—Solution concentration

Amount of sucrose $= (10.0\ \text{g}) / (342.3\ \text{g mol}^{-1}) = 0.0292\ \text{mol}$
Amount of water $= (250\ \text{g}) / (18.02\ \text{g mol}^{-1}) = 13.9\ \text{mol}$
The mole fraction of sucrose $=$ (amount of sucrose) / (total amount)
  $x = (0.0292\ \text{mol}) / (0.0292\ \text{mol} + 13.9\ \text{mo}) = 0.00210$
The molality of sucrose solution, $m =$ (amount of sucrose) / (mass of water in

kg) $= \dfrac{0.0292\ \text{mol}}{0.250\ \text{kg}} = 0.117\ \text{mol kg}^{-1}$

The mass percent of sucrose $= 100\% \times$ (mass of sucrose) / (total mass)
  $= 100\% \times (10.0\ \text{g}) / (260\ \text{g}) = 3.85\%$

Exercise 12.8—Vapour pressure of solutions, non-volatile solute

$$n(\text{H}_2\text{O}) = \frac{225\ \text{g}}{18.02\ \text{g mol}^{-1}} = 12.5\ \text{mol}$$

$$n(\text{sucrose}) = \frac{10.0\ \text{g}}{342.3\ \text{g mol}^{-1}} = 0.0291\ \text{mol}$$

$$\therefore x_{\text{H}_2\text{O}} = \frac{12.5\ \text{mol}}{12.5\ \text{mol} + 0.0291\ \text{mol}} = 0.998$$

From Raoult's law,
  $p_{\text{H}_2\text{O}} = x_{\text{H}_2\text{O}} \times p^\circ_{\text{H}_2\text{O}} = (0.998)(19.9\ \text{kPa}) = 19.85\ \text{kPa}$

The vapour pressure above the solution is less than that above pure water at the same temperature, but the values are not different to the given number of significant figures.

Exercise 12.10—Vapour pressure of solutions, non-volatile solute

$$n(\text{H}_2\text{O}) = \frac{500.0\ \text{g}}{18.015\ \text{g mol}^{-1}} = 27.75\ \text{mol}$$

$$n(\text{HOCH}_2\text{CH}_2\text{OH}) = \frac{35.0\ \text{g}}{62.07\ \text{g mol}^{-1}} = 0.564\ \text{mol}$$

$$\therefore x_{\text{H}_2\text{O}} = \frac{27.75\ \text{mol}}{27.75\ \text{mol} + 0.564\ \text{mol}} = 0.980$$

From Raoult's law,
  $p_{\text{H}_2\text{O}} = x_{\text{H}_2\text{O}} \times p^\circ_{\text{H}_2\text{O}} = (0.980)(4.76\ \text{kPa}) = 4.66\ \text{kPa}$

Exercise 12.12—Freezing point depression

Determine the freezing point of the ethylene glycol–water solution.
Amount of $HOCH_2CH_2OH$, n = (525 g) / (62.07 g mol^{-1}) = 8.46 mol
The molality of $HOCH_2CH_2OH$ = (amount of $HOCH_2CH_2OH$) / (mass of water in kg)

$$= \frac{8.46 \text{ mol}}{3.00 \text{ kg}} = 2.82 \text{ mol kg}^{-1}$$

Freezing point depression, $\Delta T_f = K_f \times m$ = (1.86 K kg mol^{-1}) (2.82 mol kg^{-1})

$$= 5.25 \text{ K} = 5.25 \text{ Celsius degrees}$$

Note that the freezing point depression constant, K_f is characteristic of the solvent—it does not depend on the solute, ethylene glycol.
Therefore, the freezing point of the solution is −5.25 °C. The concentration of ethylene glycol is not enough to prevent freezing at −25 °C.

Exercise 12.14—Molar mass from freezing point depression

Molality of aluminon in water =

$$m = \frac{\Delta T_f}{K_f} = \frac{0.197 \text{ K}}{1.86 \text{ K kg mol}^{-1}} = 0.106 \text{ mol kg}^{-1}$$

The amount of aluminon in 50.0 g water, n, is (0.0500 kg) × (0.106 mol kg^{-1}) = 0.00530 mol
Molar mass of aluminon, M = (mass of aluminon) / (amount of aluminon)

$$= (2.50 \text{ g}) / (0.00530 \text{ mol}) = 472 \text{ g mol}^{-1}$$

Exercise 12.16—Freezing points of solutions of electrolytes

Amount of NaCl, n = (25.0 g) / (58.44 g mol^{-1}) = 0.428 mol
Molality of NaCl = (amount of NaCl) / (mass of water, in kg)

$$= \frac{0.428 \text{ mol}}{0.525 \text{ kg}} = 0.815 \text{ mol kg}^{-1}$$

Assuming that the solute is a non-electrolyte

$$\Delta T_f \text{ (non-electrolyte)} = K_f \times m = (1.86 \text{ K kg mol}^{-1})(0.815 \text{ mol kg}^{-1}) = 1.52 \text{ K}$$

$$\Delta T_f \text{ (actual)} = i \times \Delta T_f \text{ (non-electrolyte)} = 1.85 \times 1.52 \text{ K} = 2.81 \text{ K}$$

The freezing point of this solution is −2.81 °C.

Exercise 12.18—Molar mass from osmotic pressure

$$\text{Molar concentration, } c = \frac{\Pi}{RT} = \frac{0.248 \text{ kPa}}{(8.314 \text{ L kPa K}^{-1} \text{ mol}^{-1})(298 \text{ K})} = 1.00 \times 10^{-4} \text{ mol L}^{-1}$$

In 100 mL of solution, amount of polyethylene, n = (1.00 × 10^{-4} mol L^{-1}) × (0.100 L) = 1.00 × 10^{-5} mol
Average molar mass of polyethylene = (1.40 g) / (1.00 × 10^{-5} mol) = 1.40 × 10^{5} g mol^{-1}

Exercise 12.19—Osmotic pressure of electrolyte solutions

The osmotic pressure of human blood (37 °C), treated as a 0.154 mol L^{-1} NaCl solution, is

$$\Pi = i \times cRT = 1.9 \times (0.154 \ \cancel{mol} \ \cancel{L}^{-1}) (8.314 \ \cancel{L} \ kPa \ \cancel{K}^{-1} \ \cancel{mol}^{-1}) (310 \ \cancel{K})$$

$$= 754 \ kPa$$

This is a very high pressure. Human cells burst from the osmotic pressure of water entering the cell —if the cell is exposed to pure water.

Exercise 12.20—Colloidal dispersions

(a) The volume of each sphere is

$$V = \frac{4}{3} \pi r^3 = \frac{4}{3} \pi (100 \ nm)^3 = 4.19 \times 10^6 \ nm^3$$

The surface area of each sphere is

$$A = 4\pi r^2 = 4\pi (100 \ nm)^2 = 1.26 \times 10^5 \ nm^2$$

(b) To give a total volume of 1.0 mL = 1.0 cm^3 = 1.0×10^{21} nm^3
requires $(1.0 \times 10^{21} \ nm^3) / (4.19 \times 10^6 \ nm^3) = 2.4 \times 10^{14}$ spheres
The total surface area of these spheres (in m^2) is
$(2.4 \times 10^{14} \ spheres) \times (1.26 \times 10^5 \ nm^2 \ sphere^{-1}) = 3.0 \times 10^{19} \ nm^2 = 30 \ m^2$

REVIEW QUESTIONS

Section 12.4 Factors affecting solubility: Pressure and temperature

12.21
The Henry's law constant for O_2 in water at 25 °C is 1.24×10^{-5} mol L^{-1} kPa^{-1}. At 50 °C, its value is
(a) 6.20×10^{-6} mol L^{-1} kPa^{-1}
The other values listed are either the same (c), or larger values (b) and (d).
Because O_2 is a gas, it is less soluble in water at higher temperatures—dissolution is an exothermic process.

12.23
For a saturated NaCl solution at 25 °C with no solid in the beaker, the amount of dissolved NaCl in the solution is increased by
(c) Raising the temperature of the solution and adding some NaCl.
The solubility of NaCl in water increases with temperature—not a lot, but some.
(a) Adding more solid NaCl does not increase NaCl in solution because the solution is saturated.
(b) Raising the temperature of the solution does not increase NaCl in solution because there is no extra solid NaCl in the beaker to dissolve.
(d) Lowering the temperature of the solution and adding some NaCl decreases the amount of NaCl in solution because the solubility goes down.

Section 12.5 More units of solute concentration

12.25
Amount of $C_2H_4(CO_2H)_2$, n = (2.56 g) / (118.09 g mol^{-1}) = 0.0217 mol
Mass of water, m = 500 mL × 1.00 g mL^{-1} = 500 g
Amount of water, $n(H_2O)$ = (500 g) / (18.02 g mol^{-1}) = 27.8 mol
The mole fraction of succinic acid, x = (amount of succinic acid) / (total amount)
    = (0.0217 mol) / (0.0217 + 27.8 mol) = 0.000780
The molality of succinic acid, m = (amount of succinic acid) / (mass of water, in kg)

$$= \frac{0.0217 \text{ mol}}{0.500 \text{ kg}} = 0.0434 \, \text{mol} \, \text{kg}^{-1}$$

The mass (weight) percent of succinic acid
    = 100% × (mass of succinic acid) / (total mass) = 100% × (2.56 g) / (502.56 g) = 0.509%

12.27
Amount of Na_2CO_3 required, n = (0.200 mol kg^{-1}) × (0.125 kg) = 0.0250 mol
Mass of Na_2CO_3 required, m = (0.0250 mol) × (105.99 g mol^{-1}) = 2.65 g
Amount of water in solution, n = (125 g) / (18.015 g mol^{-1}) = 6.94 mol
Mole fraction of Na_2CO_3 in solution = (0.0250 mol) / (0.025 + 6.94 mol) = 0.174

12.29

Amount of water $= (425 \text{ g}) / (18.02 \text{ g mol}^{-1}) = 23.59 \text{ mol}$

Amount of glycerol required $=$ (desired mole fraction glycerol) $\times$ (total amount)

i.e., $n_{\text{glycerol}} = x_{\text{glycerol}} (n_{\text{glycerol}} + n_{\text{water}})$

So, $n_{\text{glycerol}} = \dfrac{x_{\text{glycerol}} n_{\text{water}}}{\left(1 - x_{\text{glycerol}}\right)} = \dfrac{0.093 \times 23.59 \text{ mol}}{0.907} = 2.4 \text{ mol}$

Mass of glycerol required $= (2.4 \text{ mol}) \times (92.09 \text{ g mol}^{-1}) = 221 \text{ g}$

Molality of the solution $=$ (amount of gycerol) / (mass of water in kg)

$\quad = 2.4 \text{ mol} / 0.425 \text{ kg} = 5.65 \text{ mol kg}^{-1}$

12.31

(a) Suppose you have 1.000 L of the concentrated solution.

Mass of solution $= (1.18 \text{ g mL}^{-1}) \times (1000 \text{ mL}) = 1180 \text{ g}$

Amount of HCl $= (12.0 \text{ mol L}^{-1}) \times (1.000 \text{ L}) = 12.0 \text{ mol}$

Mass of HCl $= (12.0 \text{ mol}) \times (36.46 \text{ g mol}^{-1}) = 438 \text{ g}$

Mass of water $= 1180 \text{ g} - 438 \text{ g} = 742 \text{ g} = 0.742 \text{ kg}$

Molality of the solution $=$ (amount of HCl) / (mass of water in kg)

$\quad = (12.0 \text{ mol}) / (0.742 \text{ kg}) = 16.2 \text{ mol kg}^{-1}$

(b) The mass percent of HCl in the solution $= 100\% \times$ (mass of HCl) / (total mass)

$\quad = 100\% \times (438 \text{ g}) / (1180 \text{ g}) = 37.1\%$

Section 12.6 Colligative properties

12.35

The 0.15 mol kg^{-1} Na$_2$SO$_4$ solution has the higher equilibrium vapour pressure of water. The more concentrated 0.30 mol kg^{-1} NH$_4$NO$_3$ has a lower mole fraction of water and, consequently, a lower vapour pressure of water. "Vapour pressure depression"—according to Raoult's law—does not depend on the nature of the solute. In this example, assuming complete dissociation of these ionic species yields 0.45 mol kg^{-1} ions from the Na$_2$SO$_4$ solution (3 mol ions per mol Na$_2$SO$_4$) and 0.60 mol kg^{-1} ions from the NH$_4$NO$_3$ solution (2 mol ions per mol NH$_4$NO$_3$). The ammonium nitrate solution has the greatest vapour pressure depression, producing the lowest vapour pressure.

12.37

(a)

The molal concentration of the alcohol, $m = \dfrac{\Delta T_{\text{f}}}{K_{\text{f}}} = \dfrac{16.0 \text{ K}}{1.86 \text{ K kg mol}^{-1}} = 8.60 \text{ mol kg}^{-1}$

(b)

Suppose we have 1.000 kg of water.

Amount of ethanol, $n = (1.000 \text{ kg}) \times (8.60 \text{ mol kg}^{-1}) = 8.60 \text{ mol}$

Mass of ethanol $= (46.07 \text{ g mol}^{-1}) \times (8.60 \text{ mol}) = 396 \text{ g}$

Mass percent of ethanol $= 100\% \times$ (mass of ethanol) / (total mass)

$\quad = 100\% \times (396 \text{ g})/(396 + 1000 \text{ g}) = 28.4\%$

12.41

In order of decreasing freezing point, we have

(a) 0.20 mol kg^{-1} ethylene glycol solution
(d) 0.12 mol kg^{-1} KBr solution $\rightarrow$ 0.24 mol kg^{-1} ions
(c) 0.10 mol kg^{-1} MgCl$_2$ solution $\rightarrow$ 0.30 mol kg^{-1} ions
(b) 0.12 mol kg^{-1} K$_2$SO$_4$ solution $\rightarrow$ 0.36 mol kg^{-1} ions
i.e., in order of increasing concentration of aquated species

12.43

The required molality of NaCl $= m = \dfrac{\Delta T_f}{i\,K_f} = \dfrac{10.\ \text{K}}{1.85 \times 1.86\ \text{K kg mol}^{-1}} = 2.9$ mol kg^{-1}

To 3.0 kg of water, you must add 3.0 kg $\times$ 2.9 mol kg^{-1} = 8.7 mol
Required mass of NaCl = (8.7 mol) $\times$ (58.44 g mol^{-1}) = 510 g

12.45

The osmotic pressure of a 0.0120 mol L^{-1} aqueous NaCl solution at 0 °C is

$\Pi = i \times cRT = 1.94 \times (0.0120\ \text{mol L}^{-1})\,(8.314\ \text{L kPa K}^{-1}\ \text{mol}^{-1})\,(273.15\ \text{K})$

$ = 52.9$ kPa

Chapter 13
Dynamic Chemical Equilibrium

IN-CHAPTER EXERCISES

Exercise 13.3—Expressing the form of the reaction quotient

(a) $PCl_5(g) \rightleftharpoons PCl_3(g) + Cl_2(g)$

$$Q = \frac{[PCl_3][Cl_2]}{[PCl_5]}$$

(b) $CO_2(g) + C(s) \rightleftharpoons 2\,CO(g)$

$$Q = \frac{[CO]^2}{[CO_2]}$$

The solid reactant, C(s), does not appear in the reaction quotient.

(c) $Cu(NH_3)_4^{2+}(aq) \rightleftharpoons Cu^{2+}(aq) + 4\,NH_3(aq)$

$$Q = \frac{[Cu^{2+}][NH_3]^4}{[Cu(NH_3)_4^{2+}]}$$

(d) $CH_3COOH(aq) + H_2O(l) \rightleftharpoons CH_3COO^-(aq) + H_3O^+(aq)$

$$Q = \frac{[CH_3COO^-][H_3O^+]}{[CH_3COOH]}$$

Remember that the liquid reactant, H₂O(l), does not appear in the reaction quotient.

Exercise 13.5—Deducing the direction of spontaneous reaction

(a) At 298 K, the reaction mixture with
[butane] = 0.97 mol L^{-1} and [isobutane] = 2.18 mol L^{-1}

has $\quad Q = \dfrac{[\text{isobutane}]}{[\text{butane}]} = \dfrac{2.18}{0.97} = 2.25$

and is, therefore, NOT at equilibrium. Because $Q < K$, the reaction proceeds in the forward direction—to make Q increase towards K. Note that when we evaluate Q, units are left out of the concentrations. This is because the concentrations appearing in a reaction quotient expression are really "activities" which, for our purposes, are concentrations relative to a standard value. In the case of Q and K, the reference concentration is exactly 1 mol L^{-1}.

(b) At 298 K, the reaction mixture with
[butane] = 0.75 mol L^{-1} and [isobutane] = 2.60 mol L^{-1}

has $\quad Q = \dfrac{[\text{isobutane}]}{[\text{butane}]} = \dfrac{2.60}{0.75} = 3.47$

and is, therefore, NOT at equilibrium. Because $Q > K$, the reaction proceeds in the reverse direction —to make Q decrease towards K.

Exercise 13.6—Magnitude of K and extent of reaction

(a) $Cu(NH_3)_4^{2+} \rightleftharpoons Cu^{2+}(aq) + 4NH_3(aq)$ $\qquad K=1.5\times10^{-13}$

If $[Cu(NH_3)_4^{2+}] = [Cu^{2+}] = 1.0$ mol L^{-1}, and the reaction mixture is at equilibrium, then

$$Q = \frac{[Cu^{2+}][NH_3]^4}{[Cu(NH_3)_4^{2+}]} = [NH_3]^4 = K = 1.5\times10^{-13}$$

$$[NH_3] = (1.5\times10^{-13})^{1/4} = 6.2\times10^{-4}$$

i.e., $[NH_3] = 6.2\times10^{-4}$ mol L^{-1}

The units, mol L^{-1}, are put back in at the end—i.e., $[NH_3]$ is expressed as a concentration rather than a relative concentration. It is a relative concentration—relative to 1 mol L^{-1}—in the reaction quotient.

(b) $Cd(NH_3)_4^{2+} \rightleftharpoons Cd^{2+}(aq) + 4NH_3(aq)$ $\qquad K=1.0\times10^{-7}$

If $[Cd(NH_3)_4^{2+}] = [Cd^{2+}] = 1.0$ mol L^{-1}, and the reaction mixture is at equilibrium, then

$$Q = \frac{[Cd^{2+}][NH_3]^4}{[Cd(NH_3)_4^{2+}]} = [NH_3]^4 = K = 1.0\times10^{-7}$$

$$[NH_3] = (1.0\times10^{-7})^{1/4} = 1.8\times10^{-2} \text{ mol } L^{-1}$$

$[NH_3]$ is higher in solution (b) than in solution (a)

Exercise 13.7—Estimation of equilibrium constants

$C_6H_{10}I_2 (\text{in } CCl_4) \rightleftharpoons C_6H_{10}(\text{in } CCl_4) + I_2(\text{in } CCl_4)$

The order of the steps taken are indicated as (1), (2), etc.

	$C_6H_{10}I_2$	$\rightleftharpoons$	C_6H_{10}	+	I_2
Initial / (mol L^{-1})	0.050 mol / 1.00 L = 0.050 mol L^{-1}		0		0
Change / (mol L^{-1})	(2) the change in $[C_6H_{10}I_2]$ = − 0.035		(3) the change in $[C_6H_{10}]$ = + 0.035		(1) the change in $[I_2]$ = + 0.035
Equilibrium / (mol L^{-1})	(4) 0.050 − 0.035 = 0.015		(5) 0 + 0.035 = 0.035		0.035

Having determined the equilibrium concentrations, we can evaluate the reaction quotient, which equals K at equilibrium.

$$Q = \frac{[C_6H_{10}][I_2]}{[C_6H_{10}I_2]} = \frac{0.035\times0.035}{0.015} = 0.082 = K$$

Exercise 13.10—Calculating equilibrium concentrations

$H_2(g) + I_2(g) \rightleftharpoons 2HI(g)$

At some temperature, $K = 33$.

The ICE table is constructed in terms of an unknown extent of reaction, x, that we must solve for using the given equilibrium constant.

	H₂	**+**	**I₂**	**⇌**	**2 HI**
Initial / (mol L⁻¹)	6.00×10^{-3}		6.00×10^{-3}		0
Change / (mol L⁻¹)	$-x$		$-x$		$+2x$
Equilibrium / (mol L⁻¹)	$6.00 \times 10^{-3} - x$		$6.00 \times 10^{-3} - x$		$2x$

The equilibrium constant is determined by the equilibrium concentrations, expressed here in terms of the unknown, x.

$$K = \frac{[HI]^2_{eqm}}{[H_2]_{eqm}[I_2]_{eqm}} = \frac{(2x)^2}{(6.00 \times 10^{-3} - x)(6.00 \times 10^{-3} - x)} = 33$$

Taking the square root of both sides of this equation gives

$$\frac{2x}{(6.00 \times 10^{-3} - x)} = \sqrt{33} = 5.74$$

$$2x = 5.74\,(6.00 \times 10^{-3} - x) = 34.5 \times 10^{-3} - 5.74x$$

$$7.74x = 34.5 \times 10^{-3}$$

$$x = 4.46 \times 10^{-3}$$

Now that we have the extent of reaction, x, we can calculate all of the reactant and product concentrations at equilibrium.

$[H_2] = [I_2] = 6.00 \times 10^{-3} - x = 6.00 \times 10^{-3} - 4.46 \times 10^{-3} = 1.54 \times 10^{-3}$ mol L⁻¹

$[HI] = 2x = 2 \times 4.46 \times 10^{-3} = 8.92 \times 10^{-3}$ mol L⁻¹

Exercise 13.11—Reaction equations and equilibrium constants

$$A + B \xrightleftharpoons{K_1} 2C$$
$$2A + 2B \xrightleftharpoons{K_2} 4C$$

The correct answer is **(b)** $K_2 = K_1^2$

Doubling all stoichiometric coefficients doubles all exponents in the reaction quotient expression—squaring the reaction quotient.

Exercise 13.13—Reaction equations and equilibrium constants

$$\tfrac{3}{2}O_2(g) \rightleftharpoons O_3(g) \quad K_1 = 2.5\times10^{-29} \text{ at } 25\ ^\circ C$$

(a) For $3O_2(g) \rightleftharpoons 2O_3(g)$
the equilibrium constant is given by

$$K_2 = \frac{[O_3]^2}{[O_2]^3} = \left(\frac{[O_3]}{[O_2]^{3/2}}\right)^2 = (K_1)^2$$

i.e.,

$$K_2 = (K_1)^2 = (2.5\times10^{-29})^2 = 6.3\times10^{-58}$$

(b) For $2O_3(g) \rightleftharpoons 3O_2(g)$
(the reverse of reaction 2—part a), we have

$$K_3 = \frac{[O_2]^3}{[O_3]^2} = \left(\frac{[O_3]^2}{[O_2]^3}\right)^{-1} = (K_2)^{-1}$$

i.e.

$$K_3 = (K_2)^{-1} = (6.3\times10^{-58})^{-1} = 1.6\times10^{57}$$

Exercise 13.14—Deriving an equilibrium constant from others

$$SnO_2(s) + 2CO(g) \rightleftharpoons Sn(s) + 2CO_2(g) \quad K_3 = ?$$

Since

$$+1\times\Big(SnO_2(s) + 2H_2(g) \rightleftharpoons Sn(s) + 2H_2O(g) \quad K_1 = 8.12\Big)$$
$$-2\times\Big(H_2(g) + CO_2(g) \rightleftharpoons H_2O(g) + CO(g) \quad K_2 = 0.771\Big)$$
$$\overline{\quad SnO_2(s) + 2CO(g) \rightleftharpoons Sn(s) + 2CO_2(g) \quad K_3 = K_1\times(K_2)^{-2}}$$

we have $K_3 = K_1\times(K_2)^{-2}$. We can check this

$$K_3 = \frac{[CO_2]^2}{[CO]^2} = \frac{[H_2O]^2}{[H_2]^2}\left(\frac{[H_2O][CO]}{[H_2][CO_2]}\right)^{-2} = K_1(K_2)^{-2}$$

So $\quad K_3 = K_1\times(K_2)^{-2} = 8.12\times(0.771)^{-2} = 13.7$

Exercise 13.15—Effect of concentration changes on equilibrium

We will need the equilibrium constant. Use the given concentrations for the initial equilibrium mixture.

$$K = \frac{[\text{isobutane}]_{eqm}}{[\text{butane}]_{eqm}} = \frac{0.50}{0.20} = 2.5$$

Now, we set up an ICE table with the initial data corresponding to after increasing [isobutane] by 2.00.

	butane	$\rightleftharpoons$	isobutane
Initial / (mol L^{-1})	0.20		$0.50 + 2.00 = 2.50$ Note that by increasing [isobutane] by 2.00, the mixture is taken away from equilibrium.
Change / (mol L^{-1})	$+x$		$-x$
Equilibrium / (mol L^{-1})	$0.20 + x$		$2.50 - x$

The equilibrium constant is determined by the equilibrium concentrations, expressed here in terms of the unknown, x.

$$K = \frac{[\text{isobutane}]_{\text{eqm}}}{[\text{butane}]_{\text{eqm}}} = \frac{2.50 - x}{0.20 + x} = 2.5$$

Solving for x:

$$2.50 - x = 2.5\,(0.20 + x) = 0.50 + 2.5\,x$$

$$3.5\,x = 2.00$$

$$x = 0.57$$

The new equilibrium concentrations are

$$[\text{butane}] = 0.20 + x = 0.20 + 0.57 = 0.77 \text{ mol L}^{-1}$$
$$[\text{isobutane}] = 2.50 - x = 2.50 - 0.57 = 1.93 \text{ mol L}^{-1}$$

Exercise 13.16—Effect of changing the volume of a gas-phase reaction

$$3\,H_2(g) + N_2(g) \rightleftharpoons 2\,NH_3(g)$$

(a) When extra H_2 is added to an equilibrium mixture, the mixture is shifted out of equilibrium by the increase in H_2 concentration, and decreasing Q so that $Q < K$. There is net reaction to form more $NH_3(g)$, increasing Q until a new equilibrium is attained when $Q = K$ once more. Some, but not all, of the additional H_2 is 'consumed.' The concentration of N_2 decreases, since it is a reactant, while the concentration of product NH_3 increases.

When extra NH_3 is added, Q is instantaneously increased, so that $Q > K$, and the system is no longer at equilibrium. Net reaction to form more $H_2(g)$ and $N_2(g)$ occurs decreasing Q until the system comes to equilibrium again when $Q = K$ once more. Some—but not all—of the added ammonia is consumed. The final equilibrium mixture has a higher concentration of both hydrogen and nitrogen (reactants) and a higher concentration of ammonia than the original equilibrium mixture.

(b) When the volume of the system is increased, all concentrations are decreased by the same factor. Since there are different coefficients of reactant and product gases in the balanced chemical equation, this will change the value of the reaction quotient, Q, shifting it away from its equilibrium value. In this case, the coefficients of the reactant gases are larger than those of the product gases, and the reaction quotient gets bigger (because the denominator decreases by more than the numerator). This causes net reaction in the direction that produces more $N_2(g)$ and $H_2(g)$, because Q must decrease to revert to its equilibrium value, generating a net increase in the amount of gases.

Net reaction proceeds in the direction that counters the decrease in total gas concentration caused by the change in volume.

A more mathematical account of this experiment is provided as follows:

If the volume is increased by a factor λ ($\lambda > 1$), then all concentrations decrease by the same factor—i.e., they are multiplied by $1/\lambda$. The reaction quotient changes as follows:

$$Q = K = \frac{[NH_3]^2_{eqm}}{[H_2]^3_{eqm}[N_2]_{eqm}} \rightarrow$$

$$Q = \frac{\frac{1}{\lambda^2}[NH_3]^2_{eqm}}{\frac{1}{\lambda^3}[H_2]^3_{eqm}\frac{1}{\lambda}[N_2]_{eqm}} = \lambda^2 \frac{[NH_3]^2_{eqm}}{[H_2]^3_{eqm}[N_2]_{eqm}} = \lambda^2 K > K$$

The reaction quotient increases by the square of the factor applied to the volume. This results because, in the reaction equation, the sum of the coefficients of the reactants is greater than the sum of the coefficients of the products.

Exercise 13.17—Dependence of equilibrium constant on temperature

(a) The equilibrium concentration of NOCl decreases if the temperature of the system is increased.

$$2\, NOCl(g) \rightleftharpoons 2\, NO(g) + Cl_2(g) \qquad \Delta_{rxn}H^\circ = +77.1\, kJ\, mol^{-1}$$

(i) Initially, the mixture is at equilibrium. So, $Q = K$. When temperature increases, the equilibrium constant generally changes. For an endothermic reaction such as this, the equilibrium constant increases. If there were no net reaction, we would have $Q < K'$ where K' is the new equilibrium constant. *Net* reaction must proceed in the direction that forms more NO(g) and Cl_2(g) to return to equilibrium—by increasing Q up to K'.

(ii) The reaction proceeds in the forward—endothermic—direction after the temperature is increased. The endothermic direction is the direction that consumes heat and lowers temperature—i.e., countering the increase in temperature in accord with Le Chatelier's principle.

(b) The equilibrium concentration of SO_3 decreases if the temperature is increased.

$$2\, SO_2(g) + O_2(g) \rightleftharpoons 2\, SO_3(g) \qquad \Delta_{rxn}H^\circ = -198\, kJ\, mol^{-1}$$

(i) Initially, the mixture is at equilibrium, so, $Q = K$. For an exothermic reaction such as this, the equilibrium constant decreases. Now, $Q > K'$ where K' is the new equilibrium constant. To return to equilibrium, Q must decrease—i.e., net reaction proceeds in the direction to form more SO_2(g) and O_2(g) to return to equilibrium—i.e., decreasing Q down to K'.

(ii) The reaction proceeds in the reverse—endothermic—direction after the temperature is increased. The endothermic direction is the direction that absorbs heat and lowers temperature—i.e., countering the increase in temperature in accord with Le Chatelier's principle.

REVIEW QUESTIONS

Section 13.3 The reaction quotient and the equilibrium constant

13.19

$$N_2(g) + O_2(g) \rightleftharpoons 2\,NO(g)$$

$$Q = \frac{[NO]^2}{[N_2][O_2]} = \frac{(4.2 \times 10^{-3})^2}{(0.50)(0.25)} = 1.4 \times 10^{-4} \; < \; 4.0 \times 10^{-4} = K$$

Because $Q < K$, net reaction proceeds in the direction that forms more $NO(g)$, to increase Q up to K.

Section 13.4 Quantitative aspects of equilibrium constants

13.21

$$PCl_5(g) \rightleftharpoons PCl_3(g) + Cl_2(g)$$

$$Q = \frac{[PCl_3][Cl_2]}{[PCl_5]} = \frac{(1.3 \times 10^{-2})(3.9 \times 10^{-3})}{(4.2 \times 10^{-5})} = 1.2 = K$$

since the mixture is at equilibrium.

13.25

	$Br_2(g)$	$\rightleftharpoons$	$2\,Br(g)$
Initial / (mol L^{-1})	0.086 / 1.26 = 0.068		0
Change / (mol L^{-1})	(2) $= -x$ $= -0.00125$		(1) $+0.0025 = 2x$
Equilibrium / (mol L^{-1})	(3) $0.068 - 0.00125$ $= 0.067$		3.7% of 0.068 $= 0.037 \times 0.068$ $= 0.0025$

Having determined the equilibrium concentrations, we can evaluate the reaction quotient which equals K at equilibrium.

$$K = \frac{[Br]^2_{eqm}}{[Br_2]_{eqm}} = \frac{(0.0025)^2}{(0.067)} = 9.3 \times 10^{-5}$$

13.29

$$H_2(g) + CO_2(g) \rightleftharpoons H_2O(g) + CO(g)$$

(a) $$K = \frac{[H_2O]_{eqm}[CO]_{eqm}}{[H_2]_{eqm}[CO_2]_{eqm}} = \frac{(0.11/1.0)(0.11/1.0)}{(0.087/1.0)(0.087/1.0)} = 1.6$$

(b)

	H₂(g)	+	CO₂(g)	⇌	H₂O(g)	+	CO(g)
Initial / (mol L⁻¹)	0.050 / 2.0 = 0.025		0.050 / 2.0 = 0.025		0		0
Change / (mol L⁻¹)	−x		−x		+x		+x
Equilibrium / (mol L⁻¹)	0.025 − x		0.025 − x		x		x

The equilibrium constant is determined by the equilibrium concentrations, expressed here in terms of the unknown, x.

$$K = \frac{[H_2O]_{eqm}[CO]_{eqm}}{[H_2]_{eqm}[CO_2]_{eqm}} = \frac{x\,x}{(0.025-x)(0.025-x)} = 1.6$$

Taking the square root of both sides of this equation gives

$$\frac{x}{(0.025-x)} = \sqrt{1.6} = 1.26$$

$$x = 1.26\,(0.025-x) = 0.0316 - 1.26x$$

$$2.26x = 0.0316$$

$$x = 0.014$$

The new equilibrium concentrations of H₂O(g) and CO(g) are both 0.014 mol L⁻¹

13.31

	I₂(aq)	⇌	I₂(CCl₄)
Initial / (mol L⁻¹)	amount = 0.0340 g / 253.81 g mol⁻¹ = 0.000134 mol concentration = 0.000134 mol / 0.1000 L = 0.00134		0
Change / (mol L⁻¹)	−x		+x
Equilibrium / (mol L⁻¹)	0.00134 − x		x

The equilibrium constant is given by

$$K = \frac{[I_2(CCl_4)]_{eqm}}{[I_2(aq)]_{eqm}} = \frac{x}{0.00134-x} = 85.0$$

Solve this equation for x.
$$x = 85.0 \ (0.00134 - x) = 0.114 - 85.0 \ x$$

$$86.0 \ x = 0.114$$

$$x = 0.00133$$

The equilibrium concentrations are
$$[I_2(aq)]_{eqm} = 0.00134 - x = 0.00001 \text{ mol L}^{-1}$$
$$[I_2(CCl_4)]_{eqm} = x = 0.00133 \text{ mol L}^{-1}$$
To the precision of the given data, only 0.00001 mol L^{-1} concentration of I_2 remains in the aqueous layer.
The corresponding fraction of the total concentration of I_2 is $0.00001 / 0.00134 = 0.007$.
This same fraction applies to the initial mass of I_2, to give the mass of I_2 remaining in the aqueous layer. Note that it is unnecessary to convert the aqueous concentration back into an amount and then a mass.
Mass of I_2 remaining in the aqueous layer $= 0.007 \times 0.0340 \text{ g} = 0.0002 \text{ g}$.

13.33

	2 NH$_3$(g)	$\rightleftharpoons$	N$_2$(g)	+	3 H$_2$(g)
Initial / (mol L^{-1})	3.60 / 2.00 = 1.80		0		0
Change / (mol L^{-1})	$-2x$		$+x$		$+3x$
Equilibrium / (mol L^{-1})	$1.80 - 2x$		x		$3x$

The equilibrium constant is given by
$$K = \frac{[N_2]_{eqm}[H_2]_{eqm}^3}{[NH_3]_{eqm}^2} = \frac{x \ (3x)^3}{(1.80 - 2x)^2} = 6.3$$
Solve this equation for x. First take the square root of both sides
$$\frac{\sqrt{27}x^2}{1.80 - 2x} = \frac{5.196 \ x^2}{1.80 - 2x} = \sqrt{6.3} = 2.5$$

$$5.196 \ x^2 = 2.5 \ (1.80 - 2x)$$

$$5.196 \ x^2 + 5.0x - 4.5 = 0$$

$$x = \frac{-5.0 \pm \sqrt{5.0^2 + 4 \times 5.196 \times 4.5}}{2 \times 5.196}$$

$$x = 0.57 \quad \text{or} \quad -1.5$$

The latter solution gives negative $N_2(g)$ and $H_2(g)$ concentrations. Only the former solution gives admissible concentrations:

$[NH_3]_{eqm} = 1.80 - 2x = 1.80 - 1.14 = 0.66 \text{ mol L}^{-1}$

$[N_2]_{eqm} = x = 0.57 \text{ mol L}^{-1}$

$[H_2]_{eqm} = 3x = 1.71 \text{ mol L}^{-1}$

The total concentration of gases is the sum of the three gas concentrations.

Total gas concentration $= 0.66 + 0.57 + 1.71 = 2.94 \text{ mol L}^{-1}$

Pressure is determined from concentration using the ideal gas law:

$$p = \frac{n}{V} RT = \left(2.94 \; \cancel{mol} \; \cancel{L}^{-1} \right) \left(8.314 \; \cancel{L} \; kPa \, \cancel{K}^{-1} \, \cancel{mol}^{-1} \right) \left(723 \; \cancel{K} \right)$$

$$= 17670 \text{ kPa} = 17.7 \text{ MPa} \text{ (to three significant digits)}$$

13.5 Reaction equations and equilibrium constants

13.35

The second equation is double the inverse of the first. Therefore, K for the second reaction is the inverse of the square of K for the first reaction.

$K = 1/(6.66 \times 10^{12})^2 = 1/4.44 \times 10^{25} = 2.25 \times 10^{-26}$

13.37

$N_2(g) + O_2 \rightleftharpoons 2 NO(g) \qquad K = 1.7 \times 10^{-3}$ at 2300 K

(a) For $\frac{1}{2}N_2(g) + \frac{1}{2}O_2(g) \rightleftharpoons NO(g) \qquad K = (1.7 \times 10^{-3})^{1/2} = 0.041$

(b) For $2 NO(g) \rightleftharpoons N_2(g) + O_2(g) \qquad K = (1.7 \times 10^{-3})^{-1} = 588 = 590$ (to two significant digits)

13.38

$+\left(H_2O(g) + CO(g) \rightleftharpoons H_2(g) + CO_2(g) \qquad K_1 = 1.6 \right)$

$-\left(FeO(s) + CO(g) \rightleftharpoons Fe(s) + CO_2(g) \qquad K_2 = 0.67 \right)$

$\overline{\qquad Fe(s) + H_2O(g) \rightleftharpoons FeO(s) + H_2(g) \qquad K_3 = K_1 / K_2 \qquad}$

$K_3 = K_1 / K_2 = 1.6 / 0.67 = 2.4$

13.6 Disturbing reaction mixtures at equilibrium

13.39

We will need the equilibrium constant. Use the given concentrations for the initial equilibrium mixture.

$$K = \frac{[\text{isobutane}]_{eqm}}{[\text{butane}]_{eqm}} = \frac{2.5 \, \text{mol L}^{-1}}{1.0 \, \text{mol L}^{-1}} = 2.5$$

(a) Now, we set up an ICE table with the initial data corresponding to just after increasing [isobutane] by 0.50.

	butane	$\rightleftharpoons$	isobutane
Initial / (mol L^{-1})	1.0		2.5 + 0.50
Change / (mol L^{-1})	$-x$		$+x$
Equilibrium / (mol L^{-1})	$1.0 - x$		$3.0 + x$

We have used $+x$ even though we expect the reaction to shift in the other direction. In this case, solving for x will give a negative value. This shows that you can get the right answer even if you do not anticipate correctly the direction of net reaction.

The equilibrium constant is determined by the equilibrium concentrations, expressed here in terms of the unknown, x.

$$K = \frac{[\text{isobutane}]_{\text{eqm}}}{[\text{butane}]_{\text{eqm}}} = \frac{3.0 + x}{1.0 - x} = 2.5$$

Solving for x:

$$3.0 + x = 2.5\,(1.0 - x) = 2.5 - 2.5\,x$$

$$3.5\,x = -0.5$$

$$x = -0.14$$

The new equilibrium concentrations are

$$[\text{butane}] = 1.0 - x = 1.0 + 0.14 = 1.1 \text{ mol L}^{-1}$$
$$[\text{isobutane}] = 3.0 + x = 3.0 - 0.14 = 2.9 \text{ mol L}^{-1}$$

(b) Now, we set up an ICE table with the initial data corresponding to just after increasing [butane] by 0.50.

	butane	$\rightleftharpoons$	isobutane
Initial / (mol L^{-1})	1.0 + 0.50 = 1.5		2.5
Change / (mol L^{-1})	$-x$		$+x$
Equilibrium / (mol L^{-1})	$1.5 - x$		$2.5 + x$

$$K = \frac{[\text{isobutane}]_{\text{eqm}}}{[\text{butane}]_{\text{eqm}}} = \frac{2.5 + x}{1.5 - x} = 2.5$$

Solving for x:

$$2.5 + x = 2.5\,(1.5 - x) = 3.75 - 2.5\,x$$

$$3.5\,x = 1.25$$

$$x = 0.36$$

The new equilibrium concentrations are

$$[\text{butane}] = 1.5 - x = 1.5 - 0.36 = 1.1 \text{ mol L}^{-1}$$
$$[\text{isobutane}] = 2.5 + x = 2.5 + 0.36 = 2.9 \text{ mol L}^{-1}$$

We end up with the same concentrations as in part (a) at the level of significant figures used.

13.41

$$N_2O_3(g) \rightleftharpoons NO(g) + NO_2(g) \qquad \Delta_r H° = 40.5 \text{ kJ mol}^{-1}$$

(a) If more N_2O_3 (a reactant gas) is added, then the reaction quotient Q decreases—the concentration of $N_2O_3(g)$ is in the denominator of the reaction quotient. Therefore, there is net reaction in the direction that forms more $NO(g)$ and $NO_2(g)$, increasing Q until again $Q = K$.

The net reaction does not consume all of the added $N_2O_3(g)$. So, all three gases increase in concentration, compared with the concentrations before more $N_2O_3(g)$ was added.

(b) If more NO_2 (a product gas) is added, then the reaction quotient Q increases—the concentration of $NO_2(g)$ appears in the numerator of the reaction quotient. Therefore, there is net reaction in the direction that produces more $N_2O_3(g)$, decreasing Q until again $Q = K$.

Again, all three gases increase in concentration, compared with the concentrations before more $NO_2(g)$ was added. This is the case whenever reactant or product gas is added.

(c) If the volume of the reaction flask is increased, the concentrations of all gases decrease by the same factor. Since the sum of the coefficients in the equation of the two product gases, is bigger than the coefficient of the one reactant gas, the reaction quotient Q decreases (the numerator has more decreasing factors—by one). Therefore, there is net reaction in the direction that forms more $NO(g)$ and $NO_2(g)$, increasing Q until again $Q = K$.

Reactant gas concentration decreases in this case—because of the increase in volume and again because of the net reaction. Product gas concentrations decrease due to the volume increase, but then increase upon net forward reaction. They do not, however, return to their original concentrations. Altogether, there is a net decrease in all gas concentrations.

(d) If the temperature is lowered, the equilibrium constant K decreases because the reaction is endothermic. Upon decreasing K, by lowering the temperature, we have $Q > K$. Therefore, there is net reaction in the direction that produces more $N_2O_3(g)$, decreasing Q until $Q =$ (the new) K.

Here, the only change in gas concentrations is due to the net reaction. Product gases decrease in concentration, while the reactant gas increases in concentration.

13.43

$$BaCO_3(s) \rightleftharpoons BaO(s) + CO_2(g) \qquad \text{is endothermic}$$

(a) More $BaCO_3(s)$ is added.   (i) no net reaction, the system remains at equilibrium.
Adding a solid does not change Q.

(b) More $CO_2(g)$ is added.   (ii) net reaction to form more $BaCO_3(s)$
Adding $CO_2(g)$ brings about an increase of Q, so net reaction happens to decrease Q till $Q = K$ again.

(c) More $BaO(s)$ is added.   (i) no net reaction, the system remains at equilibrium
Adding a solid does not change Q.

(d) The temperature is raised.   (iii) net reaction to form more $BaO(s)$ and $CO_2(g)$.
Increasing temperature increases K for the endothermic direction. There will be net reaction in the direction that increases Q until $Q = $ new K.

(e) The volume of the reaction vessel is increased.   (iii) net reaction to form more $BaO(s)$ and $CO_2(g)$.
Increasing volume decreases the concentration of $CO_2(g)$ —the only gas in the reaction—and so decreases Q.

There will be net reaction in the direction that increases $Q = [CO_2(g)]$ back to its equilibrium value

13.45 $PCl_5(g) \rightleftharpoons PCl_3(g) + Cl_2(g)$

First, get the equilibrium constant.

Amount of $PCl_5(g)$ = 3.120 g / 208.24 g mol^{-1} = 0.01498 mol

Amount of $PCl_3(g)$ = 3.845 g / 137.33 g mol^{-1} = 0.02800 mol

Amount of $Cl_2(g)$ = 1.787 g / 70.91 g mol^{-1} = 0.02520 mol

Extra amount of $Cl_2(g)$ = 1.418 g / 70.91 g mol^{-1} = 0.02000 mol

These amounts are within a 1.00 L flask—the concentrations have the same numerical values except that they have 3 significant figures. We carry 4 figures because these calculations can amplify round-off errors.

$$K = \frac{[PCl_3]_{eqm}[Cl_2]_{eqm}}{[PCl_5]_{eqm}} = \frac{0.02800 \times 0.02520}{0.01498} = 0.04710$$

	PCl₅(g)	$\rightleftharpoons$	**PCl₃(g)**	+	**Cl₂(g)**
Initial / (mol L^{-1})	0.01498		0.02800		0.02520 + 0.02000 = 0.04520
Change / (mol L^{-1})	$+x$		$-x$		$-x$
Eqm / (mol L^{-1})	$0.01498 + x$		$0.02800 - x$		$0.04520 - x$

The equilibrium constant is given by

$$K = \frac{[PCl_3]_{eqm}[Cl_2]_{eqm}}{[PCl_5]_{eqm}} = \frac{(0.02800 - x)(0.04520 - x)}{(0.01498 + x)} = 0.04710$$

Solve this equation for x.

$$x^2 - (0.02800 + 0.04520)\,x + 0.02800 \times 0.04520 = 0.04710\,(0.01498 + x)$$

$$x^2 - 0.07320\,x + 0.001266 = 0.0007056 + 0.04710\,x$$

$$x^2 - 0.12030\,x + 0.000560 = 0$$

$$x = \frac{0.12030 \pm \sqrt{0.12030^2 - 4 \times 0.000560}}{2}$$

$$x = 0.1154 \quad \text{or} \quad 0.004851$$

The former solution gives negative $PCl_3(g)$ and $Cl_2(g)$ concentrations. Only the latter solution gives admissible concentrations:

$[PCl_5]_{eqm}$ = $0.01498 + x$ = 0.01498 + 0.00485 = 0.0198 mol L^{-1}

$[PCl_3]_{eqm}$ = $0.02800 - x$ = 0.02800 - 0.00485 = 0.0232 mol L^{-1}

$[Cl_2]_{eqm}$ = $0.04520 - x$ = 0.04520 - 0.00485 = 0.0404 mol L^{-1}

SUMMARY AND CONCEPTUAL QUESTIONS

13.47

	2 CH$_3$COOH(g)	$\rightleftharpoons$	(CH$_3$COOH)$_2$(g)
Initial / (mol L^{-1})	5.4×10^{-4}		0
Change / (mol L^{-1})	$-2x$		$+x$
Eqm / (mol L^{-1})	$5.4 \times 10^{-4} - 2x$		X

$$K = \frac{[\text{dimer}]_{eqm}}{[\text{monomer}]^2_{eqm}} = \frac{x}{(5.4 \times 10^{-4} - 2x)^2} = 3.2 \times 10^4$$

Solving for x:

$$\frac{1}{3.2 \times 10^4}x = (5.4 \times 10^{-4})^2 - 2 \times 2 \times 5.4 \times 10^{-4}x + 4x^2$$

$$3.1 \times 10^{-5}x = 2.9 \times 10^{-7} - 2.16 \times 10^{-3}x + 4x^2$$

$$4x^2 - 2.19 \times 10^{-3}x + 2.9 \times 10^{-7} = 0$$

$$x = \frac{2.19 \times 10^{-3} \pm \sqrt{(2.19 \times 10^{-3})^2 - 4 \times 4 \times 2.9 \times 10^{-7}}}{2 \times 4}$$

$$x = 3.2 \times 10^{-4} \quad \text{or} \quad 2.2 \times 10^{-4}$$

The former solution produces a negative monomer concentration. We use the latter solution. The new equilibrium concentrations are

[monomer] $= 5.4 \times 10^{-4} - 2x = 5.4 \times 10^{-4} - 2 \times 2.2 \times 10^{-4} = 1.0 \times 10^{-4}$ mol L^{-1}

[dimer] $= x = 2.2 \times 10^{-4}$ mol L^{-1}

(a) Percentage of monomer converted to dimer $=$

100% $\times$ (initial amount of monomer $-$ final amount of monomer) / (initial amount of monomer)

$= 100\% \times (5.4 \times 10^{-4} - 1.0 \times 10^{-4}) / (5.4 \times 10^{-4}) = 81\%$

(b) The reaction producing the dimer involves only the making of bonds—it must be exothermic. If the temperature is increased, the equilibrium constant will decrease and there will be net reaction in the direction that decreases Q—by formation of more monomer.

13.49

(a) SO$_2$Cl$_2$(g) $\rightleftharpoons$ SO$_2$(g) + Cl$_2$(g) $K = 0.045$ at 375 °C

Amount of SO$_2$Cl$_2$ $= 6.70$ g / 134.97 g mol^{-1} $= 0.0496$ mol

placed in a 1.00 L flask $\rightarrow$ 0.0496 mol L^{-1}

	SO$_2$Cl$_2$(g)	$\rightleftharpoons$	SO$_2$(g)	+	Cl$_2$(g)
Initial / (mol L^{-1})	0.0496		0		0
Change / (mol L^{-1})	$-x$		$+x$		$+x$
Eqm / (mol L^{-1})	$0.0496 - x$		x		x

The equilibrium constant is given by

$$K = \frac{[SO_2]_{eqm}[Cl_2]_{eqm}}{[SO_2Cl_2]_{eqm}} = \frac{x\,x}{(0.0496 - x)} = 0.045$$

Solve this equation for x.

$$x^2 = 0.045\,(0.0496 - x) = 0.0022 - 0.045\,x$$

$$x^2 + 0.045\,x - 0.0022 = 0$$

$$x = \frac{-0.045 \pm \sqrt{0.045^2 + 4 \times 0.0022}}{2}$$

$$x = 0.030 \quad \text{or} \quad -0.075$$

The latter solution gives negative SO$_2$(g) and Cl$_2$(g) concentrations. We use the former solution.
$[SO_2Cl_2]_{eqm} = 0.0496 - x = 0.0496 - 0.030 = 0.020$ mol L^{-1}
$[SO_2]_{eqm} = x = 0.030$ mol L^{-1}
$[Cl_2]_{eqm} = x = 0.030$ mol L^{-1}
Fraction of SO$_2$Cl$_2$(g) dissociated
    = (initial amount of SO$_2$Cl$_2$(g) − final amount of SO$_2$Cl$_2$(g)) / (initial amount of SO$_2$Cl$_2$(g))
    = (0.0496 − 0.020) / 0.0496 = 0.60

(b)
Initial concentration of Cl$_2$(g) =

$$\frac{n}{V} = \frac{p}{RT} = \frac{101.3\ \text{kPa}}{(8.314\ \text{L kPa mol}^{-1}\ \text{K}^{-1})(648\ \text{K})} = 0.0188\ \text{mol L}^{-1}$$

	SO$_2$Cl$_2$(g)	$\rightleftharpoons$	SO$_2$(g)	+	Cl$_2$(g)
Initial / (mol L^{-1})	0.0496		0		0.0188
Change / (mol L^{-1})	$-x$		$+x$		$+x$
Eqm / (mol L^{-1})	$0.0496 - x$		x		$0.0188 + x$

The equilibrium constant is given by

$$K = \frac{[SO_2]_{eqm}[Cl_2]_{eqm}}{[SO_2Cl_2]_{eqm}} = \frac{x\,(0.0188 + x)}{(0.0496 - x)} = 0.045$$

Solve this equation for x.

$$x^2 + 0.0188\,x = 0.045\,(0.0496 - x) = 0.0022 - 0.045\,x$$

$$x^2 + 0.0638\,x - 0.0022 = 0$$

$$x = \frac{-0.0638 \pm \sqrt{0.0638^2 + 4 \times 0.0022}}{2}$$

$$x = 0.025 \quad \text{or} \quad -0.089$$

The latter solution gives negative $SO_2(g)$ and $Cl_2(g)$ concentrations. We use the former solution.

$[SO_2Cl_2]_{eqm} = 0.0496 - x = 0.0496 - 0.025 = 0.025$ mol L^{-1}

$[SO_2]_{eqm} = x = 0.025$ mol L^{-1}

$[Cl_2]_{eqm} = 0.0188 + x = 0.0188 + 0.025 = 0.044$ mol L^{-1}

Fraction of $SO_2Cl_2(g)$ dissociated

= (initial amount of $SO_2Cl_2(g)$ − final amount of $SO_2Cl_2(g)$) / (initial amount of $SO_2Cl_2(g)$)

= (0.0496 − 0.025) / 0.0496 = 0.50

(c) The fractions of $SO_2Cl_2(g)$ dissociated in part (b) is less than that in part (b), in agreement with Le Chatelier's principle. There was net reaction in the direction to counter the increase in chlorine gas concentration.

Chapter 14
Acid-Base Equilibria in Aqueous Solution

IN-CHAPTER EXERCISES

Exercise 14.1—Brønsted-Lowry model of acids and bases

(a) $HCOOH(aq) + H_2O(l) \rightleftharpoons HCOO^-(aq) + H_3O^+(aq)$

     acid   base   conj. base   conj. acid

(b) $NH_3(aq) + H_2S(aq) \rightleftharpoons NH_4^+(aq) + HS^-(aq)$

      base   acid   conj. acid   conj. base

(c) $HSO_4^-(aq) + OH^-(aq) \rightleftharpoons SO_4^{2-}(aq) + H_2O(l)$

     acid   base   conj. base   conj. acid

Exercise 14.3—Brønsted-Lowry model of acids and bases

$HC_2O_4^-$(aq) ions acting as a Brønsted-Lowry acid:

$HC_2O_4^-(aq) + OH^-(aq) \rightleftharpoons C_2O_4^{2-}(aq) + H_2O(l)$

$HC_2O_4^-$(aq) ions acting as a Brønsted-Lowry base:

$HC_2O_4^-(aq) + H_3O^+(aq) \rightleftharpoons H_2C_2O_4(aq) + H_2O(l)$

Exercise 14.5—Hydronium and hydroxide ion concentrations

HCl is a strong acid that is essentially completely dissociated in water.
Therefore, $c(HCl) = 4.0 \times 10^{-3}$ mol L^{-1} means $[H_3O^+] = 4.0 \times 10^{-3}$ mol L^{-1} and $[HCl] = 0$ mol L^{-1}
Since,

$$K_w = [H_3O^+][OH^-] = 1.0 \times 10^{-14},$$

$$[OH^-] = \frac{1.0 \times 10^{-14}}{[H_3O^+]} = \frac{1.0 \times 10^{-14}}{4.0 \times 10^{-3}} = 2.5 \times 10^{-12} \text{ mol } L^{-1}$$

Exercise 14.7—pH and pOH of solutions

(a) Because NaOH is a strong base, we have $[OH^-] = 0.0012$ mol L^{-1}.

$pOH = -\log_{10}[OH^-] = -\log_{10}(1.2 \times 10^{-3}) = -(-2.9) = 2.9$

$pH = 14.0 - 2.9 = 11.1$

(b)
$$pOH = 14.0 - pH = 14.00 - 4.32 = 9.68$$

$$[H_3O^+] = 10^{-pH} = 10^{-4.32} = 4.79 \times 10^{-5} \text{ mol L}^{-1}$$

$$[OH^-] = 10^{-pOH} = 10^{-9.68} = 2.09 \times 10^{-10} \text{ mol L}^{-1}$$
(c)
$$pOH = 14.0 - pH = 14.00 - 10.46 = 3.54$$

$$[OH^-] = 10^{-pOH} = 10^{-3.54} = 2.88 \times 10^{-4} \text{ mol L}^{-1}$$
Each mol of dissolved $Sr(OH)_2$ gives rise to 2 mol $OH^-(aq)$ ions. Therefore,
$$c(Sr(OH)_2) = 1.44 \times 10^{-4} \text{ mol L}^{-1}$$

Exercise 14.9—Acid and base ionization constant

(a) pK_a for aquated benzoic acid, $C_6H_5COOH(aq)$, is
$$pK_a = -\log_{10} K_a = -\log_{10}(6.3 \times 10^{-5}) = -(-4.2) = 4.2$$
(b) $pK_a = 2.87$ for aquated chloroacetic acid, $ClCH_2COOH(aq)$.
Since $2.87 < 4.2$ and (like pH) a smaller pK_a corresponds to a stronger acid, aquated chloroacetic acid is a stronger acid than aquated benzoic acid.

Exercise 14.10—Acid and base ionization constants

Epinephrine hydrochloride has
$$K_a = 10^{-pK_a} = 10^{-9.53} = 2.95 \times 10^{-10}$$
It fits between benzoic and acetic acids in Table 14.5.

Exercise 14.12—Ionization constants of weak acids and their conjugate bases

The conjugate base of lactic acid—i.e., aquated lactate ions—has
$$K_b = \frac{K_w}{K_a} = \frac{10^{-14}}{1.4 \times 10^{-4}} = 7.1 \times 10^{-11}$$
It fits between dihydrogenphosphate, $H_2PO_4^-(aq)$, ions and fluoride, $F^-(aq)$, ions in Table 14.5.

Exercise 14.13—Acid-base character of solutions of salts

(a) An aqueous KBr solution has a pH of 7. KOH is a strong base—$K^+(aq)$ ions have no appreciable reaction with water. HBr is a strong acid—its conjugate base, $Br^-(aq)$ ion, is a *very* weak base with insignificant reaction with water.
(b) An aqueous NH_4NO_3 solution has a $pH < 7$—it is acidic. $NO_3^-(aq)$ ions undergo negligible reaction as a base with water—they are the conjugate base of a strong acid (HNO_3). $NH_4^+(aq)$ ions, on the other hand, are a weak acid.

(c) An aqueous $AlCl_3$ solution has a pH < 7—it is acidic. $Cl^-(aq)$ ions undergo negligible basic reaction with water—they are the conjugate base of a strong acid (HCl). Although $Al^{3+}(aq)$ ions have no protons, they are a weak acid because the Al^{3+} ions strongly attract the six water molecules bound to each, (because of the large +3 charge and the small size of the cation), withdrawing electrons from the water molecules, so that these more easily have a proton removed than do the water molecules not bound to the Al^{3+} cations

(d) An aqueous Na_2HPO_4 solution has pH > 7—it is basic. $Na^+(aq)$ ions have no appreciable reaction with water, whereas $HPO_4^{2-}(aq)$ ions are a weak base. $HPO_4^{2-}(aq)$ ions can also act as an acid. However, the K_a value of 3.6×10^{-13} is much smaller than its K_b value of 1.6×10^{-7}.

Exercise 14.15—Lewis acids and bases

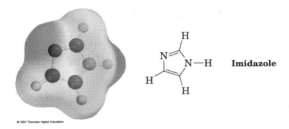

The electrostatic potential map of imidazole, on the left, shows negative charge as red and positive charge as blue.

(a) The H atom attached to an N atom has the greatest concentration of positive charge. As such, this is the most acidic hydrogen atom.

(b) The unprotonated ring nitrogen has the greatest concentration of negative charge. It is the most basic nitrogen atom in imidazole. This N atom is the site most likely to attract $H^+(aq)$ ions.

Exercise 14.16—Lewis acids and bases

CO is a Lewis base. It binds to metal ions by donating a lone pair of electrons to the formation of a bond with metal ions.

Exercise 14.17—Lewis acids and bases

(a) CH_3CH_2OH

The Lewis structure of ethanol shows lone pairs of electrons on the O atom that can be donated to bond formation with a Lewis acid. It is a Lewis base. There are no places in this structure that can accept a lone pair of electrons, and the H atoms bonded to alkyl carbons are not very acidic.

(b) $(CH_3)_2NH$

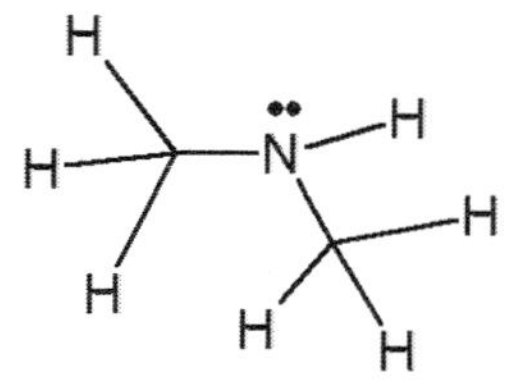

The Lewis structure of dimethyl amine shows a lone pair of electrons on the N atom that can be donated to bond formation. It is a Lewis base.

(c) Br^-

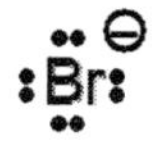

The bromide ion clearly has lone pairs that it can donate to bond formation. It is a Lewis base. However, because it is the conjugate base of a strong acid, we know that it is a very weak base in the Brønsted-Lowry sense.

(d) $(CH_3)_3B$

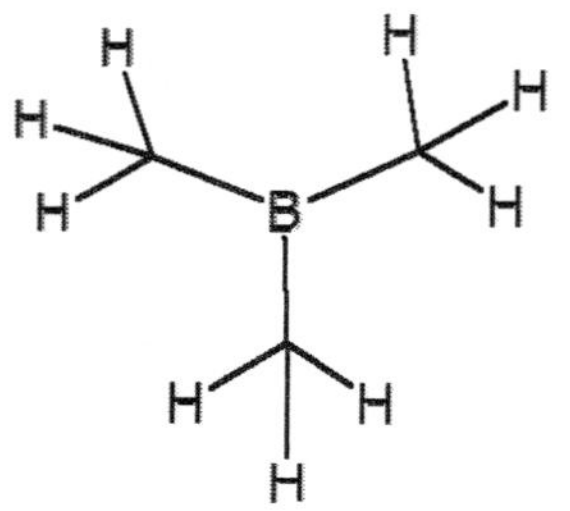

Trimethyl borane is a Lewis acid. The boron atom has only 6 electrons in its valence shell. It can accept a lone pair of electrons from a Lewis base. Trimethyl borane has no lone pairs available to be donated—it cannot act as a Lewis base.

(e) H_3CCl

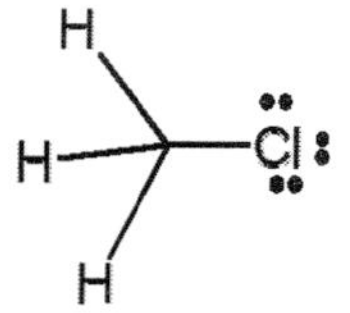

Methyl chloride has lone pairs on the Cl atom, so has the potential to react as a Lewis base.

(f) $(CH_3)_3P$

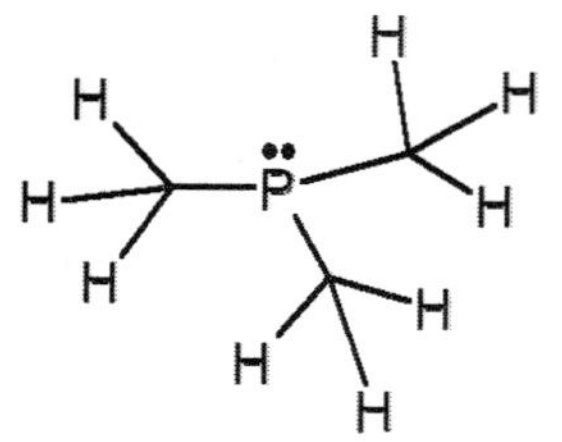

Trimethylamine has a lone pair of electrons on the P atom that can be donated to bond formation. It is a Lewis base.

Exercise 14.19—Calculating K_a from a measured pH

In an imaginary situation before any ionization occurred
[butanoic acid]$_{initial}$ = 0.055 mol / 1.0 L = 0.055 mol L^{-1}
[butanoate ions]$_{initial}$ = 0
At equilibrium, if we assume that x mol L^{-1} of butanoic acid has ionized:
[butanoic acid]$_{eqm}$ = 0.055 − x mol L^{-1}
[butanoate ions]$_{eqm}$ = x mol L^{-1}
[H$_3$O$^+$]$_{eqm}$ = x mol L^{-1}

$$CH_3CH_2CH_2COOH(aq) + H_2O(l) \rightleftharpoons CH_3CH_3CH_2COO^-(aq) + H_3O^+(aq)$$

Since the pH of the equilibrium solution is 2.72,
[H$_3$O$^+$]$_{eqm}$ = $10^{-2.72}$ mol L^{-1} = 1.91×10^{-3} mol L^{-1}
Therefore, x = 1.91×10^{-3} mol L^{-1}
and

$$K_a = \frac{[CH_3CH_3CH_2COO^-]_{eqm}[H_3O^+]_{eqm}}{[CH_3CH_3CH_2COOH]_{eqm}}$$

$$= \frac{(1.91\times10^{-3})(1.91\times10^{-3})}{(0.055 - 1.91\times10^{-3})} = 6.87\times10^{-5}$$

Exercise 14.22—Equilibrium concentrations, pH, and % ionized from K_a

K_a for acetic acid is 1.8×10^{-5}

Concentrations (mol L^{-1})	CH$_3$COOH(aq)	+	H$_2$O(l)	$\rightleftharpoons$	CH$_3$COO$^-$(aq)	+	H$_3$O$^+$(aq)
Initial	0.10				0		0
Change	−x				+x		+x
Equilibrium	0.10 − x				x		x

At equilibrium the condition $Q = K$ must be satisfied:

$$Q = \frac{[CH_3COO^-][H_3O^+]}{[CH_3COOH]} = \frac{(x)(x)}{0.10-x} = K_a = 1.8\times10^{-5}$$

From this we obtain the quadratic equation
$$x^2 + (1.8\times10^{-5})\,x - (1.8\times10^{-6}) = 0$$
Solving the quadratic equation, gives $x = 0.0013$
We now have the equilibrium concentrations
[C$_6$H$_5$COO$^-$] = [H$_3$O$^+$] = x = 0.0013 mol L^{-1}
[C$_6$H$_5$COOH] = 0.10 − x = 0.10 − 0.0013 = 0.0987 mol L^{-1}
and   pH = $-\log_{10}$[H$_3$O$^+$] = $-\log_{10}$(0.0013) = 2.89

and   % of acetic acid ionized = 100% $\times \dfrac{x}{0.10}$ = 100% $\times \dfrac{0.0013}{0.10}$ = 1.3%

Exercise 14.24—Magnitude of ionization constant and % ionization

For benzoic acid, $K_a = 6.3 \times 10^{-5}$. The quadratic equation derived from assuming that x mol L^{-1} of the weak acid ionizes before equilibrium is achieved is

$$x^2 + (6.3 \times 10^{-5})\, x - (6.3 \times 10^{-6}) = 0$$

Solving the quadratic equation, gives $x = 0.0025$
The equilibrium H_3O^+ concentration is

$$[H_3O^+] = x = 0.0025 \text{ mol } L^{-1}$$

and $pH = -\log_{10}[H_3O^+] = -\log_{10}(0.0025) = 2.61$

and   % of benzoic acid ionized $= 100\% \times \dfrac{x}{0.10} = 100\% \times \dfrac{0.0025}{0.10} = 2.5\%$

Note that the pH comes out as 2.60 if computed as shown above. The value 2.61 results if you carry an extra digit from the previous calculation—i.e., use $x = 0.00248$ mol L^{-1}.
For hypochlorous acid, $K_a = 3.5 \times 10^{-8}$. The quadratic equation derived is

$$x^2 + (3.5 \times 10^{-8})\, x - (3.5 \times 10^{-9}) = 0$$

Solving the quadratic equation, gives $x = 5.9 \times 10^{-5}$
The equilibrium H_3O^+ concentration is

$$[H_3O^+] = x = 5.9 \times 10^{-5} \text{ mol } L^{-1}$$

and $pH = -\log_{10}[H_3O^+] = -\log_{10}(5.9 \times 10^{-5}) = 4.23$

and   % of hypochlorous acid ionized $= 100\% \times \dfrac{x}{0.10} = 100\% \times \dfrac{5.9 \times 10^{-5}}{0.10} = 5.9 \times 10^{-2}\%$

For ammonium ions, $K_a = 5.6 \times 10^{-10}$. The quadratic equation derived is

$$x^2 + (5.6 \times 10^{-10})\, x - (5.6 \times 10^{-11}) = 0$$

Solving the quadratic equation, gives $x = 7.5 \times 10^{-6}$
The equilibrium $H_3O^+(aq)$ ion concentration is

$$[H_3O^+] = x = 7.5 \times 10^{-6} \text{ mol } L^{-1}$$

and $pH = -\log_{10}[H_3O^+] = -\log_{10}(7.5 \times 10^{-6}) = 5.13$

and   % of ammonium reacted $= 100\% \times \dfrac{x}{0.10} = 100\% \times \dfrac{7.5 \times 10^{-6}}{0.10} = 7.5 \times 10^{-3}\%$

Exercise 14.25—Dependence of % ionization on solution concentration

Here, only the initial acid concentration is varied. The acid is propanoic acid with $K_a = 1.3 \times 10^{-5}$ at 25 °C.
For initial acid concentration, 1.00 mol L^{-1}, the quadratic equation derived from assuming x mol L^{-1} ionization takes the form

$$x^2 + K_a\, x - c_{initial} K_a = 0$$

$$x^2 + 1.3 \times 10^{-5}\, x - 1.3 \times 10^{-5} = 0$$

Solving the quadratic equation, gives $x = 3.6 \times 10^{-3}$
The equilibrium $H_3O^+(aq)$ ion concentration is

$$[H_3O^+] = x = 3.6 \times 10^{-3} \text{ mol } L^{-1}$$

and $pH = -\log_{10}[H_3O^+] = -\log_{10}(3.6 \times 10^{-3}) = 2.44$

and $\quad$ % of propanoic acid ionized $= 100\% \times \dfrac{x}{1.00} = 100\% \times \dfrac{3.6\times10^{-3}}{1.00} = 0.36\%$

For initial acid concentration, 1.00×10^{-2} mol L^{-1}, the quadratic equation is of the form
$$x^2 + 1.3 \times 10^{-5}\, x - 1.3 \times 10^{-7} = 0$$
Solving the quadratic equation, gives $x = 3.5 \times 10^{-4}$
The equilibrium $H_3O^+(aq)$ ion concentration is
$$[H_3O^+] = x = 3.5 \times 10^{-4} \text{ mol L}^{-1}$$
and $\quad$ pH $= -\log_{10}[H_3O^+] = -\log_{10}(3.5 \times 10^{-4}) = 3.45$

and $\quad$ % of propanoic acid ionized $= 100\% \times \dfrac{x}{0.0100} = 100\% \times \dfrac{3.5\times10^{-4}}{0.0100} = 3.5\%$

For initial acid concentration, 1.00×10^{-4} mol L^{-1}, the quadratic equation derived is
$$x^2 + 1.3 \times 10^{-5}\, x - 1.3 \times 10^{-9} = 0$$
Solving the quadratic equation, gives $x = 3.0 \times 10^{-6}$
The equilibrium $H_3O^+(aq)$ ion concentration is
$$[H_3O^+] = x = 3.0 \times 10^{-5} \text{ mol L}^{-1}$$
and $\quad$ pH $= -\log_{10}[H_3O^+] = -\log_{10}(3.0 \times 10^{-5}) = 4.52$

and $\quad$ % of propanoic acid ionized $= 100\% \times \dfrac{x}{0.000100} = 100\% \times \dfrac{3.0\times10^{-5}}{0.000100} = 30\%$

Exercise 14.27—The common ion effect

(a) The quadratic equation for the extent of reaction, x, in the case of a 0.30 mol L^{-1} aqueous formic acid solution ($K_a = 1.8 \times 10^{-4}$) is
$$x^2 + 1.8 \times 10^{-4}\, x - 5.4 \times 10^{-5} = 0$$
Solving the quadratic equation, gives $x = 7.3 \times 10^{-3}$
The equilibrium $H_3O^+(aq)$ ion concentration is
$$[H_3O^+] = x = 7.3 \times 10^{-3} \text{ mol L}^{-1}$$
and $\quad$ pH $= -\log_{10}[H_3O^+] = -\log_{10}(7.3 \times 10^{-3}) = 2.14$

and $\quad$ % of formic acid ionized $= 100\% \times \dfrac{x}{0.30} = 100\% \times \dfrac{7.3\times10^{-3}}{0.30} = 2.4\%$

In the presence of a 0.10 mol L^{-1} initial concentration of formate ions, the reaction quotient in terms of extent of reaction takes the form,
$$Q = \frac{[HCOO^-][H_3O^+]}{[HCOOH]} = \frac{(0.10+x)(x)}{(0.30-x)} = K_a = 1.8 \times 10^{-4}$$
The quadratic equation that results is
$$x^2 + (1.8 \times 10^{-4} + 0.10)\, x - 5.4 \times 10^{-5} = 0$$

$$x^2 + 0.10018\, x - 5.4 \times 10^{-5} = 0$$
Solving the quadratic equation, gives $x = 5.4 \times 10^{-4}$

The equilibrium $H_3O^+(aq)$ ion concentration is

$$[H_3O^+] = x = 5.4 \times 10^{-4} \text{ mol L}^{-1}$$

and $\quad$ $pH = -\log_{10}[H_3O^+] = -\log_{10}(5.4 \times 10^{-4}) = 3.27$

and $\quad$ % of formic acid ionized $= 100\% \times \dfrac{x}{0.30} = 100\% \times \dfrac{5.4 \times 10^{-4}}{0.30} = 0.18\%$

Adding formate ions in an amount comparable to the amount of formic acid significantly reduces the degree of ionization of the acid.

(b) If the initial solution contained 0.10 mol L^{-1} $H_3O^+(aq)$ ions in addition to 0.30 mol L^{-1} formic acid, then the reaction quotient at equilibrium becomes

$$Q = \frac{[HCOO^-][H_3O^+]}{[HCOOH]} = \frac{(x)(0.10+x)}{(0.30-x)} = K_a = 1.8 \times 10^{-4}$$

This equation gives the same quadratic equation as in part (a). Therefore, $x = 5.4 \times 10^{-4}$ as above. The difference here is the equilibrium $H_3O^+(aq)$ ion concentration.

$$[H_3O^+] = 0.10 + x = 0.10 + 5.4 \times 10^{-4} = 0.10 \text{ mol L}^{-1}$$

and $\quad$ $pH = -\log_{10}[H_3O^+] = -\log_{10}(0.10) = 1.00$

However, the

$$\text{% of formic acid ionized} = 100\% \times \frac{x}{0.30} = 100\% \times \frac{5.4 \times 10^{-4}}{0.30} = 0.18\%$$

is the same as in part (a), and shows the effect of a common ion (in this case, $H^+(aq)$ ions).

Exercise 14.28—pH of a solution of a weak base

Concentrations (mol L^{-1})	OCl$^-$(aq)	+	H$_2$O(l)	$\rightleftharpoons$	HOCl(aq)	+	OH$^-$(aq)
Initial	0.015				0		0
Change	$-x$				$+x$		$+x$
Equilibrium	0.015 $- x$				x		x

Substituting equilibrium concentrations into the reaction quotient:

$$Q = \frac{[HOCl][OH^-]}{[OCl^-]} = \frac{(x)(x)}{(0.015-x)} = K_b(OCl^-) = 2.9 \times 10^{-7}$$

From this we obtain a quadratic equation: $x^2 + (2.9 \times 10^{-7})x - 4.4 \times 10^{-9} = 0$ with solution $x = 6.6 \times 10^{-6} \text{ mol L}^{-1}$

Now we can write the equilibrium concentrations:

$$[HOCl] = [OH^-] = x = 6.6 \times 10^{-6} \text{ mol L}^{-1}$$
$$[OCl^-] = 0.015 - x = 0.015 - 6.6 \times 10^{-6} = 0.015 \text{ mol L}^{-1}$$

$$[H_3O^+] = \frac{K_w}{[OH^-]} = \frac{1.0 \times 10^{-14}}{6.6 \times 10^{-6}} = 1.5 \times 10^{-9} \text{ mol L}^{-1}$$

Finally, $\quad pH = -\log_{10}(1.5 \times 10^{-9}) = 8.82$

Exercise 14.30—pH of a solution of a polyprotic acid

Oxalic acid is a diprotic acid with $K_{a1} = 5.9 \times 10^{-2}$ and $K_{a2} = 6.4 \times 10^{-5}$. Because the second ionization constant is 1000 times smaller than the first, we can neglect the second ionization for the purpose of computing pH of solution.

ICE table for stage one ionization of oxalic acid.

Concentrations (mol L^{-1})	(COOH)$_2$(aq)	+	H$_2$O(l)	$\rightleftharpoons$	HOOCCOO$^-$(aq)	+	H$_3$O$^+$(aq)
Initial	0.10				0		0
Change	$-x$				$+x$		$+x$
Equilibrium	$0.10 - x$				x		x

At equilibrium the condition $Q = K$ must be satisfied:

$$Q = \frac{[\text{oxalate}^-][\text{H}_3\text{O}^+]}{[\text{oxalic acid}]} = \frac{(x)(x)}{0.10 - x} = K_{a1} = 5.9 \times 10^{-2}$$

From this we obtain the quadratic equation

$$x^2 + (5.9 \times 10^{-2})\,x - (5.9 \times 10^{-3}) = 0$$

Solving the quadratic equation, gives $x = 0.053$
We now have the equilibrium concentrations

$$[\text{oxalate}^-] = [\text{H}_3\text{O}^+] = x = 0.053 \text{ mol L}^{-1}$$
$$[\text{oxalic acid}] = 0.10 - x = 0.10 - 0.053 = 0.047 \text{ mol L}^{-1}$$

and $\quad$ pH $= -\log_{10}[\text{H}_3\text{O}^+] = -\log_{10}(0.053) = 1.28$

Exercise 14.31—Acidity of solutions

(a) To get the pH in cases A and C, requires equilibrium calculations. B requires no equilibrium calculation because HCl is a strong acid.
A. 0.005 mol L^{-1} formic acid ($K_a = 1.8 \times 10^{-4}$) has pH $= 3.06$. This is derived from the quadratic equation

$$x^2 + (1.8 \times 10^{-4})\,x - (9.0 \times 10^{-7}) = 0$$
$$\rightarrow x = 0.00086$$
$$[\text{H}_3\text{O}^+] = x = 0.00086 \text{ mol L}^{-1}$$
$$\text{pH} = -\log_{10}[\text{H}_3\text{O}^+] = -\log_{10}(0.00086) = 3.06$$

B. 0.001 mol L^{-1} hydrochloric acid (strong acid) has $[\text{H}^+] = 0.001$ mol L^{-1} and pH $= 3$
C. 0.003 mol L^{-1} carbonic acid solution ($K_a = 4.2 \times 10^{-7}$) has pH $= 4.43$
Consideration of only the first step of ionization leads to the quadratic equation:

$$x^2 + (4.7 \times 10^{-7})\,x - (1.41 \times 10^{-9}) = 0$$
$$\rightarrow x = 3.7 \times 10^{-5}$$
$$[\text{H}_3\text{O}^+] = x = 3.7 \times 10^{-5} \text{ mol L}^{-1}$$
$$\text{pH} = -\log_{10}[\text{H}_3\text{O}^+] = -\log_{10}(3.7 \times 10^{-5}) = 4.43$$

According to increasing acidity in terms of pH (i.e., decreasing pH) we have
C < A < B
The pH calculation for case C could have been skipped because there are fewer moles of carbonic acid and it is a weaker acid—also, the second dissociation occurs to a negligible extent unless base is added.

(b) To rank the three solutions according to amount of NaOH solution that can be consumed, it is sufficient to compare the amounts of ionizable protons available for reaction with OH⁻(aq) ions—regardless of the amounts of H⁺(aq) ions that are in solution before any base is added. Therefore, according to amount of OH⁻(aq) ions that will react with 1 L of each solution, we have
B (0.001 mol) < A (0.005 mol) < C (0.006 mol)

Exercise 14.34—pH-dependent weak acid-conjugate base speciation

(a) At pH 4

$$\frac{[HOCl]}{[OCl^-]} = \frac{[H_3O^+]}{K_a} = \frac{10^{-4}}{10^{-7.46}} = \frac{10^{-4}}{3.47 \times 10^{-8}} = \frac{2.88 \times 10^3}{1}$$

(b) At pH 7

$$\frac{[HOCl]}{[OCl^-]} = \frac{[H_3O^+]}{K_a} = \frac{10^{-7}}{10^{-7.46}} = 10^{0.46} = \frac{2.88}{1}$$

(c) At pH 10

$$\frac{[HOCl]}{[OCl^-]} = \frac{[H_3O^+]}{K_a} = \frac{10^{-10}}{10^{-7.46}} = 10^{-2.54} = \frac{2.88 \times 10^{-3}}{1} = \frac{2.88}{1000}$$

Exercise 14.35—pH-dependent speciation of polyprotic acid species

(a) At pH 6.0

(i) $$\frac{[H_3PO_4]}{[H_2PO_4^-]} = \frac{[H_3O^+]}{K_{a1}} = \frac{[H_3O^+]}{7.5 \times 10^{-3}} = \frac{10^{-6}}{7.5 \times 10^{-3}} = \frac{1.33 \times 10^{-4}}{1}$$

(ii) $$\frac{[H_2PO_4^-]}{[HPO_4^{2-}]} = \frac{[H_3O^+]}{K_{a2}} = \frac{[H_3O^+]}{6.2 \times 10^{-8}} = \frac{10^{-6}}{6.2 \times 10^{-8}} = \frac{16}{1}$$

(iii) $$\frac{[HPO_4^{2-}]}{[PO_4^{3-}]} = \frac{[H_3O^+]}{K_{a3}} = \frac{[H_3O^+]}{3.6 \times 10^{-13}} = \frac{10^{-6}}{3.6 \times 10^{-13}} = \frac{2.78 \times 10^6}{1}$$

At pH = 6.0, the dominant species is H₂PO₄⁻(aq) ions

(b) At pH 9.0

(i) $$\frac{[H_3PO_4]}{[H_2PO_4^-]} = \frac{[H_3O^+]}{K_{a1}} = \frac{[H_3O^+]}{7.5 \times 10^{-3}} = \frac{10^{-9}}{7.5 \times 10^{-3}} = \frac{1.33 \times 10^{-7}}{1}$$

(ii) $$\frac{[H_2PO_4^-]}{[HPO_4^{2-}]} = \frac{[H_3O^+]}{K_{a2}} = \frac{[H_3O^+]}{6.2 \times 10^{-8}} = \frac{10^{-9}}{6.2 \times 10^{-8}} = \frac{1.6 \times 10^{-2}}{1}$$

(iii) $$\frac{[HPO_4^{2-}]}{[PO_4^{3-}]} = \frac{[H_3O^+]}{K_{a3}} = \frac{[H_3O^+]}{3.6 \times 10^{-13}} = \frac{10^{-9}}{3.6 \times 10^{-13}} = \frac{2.78 \times 10^3}{1}$$

At pH = 9.0, the dominant species is HPO₄²⁻(aq) ions

Exercise 14.36—pH-dependent distribution among amino acid species in solution

(a) phenylalanine

$$H_2Phe^+(aq) + H_2O(l) \rightleftharpoons HPhe(aq) + H_3O^+(aq) \qquad pK_{a1} = 1.83$$

$$HPhe(aq) + H_2O(l) \rightleftharpoons Phe^-(aq) + H_3O^+(aq) \qquad pK_{a2} = 9.13$$

$$\frac{[H_2Phe^+]}{[HPhe]} = \frac{[H_3O^+]}{K_{a1}} = \frac{10^{-7.40}}{10^{-1.83}} = 10^{-5.57} = \frac{2.7 \times 10^{-6}}{1}$$

$$\frac{[HPhe]}{[Phe^-]} = \frac{[H_3O^+]}{K_{a2}} = \frac{10^{-7.40}}{10^{-9.13}} = 10^{1.73} = \frac{54}{1}$$

The dominant species at pH = 7.40 is HPhe(aq) (the zwitterion).

(b) glutamic acid

$$H_3Glu^+(aq) + H_2O(l) \rightleftharpoons H_2Glu(aq) + H_3O^+(aq) \qquad pK_{a1} = 2.19$$

$$H_2Glu(aq) + H_2O(l) \rightleftharpoons HGlu^-(aq) + H_3O^+(aq) \qquad pK_{a2} = 4.25$$

$$HGlu^-(aq) + H_2O(l) \rightleftharpoons Glu^{2-}(aq) + H_3O^+(aq) \qquad pK_{a3} = 9.67$$

$$\frac{[H_3Glu^+]}{[H_2Glu]} = \frac{[H_3O^+]}{K_{a1}} = \frac{10^{-7.40}}{10^{-2.19}} = 10^{-5.21} = \frac{6.2 \times 10^{-6}}{1}$$

$$\frac{[H_2Glu]}{[HGlu^-]} = \frac{[H_3O^+]}{K_{a2}} = \frac{10^{-7.40}}{10^{-4.25}} = 10^{-3.15} = \frac{7.1 \times 10^{-4}}{1}$$

$$\frac{[HGlu^-]}{[Glu^{2-}]} = \frac{[H_3O^+]}{K_{a3}} = \frac{10^{-7.40}}{10^{-9.67}} = 10^{2.27} = \frac{190}{1}$$

The dominant species at pH = 7.40 is HGlu⁻(aq) ions

(c) lysine

$$H_3Lys^{2+}(aq) + H_2O(l) \rightleftharpoons H_2Lys^+(aq) + H_3O^+(aq) \qquad pK_{a1} = 2.18$$

$$H_2Lys^+(aq) + H_2O(l) \rightleftharpoons HLys(aq) + H_3O^+(aq) \qquad pK_{a2} = 8.95$$

$$HLys(aq) + H_2O(l) \rightleftharpoons Lys^-(aq) + H_3O^+(aq) \qquad pK_{a3} = 10.53$$

$$\frac{[H_3Lys^{2+}]}{[H_2Lys^+]} = \frac{[H_3O^+]}{K_{a1}} = \frac{10^{-7.40}}{10^{-2.18}} = 10^{-5.22} = \frac{6.0 \times 10^{-6}}{1}$$

$$\frac{[H_2Lys^+]}{[HLys]} = \frac{[H_3O^+]}{K_{a2}} = \frac{10^{-7.40}}{10^{-8.95}} = 10^{1.55} = \frac{35}{1}$$

$$\frac{[\text{HLys}]}{[\text{Lys}^-]} = \frac{[\text{H}_3\text{O}^+]}{K_{a3}} = \frac{10^{-7.40}}{10^{-10.53}} = 10^{3.13} = \frac{1350}{1}$$

The dominant species at pH = 7.40 is $\text{H}_2\text{Lys}^+(\text{aq})$ ions

Exercise 14.40—pH of buffer solutions

Amount of benzoic acid, $n(\text{HA}) = 2.00\ \text{g} / 122.12\ \text{g mol}^{-1} = 0.0164\ \text{mol}$
Amount of sodium benzoate, $n(\text{A}^-) = 2.00\ \text{g} / 144.11\ \text{g mol}^{-1} = 0.0139\ \text{mol}$
dissolved to make 1.00 L of solution
$[\text{C}_6\text{H}_5\text{COOH}] = 0.0164\ \text{mol L}^{-1}$
$[\text{C}_6\text{H}_5\text{COO}^-] = 0.0139\ \text{mol L}^{-1}$
Recognizing that we have a buffer solution, we can apply the buffer equation 14.1:

$$[\text{H}_3\text{O}^+] = \frac{[\text{C}_6\text{H}_5\text{COOH}]}{[\text{C}_6\text{H}_5\text{COO}^-]} \times K_a = \frac{0.0164\,\text{mol L}^{-1}}{0.0139\,\text{mol L}^{-1}} \times 6.3 \times 10^{-5} = 7.4 \times 10^{-5}$$

pH = 4.13
Check that you necessarily should obtain the same answer by using the alternative buffer equation 14.4.

Exercise 14.42—Designing buffer solutions

To make a buffer solution at a pH near 9, we use a weak acid (and its conjugate base) with pK_a near 9.
(a) HCl and NaCl would not make a buffer solution since HCl is a strong acid.
(b) NH_3 and NH_4Cl would make a buffer solution near pH = 9
$K_a(\text{HA}) = K_a(\text{NH}_4^+) = 5.6 \times 10^{-10}$ and $pK_a = 9.25$
(c) CH_3COOH and NaCH_3COO would not make a buffer solution near pH = 9
$K_a(\text{HA}) = K_a(\text{CH}_3\text{COOH}) = 1.8 \times 10^{-5}$ and $pK_a = 4.74$
Acetic acid is essentially fully ionized this far above pH = pK_a, so has no capacity to buffer against the addition of bases.

Exercise 14.43—Designing buffer solutions

Solve for the required ratio of $\text{H}_2\text{PO}_4^-(\text{aq})$ and $\text{HPO}_4^{2-}(\text{aq})$ ions using equation 14.1, and the target $[\text{H}^+] = 10^{-7.5} = 3.16 \times 10^{-8}\ \text{mol L}^{-1}$. From Table 14.11 we see that $K_a(\text{H}_2\text{PO}_4^-) = 6.2 \times 10^{-8}$.

$$[\text{H}_3\text{O}^+] = 3.16 \times 10^{-8} = \frac{[\text{H}_2\text{PO}_4^-]}{[\text{HPO}_4^{2-}]} \times K_a(\text{H}_2\text{PO}_4^-) = \frac{[\text{H}_2\text{PO}_4^-]}{[\text{HPO}_4^{2-}]} \times 6.2 \times 10^{-8}$$

$$\therefore \frac{[\text{H}_2\text{PO}_4^-]}{[\text{HPO}_4^{2-}]} = \frac{3.16 \times 10^{-8}}{6.2 \times 10^{-8}} = \frac{0.51}{1}$$

So if we make a (not too dilute) solution using amounts (mol) of $\text{NaH}_2\text{PO}_4(\text{s})$ and $\text{Na}_2\text{HPO}_4(\text{s})$ in the ratio 0.51:1, it will function as a buffer solution with pH = 7.5.

Exercise 14.45—pH change of a buffer solution

(a) In the 0.500 L of buffer solution
$n(\text{HCOOH}) = 0.25$ mol and $n(\text{HCOO}^-) = 0.35$ mol
We can directly calculate the pH using the buffer equation 14.4.

$$[\text{H}_3\text{O}^+] = \frac{n(\text{HCOOH})}{n(\text{HCOO}^-)} \times K_a = \frac{0.25\,\text{mol}}{0.35\,\text{mol}} \times 1.8 \times 10^{-4} = 1.29 \times 10^{-4}$$

pH = 3.89

(b) By adding 10.0 mL of 1.0 mol L^{-1} HCl solution, we add
0.010 L × 1.0 mol L^{-1} = 0.010 mol of H_3O^+(aq) ions
These are essentially "mopped up" by HCOO⁻(aq) ions
H_3O^+(aq) + HCOO⁻(aq) → HCOOH(aq) + H_2O(l)
Then $n(\text{HCOOH}) = (0.25 + 0.01)$ mol = 0.26 mol
$n(\text{HCOO}^-) = (0.35 - 0.01)$ mol = 0.34 mol

$$[\text{H}_3\text{O}^+] = \frac{n(\text{HCOOH})}{n(\text{HCOO}^-)} \times K_a = \frac{0.26\,\text{mol}}{0.34\,\text{mol}} \times 1.8 \times 10^{-4} = 1.38 \times 10^{-4}$$

pH = 3.86
ΔpH = 0.03 only. Clearly the buffering action is quite effective. By how much would the pH change if the same amount of HCl solution were added to 0.50 L of water?

Exercise 14.46—Buffer capacity

(a) 100 mL of solution with $c(\text{HCO}_3^-) = 0.50$ mol L^{-1} and $c(\text{CO}_3^{2-}) = 0.25$ mol L^{-1}

$$[\text{H}_3\text{O}^+] = \frac{[\text{HCO}_3^-]}{[\text{CO}_3^{2-}]} \times K_a = \frac{0.50\,\text{mol L}^{-1}}{0.25\,\text{mol L}^{-1}} \times 4.8 \times 10^{-11} = 9.6 \times 10^{-11}$$

pH = 10.02
Upon adding (0.40 g) / (40.00 g mol⁻¹) = 0.010 mol of OH⁻, the amounts of HCO_3^-(aq) and CO_3^{2-}(aq) ions change to
Amount of HCO_3^-(aq) ions = 0.50 mol L^{-1} × 0.100 L − 0.010 mol = 0.040 mol
Amount of CO_3^{2-}(aq) ions = 0.25 mol L^{-1} × 0.100 L + 0.010 mol = 0.035 mol
Now to find the pH of the new buffer solution we can use:

$$[\text{H}_3\text{O}^+] = \frac{n(\text{HCO}_3^-)}{n(\text{CO}_3^{2-})} \times K_a = \frac{0.040\,\text{mol}}{0.035\,\text{mol}} \times 4.8 \times 10^{-11} = 5.49 \times 10^{-11}$$

The new pH is 10.26
ΔpH = 10.26 − 10.02 = 0.24

(b) 100 mL of solution with $c(\text{HCO}_3^-) = 0.080$ mol L^{-1} and $c(\text{CO}_3^{2-}) = 0.40$ mol L^{-1}

$$[\text{H}_3\text{O}^+] = \frac{[\text{HCO}_3^-]}{[\text{CO}_3^{2-}]} \times K_a = \frac{0.080\,\text{mol L}^{-1}}{0.40\,\text{mol L}^{-1}} \times 4.8 \times 10^{-11} = 9.6 \times 10^{-12}$$

pH of buffer solution = 11.02
We add (0.40 g) / (40.00 g mol⁻¹) = 0.010 mol of OH⁻ ions. These react with the HCO_3^-(aq) ions, but there is not enough HCO_3^-(aq) ions to react with all of the added OH⁻(aq) ions. Instead, the
0.080 mol L^{-1} × 0.100 L = 0.008 mol HCO_3^-(aq) ions react, leaving
(0.010 − 0.008) mol = 0.002 mol remaining OH⁻(aq) ions

The $H_3O^+(aq)$ ions concentration is $(1 \times 10^{-14}) / 0.002 = 5 \times 10^{-12}$ mol L^{-1}
and pH = 11.30

ΔpH $= 11.30 - 11.02 = 0.28$

Although the buffer failed, the pH did not change too much because the buffer solution already has a significant $OH^-(aq)$ ion concentration—the HCO_3^-/CO_3^{2-} buffer system has a quite high pH (approaching 14)—and there was enough $HCO_3^-(aq)$ ions to consume most of the added $OH^-(aq)$ ions. Another way of considering this is to recognize that the absolute change of $[OH^-]$ was considerable, but at the already high $[OH^-]$, the fractional change was relatively small.

Exercise 14.47—Acid-base titration

$$CH_3COOH(aq) + OH^-(aq) \rightleftharpoons CH_3COO^-(aq) + H_2O(l)$$

Amount of $OH^-(aq)$ ions consumed in titration $= 0.02833$ L $\times 0.953$ mol $L^{-1} = 0.0270$ mol
$\quad$ = total amount of $CH_3COOH(aq)$ in the vinegar solution

Therefore, the concentration of the vinegar sample, $c(CH_3COOH) = (0.0270$ mol$) / (0.0250$ L$) = 1.08$ mol L^{-1}

and the mass of acetic acid in vinegar sample $= 0.0270$ mol $\times 60.05$ g mol$^{-1} = 1.62$ g

Exercise 14.49—Standardizing solution concentration by titration

$$H_3O^+(aq) + OH^-(aq) \rightleftharpoons 2 H_2O(l)$$

Amount of $H_3O^+(aq)$ ions consumed in titration $= 0.02967$ L $\times 0.100$ mol $L^{-1} = 0.00297$ mol
$\quad$ = amount of $OH^-(aq)$ ions

Therefore, the concentration of the NaOH solution, $c(NaOH) = (0.00297$ mol$) / 0.0250$ L $= 0.119$ mol L^{-1}

Exercise 14.51—Titration of a solution of a weak acid with a solution of a strong base

(a) Adding 35.0 mL of 0.100 mol L^{-1} NaOH solution to 100.0 mL of 0.100 mol L^{-1} acetic acid solution

$\quad$ Amount of $OH^-(aq)$ ions added $= 0.035$ L $\times 0.100$ mol $L^{-1} = 0.0035$ mol
$\quad$ Total amount of $CH_3COOH(aq) = 0.100$ L $\times 0.100$ mol $L^{-1} = 0.0100$ mol

Since the amount of added base does not exceed the amount of weak acid, we end up with a buffer solution. The acetic acid will consume essentially all of the added $OH^-(aq)$ ions.

Upon adding 0.0035 mol OH^-, the amounts of $CH_3COOH(aq)$ and $CH_3COO^-(aq)$ ions change to

$\quad$ Amount of $CH_3COOH(aq) = n(CH_3COOH) = 0.0100 - 0.0035$ mol $= 0.0065$ mol
$\quad$ Amount of $CH_3COO^-(aq)$ ions $= n(CH_3COO^-) = 0.0035$ mol

So we have a buffer solution in which

$$[H_3O^+] = \frac{n(CH_3COOH)}{n(CH_3COO^-)} \times K_a = \frac{0.0065 \text{ mol}}{0.0035 \text{ mol}} \times 1.8 \times 10^{-5} = 3.34 \times 10^{-5}$$

and pH $= 4.48$

Alternatively we could have arrived at the same answer (except for rounding-off errors) by:

$$pH = pK_a + \log_{10} \frac{n(CH_3COO^-)}{n(CH_3COOH)} = -\log_{10}(1.8 \times 10^{-5}) + \log_{10} \frac{0.0035}{0.0065}$$

$$= 4.74 - 0.27 = 4.47$$

(b) Adding 100.0 mL of 0.100 mol L^{-1} NaOH to 100.0 mL of 0.100 mol L^{-1} acetic acid solution
Amount of $OH^-(aq)$ ions $= 0.100$ L $\times 0.100$ mol $L^{-1} = 0.0100$ mol
Total amount of $CH_3COOH(aq) = 0.0100$ mol
Since the amount of added acid equals the amount of weak acid, using the approximation of part (a) yields an apparently nonsensical result. We can recognize that the final solution corresponds with the exact equivalence point of titration, where the solution is identical to a sodium acetate solution. The pH of this solution depends upon reaction of $CH_3COO^-(aq)$ ions as a weak base

$$CH_3COO^-(aq) + H_2O(l) \rightleftharpoons CH_3COOH(aq) + OH^-(aq)$$

Since we go back to working with concentrations, we must take account of the change in total volume of the combined solutions. Total volume $= 100.0$ mL $+ 100.0$ mL $= 200.0$ mL.

Concentrations (mol L^{-1})	$CH_3COO^-(aq)$	+	$H_2O(l)$	$\rightleftharpoons$	$CH_3COOH(aq)$	+	$OH^-(aq)$
Initial	0.0100 mol / 0.2000 L = 0.0500				0		0
Change	$-x$				$+x$		$+x$
Equilibrium	$0.0500 - x$				x		x

At equilibrium the condition $Q = K$ must be satisfied:

$$Q = \frac{[CH_3COOH][OH^-]}{[CH_3COO^-]} = K_b = \frac{K_w}{K_a} = \frac{10^{-14}}{1.8 \times 10^{-5}} = 5.6 \times 10^{-10} = \frac{x^2}{0.0500 - x}$$

This leads to the quadratic equation: $x^2 + (5.6 \times 10^{-10})x - 2.8 \times 10^{-11} = 0$
for which the solution is $x = 5.3 \times 10^{-6}$
We get the small residual $OH^-(aq)$ concentration:

$$[OH^-] = x = 5.3 \times 10^{-6} \text{ mol } L^{-1}$$

From which we get

$$[H_3O^+] = 10^{-14} / [OH^-] = 10^{-14} / 5.3 \times 10^{-6} = 1.9 \times 10^{-9} \text{ mol } L^{-1}$$

and $\quad pH = -\log_{10}[H_3O^+] = -\log_{10}(1.9 \times 10^{-9}) = 8.72$

Exercise 14.52—Titration of a diprotic weak acid with a strong base

100.0 mL of 0.010 mol L^{-1} solution of carbonic acid (H_2CO_3) is titrated against 0.010 mol L^{-1} NaOH solution.
(a) At the beginning of the titration, we have 100.0 mL of 0.010 mol L^{-1} carbonic acid solution. K_a for carbonic acid is 4.2×10^{-7}

Concentrations (mol L^{-1})	H$_2$CO$_3$(aq)	+	H$_2$O(l)	⇌	HCO$_3^-$(aq)	+	H$_3$O$^+$(aq)
Initial	0.010				0		0
Change	$-x$				$+x$		$+x$
Equilibrium	$0.010 - x$				x		x

At equilibrium the condition $Q = K$ must be satisfied:
$$Q = \frac{[\text{HCO}_3^-][\text{H}_3\text{O}^+]}{[\text{H}_2\text{CO}_3]} = \frac{(x)(x)}{0.010-x} = K_a = 4.2\times10^{-7}$$
From this we obtain the quadratic equation
$$x^2 + (4.2\times10^{-7})x - (4.2\times10^{-9}) = 0$$
Solving the quadratic equation, gives $x = 6.5 \times 10^{-5}$
So $[\text{H}_3\text{O}^+] = x = 6.5 \times 10^{-5}$ mol L^{-1}
and $\text{pH} = -\log_{10}[\text{H}_3\text{O}^+] = -\log_{10}(6.5 \times 10^{-5}) = 4.19$
(b) When the volume of NaOH solution added is half that needed to arrive at the first equivalence point, half of the amount of H$_2$CO$_3$(aq) molecules have been converted into HCO$_3^-$(aq) ions—the conjugate base. So we have a buffer solution in which [weak acid] = [conjugate base]. So
$$\text{pH} = \text{p}K_{a1} = -\log_{10}(4.2 \times 10^{-7}) = 6.38$$
(c) At the first equivalence point, all of the H$_2$CO$_3$(aq) molecules have been converted into HCO$_3^-$(aq) ions. We have a 0.010 mol L^{-1} sodium hydrogencarbonate solution.
HCO$_3^-$(aq) ions are both a weak acid (to form CO$_3^{2-}$(aq) ions and H$_3$O$^+$(aq) ions) and a weak base (to form H$_2$CO$_3$(aq) molecules and OH$^-$(aq) ions). Accurate calculation of pH is possible, but (page 541) we can take advantage of the approximation that applies at the first equivalence point:
$$\text{pH} = \tfrac{1}{2}(\text{p}K_{a1} + \text{p}K_{a2}) = \tfrac{1}{2}(6.38 + 10.32) = 8.35$$
which tells us that the reaction of HCO$_3^-$(aq) ions as a base occurs a little more than its reaction as an acid (which we could also deduce by comparison of its K_a (K_{a2} of H$_2$CO$_3$(aq)) and its K_b (=K_w / K_{a1}).
(d) When the volume of NaOH solution added is exactly midway between the first and second equivalence points, half of the HCO$_3^-$(aq) ions have been converted into CO$_3^{2-}$(aq) ions. Then [HCO$_3^-$] = [CO$_3^{2-}$], and
$$\text{pH} = \text{p}K_{a2} = -\log_{10}(4.8 \times 10^{-11}) = 10.32$$
(e) At the second equivalence point, all of the HCO$_3^-$(aq) ions have been converted into CO$_3^{2-}$(aq) ions. With dilution by the added base solution, we have a 0.0033 mol L^{-1} sodium carbonate solution. K_{b1} for CO$_3^{2-}$(aq) ions is 2.1×10^{-4}

Concentrations (mol L^{-1})	CO$_3^{2-}$	+	H$_2$O	⇌	HCO$_3^-$	+	OH$^-$
Initial	0.0033				0		0
Change	$-x$				$+x$		$+x$
Equilibrium	$0.0033 - x$				x		x

At equilibrium the condition $Q = K$ must be satisfied:
$$Q = \frac{[\text{HCO}_3^-][\text{OH}^-]}{[\text{CO}_3^{2-}]} = \frac{(x)(x)}{0.0033-x} = K_{b1} = 2.1\times10^{-4}$$

From this we obtain the quadratic equation
$$x^2 + (2.1\times10^{-4})\,x - (6.93\times10^{-7}) = 0$$
Solving the quadratic equation, gives $x = 7.3 \times 10^{-4}$
So $\quad [OH^-] = x = 7.3 \times 10^{-4}$ mol L^{-1}
and $\quad [H_3O^+] = (1 \times 10^{-14})/[OH^-] = (1 \times 10^{-14})/(7.3 \times 10^{-4}) = 1.4 \times 10^{-11}$
and $\quad pH = -\log_{10}[H_3O^+] = -\log_{10}(1.4 \times 10^{-11}) = 10.86$

Exercise 14.53—Titration of a weak base with a strong acid

Adding 75.0 mL of 0.100 mol L^{-1} HCl solution to 100.0 mL of 0.100 mol L^{-1} ammonia solution
$\quad$ Amount of H$_3$O$^+$(aq) ions added $= 0.075$ L $\times 0.100$ mol L^{-1} $= 0.0075$ mol
$\quad$ Initial amount of NH$_3$(aq) $= 0.100$ L $\times 0.100$ mol L^{-1} $= 0.0100$ mol
Since the amount of added acid does not exceed the amount of weak base, we end up with a buffer solution. NH$_3$(aq) molecules will react with essentially all of the added H$_3$O$^+$(aq) ions.
Upon adding 0.0075 mol H$_3$O$^+$(aq) ions, the amounts of NH$_3$(aq) molecules and NH$_4^+$(aq) ions change to
$\quad$ Amount of NH$_3$(aq) molecules $= 0.0100 - 0.0075$ mol $= 0.0025$ mol
$\quad$ Amount of NH$_4^+$(aq) ions $= 0.0075$ mol

$$[H_3O^+] = \frac{[NH_4^+]}{[NH_3]} \times K_a(NH_4^+) = \frac{0.0075}{0.0025} \times (5.6\times10^{-10}) = 1.68\times10^{-9}$$

pH = 8.77

REVIEW EXERCISES

14.3 Water and the pH scale

14.55
(a) A solution with $[H_3O^+] = 5 \times 10^{-10}$ mol L^{-1} is basic ($[H_3O^+] < 10^{-7}$ mol L^{-1})
(b) A solution with $[H_3O^+] = 7 \times 10^{-7}$ mol L^{-1} is acidic ($[H_3O^+] > 10^{-7}$ mol L^{-1})
(c) A solution with $[OH^-] = 3 \times 10^{-9}$ mol L^{-1} is acidic ($[OH^-] < 10^{-7}$ mol L^{-1})

14.57
$[H_3O^+] = 10^{-pH} = 10^{-3.75} = 1.78 \times 10^{-4}$ mol L^{-1}
The solution is acidic (pH < 7).

14.59

$$K_b\left((CH_3)_3 N(aq)\right) = \frac{[(CH_3)_3 NH^+][OH^-]}{[(CH_3)_3 N]} = \left(\frac{[(CH_3)_3 N][H_3O^+]}{[(CH_3)_3 NH^+]} \frac{1}{[H_3O^+][OH^-]}\right)^{-1}$$

$$= \left(K_a\left((CH_3)_3 NH^+\right)\frac{1}{K_w}\right)^{-1} = \frac{K_w}{K_a\left((CH_3)_3 NH^+\right)}$$

$$= \frac{10^{-14}}{10^{-pK_a}} = \frac{10^{-14}}{10^{-9.80}} = 10^{-4.20} = 6.31 \times 10^{-5}$$

14.4 Relative strengths of weak acids and bases

14.61
HPO_4^{2-}(aq) ions acting as an acid:
$$HPO_4^{2-}(aq) + H_2O(l) \rightleftharpoons PO_4^{3-}(aq) + H_3O^+(aq)$$
HPO_4^{2-}(aq) ions acting as a base:
$$HPO_4^{2-}(aq) + H_2O(l) \rightleftharpoons H_2PO_4^-(aq) + OH^-(aq)$$

14.63
(b) ClC_6H_4COOH ($pK_a = 2.88$) is the stronger acid, stronger than benzoic acid ($pK_a = 4.20$). The acid with the lower pK_a value has the higher K_a value, and is the stronger acid.

14.65

$$C_6H_5OH(aq) + H_2O(l) \rightleftharpoons H_3O^+(aq) + C_6H_5O^-(aq) \qquad K_a = 1.3 \times 10^{-10}$$

$$HCOOH(aq) + H_2O(l) \rightleftharpoons H_3O^+(aq) + HCOO^-(aq) \qquad K_a = 1.8 \times 10^{-4}$$

$$HC_2O_4^-(aq) + H_2O(l) \rightleftharpoons H_3O^+(aq) + C_2O_4^{2-}(aq) \qquad K_a = 6.4 \times 10^{-5}$$

(a) HCOOH(aq) (formic acid) is the strongest acid. C_6H_5OH(aq) (phenol) is the weakest acid.
(b) HCOOH(aq) (formic acid) has the weakest conjugate base.
(c) C_6H_5OH(aq) (phenol) has the strongest conjugate base.

14.69

The 0.10 mol L^{-1} solution of **(a)** $Na_2S(s)$ has the highest pH. S^{2-} is a stronger base than PO_4^{3-} which is stronger than acetate which is stronger than F^-. Al^{3+} forms the acid, $Al(H_2O)_6^{3+}$. The 0.10 mol L^{-1} solution of **(f)** $AlCl_3(s)$ has the lowest pH.

14.5 The Lewis model of acids and bases

14.71

(a)

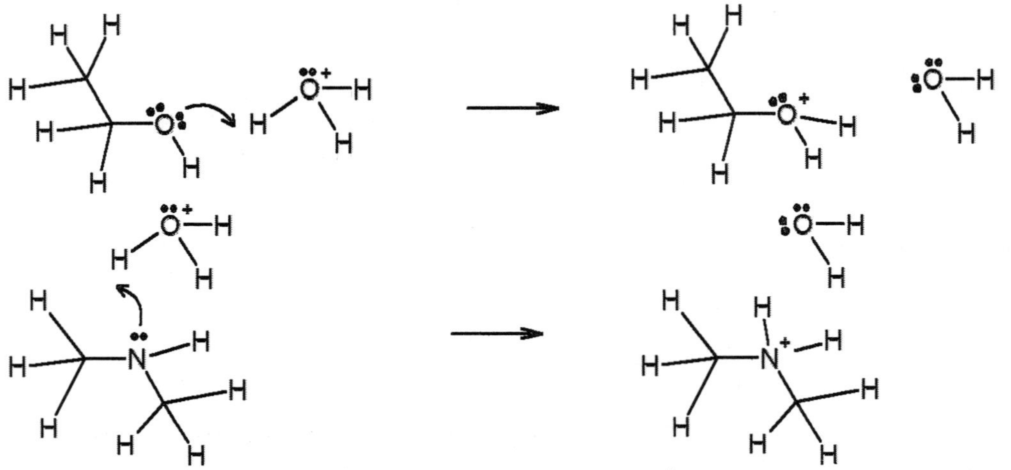

The reaction of trimethyl phosphine, $(CH_3)_3P$, with H_3O^+ is similar to that of dimethylamine, $(CH_3)_2NH$, with H_3O^+

(b)

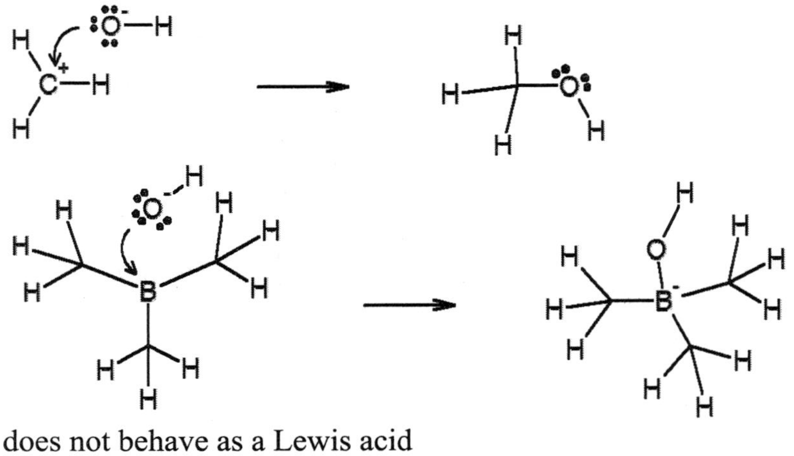

Br^- does not behave as a Lewis acid

14.6 Equilibria in aqueous solutions of weak acids or bases

14.73
(a) pH = 3.80
$[H_3O^+] = 10^{-3.80} = 1.58 \times 10^{-4}$ mol L^{-1}
(b) Compute K_a

$$HA + H_2O \rightleftharpoons A^- + H_3O^+$$

$[A^-] = [H_3O^+] = x = 1.58 \times 10^{-4}$ mol L^{-1}
$[HA] = 2.5 \times 10^{-3} - x = 2.5 \times 10^{-3} - 1.58 \times 10^{-4} = 2.3 \times 10^{-3}$ mol L^{-1}
Therefore,

$$K_a = \frac{[A^-][H_3O^+]}{[HA]} = \frac{(1.58 \times 10^{-4})^2}{2.3 \times 10^{-3}} = 1.1 \times 10^{-5}$$

i.e., this is a moderately weak acid.

14.75
(a) Dissolving solid ammonium chloride in a dilute aqueous ammonia solution decreases pH. $NH_4^+(aq)$ ions are a weak acid. Alternatively, because $NH_4^+(aq)$ ions are the conjugate acid to ammonia, $NH_3(aq)$, we can think of the change in the ratio of $NH_3(aq)$ and $NH_4^+(aq)$ ions in the buffer equation 14.1 and the corresponding change in $[H_3O^+]$.
(b) Dissolving solid sodium acetate in a dilute aqueous acetic acid solution increases pH.
(c) Dissolving solid NaCl in a dilute aqueous NaOH solution has no affect on the pH.

14.77

Concentrations (mol L^{-1})	NH$_3$(aq)	+	H$_2$O(l)	$\rightleftharpoons$	NH$_4^+$(aq)	+	OH$^-$(aq)
Initial	0.15				0		0
Change	$-x$				$+x$		$+x$
Equilibrium	$0.15 - x$				x		x

Substituting equilibrium concentrations into the reaction quotient:

$$Q = \frac{[NH_4^+][OH^-]}{[NH_3]} = \frac{(x)(x)}{(0.15-x)} = K_b(NH_3) = 1.8 \times 10^{-5}$$

From this we obtain a quadratic equation: $x^2 + (1.8 \times 10^{-5})x - 2.7 \times 10^{-6} = 0$
with solution $x = 1.6 \times 10^{-3}$ mol L^{-1}
Now we can write the equilibrium concentrations:

$$[NH_4^+] = [OH^-] = x = 1.6 \times 10^{-3} \text{ mol L}^{-1}$$
$$[NH_3] = 0.15 - x = 0.15 - 1.6 \times 10^{-3} = 0.015 \text{ mol L}^{-1}$$
$$[H_3O^+] = \frac{K_w}{[OH^-]} = \frac{1.0 \times 10^{-14}}{1.6 \times 10^{-3}} = 6.25 \times 10^{-12} \text{ mol L}^{-1}$$

Finally, pH $= -\log_{10}(6.25 \times 10^{-12}) = 11.20$

$$\% \text{ of } NH_3 \text{ ionized} = 100\% \times \frac{x}{0.15} = 100\% \times \frac{1.6 \times 10^{-3}}{0.15} = 1.1\%$$

14.79

$K_b = 2.9 \times 10^{-7}$ for aquated hypochlorite ions, $ClO^-(aq)$, at 25 °C.

(a) For initial base concentration, 1.00 mol L^{-1}, the quadratic equation for the extent of reaction (in this case, abstraction of a proton from water by hypochlorite to form hypochlorous acid) takes the form

$$x^2 + K_b\, x - c_{initial} K_b = 0$$

$$x^2 + 2.9 \times 10^{-7}\, x - 2.9 \times 10^{-7} = 0$$

Solving the quadratic equation, gives $x = 5.4 \times 10^{-4}$

The equilibrium $OH^-(aq)$ ion concentration is

$[OH^-] = x = 5.4 \times 10^{-4}$ mol L^{-1}

and $\quad pOH = -\log_{10}[OH^-] = -\log_{10}(5.4 \times 10^{-4}) = 3.27$

% of hypochlorite converted to hypochlorous acid $= 100\% \times \dfrac{x}{1.00} = 100\% \times \dfrac{5.4 \times 10^{-4}}{1.00} = 0.054\%$

(b) For initial base concentration, 0.100 mol L^{-1}, the quadratic equation takes the form

$$x^2 + 2.9 \times 10^{-7}\, x - 2.9 \times 10^{-8} = 0$$

Solving the quadratic equation, gives $x = 1.7 \times 10^{-4}$

The equilibrium $OH^-(aq)$ ion concentration is

$[OH^-] = x = 1.7 \times 10^{-4}$ mol L^{-1}

and $\quad pOH = -\log_{10}[OH^-] = -\log_{10}(3.5 \times 10^{-4}) = 3.77$

% of hypochlorite converted to hypochlorous acid $= 100\% \times \dfrac{x}{0.100} = 100\% \times \dfrac{1.7 \times 10^{-4}}{0.100} = 0.17\%$

(c) For initial base concentration, 0.0100 mol L^{-1}, the quadratic equation takes the form

$$x^2 + 2.9 \times 10^{-7}\, x - 2.9 \times 10^{-9} = 0$$

Solving the quadratic equation, gives $x = 5.4 \times 10^{-5}$

The equilibrium $OH^-(aq)$ ion concentration is

$[OH^-] = x = 5.4 \times 10^{-5}$ mol L^{-1}

and $\quad pOH = -\log_{10}[OH^-] = -\log_{10}(5.4 \times 10^{-5}) = 4.27$

% of hypochlorite converted to hypochlorous acid $= 100\% \times \dfrac{x}{0.00100} = 100\% \times \dfrac{5.4 \times 10^{-5}}{0.00100} = 5.4\%$

The trend here is as for all weak acids: the percentage of ClO–(aq) ions reacted increases with decreasing concentration.

14.81

$$N_2H_4(aq) + H_2O(l) \rightleftharpoons N_2H_5 + OH^-(aq) \qquad\qquad K_{b1} = 8.5 \times 10^{-7}$$

$$N_2H_5^+(aq) + H_2O(l) \rightleftharpoons N_2H_6^{2+}(aq) + OH^-(aq) \qquad K_{b2} = 8.9 \times 10^{-16}$$

(a) In a 0.010 mol L^{-1} aqueous hydrazine solution, we can neglect the second ionization reaction. Hydrazine being a weak base is only slightly ionized when added to water at this concentration. Of the relatively few $N_2H_5^+$(aq) ions that are formed, only a negligible number of them react further with water to form $N_2H_6^{2+}$(aq) ions.

For initial base concentration $c(N_2H_4) = 0.010$ mol L^{-1}, the quadratic equation takes the form

$$x^2 + 8.5 \times 10^{-7} x - 8.5 \times 10^{-9} = 0$$

Solving the quadratic equation, gives $x = 9.2 \times 10^{-5}$

The equilibrium OH$^-$(aq) ion concentration is

$$[OH^-] = x = 9.2 \times 10^{-5} \text{ mol L}^{-1}$$

and $\quad$ pOH $= -\log_{10}[OH^-] = -\log_{10}(9.2 \times 10^{-5}) = 4.04$

$$[N_2H_5^+] = x = 9.2 \times 10^{-5} \text{ mol L}^{-1}$$

The concentration of $N_2H_6^{2+}$(aq) ions can be computed by considering that the second stage of weak acid ionization occurs in an environment dominated by the concentrations of OH$^-$(aq) and $N_2H_5^+$(aq) ions produced in the first stage—and which will be negligibly altered as a result of the very weak second stage.

$$K_{b2} = \frac{[N_2H_6^{2+}][OH^-]}{[N_2H_5^+]} = \frac{[N_2H_6^{2+}](\cancel{9.2 \times 10^{-5}})}{\cancel{9.2 \times 10^{-5}}} = [N_2H_6^{2+}] = 8.9 \times 10^{-16} \text{ mol L}^{-1}$$

This very low concentration justifies our earlier assumption that the extent of the second ionization step is very small, and negligible in terms of calculating the concentrations of species from the first step.

(b) pH $= 14.00 - $ pOH $= 14.00 - 4.04 = 9.96$

14.7 Speciation: Relative concentrations of species

14.83

$$CH_3NH_3^+(aq) + H_2O(l) \xrightleftharpoons{pK_a = 3.38} CH_3NH_2(aq) + H_3O^+(l)$$

The relative equilibrium concentrations of the weak acid and its conjugate base are given by

$$\frac{[CH_3NH_3^+]}{[CH_3NH_2]} = \frac{[H_3O^+]}{K_a} = \frac{[H_3O^+]}{10^{-3.38}} = \frac{[H_3O^+]}{4.17 \times 10^{-4}}$$

At pH = 3, $[H_3O^+] = 1 \times 10^{-3}$ mol L^{-1}: $\quad \dfrac{[CH_3NH_3^+]}{[CH_3NH_2]} = \dfrac{[H_3O^+]}{4.17 \times 10^{-4}} = \dfrac{1 \times 10^{-3}}{4.17 \times 10^{-4}} = \dfrac{2.4}{1}$

At pH = 7, $[H_3O^+] = 1 \times 10^{-7}$ mol L^{-1}: $\quad \dfrac{[CH_3NH_3^+]}{[CH_3NH_2]} = \dfrac{[H_3O^+]}{4.17 \times 10^{-4}} = \dfrac{1 \times 10^{-7}}{4.17 \times 10^{-4}} = \dfrac{2.4 \times 10^{-4}}{1}$

At pH = 11, $[H_3O^+] = 1 \times 10^{-11}$ mol L^{-1}: $\quad \dfrac{[CH_3NH_3^+]}{[CH_3NH_2]} = \dfrac{[H_3O^+]}{4.17 \times 10^{-4}} = \dfrac{1 \times 10^{-11}}{4.17 \times 10^{-4}} = \dfrac{2.4 \times 10^{-8}}{1}$

$\dfrac{[CH_3NH_3^+]}{[CH_3NH_2]} = \dfrac{1}{1} \quad$ when $[H_3O^+] = K_a = 10^{-3.38}$. That is, when pH = 3.38.

14.85
(a) At
  (i) pH = 6, $H_2A(aq)$ is at the highest concentration
  (ii) pH = 8, $HA^-(aq)$ ions are at the highest concentration
  (iii) pH = 10, $HA^-(aq)$ ions are at the highest concentration—but not by as wide a margin
(b) pK_{a1} and pK_{a2} can be seen in the plot.
$[H_2A(aq)] = [HA^-(aq)]$ when pH = pK_{a1}. So, pK_{a1} = about 6.4
$[HA^-(aq)] = [A^{2-}(aq)]$ when pH = pK_{a2}. So, pK_{a1} = about 10.3
(c) Since $pK_{a1} = 6.38$ and $pK_{a2} = 10.32$ for carbonic acid, the acid is probably carbonic acid.

14.8 Acid-base properties of amino acids and proteins

14.87

$$H_3Glu^+(aq) + H_2O(l) \rightleftharpoons H_2Glu(aq) + H_3O^+(aq) \qquad pK_{a1} = 2.19$$

$$H_2Glu(aq) + H_2O(l) \rightleftharpoons HGlu^-(aq) + H_3O^+(aq) \qquad pK_{a2} = 4.25$$

$$HGlu^-(aq) + H_2O(l) \rightleftharpoons Glu^{2-}(aq) + H_3O^+(aq) \qquad pK_{a3} = 9.67$$

(a) At pH = 1.0
H_3Glu^+ is the dominant glutamic acid species. pH = 1.0 < pK_{a1}
H_3Glu^+

(b) At pH = 3.0
H_2Glu is the dominant glutamic acid species. pK_{a1} < pH = 3.0 < pK_{a2}
H_2Glu

(c) At pH = 8.0
$HGlu^-$ is the dominant glutamic acid species. pK_{a2} < pH = 8.0 < pK_{a3}
$HGlu^-$

(d) pH 12.0
Glu^{2-} is the dominant glutamic acid species. $pK_{a3} < pH = 12.0$
Glu^{2-}

14.9 Controlling pH: buffer solutions

14.89
With known labelled concentrations of the weak acid HCOOH(aq) and its conjugate base HCOO⁻(aq) ions, we can use the form of the buffer equation 14.3

$$[H_3O^+] = \frac{c(HCOOH)}{c(HCOO^-)} \times K_a = \frac{0.50\,mol\,L^{-1}}{0.70\,mol\,L^{-1}} \times (1.8 \times 10^{-4}) = 1.29 \times 10^{-4}\,mol\,L^{-1}$$

$$pH = 3.89$$

14.91
(a) We can use buffer equation 14.3, based on solute concentrations:

$$[H_3O^+] = \frac{c(HCOOH)}{c(HCOO^-)} \times K_a = \frac{0.050\,mol\,L^{-1}}{0.035\,mol\,L^{-1}} \times (1.8 \times 10^{-4}) = 2.57 \times 10^{-4}\,mol\,L^{-1}$$

$$pH = 3.59$$

(b) To raise the pH by 0.5, to 4.09, we must have (equation 14.1):

$$[H_3O^+] = 10^{-4.09} = 8.13 \times 10^{-5} = \frac{[HCOOH]}{[HCOO^-]} \times (1.8 \times 10^{-4})$$

$$\frac{[HCOOH]}{[HCOO^-]} = \frac{8.13 \times 10^{-5}}{1.8 \times 10^{-4}} = \frac{0.452}{1}$$

14.93
To make a hydrogencarbonate/carbonate buffer solution with pH 10.00, we would dissolve in the same sample of water amounts of a hydrogencarbonate salt and a carbonate salt such that

$$[H_3O^+] = 1.0 \times 10^{-10} = \frac{n(HCO_3^-)}{n(CO_3^{2-})} \times K_{a2}(H_2CO_3) = \frac{n(HCO_3^-)}{n(CO_3^{2-})} \times (4.8 \times 10^{-11})$$

$$\frac{n(HCO_3^-)}{n(CO_3^{2-})} = \frac{1.0 \times 10^{-10}}{4.8 \times 10^{-11}} = \frac{2.08}{1}$$

So we could make the buffer solution by dissolving 2.08 mol of NaHCO₃(s) for every 1.00 mol of Na₂CO₃(s) in a sample of water. The actual amounts would depend on the amounts of added acid or base against which the solution is required to buffer (that is, on the buffer capacity required).

14.95

(a)

Amount of NaCH$_3$COO(s) dissolved = n(CH$_3$COO$^-$) = 4.95 g / 82.03 g mol^{-1} = 0.0603 mol

n(CH$_3$COOH) = 0.250 L × 0.150 mol L^{-1} = 0.0375 mol

Using buffer equation 14.4, expressed in amounts of solute species:

$$[\text{H}_3\text{O}^+] = \frac{n(\text{CH}_3\text{COOH})}{n(\text{CH}_3\text{COO}^-)} \times K_a = \frac{0.0375\,\text{mol}}{0.0603\,\text{mol}} \times (1.8\times10^{-5}) = 1.12\times10^{-5}\,\text{mol L}^{-1}$$

pH = 4.95

(b) In 100 mL of buffer solution

n(CH$_3$COOH) = 0.0375 mol × 100/250 = 0.0150 mol

= n(CH$_3$COO$^-$) = 0.0603 mol × 100/250 = 0.0241 mol

If 82 mg of NaOH(s) is dissolved in 100 mL of the buffer solution, CH$_3$COOH(aq) "mops up" the added OH$^-$(aq) ions.

Amount of NaOH(s) dissolved = n(OH$^-$) added = 0.082 g / 40.00 g mol^{-1} = 0.00205 mol

After addition of the NaOH(s):

n(CH$_3$COOH) = (0.0150 − 0.00205) mol = 0.0129 mol

n(CH$_3$COO$^-$) = (0.0150 + 0.00205) mol = 0.0262 mol

$$[\text{H}_3\text{O}^+] = \frac{n(\text{CH}_3\text{COOH})}{n(\text{CH}_3\text{COO}^-)} \times K_a = \frac{0.0129\,\text{mol}}{0.0262\,\text{mol}} \times (1.8\times10^{-5}) = 8.86\times10^{-6}\,\text{mol L}^{-1}$$

pH = 5.05

14.10 Acid-base titrations

14.99

Amount of potassium hydrogenphthalate = 0.902 g / 204.22 g mol^{-1} = 0.00442 mol

To neutralize the aquated hydrogenphthalate ions requires an equal amount of OH$^-$(aq) ions. If this is achieved with 26.45 mL of NaOH solution, the concentration of OH$^-$(aq) ions in the solution is

0.00442 mol / 0.02645 L = 0.167 mol L^{-1}

14.101

(a) Adding 100.0 mL of 0.100 mol L^{-1} NaOH solution to 100.0 mL of 0.200 mol L^{-1} hypochlorous acid solution

Total amount of HOCl = 0.100 L × 0.200 mol L^{-1} = 0.0200 mol

Amount of OH$^-$(aq) ions added = 0.100 L × 0.100 mol L^{-1} = 0.0100 mol

Since the amount of added base does not exceed the amount of weak acid, we end up with a buffer solution. The hypochlorous acid will consume all of the added hydroxide.

Upon adding 0.0100 mol OH$^-$(aq) ions, the amounts of HOCl(aq) and OCl$^-$(aq) ions change to

n(HOCl) = (0.0200 − 0.0100) mol = 0.0100 mol

n(OCl$^-$) = (0 + 0.0100) mol

The new pH is given by

$$[\text{H}_3\text{O}^+] = \frac{n(\text{HOCl})}{n(\text{OCl}^-)} \times K_a(\text{HOCl}) = \frac{0.0100\,\text{mol}}{0.0100\,\text{mol}} \times (3.5\times10^{-8}) = 3.5\times10^{-8}\,\text{mol L}^{-1}$$

pH = 7.46

(b) Adding 80.0 mL of 0.100 mol L^{-1} NaOH solution to 100.0 mL of 0.200 mol L^{-1} hypochlorous acid solution

    amount of OH$^-$ = 0.0800 L × 0.100 mol L^{-1} = 0.00800 mol

    total amount of HOCl = 0.0200 mol

Upon adding 0.00800 mol OH$^-$(aq) ions, the amounts of HOCl(aq) and OCl$^-$(aq) ions change to

    n(HOCl) = (0.0200 − 0.00800) mol = 0.0120 mol

    n(OCl$^-$) = 0.00800 mol

The new pH is given by

$$[H_3O^+] = \frac{n(\text{HOCl})}{n(\text{OCl}^-)} \times K_a(\text{HOCl}) = \frac{0.0120\,\text{mol}}{0.0080\,\text{mol}} \times (3.5\times10^{-8}) = 5.25\times10^{-8}\,\text{mol L}^{-1}$$

    pH = 7.28

14.103

(a) Amount of aniline = amount of HCl required to reach equivalence

    = 0.02567 L × 0.175 mol L^{-1} = 0.00449 mol

c(aniline) = 0.00449 mol / 0.0250 L = 0.180 mol L^{-1}

(b) At the equivalence point, the solution is the same as a solution of aniline hydrochloride (C$_6$H$_5$NH$_3^+$).

Volume = (25.0 + 25.67) mL = 50.67 mL

So c(C$_6$H$_5$NH$_3^+$) = 0.180 mol L^{-1} × (25.0/50.67) = 0.0888 mol L^{-1}

K_a for C$_6$H$_5$NH$_3^+$ is 10^{-14} / 4.0 × 10^{-10} = 2.5 × 10^{-5}

Concentrations (mol L^{-1})	C$_6$H$_5$NH$_3^+$(aq)	+	H$_2$O(l)	⇌	C$_6$H$_5$NH$_2$(aq)	+	H$_3$O$^+$(aq)
Initial	0.0888				0		0
Change	−x				+x		+x
Equilibrium	0.0888 − x				x		x

Setting Q (at equilibrium) = K gives the quadratic equation,

    $x^2 + (2.5\times10^{-5})x - (2.22\times10^{-6}) = 0$

Solving the quadratic equation, gives $x = 1.48 \times 10^{-3}$

So   [H$_3$O$^+$] = x = 1.48 × 10^{-3} mol L^{-1}

    [C$_6$H$_5$NH$_3^+$] = (0.0888 − 1.48 × 10^{-3}) mol L^{-1} = 0.0873 mol L^{-1}

    [OH$^-$] = (1 × 10^{-14}) / (1.48 × 10^{-3}) = 6.76 × 10^{-12} mol L^{-1}

(c) pH = −log$_{10}$(1.48 × 10^{-3}) = 2.83

14.105

(a) We have a 0.050 mol L^{-1} HCN solution

K_a for HCN is 4.0×10^{-10}

Concentrations (mol L^{-1})	HCN(aq)	+	H$_2$O(l)	$\rightleftharpoons$	CN$^-$(aq)	+	H$_3$O$^+$(aq)
Initial	0.050				0 -		0
Change	$-x$				$+x$		$+x$
Equilibrium	0. 050 $- x$				X		x

Setting Q (at equilibrium) $= K_a$ gives the quadratic equation,

$$x^2 + (4.0 \times 10^{-10})\, x - (2.0 \times 10^{-11}) = 0$$

Solving the quadratic equation, gives $x = 4.5 \times 10^{-6}$

So $\quad$ [H$_3$O$^+$] $= x = 4.5 \times 10^{-6}$ mol L^{-1}

$\quad$ [CN$^-$] $= 4.5 \times 10^{-6}$ mol L^{-1}

$\quad$ [HCN] $= 0.050 - 4.5 \times 10^{-6}$ mol L^{-1} $= 0.050$ mol L^{-1}

$\quad$ pH $= -\log_{10}(4.5 \times 10^{-6}) = 5.35$

(b) At the halfway point of the titration, the HCN$^{\square}$ and CN$^-$ concentrations are equal and

$\quad$ pH $= pK_a$(HCN) $= 9.40$

(c) Initial amount of HCN$^{\square}$ $= 0.025$ L $\times 0.050$ mol L^{-1} $= 0.00125$ mol

$\quad$ amount of NaOH added to reach 95% of equivalence value $= 0.95 \times 0.00125$ mol $= 0.00119$ mol

Normally, at this point we would assume that all of this added NaOH is consumed by the HCN. However, the numbers are such that this does not give the correct answer. To be accurate, we need an ICE table and associated quadratic equation. This ICE table is constructed in terms of amounts. The ICE table is constructed with the initial amounts, starting with the added OH$^-$ completely consumed. We construct the ICE table for the reverse reaction which corresponds to some of the hydroxide being regenerated—i.e., we treat the reaction of cyanide with water.

If the OH$^-$ were completely consumed, we would have

$\quad$ Final amount of CN$^-$ $= 0.00119$ mol

$\quad$ Final amount of HCN $= 0.00125 - 0.00119$ mol $= 0.00006$ mol

Use these values as the initial amounts in the ICE table. K_b for cyanide is 2.5×10^{-5}

Amounts / mol	CN$^-$	+	H$_2$O	$\rightleftharpoons$	HCN	+	OH$^-$
Initial	0.00119				0.00006		0
Change	$-x$				$+x$		$+x$
Equilibrium	0. 00119 $-$ x				0.00006 $+ x$		x

Setting Q (at equilibrium) $= K_b$ gives the quadratic equation,

$$x^2 + (2.5 \times 10^{-5} + 6 \times 10^{-5})\, x - (3.0 \times 10^{-8}) = 0$$

$$x^2 + (8.5 \times 10^{-5})\, x - (3.0 \times 10^{-8}) = 0$$

Solving the quadratic equation, gives $x = 0.00014$

So $\quad$ n(CN$^-$) $= 0.00119 - 0.00014$ mol $= 0.00105$ mol

$\quad$ n(HCN) $= 0.00006 + x = 0.00006 + 0.00014 = 0.00020$ mol

Use the Henderson-Hasselbalch equation to get the pH

$$pH = pK_a(HCN) + \log_{10}\left(\frac{n(CN^-)}{n(HCN)}\right) = 9.40 + \log_{10}\left(\frac{0.00105}{0.00020}\right)$$

$$= 9.40 + 0.72 = 10.12$$

(d) The volume of NaOH solution added to reach equivalence
= 25.0 mL × (0.050 mol L^{-1} / 0.075 mol L^{-1}) = 16.7 mL

(e) At the equivalence point, we have a sodium cyanide solution with a total concentration of CN^-
= (0.0250 L × 0.050 mol) / total volume of solution after titration
= 0.00125 mol / (0.0250 + 0.0167 L)
= 0.030 mol L^{-1}

K_b for cyanide is 2.5×10^{-5}

Concentrations (mol L^{-1})	CN^-(aq)	+	H_2O(l)	⇌	HCN(aq)	+	OH^-(aq)
Initial	0.030				0		0
Change	$-x$				$+x$		$+x$
Equilibrium	$0.030 - x$				x		x

Setting Q (at equilibrium) = K_b gives the quadratic equation,
$$x^2 + (2.5\times10^{-5})\,x - (7.5\times10^{-7}) = 0$$

Solving the quadratic equation, gives $x = 8.5 \times 10^{-4}$

So $[OH^-] = x = 8.5 \times 10^{-4}$ mol L^{-1}
$[CN^-] = 0.030 - 8.5 \times 10^{-4}$ mol L^{-1} = 0.030 mol L^{-1}
$[HCN] = 8.5 \times 10^{-4}$ mol L^{-1}
$[H_3O^+] = 10^{-14} / [OH^-] = 1.18 \times 10^{-11}$ mol L^{-1}
$pH = -\log_{10}(1.18 \times 10^{-11}) = 10.93$

(f) We need an indicator with pK_a near the equivalence point pH—i.e., $pK_a \approx 10.93$.
Indicators which change colour in the neighbourhood of pH = 11 include
alizarine yellow R

We have the following data with which to plot the titration curve:

pH	$n(\text{HCN})$ / mol	$n(\text{CN}^-)$ / mol
5.35	$0.050 \text{ mol L}^{-1} \times 0.025 \text{ L}$ $= 0.00125$	$4.5 \times 10^{-6} \text{ mol L}^{-1} \times 0.025 \text{ L}$ $= 1.1 \times 10^{-7}$
9.40	$0.00125 / 2$ $= 0.000625$ here, concentrations of HCN and CN^- are equal —i.e., they equal half the total amount	0.000625
10.41	0.00032	0.00093
10.93	$8.5 \times 10^{-4} \text{ mol L}^{-1} \times$ 0.0417 L $= 3.5 \times 10^{-5}$	$0.030 \text{ mol L}^{-1} \times 0.0417 \text{ L}$ $= 0.00125$

Note that we use amounts here for consistency. The volumes are different, so concentrations cannot be compared directly.

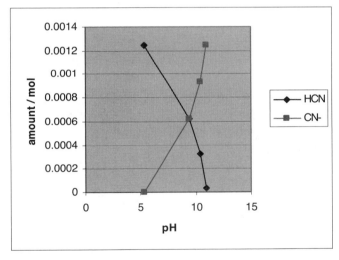

SUMMARY AND CONCEPTUAL PROBLEMS

14.107

(a) If the pH were 2, the concentration of hydronium ions would be 10 times smaller—i.e., there would be about 10 H_3O^+(aq) ions.

(b) If the pH were 3, the concentration of hydronium ions would be another 10 times smaller—i.e., there would just be about 1 H_3O^+(aq) ion.

(c) If the pH were 4, then there would likely not be any hydronium ions in the cube. To visualize pH = 4, you would have to imagine the cube as a snapshot taken from a larger quantity of water. If many such snapshots were considered, then about 1 in 10 of them would have a single hydronium ion.

(d) If the H_3O^+(aq) ion to water molecule ratio is 100:55000 = 1:550 when pH = 1.0, then this ratio is 1:55000000 when pH = 6.0 (i.e., a hydronium ion concentration 10^5 times smaller).

14.109

A buffer solution contains a not-too-dilute solution of both a weak acid and a weak base. When a strong base is added, the weak acid reacts with it and consumes most of it. In the case of a carbonate/hydrogencarbonate buffer, the hydrogencarbonate reacts with the added base.

$$HCO_3^-(aq) + OH^-(aq) \rightarrow CO_3^{2-}(aq) + H_2O(l)$$

14.111

(a) pH can only be increased by addition of base. If the added base is OH^-(aq), then a reaction occurs as base is added:

$$CH_3COOH(aq) + OH^-(aq) \rightarrow CH_3COO^-(aq) + H_2O(l)$$

So [CH_3COOH] decreases, and [CH_3COO^-] increases.

Quantitatively, we see this through the speciation equation

$$\frac{[CH_3COOH]}{[CH_3COO^-]} = \frac{[H_3O^+]}{K_a}$$

As pH increases, [H_3O^+] decreases, and proportionately so does the ratio [acid]/[base].

(b) From the graph, we can see that at pH 4 (indeed at all pH < pK_a (= 4.7)), there is a higher concentration of CH_3COOH(aq) than CH_3COO^-(aq) ions.

We can also see that at pH 6 (and all pH > pK_a (= 4.7)), there is a higher concentration of CH_3COO^-(aq) ions than CH_3COOH(aq) molecules.

(c) From the $Q = K_a$ condition, we can derive the speciation equation

$$\frac{[CH_3COOH]}{[CH_3COO^-]} = \frac{[H_3O^+]}{K_a}$$

We can see from this equation that [CH_3COOH] = [CH_3COO^-] when [H_3O^+] = K_a—i.e., when pH = pK_a .

14.113

(a)
Amount of salicylic acid, $n(C_6H_4(OH)COOH) = 1.00 \text{ g} / 138.12 \text{ g mol}^{-1} = 0.00724 \text{ mol}$
For the solution made up, $c(C_6H_4(OH)COOH) = 0.00724 \text{ mol} / 0.460 \text{ L} = 0.0157 \text{ mol L}^{-1}$
Ionization of salicylic acid produces $H_3O^+(aq)$ ions
Extent of reaction before equilibrium, $x = [H_3O^+] = 10^{-pH} = 10^{-2.4} = 0.00398 \text{ mol L}^{-1}$
At equilibrium $[C_6H_4(OH)COOH] = 0.0157 - 0.00398 \text{ mol L}^{-1} = 0.0117 \text{ mol L}^{-1}$
and salicylate concentration $[C_6H_4(OH)COO^-] = 0.00398 \text{ mol L}^{-1}$

$$K_a = \frac{[C_6H_4(OH)COO^-][H_3O^+]}{[C_6H_4(OH)COOH]} = \frac{(0.00398)^2}{(0.0117)} = 0.00135$$

(b) We can use the speciation equation 14.1

$$\frac{[C_6H_4(OH)COOH]}{[C_6H_4(OH)COO^-]} = \frac{[H_3O^+]}{K_a} = \frac{1.0 \times 10^{-2}}{1.35 \times 10^{-3}} = \frac{7.4}{1}$$

So % as $C_6H_4(OH)COO^-(aq)$ ions $= (1 \times 100\%)(1/(1 + 7.4) = 12\%$

(c) At the halfway point of the titration, $[C_6H_4(OH)COOH] = [C_6H_4(OH)COO^-]$ and pH = $pK_a = 2.86$.
At the equivalence point, the solution is identical to a solution of sodium salicylate. What is its concentration?

Initial amount of salicylic acid $= 0.043 \text{ L} \times 0.014 \text{ mol L}^{-1} = 0.00035 \text{ mol}$
Amount of $OH^-(aq)$ ions to neutralize acid $= 0.00035 \text{ mol}$
Volume of sodium hydroxide solution $= 0.00035 \text{ mol} / 0.010 \text{ mol L}^{-1} = 0.035 \text{ L}$
Total amount of salicylate $= 0.00035 \text{ mol}$
$c(C_6H_4(OH)COO^-) = 0.00035 \text{ mol} / 0.060 \text{ L} = 0.0058 \text{ mol L}^{-1}$
The solution is basic because of the basic hydrolysis reaction of $C_6H_4(OH)COO^-(aq)$ ions.
Substituting equilibrium concentrations into the reaction quotient:

$$Q = \frac{[C_6H_4(OH)COOH][OH^-]}{[C_6H_4(OH)COO^-]} = \frac{(x)(x)}{(0.0058 - x)} = K_b(C_6H_4(OH)COO^-)$$

$$= \frac{K_w}{K_a(C_6H_4(OH)COOH)} = \frac{1 \times 10^{-14}}{0.00137} = 7.30 \times 10^{-12}$$

From this we obtain a quadratic equation: $x^2 + (7.30 \times 10^{-12}) x - 4.26 \times 10^{-14} = 0$
with solution $x = 2.1 \times 10^{-7} \text{ mol L}^{-1}$
Now we can write the equilibrium concentrations of $OH^-(aq)$ ions:

$$[OH^-] = x = 2.1 \times 10^{-7} \text{ mol L}^{-1}$$

$$[H_3O^+] = \frac{K_w}{[OH^-]} = \frac{1.0 \times 10^{-14}}{2.1 \times 10^{-7}} = 4.8 \times 10^{-8} \text{ mol L}^{-1}$$

Finally, $pH = -\log_{10}(4.8 \times 10^{-8}) = 7.32$
This is only very slightly alkaline, as we might expect for such a weak base as $C_6H_4(OH)COO^-(aq)$ ions.

Chapter 15
Solubility, Precipitation, and Complexation

IN-CHAPTER EXERCISES

Exercise 15.1—Reaction quotients for slightly soluble salts in aqueous solution

(a) $AgI(s) \rightleftharpoons Ag^+(aq) + I^-(aq)$

$Q = [Ag^+][I^-] = K_{sp} = 1.5 \times 10^{-16}$ at equilibrium

(b) $BaF_2(s) \rightleftharpoons Ba^{2+}(aq) + 2\,F^-(aq)$

$Q = [Ba^{2+}][F^-]^2 = K_{sp} = 1.7 \times 10^{-6}$ at equilibrium

(c) $Ag_2CO_3(s) \rightleftharpoons 2\,Ag^+(aq) + CO_3^{2-}(aq)$

$Q = [Ag^+]^2[CO_3^{2-}] = K_{sp} = 8.1 \times 10^{-12}$ at equilibrium

Exercise 15.4—K_{sp} from ion concentrations in saturated solution

$BaF_2(s) \rightleftharpoons Ba^{2+}(aq) + 2\,F^-(aq)$
If $[Ba^{2+}] = 3.6 \times 10^{-3}$ mol L^{-1},
$[F^-] = 2 \times [Ba^{2+}] = 7.2 \times 10^{-3}$ mol L^{-1}
and

$$Q = [Ba^{2+}][F^-]^2 = (3.6 \times 10^{-3})(7.2 \times 10^{-3})^2 = 1.9 \times 10^{-7} = K_{sp}$$

at the temperature of the measurement, since the solid is in equilibrium with the solution containing Ba^{2+} (aq) and F^-(aq) ions.

Exercise 15.6—Solubility from K_{sp}

$$Ca(OH)_2(s) \rightleftharpoons Ca^{2+}(aq) + 2\,OH^-(aq)$$

At equilibrium, if the concentration of Ca^{2+}(aq) ions is x mol L^{-1}, then the concentration of OH^-(aq) ions is $2\,x$ mol L^{-1}, and

$$Q = [Ca^{2+}][OH^-]^2 = x\,(2x)^2 = 4x^3$$

$$= K_{sp} = 7.9 \times 10^{-6}$$

(a) Therefore,

$$\text{Solubility of Ca(OH)}_2 = x = \sqrt[3]{\frac{K_{sp}}{4}} = \sqrt[3]{\frac{7.9 \times 10^{-6}}{4}} = 1.3 \times 10^{-2} \text{ mol } L^{-1}$$

(b) This corresponds to
$$1.3 \times 10^{-2} \text{ mol } L^{-1} \times 74.09 \text{ g mol}^{-1} = 9.6 \times 10^{-1} \text{ g } L^{-1}$$

Exercise 15.8—Comparing solubilities from solubility products

(a) At 25 °C, $K_{sp}(AgCl) = 1.8 \times 10^{-10}$ and $K_{sp}(AgCN) = 1.2 \times 10^{-16}$
Because the reaction quotients have the same form for the dissolution of these species (both salts are 1:1 salts), we can compare solubilities simply by comparing K_{sp} values. Therefore, AgCl(s) is more soluble than AgCN(s) at 25 °C.
(b) At 25 °C, $K_{sp}(Mg(OH)_2) = 1.5 \times 10^{-11}$ and $K_{sp}(Ca(OH)_2) = 7.9 \times 10^{-6}$
Both are 1:2 hydroxides, so their relative solubilities can be deduced from the relative magnitudes of the solubility products.
$Ca(OH)_2$ is more soluble than $Mg(OH)_2$ at 25 °C.
(c) At 25 °C, $K_{sp}(Ca(OH)_2) = 7.9 \times 10^{-6}$ and $K_{sp}(CaSO_4) = 2.4 \times 10^{-5}$
Since $Ca(OH)_2(s)$ is a 1:2 compound, and $CaSO_4(s)$ is a 1:1 salt, we cannot deduce relative solubilities directly by comparison of the magnitudes of the K_{sp} values. Solubilities must be computed.
The solubility of $Ca(OH)_2(s)$ was determined in Exercise 15.6.

$$\text{Solubility of } Ca(OH)_2 = x = \sqrt[3]{\frac{K_{sp}}{4}} = \sqrt[3]{\frac{7.9 \times 10^{-6}}{4}} = 1.3 \times 10^{-2} \text{ mol L}^{-1}$$

$$CaSO_4(s) \rightleftharpoons Ca^{2+}(aq) + SO_4^{2-}(aq)$$

So at equilibrium, $[Ca^{2+}] = [SO_4^{2-}]$
The solubility of $CaSO_4(s)$ is $\{K_{sp}(CaSO_4)\}^{1/2} = 4.9 \times 10^{-3}$ mol L^{-1}
So we see that $Ca(OH)_2(s)$ is more soluble than $CaSO_4(s)$ at 25 °C, even though its solubility product is smaller.

Exercise 15.10—Participation of ions in other reactions

$$Ag_3PO_4(s) \rightleftharpoons 3 Ag^+(aq) + PO_4^{3-}(aq) \qquad K_{sp} = [Ag^+]^3 [PO_4^{3-}]$$

The solubility of $Ag_3PO_4(s)$ is larger than that predicted using K_{sp} of the salt because the aquated phosphate ion, $PO_4^{3-}(aq)$, is a weak base and reacts with water to form $HPO_4^{2-}(aq)$ ions (as well as smaller concentrations of $H_2PO_4^-(aq)$ ions and $H_3PO_4(aq)$ upon further reaction with water). If we ignore the reaction of $PO_4^{3-}(aq)$ ions as a base, we would presume that $[Ag^+] = 3 \times [PO_4^{3-}]$. In fact, $[PO_4^{3-}]$ is less than this assumption implies, so $[Ag^+]$ must be larger to maintain the $Q = K_{sp}$ condition.

Exercise 15.13—The common ion effect and salt solubility

(a) In pure water

	$[Ba^{2+}]$ / (mol L^{-1})	$[SO_4^{2-}]$ / (mol L^{-1})
Initial	0	0
Change in concentration	$+x$	$+x$
Equilibrium concentration	x	X

So $x = \{K_{sp}(BaSO_4)\}^{1/2} = \{1.1 \times 10^{-10}\}^{1/2} = 1.0 \times 10^{-5}$ mol L^{-1}

(b) In the presence of 0.010 mol L^{-1} Ba(NO$_3$)$_2$.

	$[Ba^{2+}]$ / (mol L^{-1})	$[SO_4^{2-}]$ / (mol L^{-1})
Initial (before adding BaSO$_4$)	0.010	0
Change in concentration	$+x$	$+x$
Equilibrium concentration	$0.010 + x$	X

Now,

$$Q = [Ba^{2+}][SO_4^{2-}] = K_{sp} = 1.1 \times 10^{-10}$$

$$= (0.010 + x)x$$

which gives the following quadratic equation

$$x^2 + 0.010\,x - 1.1 \times 10^{-10} = 0$$

with (sensible) solution,

$$x = 1.1 \times 10^{-8} \text{ mol L}^{-1}$$

Exercise 15.15—pH-dependence of solubility of salts

(a) PbS(s) is more soluble in nitric acid solution than in pure water because acidic conditions increase the extent of the reactions of S^{2-}(aq) ions as a moderately weak base

$$S^{2-}(aq) + H_3O^+(aq) \rightleftharpoons HS^-(aq) + H_2O(l)$$

and $HS^-(aq) + H_3O^+(aq) \rightleftharpoons H_2S(aq) + H_2O(l)$
decreasing the concentration of S^{2-}(aq) ions , so that more PbS(s) must dissolve to achieve the condition $Q = K_{sp}$(PbS). Cl$^-$(aq) ions undergo negligible reaction as a base, because HCl(aq) is a strong base. Their concentration is not reduced by acidifying the solution. So, the solubility of PbS(s) increases more than the solubility of PbCl$_2$(s) does.

(b) The solubility of Ag$_2$CO$_3$(s) is greater in acidic solution than in pure water because, like sulfide in part (a), CO$_3^{2-}$(aq) ions are a weak base, so their concentration is reduced as a result of reaction with H$_3$O$^+$(aq) ions. HI(aq) is a strong acid, so I$^-$(aq) ions undergo negligible reaction as a base, and the solubility of AgI(s) is not changed by acidifying of the solution.

(c) The solubility of Al(OH)$_3$(s)—as measured by the concentration of Al^{3+}(aq) ions in solution—is greater in acidic solutions than in pure water because aquated hydroxide ions, OH$^-$(aq), are a strong base neutralized by H$_3$O$^+$(aq) ions.

Exercise 15.16—Is the solution saturated?

$$PbI_2(s) \rightleftharpoons Pb^{2+}(aq) + 2\,I^-(aq)$$
$$Q = [Pb^{2+}][I^-]^2 = (1.1 \times 10^{-3})(2 \times 1.1 \times 10^{-3})^2 = 5.3 \times 10^{-9}$$

$$< K_{sp} = 9.8 \times 10^{-9}$$

The solution is not saturated. More PbI$_2$(s) can dissolve.

Exercise 15.19—Mixing solutions: Is the mixture saturated?

$$Q = [Sr^{2+}][SO_4{}^{2-}] = (2.5 \times 10^{-4})(2.5 \times 10^{-4}) = 6.3 \times 10^{-8}$$

$$< K_{sp} = 3.4 \times 10^{-7}$$

The mixture is not saturated. $SrSO_4(s)$ will not precipitate.

Exercise 15.20—Adjusting the concentration of one ion

$PbI_2(s)$ will precipitate if $[I^-]$ is such that Q exceeds K_{sp}. $Q = K_{sp}$ for the minimum such concentration—i.e., when the solution is <u>just</u> saturated.

$$Q = [Pb^{2+}][I^-]^2 = (0.050)\,[I^-]^2 = K_{sp} = 9.8 \times 10^{-9}$$

Solving for $[I^-]$, we get

$$[I^-] = (\,9.8 \times 10^{-9} / 0.050\,)^{1/2} = 4.4 \times 10^{-4} \text{ mol L}^{-1}$$

When the concentration of I^- is 0.0015 mol L^{-1}, which exceeds the minimum value needed, $PbI_2(s)$ precipitates until the lead concentration is such that $Q = K_{sp}$.

$$Q = [Pb^{2+}][I^-]^2 = [Pb^{2+}](0.0015)^2 = K_{sp} = 9.8 \times 10^{-9}$$

i.e., $[Pb^{2+}] = 9.8 \times 10^{-9} / (0.0015)^2 = 4.4 \times 10^{-3} \text{ mol L}^{-1}$

Exercise 15.22—Separation of metals by selective precipitation

Because K_{sp} for copper hydroxide is much smaller than that for magnesium hydroxide, it is possible to separate the metals via precipitation of $Cu(OH)_2(s)$ before any significant precipitation of $Mg(OH)_2(s)$. The maximum separation is achieved by adjusting the concentration of OH$^-$(aq) ions that just makes the solution a saturated solution of $Mg(OH)_2$. If $[OH^-]$ is not increased beyond this, no $Mg(OH)_2(s)$ actually precipitates. At the same time, this concentration of hydroxide produces the lowest concentration of Cu^{2+}(aq) ions remaining in solution.
To achieve a solution saturated with respect to $Mg(OH)_2$:

$$Q = [Mg^{2+}][OH^-]^2 = (0.200)\,[OH^-]^2 = K_{sp} = 5.6 \times 10^{-12}$$

i.e., $[OH^-] = (\,5.6 \times 10^{-12} / 0.200\,)^{1/2} = 5.3 \times 10^{-6} \text{ mol L}^{-1}$

Exercise 15.24—Competition: Precipitation vs. complexation

$$Cu(OH)_2(s) \rightleftharpoons Cu^{2+}(aq) + 2\,OH^-(aq) \qquad K_{sp} = 1.6 \times 10^{-19}$$

$$\underline{Cu^{2+}(aq) + 4\,NH_3(aq) \rightleftharpoons Cu(NH_3)_4{}^{2+}(aq) \qquad K_f = 6.8 \times 10^{12}}$$

$$Cu(OH)_2(s) + 4\,NH_3(aq) \rightleftharpoons Cu(NH_3)_4{}^{2+}(aq) + 2\,OH^-(aq)$$

$$K_{net} = K_{sp} \times K_f = 1.6 \times 10^{-19} \times 6.8 \times 10^{12} = 1.1 \times 10^{-6}$$

Exercise 15.25—Complexation vs. Lewis base protonation

At pH = 1, the cyanide is entirely present as aquated hydrogen cyanide, HCN(aq). Silver appears only as Ag^+(aq) or $[Ag(OH_2)_n]^+$(aq). As pH of solution is increased, an increasing fraction of HCN(aq) molecules are de-protonated, and at pH = pK_a, [HCN] = [CN⁻]. As pH increases past pK_a = $-\log_{10}(3.5 \times 10^{-4}) = 3.5$, [CN⁻] > [HCN].

Depending on the relative amounts of cyanide and silver, some or all of the cyanide ions form complexes with silver ions, Ag^+(aq) – the Ag^+(aq) concentration decreases and the $[Ag(CN)_2]^-$ increases.

The concentrations remain constant from pH about 6 (i.e., when very little HCN(aq) remains) all the way to 13.

REVIEW PROBLEMS

15.2 Solubility and precipitation of ionic salts

15.25

(a) $PbSO_4(s) \rightleftharpoons Pb^{2+}(aq) + SO_4^{2-}(aq)$

$Q = [Pb^{2+}][SO_4^{2-}] = K_{sp} = 1.8 \times 10^{-8}$ at equilibrium.

(b) $NiCO_3(s) \rightleftharpoons Ni^{2+}(aq) + CO_3^{2-}(aq)$

$Q = [Ni^{2+}][CO_3^{2-}] = K_{sp} = 1.4 \times 10^{-7}$ at equilibrium.

(c) $Ag_3PO_4(s) \rightleftharpoons 3Ag^+(aq) + PO_4^{3-}(aq)$

$Q = [Ag^+]^3[PO_4^{3-}] = K_{sp} = 1.3 \times 10^{-20}$ at equilibrium.

15.27

$SrF_2(s) \rightleftharpoons Sr^{2+}(aq) + 2 F^-(aq)$

Note that $[F^-] = 2[Sr^{2+}]$.

$K_{sp} = [Sr^{2+}][F^-]^2 = (1.0 \times 10^{-3})(2 \times 1.0 \times 10^{-3})^2 = 4.0 \times 10^{-9}$

15.29

Amount of lead sulfate $= 0.055\ g\ /\ 303.26\ g\ mol^{-1} = 1.8 \times 10^{-4}\ mol$

If all of the lead sulfate dissolved, its concentration would be $1.8 \times 10^{-4}\ mol\ /\ 0.100\ L = 1.8 \times 10^{-3}\ mol\ L^{-1}$.

The appropriate reaction quotient is

$Q = [Pb^{2+}][SO_4^{2-}] = (1.8 \times 10^{-3})(1.8 \times 10^{-3}) = 3.2 \times 10^{-6}$

$> K_{sp} = 1.8 \times 10^{-8}$

Therefore, not all of the lead sulfate dissolves. The concentration of a saturated solution of $PbSO_4$ is

$(K_{sp})^{1/2} = 1.3 \times 10^{-4}\ mol\ L^{-1}$

This corresponds to $1.3 \times 10^{-4}\ mol\ L^{-1} \times 0.100\ L = 1.3 \times 10^{-5}\ mol$

and $1.3 \times 10^{-5}\ mol \times 303.26\ g\ mol^{-1} = 4.1 \times 10^{-3}\ g = 4.1\ mg$

i.e., most of the lead sulfate ($55 - 4.1 = 51$ mg) remains undissolved.

15.31

(a) $K_{sp}(PbCl_2(s)) = 1.7 \times 10^{-5}$ and $K_{sp}(PbBr_2(s)) = 6.3 \times 10^{-6}$

Since these solids ionize on dissolving with the same stoichiometry, solubility is in accord with K_{sp}. Therefore, we can say that $PbCl_2(s)$ is more soluble than $PbBr_2(s)$.

(b) $K_{sp}(HgS(s)) = 3.0 \times 10^{-53}$ and $K_{sp}(FeS(s)) = 4.9 \times 10^{-18}$

Since these are both 1:1 salts, we can say that $FeS(s)$ is more soluble than $HgS(s)$.

(c) $K_{sp}(Fe(OH)_2(s)) = 7.9 \times 10^{-15}$ and $K_{sp}(Zn(OH)_2(s)) = 4.5 \times 10^{-17}$

Both are 1:2 compounds, so we can conclude that $Fe(OH)_2(s)$ is more soluble than $Zn(OH)_2(s)$.

15.3 Precipitation Reactions

15.33
Amount of HCl added $= 0.0050 \text{ L} \times 0.025 \text{ mol L}^{-1} = 1.25 \times 10^{-4} \text{ mol}$
Concentration of $Cl^-(aq)$ ions in resulting solution $= 1.25 \times 10^{-4} \text{ mol} / 0.105 \text{ L} = 1.2 \times 10^{-3} \text{ mol L}^{-1}$
Concentration of $Ag^+(aq)$ ions in resulting solution $= 0.0010 \text{ mol L}^{-1} \times 0.100 \text{ L} / 0.105 \text{ L} = 9.5 \times 10^{-4} \text{ mol L}^{-1}$

$$Q = [Ag^+][Cl^-] = (9.5 \times 10^{-4})(1.2 \times 10^{-3}) = 1.1 \times 10^{-6}$$
$$> K_{sp} = 1.8 \times 10^{-8}$$

Therefore, AgCl will precipitate out of solution.

15.35
(a)
$$Q = [Ni^{2+}][CO_3^{2-}] = (2.4 \times 10^{-3})(1.0 \times 10^{-6}) = 2.4 \times 10^{-9}$$
$$< K_{sp} = 6.6 \times 10^{-9}$$

$NiCO_3$ will not precipitate.
(b)
$$Q = [Ni^{2+}][CO_3^{2-}] = (2.4 \times 10^{-3})(1.0 \times 10^{-4}) = 2.4 \times 10^{-7}$$
$$> K_{sp} = 6.6 \times 10^{-9}$$

$NiCO_3$ will precipitate.

15.37
In general, we cannot answer such a question simply by looking at the associated K_{sp} values, because the stoichiometry of dissociation of the lead hydroxide is different from that of iron(III) and aluminum hydroxides. We must determine the hydroxide concentration at the onset of precipitation for each of the three salts.
$Fe(OH)_3(s)$ precipitates when
$$Q = [Fe^{3+}][OH^-]^3 = (0.10)[OH^-]^3 = K_{sp} = 6.3 \times 10^{-38}$$
from which we get
$$[OH^-] = \{6.3 \times 10^{-38} / 0.10\}^{1/3} = 8.6 \times 10^{-13} \text{ mol L}^{-1}$$
$Pb(OH)_2(s)$ precipitates when
$$Q = [Pb^{2+}][OH^-]^2 = (0.10)[OH^-]^2 = K_{sp} = 2.8 \times 10^{-16}$$
from which we get
$$[OH^-] = \{2.8 \times 10^{-16} / 0.10\}^{1/2} = 5.3 \times 10^{-8} \text{ mol L}^{-1}$$
$Al(OH)_3(s)$ precipitates when
$$Q = [Al^{3+}][OH^-]^3 = (0.10)[OH^-]^3 = K_{sp} = 1.9 \times 10^{-33}$$
from which we get
$$[OH^-] = \{1.9 \times 10^{-33} / 0.10\}^{1/3} = 2.7 \times 10^{-11} \text{ mol L}^{-1}$$
We see that $Fe(OH)_3(s)$ precipitates first (i.e., at the lowest hydroxide concentration), followed by $Al(OH)_3(s)$. $Pb(OH)_2(s)$ is last to precipitate.

15.4 Solubility and complexation: competitive equilibria

15.39

$$AuCl(s) \rightleftharpoons Au^+(aq) + Cl^-(aq)$$

$$\underline{Au^+(aq) + 2\,CN^-(aq) \rightleftharpoons [Au(CN)_2]^-(aq)}$$

$$AuCl(s) + 2\,CN^-(aq) \rightleftharpoons [Au(CN)_2]^-(aq) + Cl^-(aq)$$

$$
\begin{aligned}
K_{net} &= K_{sp}\,K_f \\
&= 2.0 \times 10^{-13} \times 2.0 \times 10^{38} = 4.0 \times 10^{25}
\end{aligned}
$$

SUMMARY AND CONCEPTUAL QUESTIONS

15.41

At pH $= 12.68$, $[H^+] = 2.09 \times 10^{-13}$ mol L^{-1} and $[OH^-] = 4.79 \times 10^{-2}$ mol L^{-1}

The associated $Ca^{2+}(aq)$ ion concentration is $[Ca^{2+}] = [OH^-] / 2 = 2.40 \times 10^{-2}$ mol L^{-1} and

$$K_{sp} = [Ca^{2+}][OH^-]^2 = (2.40 \times 10^{-2})(4.79 \times 10^{-2})^2$$

$$= 5.51 \times 10^{-5}$$

15.43

The maximum concentration of $F^-(aq)$ ions that can be present without precipitating $CaF_2(s)$ is such that the reaction quotient is just equal to solubility product.

$$Q = [Ca^{2+}][F^-]^2 = (2.0 \times 10^{-3})[F^-]^2 = K_{sp} = 3.9 \times 10^{-11}$$

which gives

$[F^-] = \{3.9 \times 10^{-11} / 2.0 \times 10^{-3}\}^{1/2} = 1.4 \times 10^{-4}$ mol L^{-1}

Chapter 16
Electron Transfer Reactions and Electrochemistry

IN-CHAPTER EXERCISES

Exercise 16.2—Calculating oxidation states

(a) $\underline{Fe}_2O_3$
The sum of oxidation states equals zero for a neutral compound.
$2 \times$ Oxidation State(Fe) $+ 3 \times$ Oxidation State(O) $= 0$
$2 \times$ Oxidation State(Fe) $+ 3 \times (-2) = 0$
Oxidation State(Fe) $= 3 \times 2 /2 = +3$

(b) $H_2\underline{S}O_4$
The sum of oxidation states equals zero for a neutral compound.
$2 \times$ Oxidation State(H) $+$ Oxidation State(S) $+ 4 \times$ Oxidation State(O) $= 0$
$2 \times 1 +$ Oxidation State(S) $+ 4 \times (-2) = 0$
Oxidation State(S) $= 4 \times 2 - 2 \times 1 = +6$

(c) $\underline{C}O_3^{2-}$
The sum of oxidation states equals the charge on an ionic species.
Oxidation State(C) $+ 3 \times$ OxidationState(O) $= -2$
Oxidation State(C) $+ 3 \times (-2) = -2$
Oxidation State(C) $= -2 + 3 \times 2 = +4$

(d) $\underline{N}O_2^{+}$
The sum of oxidation states equals the charge on an ionic species.
Oxidation State(N) $+ 2 \times$ Oxidation State(O) $= +1$
Oxidation State(N) $+ 2 \times (-2) = +1$
Oxidation State(C) $= 1 + 2 \times 2 = +5$

Exercise 16.4—Recognizing oxidation and reduction

$$3C_2H_5OH(aq) + 2Cr_2O_7^{2-}(aq) + 16H^+(aq) \longrightarrow 3CH_3COOH(aq) + 4Cr^{3+}(aq) + 11H_2O(l)$$

Left Side: Right Side:
$OS(C) = (0 - 6\times1 - (-2))/2 = -2$ $OS(C) = (0 - 4\times1 - 2\times(-2))/2 = 0$
-2 is is the average $OS(C)$ in ethanol. It is possible to assign separate oxidation states to the two carbon atoms in ethanol, -3 and -1 (the C atom bonded to the O atom).
0 is the average $OS(C)$ in acetic acid. It is possible to assign separate oxidation states to the two carbon atoms in acetic acid, -3 and $+3$ (the C atom bonded to two O atoms).
$OS(Cr) = (-2 - 7\times(-2))/2 = +6$ $OS(Cr) = +3$
The reaction increases the oxidation state of carbon, $OS(C)$. $CH_3CH_2OH(aq)$ is oxidized. $CH_3CH_2OH(aq)$ is the reducing agent. The reaction decreases the oxidation state of chromium, $OS(Cr)$. $Cr_2O_7^{2-}(aq)$ ions are reduced. $Cr_2O_7^{2-}(aq)$ ions are the oxidizing agent.

Exercise 16.6—Voltaic cells

Reduction occurs at the cathode. In accord with the reduction half-reaction, a silver electrode immersed in an aqueous solution containing $Ag^+(aq)$ ions (such as an $AgNO_3$ solution) provides the cathode. A nickel electrode immersed in a solution containing $Ni^{2+}(aq)$ ions (such as a $NiNO_3$ solution) provides the anode—where oxidation occurs.

The equation for the overall cell reaction is obtained by adding the half-reactions so as the number of electrons removed is the same as the number of electrons gained:

$$2 \times \left(Ag^+(aq) + e^- \longrightarrow Ag(s) \right)$$
$$\underline{Ni(s) \longrightarrow Ni^{2+}(aq) + 2\,e^-}$$
$$2\,Ag^+(aq) + Ni(s) \longrightarrow 2\,Ag(s) + Ni^{2+}(aq)$$

Anions in the salt bridge flow from cathode to anode within the salt bridge, while cations in the salt bridge flow in the opposite direction (toward the cathode).

Exercise 16.7—Electrochemical cell conventions

(a) $Zn(s)|Zn^{2+}(aq)\|Ni^{2+}(aq)|Ni(s)$

(b) $C(s)|Fe^{2+}(aq)|Fe^{3+}(aq)\|Ag^+(aq)|Ag(s)$

(c) $Mg(s)|Mg^{2+}(aq)\|Br_2(l)|Br^-(aq)|Pt(s)$

Cell diagrams:

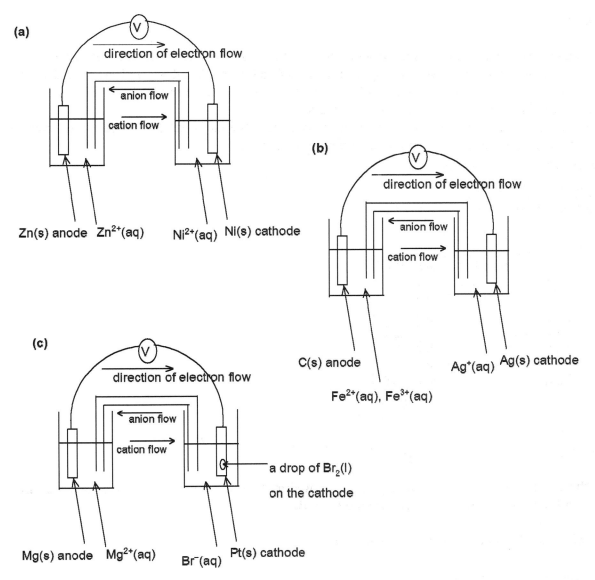

(a)

direction of electron flow

anion flow

cation flow

Zn(s) anode   Zn^{2+}(aq)   Ni^{2+}(aq) Ni(s) cathode

(b)

direction of electron flow

anion flow

cation flow

C(s) anode   Ag$^+$(aq) Ag(s) cathode

Fe^{2+}(aq), Fe^{3+}(aq)

(c)

direction of electron flow

anion flow

cation flow

a drop of Br$_2$(l)

on the cathode

Mg(s) anode   Mg^{2+}(aq)   Br$^-$(aq)   Pt(s) cathode

Exercise 16.8—Half-cell reduction potentials from experimental measurements

(a) The measured cell potential is the difference between the half-cell reduction potentials ($E_{cathode} - E_{anode}$). Here, the standard hydrogen electrode is the cathode—its half-cell reduction potential is (by convention) 0 V. The specified conditions tell us that the Ni^{2+}(aq)|Ni(s) half-cell is a standard half-cell, so we can use the symbol $E°$—for this half-cell as well as for the cell.

$$E°_{cell} = 0.25\ V = 0\ V - E°[\ Ni^{2+}(aq)|Ni(s)\ ]$$

i.e., $E°[\ Ni^{2+}(aq)|Ni(s)\ ] = -0.25\ V$

(b) In this case, we are given neither the [Cl$^-$], nor $p(Cl_2)$, so we cannot presume the Cl$_2$(g)|Cl$^-$(aq) half-cell is a standard cell, so we use the symbol E (not $E°$) for half-cell and cell.

$$E_{cell} = 1.36\ V = E[Cl_2(g)|Cl^-(aq)|Pt(s)\ ] - 0\ V$$

i.e., $E[Cl_2(g)|Cl^-(aq)|Pt(s)] = 1.36\ V$

Exercise 16.9—Standard half-cell reduction potentials

This is a similar exercise to Exercise 16.8(b), except that the conditions here specify that the $Cl_2(g)|Cl^-(aq)$ half-cell is a standard cell, so we use the symbol $E°$ for this half-cell and for the cell.

$$E°_{cell} = 1.36 \text{ V} = E°[Cl_2(g)|Cl^-(aq)|Pt(s)] - 0 \text{ V}$$

i.e., $E°[Cl_2(g)|Cl^-(aq)|Pt(s)] = 1.36 \text{ V}$

Exercise 16.13—$E°_{cell}$ from standard half-cell reduction potentials

$$Zn(s) \rightarrow Zn^{2+}(aq) + 2 \text{ e}^- \qquad \text{at the anode}$$
$$\underline{2 \times (Ag^+(aq) + e^- \rightarrow Ag(s))} \qquad \text{at the cathode}$$
$$Zn(s) + 2 \text{ Ag}^+(aq) \rightarrow Zn^{2+}(aq) + 2 \text{ Ag}(s)$$

So the standard cell emf is

$$E°_{cell} = E°[Ag^+(aq)|Ag(s)] - E°[Zn^{2+}(aq)|Zn(s)]$$
$$= 0.7994 \text{ V} - (-0.763 \text{ V}) = 1.56 \text{ V}$$

Exercise 16.14—$E°_{cell}$ from standard half-cell reduction potentials

(a) $2 \text{ I}^-(aq) + Zn^{2+}(aq) \longrightarrow I_2(s) + Zn(s)$

$$E°_{cell} = E°[Zn^{2+}(aq)|Zn(s)] - E°[I_2(s)|I^-(aq)]$$
$$= -0.763 \text{ V} - 0.535 \text{ V} = -1.298 \text{ V}$$

Because we get a negative value, we know that the spontaneous cell reaction is in the reverse direction to that written—iodine and iodide ions are at the cathode and zinc and $Zn^{2+}(aq)$ ions are at the anode.

(b) $Zn^{2+}(aq) + Ni(s) \longrightarrow Zn(s) + Ni^{2+}(aq)$

$$E°_{cell} = E°[Zn^{2+}(aq)|Zn(s)] - E°[Ni^{2+}(aq)|Ni(s)]$$
$$= -0.763 \text{ V} - (-0.25 \text{ V}) = -0.51 \text{ V}$$

Because we get a negative value, we know that the spontaneous cell reaction is in the reverse direction to that written—nickel is the cathode and zinc is the anode.

(c) $2 \text{ Cl}^-(aq) + Cu^{2+}(aq) \longrightarrow Cu(s) + Cl_2(g)$

$$E°_{cell} = E°[Cu^{2+}(aq)|Cu(s)] - E°[Cl_2(g)|Cl^-(aq)]$$
$$= 0.337 \text{ V} - (1.36 \text{ V}) = -1.023 \text{ V}$$

Because we get a negative value, we know that the spontaneous cell reaction is in the reverse direction to that written—chlorine and aquated chloride ions are at the cathode and copper and aquated copper(II) ions are at the anode.

(d) $Fe^{2+}(aq) + Ag^+(aq) \longrightarrow Fe^{3+}(aq) + Ag(s)$

$$E°_{cell} = E°[Ag^+(aq)|Ag(s)] - E°[Fe^{3+}(aq),Fe^{2+}(aq)|Pt]$$
$$= 0.799 \text{ V} - (0.771 \text{ V}) = 0.028 \text{ V}$$

Because we get a positive value, we know that the spontaneous cell reaction is in this direction—silver is the cathode and $Fe^{3+}(aq)$ and $Fe^{2+}(aq)$ ions are at the anode.

Exercise 16.17— Relative oxidizing and reducing abilities

$E°[Cu^{2+}(aq)|Cu(s)] = 0.337$ V $E°[Sn^{2+}(aq)|Sn(s)] = -0.14$ V $E°[Fe^{2+}(aq)|Fe(s)] = -0.44$ V

$E°[Zn^{2+}(aq)|Zn(s)] = -0.763$ V $E°[Al^{3+}(aq)|Al(s)] = -1.66$ V
(a)   Aluminum is the most easily oxidized of these metals.
(b)   Aluminum and zinc are capable of reducing $Fe^{2+}(aq)$ ions to Fe(s).
(c) $Fe^{2+}(aq) + Sn(s) \rightarrow Fe(s) + Sn^{2+}(aq)$
Because the reduction potential for $Sn^{2+}(aq)|Sn(s)$ is greater than that for $Fe^{2+}(aq)|Fe(s)$, this is NOT the direction of spontaneous reaction. E_{cell} is negative.
(d) $Zn^{2+}(aq) + Sn(s) \rightarrow Zn(s) + Sn^{2+}(aq)$
Because the reduction potential for $Sn^{2+}(aq)|Sn(s)$ is greater than that for $Zn^{2+}(aq)|Zn(s)$, this is NOT the direction of spontaneous reaction.

Exercise 16.20—Cell emf under non-standard conditions at 25 °C
$E°_{cell} = E°[Fe^{2+}(aq)|Fe(s)] - E°[Al^{3+}(aq)|Al(s)] = -0.44$ V $- (-1.66$ V $) = 1.22$ V
$3 \times (Fe^{2+}(aq) + 2 e^- \rightarrow Fe(s))$
$\underline{2 \times (Al(s) \rightarrow Al^{3+}(aq) + 3 e^-)}$
$3 Fe^{2+}(aq) + 2 Al(s) \rightarrow 3 Fe(s) + 2 Al^{3+}(aq)$

$$Q = \frac{[Al^{3+}]^2}{[Fe^{2+}]^3} = \frac{(0.025)^2}{(0.50)^3} = 0.0050$$

$$E_{cell} = E°_{cell} - \frac{0.0257 \text{ V}}{n} \times \ln Q = +1.22 \text{ V} - \frac{0.0257 \text{ V}}{6} \times \ln(0.0050)$$

$$= +1.22 \text{ V} - \frac{0.0257 \text{ V}}{6} \times (-5.3) = +1.24 \text{ V}$$

Exercise 16.22—Ion concentration from cell emf

The $Zn^{2+}(aq)|Zn(s)$ half-cell is a standard half-cell, but it is most unlikely that $Ag^+(aq)|Ag(s)$ is a standard half-cell, so we use the symbol E (not $E°$) for this half-cell and for the cell. First, we do need to know the value of an imagined standard cell comprised of the same components.
$E°_{cell} = E°[Ag^+(aq)|Ag(s)] - E°[Zn^{2+}(aq)|Zn(s)] = 0.799$ V $- (-0.763$ V $) = 1.562$ V
$2 \times (Ag^+(aq) + e^- \rightarrow Ag(s))$
$\underline{Zn(s) \rightarrow Zn^{2+}(aq) + 2 e^-}$
$2 Ag^+(aq) + Zn(s) \rightarrow 2 Ag(s) + Zn^{2+}(aq)$

$$Q = \frac{[Zn^{2+}]}{[Ag^+]^2} = \frac{1.0}{[Ag^+]^2}$$

$$E_{cell} = E°_{cell} - \frac{0.0257 \text{ V}}{n} \times \ln Q = +1.562 \text{ V} - \frac{0.0257 \text{ V}}{2} \times \ln\left(\frac{1.0}{[Ag^+]^2}\right) = +1.48 \text{ V}$$

i.e,

$$\frac{0.0257\,\text{V}}{2} \times \ln\left(\frac{1.0}{[Ag^+]^2}\right) = 1.562 - 1.48\,\text{V} = 0.082\,\text{V}$$

$$\ln\left(\frac{1.0}{[Ag^+]^2}\right) = \frac{2}{0.0257\,\text{V}} \times 0.082\,\text{V} = 6.4$$

$$\frac{1.0}{[Ag^+]^2} = e^{6.4} = 602$$

$$[Ag^+] = 1.0/\sqrt{602} = 0.041\,\text{mol L}^{-1}$$

Alternatively, you can use $E_{cell} = E[Ag^+(aq)|Ag(s)] - E°[Zn^{2+}(aq)|Zn(s)]$ to calculate $E[Ag^+(aq)|Ag(s)]$. Then you can apply the Nernst equation to this half-cell, using the appropriate reduction half-equation and the tabulated value of $E°[Ag^+(aq)|Ag(s)]$. Check that you get the same answer.

Exercise 16.23—pH-dependence of oxidizing power of oxoanions

$E°[MnO_4^-(aq)|Mn^{2+}(aq)] = 1.51\,\text{V}$
$MnO_4^-(aq) + 8\,H^+ + 5\,e^- \rightarrow Mn^{2+}(aq) + 4\,H_2O(l)$
(a) At pH = 0.0, $[H^+] = 1.0\,\text{mol L}^{-1}$

$$Q = \frac{[Mn^{2+}]}{[MnO_4^-][H^+]^8} = \frac{1.00}{(1.00)[H^+]^8} = \frac{1.00}{(1.00)(1.00\times10^{-0})^8} = 1.00$$

and

$$E_{cell} = E°_{cell} - \frac{0.0257\,\text{V}}{n} \times \ln Q = +1.51\,\text{V} - \frac{0.0257\,\text{V}}{5} \times \ln(1.00)$$

$$= +1.51\,\text{V} - \frac{0.0257\,\text{V}}{5} \times (0) = +1.51\,\text{V}$$

(b) At pH = 5.0, $[H^+] = 1 \times 10^{-5.0}\,\text{mol L}^{-1}$

$$Q = \frac{[Mn^{2+}]}{[MnO_4^-][H^+]^8} = \frac{1.00}{(1.00)[H^+]^8} = \frac{1.00}{(1.00)(1.00\times10^{-5.0})^8} = 1.00\times10^{40}$$

and

$$E_{cell} = E°_{cell} - \frac{0.0257\,\text{V}}{n} \times \ln Q = +1.51\,\text{V} - \frac{0.0257\,\text{V}}{5} \times \ln(1.00\times10^{40})$$

$$= +1.04\,\text{V}$$

The half-cell reduction potential is greater at pH = 0 than at pH 5.0. This means permanganate is a more powerful oxidizing agent at pH = 0 than at pH 5.0.

Exercise 16.25—$E°_{cell}$ and K for the cell reaction

$E°_{cell} = E°[Ag^+(aq)|Ag(s)] - E°[Hg^{2+}(aq)|Hg(l)] = 0.799\ V - (0.855\ V) = -0.56\ V$
This reaction is NOT the spontaneous direction of reaction under standard conditions. At 25 °C,

$$\ln K = \frac{2}{0.0257\ V} \times E°_{cell} = \frac{2}{0.0257\ V} \times (-0.56\ V)$$
$$= -43.6$$

and

$$K = 1.2 \times 10^{-19}$$

Exercise 16.27—Electrolysis

In an aqueous sodium hydroxide solution, the reduction of H_2O is the principal cathode process.
$$2\ H_2O(l) + 2\ e^- \longrightarrow H_2(g) + 2\ OH^-(aq)$$
Presuming standard conditions – $[OH^-] = 1.0\ mol\ L^{-1}$ and $p(H_2) = 1.0$ bar:
$E°[H_2O(l)|H_2(g)] = -0.8277\ V$
The oxidation of $OH^-(aq)$ ions is the principal anode process.
$$4\ OH^-(aq) \longrightarrow O_2(g) + 2\ H_2O(l) + 4\ e^-$$
$E°[O_2(g)|OH^-(aq)] = 0.40\ V$
$E°_{cell} = E°[H_2O(l)|H_2(g)] - E°[O_2(g)|OH^-(aq)] = -0.8277\ V - (0.40\ V) = -1.23\ V$
The minimum applied voltage required to make this reaction go is 1.23 V.

Exercise 16.28—Corrosion of iron

The brass fittings provide cathodes where oxygen is reduced to water. The steel provides the anode where iron is oxidized to Fe^{2+}. Of course, the steel can provide its own cathode and can corrode without brass fittings. However, because copper (the principal component of brass) is a more noble metal than iron, an electric potential is created—in essence a voltaic cell with brass the cathode and iron the anode. Electrons are driven from the iron (which is oxidized) to the brass (where oxygen is reduced).
The rate of corrosion varies from city to city because it depends upon the ionic strength—the total concentration of ions—of the city water. Ions in the water carry current from the anode to cathode—to complete the circuit. The conductivity of the water increases with ion concentration.

Exercise 16.29—Corrosion of iron

The main reason for corrosion at the water line is differential aeration—the different concentrations of O_2 in the air and dissolved in the water. This creates a difference in reduction potentials, giving a driving force for electrons (from where the concentration is least, to where it is highest). Parts of the pier out of the water have almost uniform O_2 concentrations so a cell is not set up. The deeper the water, the greater is the difference between O_2 concentrations in the air and in the water. But also the deeper the water, the greater is the resistance to movement of ions through the water. The optimum 'trade-off' between increasing potential (due to increasing difference of O_2 concentrations) and increasing resistance, occurs not far below the water line.

Exercise 16.30—Protection against corrosion of iron

(i) Tin impedes corrosion of iron by coating it, preventing oxygen from contacting the iron. If the tin coating gets scratched, the iron will corrode quickly at the scratch because the more noble metal tin provides a preferred cathode for reduction of oxygen.

(ii) Zinc impedes corrosion of iron by providing a "sacrificial anode." Because zinc is a less noble metal, it provides a preferred anodic site. The zinc is oxidized, rather than the iron. Scratching the zinc coating will not impact upon its ability to protect the iron.

(iii) Paint impedes corrosion simply by providing a coating, keeping the oxygen out of contact with the iron. If it is scratched, a situation of differential aeration is set up, with the O_2 concentration in the air higher than that under the paint. If there is moisture to allow ion movement, electrons travel through the metal from under the paint (oxidation of iron) to where the metal is exposed (reduction of oxygen).

REVIEW QUESTIONS

16.2 Oxidation-reduction reactions

16.31
(a) PF_6^-
$OS(F) = -1$
$-1 = OS(P) + 6\times(-1)$
$OS(P) = -1 - 6\times(-1) = +5$
(b) $H_2AsO_4^-$
$OS(H) = +1$ & $OS(O) = -2$
$-1 = 2\times(+1) + OS(As) + 4\times(-2)$
$OS(As) = -1 - 2\times(+1) - 4\times(-2) = +5$
(c) UO^{2+}
$OS(O) = -2$
$+2 = OS(U) + (-2)$
$OS(U) = +4$

(d) N_2O_5
$OS(O) = -2$
$0 = 2\times OS(N) + 5\times(-2)$
$OS(N) = (0 - 5\times(-2))/2 = +5$
(e) $POCl_3$
$OS(Cl) = -1$ & $OS(O) = -2$
$0 = OS(P) + 3\times(-1) + (-2)$
$OS(P) = +5$
(f) XeO_4^{2-}
$OS(O) = -2$
$-2 = OS(Xe) + 4\times(-2)$
$OS(Xe) = +6$

16.33
(a) $OH^-(aq) + H^+(aq) \longrightarrow H_2O(l)$
is NOT an oxidation-reduction reaction. Oxidation states are the same on both sides: +1 and −2 for H and O, respectively.
(b) $Cu(s) + Cl_2(g) \longrightarrow CuCl_2(s)$
is an oxidation-reduction reaction. The oxidation state of Cu increases from 0 to +2, while that of Cl decreases from 0 to −1.
(c) $CO_3^{2-}(aq) + 2H^+(aq) \longrightarrow CO_2(g) + H_2O(l)$
is NOT an oxidation-reduction reaction. The oxidation state of C equals +4 on both sides. Also, we have +1 and −2 for H and O on both sides.
(d) $2S_2O_3^{2-}(aq) + I_2(s) \longrightarrow S_4O_6^-(aq) + 2I^-(aq)$
is an oxidation-reduction reaction. The oxidation state of S increases from +2 to +11/4, while that of I decreases from 0 to −1.

16.35
(a) $Sn(s) + H^+(aq) \longrightarrow Sn^{2+}(aq) + H_2(g)$
First, we balance the electrons. 2 electrons are taken from each Sn atom, while two $H^+(aq)$ ions each gain one electron.
$$Sn(s) + 2H^+(aq) \longrightarrow Sn^{2+}(aq) + H_2(g)$$
The reaction is now balanced.
(b) $Cr_2O_7^{2-}(aq) + Fe^{2+}(aq) \longrightarrow Cr^{3+}(aq) + Fe^{3+}(aq)$
First, we balance the electrons. Each $Cr_2O_7^{2-}(aq)$ ion gains 6 electrons (write the half-equation), while one electron is taken from each $Fe^{2+}(aq)$ ion.
To balance the electrons, the 6 $Fe^{2+}(aq)$ ions must be oxidized for every $Cr_2O_7^{2-}(aq)$ ion that is reduced. This gives
$$Cr_2O_7^{2-}(aq) + 6Fe^{2+}(aq) \longrightarrow 2Cr^{3+}(aq) + 6Fe^{3+}(aq)$$

What remains is to balance the O atoms via participation of the water solvent. Balancing in acid, we balance the O atoms by adding 7 H_2O to the O-deficient side—the right-hand side. This introduces 14 H atoms on the right, which must be balanced by adding 14 H^+(aq) ions to the left. The final balanced equation is

$$Cr_2O_7^{2-}(aq) + 6\ Fe^{2+}(aq) + 14\ H^+(aq) \longrightarrow 2\ Cr^{3+}(aq) + 6\ Fe^{3+}(aq) + 7\ H_2O(l)$$

(c) $\quad MnO_2(s) + Cl^-(aq) \longrightarrow Mn^{2+}(aq) + Cl_2(g)$

First, we balance the electrons. Each mol of $MnO_2(s)$ gains 2 mol of electrons, while 1 mol of electrons is removed from each 1 mol of Cl^-(aq) ions.

To balance the electrons, 2 mol of Cl^-(aq) ions must be oxidized for every 1 mol of $MnO_2(s)$ reduced. This gives

$$MnO_2(s) + 2\ Cl^-(aq) \longrightarrow Mn^{2+}(aq) + Cl_2(g)$$

What remains is to balance the O atoms via participation of the aqueous solvent. Balancing in acid, we balance the O atoms by adding 2 H_2O molecules to the O-deficient side—the right. This introduces 4 H atoms on the right, which must be balanced by adding 4 H^+(aq) ions to the left. The final balanced equation is

$$MnO_2(s) + 2\ Cl^-(aq) + 4\ H^+(aq) \longrightarrow Mn^{2+}(aq) + Cl_2(g) + 2\ H_2O(l)$$

(d) $\quad HCHO(aq) + Ag^+(aq) \longrightarrow HCOOH(aq) + Ag(s)$

First, we balance the electrons. Each Ag^+(aq) ion gains one electron, while two electrons are removed from each HCHO(aq) molecule.

To balance the electrons, 2 mol of Ag^+(aq) ions must be reduced for every 1 mol of HCHO(aq) oxidized. This gives

$$HCHO(aq) + 2\ Ag^+(aq) \longrightarrow HCOOH(aq) + 2\ Ag(s)$$

What remains is to balance the O atoms via participation of the aqueous solvent. Balancing in acid, we balance the O atoms by adding one H_2O molecule to the O-deficient side—the left. Now we add 2 H^+(aq) ions to the H-deficient side, the right. The final balanced equation is

$$HCHO(aq) + 2\ Ag^+(aq) + H_2O(l) \longrightarrow HCOOH(aq) + 2\ Ag(s) + 2\ H^+(aq)$$

16.3—Voltaic cells: Electricity from chemical change

16.37

A voltaic cell is constructed using the reaction of chromium metal and aquated iron(II) ions

$$2\ Cr(s) + 3\ Fe^{2+}(aq) \longrightarrow 2\ Cr^{3+}(aq) + 3\ Fe(s)$$

Electrons in the external circuit flow from the $Cr^{3+}|Cr$ electrode (the <u>anode</u>) to the $Fe^{2+}|Fe$ electrode (the <u>cathode</u>). Negative ions move in the salt bridge from the $Fe^{2+}|Fe$ half-cell (the <u>reduction</u> half-cell) to the $Cr^{3+}|Cr$ half-cell (the <u>oxidation</u> half-cell). The half-reaction at the anode is <u>$Cr(s) \rightarrow$</u> <u>$Cr^{3+}(aq) + 3\ e^-$</u> and that at the cathode is <u>$Fe^{2+}(aq) + 2\ e^- \rightarrow Fe(s)$</u>.

16.39
(a) and (c)

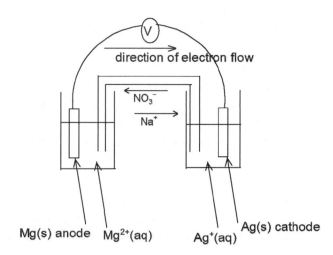

(b) $Mg(s) \longrightarrow Mg^{2+}(aq) + 2\,e^-$ is the oxidation half-reaction that occurs at the anode.

$Ag^+(aq) + e^- \longrightarrow Ag(s)$ is the reduction half-reaction that occurs at the cathode.

The salt bridge is needed to complete the circuit. Current flows only when the circuit is completed.

16.4 Cell emf, and half-cell reduction potentials

16.41
(a) $E^{\circ}_{Fe\ scale}[Cu^{2+}(aq)|Cu(s)] = E^{\circ}[Cu^{2+}(aq)|Cu(s)] - E^{\circ}[Fe^{2+}(aq)|Fe(s)]$
$= 0.34\ V - (-0.44\ V) = 0.78\ V$
(b) $E^{\circ}_{Fe\ scale}[Zn^{2+}(aq)|Zn(s)] = E^{\circ}[Zn^{2+}(aq)|Zn(s)] - E^{\circ}[Fe^{2+}(aq)|Fe(s)]$
$= -0.76\ V - (-0.44\ V) = -0.32\ V$

16.43
$$NO_3^-(aq) + 4\,H^+(aq) + 3\,e^- \longrightarrow NO(g) + 2\,H_2O(l)$$

(a) $E[\ NO_3^-(aq)|NO(g)\ ] = E^{\circ}[\ NO_3^-(aq)|NO(g)\ ]$
when $[NO_3^-] = 1.00\ mol\ L^{-1}$, $[H^+] = 1.00\ mol\ L^{-1}$ and $p(NO) = 1\ bar$,
or, more generally, whenever
$$Q = \frac{p(NO)}{[NO_3^-][H^+]^4} = 1.00$$

(b) $E^{\circ}_{cell} = E^{\circ}[\ NO_3^-(aq)|NO(g)\ ] - E^{\circ}[\ H^+(aq)|H_2(g)\ ] = 0.96\ V$
So, $E^{\circ}[\ NO_3^-(aq)|NO(g)\ ] = 0.96\ V$

16.45

(a) To make a cell with potential close to 1.1 V, using the $Zn^{2+}(aq)\,|\,Zn(s)$ half-cell, we could use the $Cu^{2+}(aq)\,|\,Cu(s)$ half-cell as the cathode and the $Zn^{2+}(aq)\,|\,Zn(s)$ half-cell as the anode.
$E°_{cell} = E°[Cu^{2+}(aq)|Cu(s)] - E°[Zn^{2+}(aq)|Zn(s)] = 0.337\ V - (-0.763\ V) = 1.10\ V$
Alternatively we could use any cell with $E°_{cell}$ approximately 1.1 V and calculate, using the Nernst equation, concentrations of species that would result in $E_{cell} = 1.1\ V$.
For example, a cell that we could use, in which the $Zn^{2+}(aq)\,|\,Zn(s)$ half-cell is the anode is
$$Zn(s)|Zn^{2+}(aq)||I_2(s)|I^-(aq) \qquad E°_{cell} = 0.535\ V - (-0.763\ V) = 1.30\ V$$
A cell in which the $Zn^{2+}(aq)\,|\,Zn(s)$ half-cell is the cathode is
$$Al(s)|Al^{3+}(aq)||Zn^{2+}(aq)\,|\,Zn(s) \qquad E°_{cell} = -0.763\ V - (-1.66\ V) = 0.90\ V$$
(b) To make a cell with potential close to 0.5 V, using the $Zn^{2+}(aq)\,|\,Zn(s)$ half-cell, we could use the $Ni^{2+}(aq)\,|\,Ni(s)$ half-cell as the cathode and the $Zn^{2+}(aq)\,|\,Zn(s)$ half-cell as the anode.
$E°_{cell} = E°[Ni^{2+}(aq)\,|\,Ni(s)] - E°[Zn^{2+}(aq)|Zn(s)] = -0.25\ V - (-0.763\ V) = 0.51\ V$
We can then use the Nernst equation to calculate concentrations of $Ni^{2+}(aq)$ and $Zn^{2+}(aq)$ ions that result in $E_{cell} = 0.50\ V$.
A cell with the $Zn^{2+}(aq)\,|\,Zn(s)$ half-cell as the cathode and which has $E°_{cell} \approx 0.5\ V$ is
$$Cd(s),\ S^{2-}(aq)|CdS(s)||\ Zn^{2+}(aq)\,|\,Zn(s) \qquad E°_{cell} = -0.763\ V - (-1.21\ V) = 0.45\ V$$

16.47

Ranked in order of half-cell reduction potentials:
$E°[F_2(g)|F^-(aq)] = 2.87\ V$
$E°[Cl_2(g)|Cl^-(aq)] = 1.358\ V$
$E°[O_2(g)|H^+(aq)|H_2O(l)] = 1.229\ V$
$E°[Br_2(l)|Br^-(aq)] = 1.066\ V$
$E°[I_2(s)|I^-(aq)] = 0.535\ V$
$E°[S(s)|H^+(aq)|H_2S(aq)] = 0.14\ V$
$E°[Se(s)|H^+(aq)|H_2Se(l)] = -0.40\ V$
(a) Se(s) is the least powerful oxidizing agent.
(b) $F^-(aq)$ ions are the least powerful reducing agent.
(c) $F_2(g)$ and $Cl_2(g)$ are capable of oxidizing $H_2O(l)$ to $O_2(g)$.
(d) $F_2(g)$, $Cl_2(g)$, $Br_2(l)$, $I_2(s)$ and $O_2(g)$ are capable of oxidizing $H_2S(g)$ to S(s).
(e) $O_2(g)$ is capable of oxidizing $I^-(aq)$ ions to $I_2(s)$ in acidic solution.
(f) S(s) is NOT capable of oxidizing $I^-(aq)$ ions to $I_2(s)$ in acidic solution.
(g) $H_2S(aq)\ \square\ Se(s) \to H_2Se(aq)\ \square\ S(s)$ is NOT the direction of spontaneous reaction when all species are at standard concentrations at 25 °C.
(h) $H_2S(aq)\ \square\ I_2(s) \to 2\ H^{\square+}(aq)\ \square + 2\ I^-(aq) + S(s)$ is the direction of spontaneous reaction when all species are at standard concentrations at 25 °C.

16.49

If $E°_{cell}$, computed for the direction of reaction as written has a positive value, then at standard conditions the equation indicates the direction of spontaneous reaction.

(a) $Ni^{2+}(aq) + H_2(g) \longrightarrow Ni(s) + 2H^+(aq)$

is NOT the direction of spontaneous reaction.

(b) $Fe^{3+}(aq) + 2I^-(aq) \longrightarrow Fe^{2+}(aq) + I_2(s)$

is the direction of spontaneous reaction.

(c) $Br_2(l) + 2Cl^-(aq) \longrightarrow 2 Br^-(aq) + Cl_2(g)$

is NOT the direction of spontaneous reaction.

(d) $Cr_2O_7^{2-}(aq) + 6 Fe^{2+}(aq) + 14 H^+(aq) \longrightarrow 2 Cr^{3+}(aq) + 6 Fe^{3+}(aq) + 7 H_2O(l)$

is the direction of spontaneous reaction.

16.53

If $E°_{cell}$, computed for the direction of reaction as written has a positive value, then at standard conditions the equation indicates the direction of spontaneous reaction.

(a) $Zn(s) + I_2(s) \longrightarrow Zn^{2+}(aq) + 2 I^-(aq)$

is the direction of spontaneous reaction.

(b) $2 Cl^-(aq) + I_2(s) \longrightarrow Cl_2(g) + 2 I^-(aq)$

is NOT the direction of spontaneous reaction.

(c) $2 Na^+(aq) + 2 Cl^-(aq) \longrightarrow 2 Na(s) + Cl_2(g)$

is NOT the direction of spontaneous reaction.

(d) $2 K(s) + H_2O(l) \longrightarrow 2 K^+(aq) + H_2(g) + 2 OH^-(aq)$

is the direction of spontaneous reaction.

16.55

Adding a KI solution to a standard acidic $Cu(NO_3)_2$ solution causes a brown colour and a precipitate to form because the nitrate ions oxidize iodide ions under these conditions. The $I_2(s)$ that is formed first reacts with $I^-(aq)$ ions to form brown $I_3^-(aq)$ ions. When all of the $I^-(aq)$ ions have reacted, further iodine formed remains as the insoluble solid precipitate. A similar response is not observed when KCl or KBr solutions are added because nitrate cannot oxidize chloride or bromide under standard conditions—there is no reaction.

$6 I^-(aq) + 2 NO_3^-(aq) + 8 H^+(aq) \rightarrow 3 I_2(s) + 2 NO(g) + 4 H_2O(l)$

is the balanced equation for reaction with $I^-(aq)$ ions.

16.5 Electrochemical cells under non-standard conditions

16.57

$E^\circ_{cell} = E^\circ[Ag^+(aq)|Ag(s)] - E^\circ[Zn^{2+}(aq)|Zn(s)] = 0.7994\ V - (-0.763\ V) = +1.562\ V$

$Zn(s) \rightarrow Zn^{2+}(aq) + 2\ e^-$

$\underline{2\times(\ Ag^+(aq) + e^- \rightarrow Ag(s)\ )}$

$Zn(s) + 2\ Ag^+(aq) \rightarrow Zn^{2+}(aq) + 2\ Ag(s)$

$$Q = \frac{[Zn^{2+}]}{[Ag^+]^2} = \frac{(0.010)}{(0.25)^2} = 0.16$$

$$E_{cell} = E^\circ_{cell} - \frac{0.0257\ V}{n} \times \ln Q = +1.562\ V - \frac{0.0257\ V}{2} \times \ln(0.16)$$

$$= +1.562\ V - \frac{0.0257\ V}{2} \times (-1.8) = +1.585\ V$$

16.59

(a)

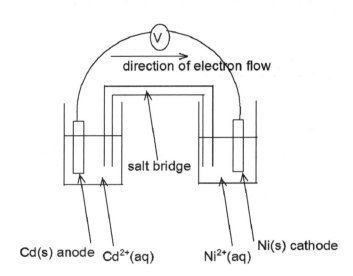

direction of electron flow

salt bridge

Cd(s) anode Cd^{2+}(aq)   Ni^{2+}(aq)   Ni(s) cathode

(b)

$$Ni^{2+}(aq) + 2\ e^- \longrightarrow Ni(s)$$

$$\underline{Cd(s) \longrightarrow Cd^{2+}(aq) + 2\ e^-}$$

$$Ni^{2+}(aq) + Cd(s) \longrightarrow Ni(s) + Cd^{2+}(aq)$$

(c) Since Ni^{2+}(aq)|Ni(s) has the higher reduction potential, the Ni(s) electrode is the cathode—it is labelled (+). The Cd^{2+}(aq)|Cd(s) half-cell is the anode compartment, labelled (−).

(d) $E^\circ_{cell} = E^\circ[Ni^{2+}(aq)|Ni(s)] - E^\circ[Cd^{2+}(aq)|Cd(s)] = -0.25\ V - (-0.403\ V) = +0.15\ V$

(e) Electrons flow from the anode to cathode in the external circuit.

(f) The Na$^+$(aq) ions (cations) in the salt bridge move from the anode compartment toward the cathode compartment. The NO$_3^-$(aq) ions (anions) move from the cathode compartment toward the anode compartment.

(g) The cell is at equilibrium when $E_{cell} = 0$.

$$E_{cell} = E^\circ_{cell} - \frac{0.0257\ V}{n} \times \ln Q = 0\ V$$

when $Q = K$. So

$$\ln K \;=\; \frac{2}{0.0257\,\text{V}} \times E^{\circ}{}_{\text{cell}} \;=\; \frac{2}{0.0257\,\text{V}} \times (+0.15\,\text{V})$$

$$=\; 11.7$$

and

$$K \;=\; 1.2 \times 10^{5}$$

(h) If the concentration of Cd^{2+}(aq) ions is reduced to 0.010 mol L^{-1}, keeping $[Ni^{2+\square}] = 1.0$ mol L^{-1}, then

$$E_{\text{cell}} = E^{\circ}{}_{\text{cell}} - \frac{0.0257\,\text{V}}{n} \times \ln Q = +0.15\,\text{V} - \frac{0.0257\,\text{V}}{2} \times \ln\left(\frac{[Cd^{2+}]}{[Ni^{2+}]}\right)$$

$$= +0.15\,\text{V} - \frac{0.0257\,\text{V}}{2} \times \ln\left(\frac{0.010}{1.0}\right) = +0.21\,\text{V}$$

Since E_{cell} is still positive, the net reaction is in the same direction given in part (b). In fact, the cell emf is increased, as we might have expected as a result of decreasing the concentration of a product species.

16.6 Standard cell emf and equilibrium constant

16.61

(a) $2\,Fe^{3+}(aq) + 2\,I^{-}(aq) \longrightarrow 2\,Fe^{2+}(aq) + I_{2}(s)$

$E^{\circ}{}_{\text{cell}} = E^{\circ}[Fe^{3+}(aq)|\,Fe^{2+}(aq)] - E^{\circ}[I_2(s)|I^{-}(aq)] = 0.771\,\text{V} - (0.535\,\text{V}) = +0.236\,\text{V}$

$$\ln K \;=\; \frac{2}{0.0257\,\text{V}} \times E^{\circ}{}_{\text{cell}} \;=\; \frac{2}{0.0257\,\text{V}} \times (+0.236\,\text{V})$$

$$=\; 18.4$$

and

$$K \;=\; 9.5 \times 10^{7}$$

(b) $I_{2}(s) + 2\,Br^{-}(aq) \longrightarrow 2\,I^{-}(aq) + Br_{2}(l)$

$E^{\circ}{}_{\text{cell}} = E^{\circ}[\,I_2(s)|I^{-}(aq)] - E^{\circ}[Br_2(l)|Br^{-}(aq)] = 0.535\,\text{V} - (1.066\,\text{V}) = -0.531\,\text{V}$

$$\ln K \;=\; \frac{2}{0.0257\,\text{V}} \times E^{\circ}{}_{\text{cell}} \;=\; \frac{2}{0.0257\,\text{V}} \times (-0.531\,\text{V})$$

$$=\; -41.3$$

and

$$K \;=\; 1.1 \times 10^{-18}$$

16.65

$$Au^{3+}(aq) + 3\,e^- \longrightarrow Au(s)$$
$$\underline{Au(s) + 4\,Cl^-(aq) \longrightarrow [AuCl_4]^-(aq) + 3\,e^-}$$
$$Au^{3+}(aq) + 4\,Cl^-(aq) \longrightarrow [AuCl_4]^-(aq)$$

The formation reaction is expressed here as the sum of reduction and oxidation half-reactions.

$$E^\circ{}_{cell} = E^\circ[Au^{3+}(aq)|Au(s)] - E^\circ\{[AuCl_4]^-(aq)|Au(s)\} = 1.50\ V - (1.00\ V) = +0.50\ V$$

$$\ln K_f = \frac{3}{0.0257\ V} \times E^\circ{}_{cell} = \frac{3}{0.0257\ V} \times (0.50\ V)$$

$$= 58.4$$

and

$$K_f = 2.3 \times 10^{25}$$

16.7 Electrolysis: Chemical change using electrical energy

16.67

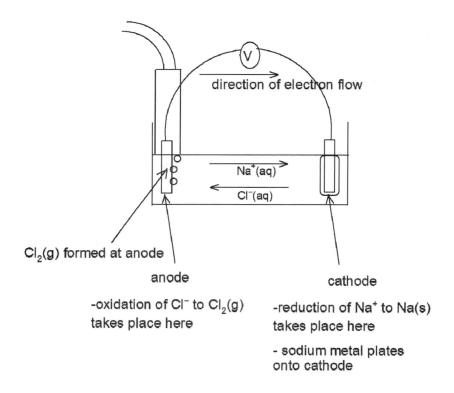

Cl$_2$(g) formed at anode

anode

-oxidation of Cl⁻ to Cl$_2$(g)
takes place here

cathode

-reduction of Na⁺ to Na(s)
takes place here

- sodium metal plates
onto cathode

16.69
Fluorine, $F_2(g)$, has the largest reduction potential of all species—it is easily reduced to $F^-(aq)$ ions. This means that $F^-(aq)$ ions are difficult to oxidize to $F_2(g)$. Oxygen gas forms at the anode in the electrolysis of an aqueous KF solution.

16.71

The bumper on an off-road vehicle is likely to get scratched. The chrome plated steel will corrode at the scratch faster than it would without the plating. This is because chromium is a (slightly) more noble metal than iron, and a voltaic cell is set up with the iron as the anode area. The galvanized steel stays protected, even with scratches, until all of the zinc plating is oxidized by corrosion. Note that corrosion of zinc is slower than that for iron because zinc oxide coats the zinc protecting it from atmospheric oxygen. When iron oxidizes, its oxide does not coat the metal, so it is vulnerable to corrosion.

16.8 Corrosion of iron

16.73

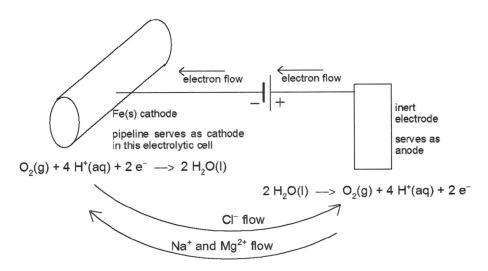

$$O_2(g) + 4\ H^+(aq) + 2\ e^- \longrightarrow 2\ H_2O(l)$$

$$2\ H_2O(l) \longrightarrow O_2(g) + 4\ H^+(aq) + 2\ e^-$$

Fe(s) cathode

pipeline serves as cathode in this electrolytic cell

inert electrode

serves as anode

Cl⁻ flow

Na⁺ and Mg²⁺ flow

electron flow

electron flow

By keeping the pipeline at a higher potential than the inert electrode, the pipeline behaves like a metal even more noble - it does not corrode.

SUMMARY AND CONCEPTUAL QUESTIONS

16.75
Both of the problems encountered are examples of a more noble metal polarizing a less noble metal, making it more susceptible to corrosion.
(a) In this case, the iron screws in contact with the more noble copper alloy corrode rapidly. They crumble, leaving holes in the copper alloy sheets where water leaks in.
(b) In this case, the iron wrench is the more noble metal. It accelerates the corrosion of the aluminum in contact with it by creating a voltaic cell in which the aluminium is the anode area. The aluminum corrodes rapidly leaving a wrench-shaped hole in the hull.

16.77
If the emf is applied in the wrong direction, the corrosion of the pipeline will be accelerated. Electrons would be taken away from the pier, causing oxidation of the iron—i.e., rendering the pier the anode of a cell. The electrons would be directed to the platinum-coated titanium, where the only possible reaction would be reduction of $O_2(g)$ at its surface.

Chapter 17
Spontaneous Change: How Far?

IN-CHAPTER EXERCISES

Exercise 17.1—Entropy change

The change in entropy for a reversible process (i.e., add the heat very slowly) is just heat flow into the system divided by temperature. Therefore, the entropy of the water and hexane change by the same amount—it depends only on the temperature of the system and the amount of heat added.

$$\Delta S = q / T = 1 \times 10^6 \text{ J} / 323.15 \text{ K} = 3095 \text{ J K}^{-1}$$

Exercise 17.3—Entropy change

$$\Delta S = q / T = 29.1 \times 10^3 \text{ J} / 239.7 \text{ K} = 121 \text{ J K}^{-1}$$

i.e., the entropy of 1 mol of ammonia vapour is 121 J K^{-1} greater than the entropy of 1 mol of the liquid.

Exercise 17.5—Comparison of entropies of substances

(a) 1 mol of $O_3(g)$ has higher entropy than 1 mol of $O_2(g)$.
There are 1.5 times as many O atoms in 1 mol of $O_3(g)$ than there are in 1 mol of $O_2(g)$ so there are many more possibilities for different arrangements of the atoms. Ozone has more internal degrees of freedom. Ozone molecules have more modes of vibration (two stretching vibrations and a bending mode), as well as three distinct rotational modes. Oxygen molecules have only one stretching vibrational mode and two rotational modes (rotation about the bond does not constitute rotation of the molecule). Ozone has more ways of distributing its energy which corresponds to greater disorder —i.e., greater entropy.
(b) 1 mol of $SnCl_4(g)$ has more entropy than 1 mol of $SnCl_4(l)$.
Gases have much more entropy per mole than the corresponding liquid. The gas molecules are free to explore the entire container—liquid molecules have less volume available and reduced mobility within the bulk.
(c) 3 mol of $O_2(g)$ has more entropy than 2 mol of $O_3(g)$.
There is the same number of O atoms—i.e., 6 mol. 2 mol of the larger molecules is a more ordered/constrained arrangement of those 6 mol of O atoms than is 3 mol of O_2 molecules.

Exercise 17.8—Standard entropy change of reaction

(a) $CaCO_3(s) \longrightarrow CaO(s) + CO_2(g)$

$$\begin{aligned}
\Delta_r S^\circ &= [1\times S^\circ(CO_2,g) + 1\times S^\circ(CaO,s)] - 1\times S^\circ(CaCO_3,s) \\
&= [213.74\,J\,K^{-1}\,mol^{-1} + 39.75\,J\,K^{-1}\,mol^{-1})] - 92.9\,J\,K^{-1}\,mol^{-1} \\
&= +160.59\ J\,K^{-1}\,mol^{-1}
\end{aligned}$$

(b) $N_2(g) + 3H_2(g) \longrightarrow 2NH_3(g)$

$$\begin{aligned}
\Delta_r S^\circ &= 2\times S^\circ(NH_3,g) - [1\times S^\circ(N_2,g) + 3\times S^\circ(H_2,g)] \\
&= 2\times192.45\,J\,K^{-1}\,mol^{-1} - [191.61\,J\,K^{-1}\,mol^{-1} + 3\times130.684\,J\,K^{-1}\,mol^{-1}] \\
&= -198.76\ J\,K^{-1}\,mol^{-1}
\end{aligned}$$

Exercise 17.10—Predicting spontaneity from ΔS°_{univ}

$Si(s) + 2Cl_2(g) \longrightarrow SiCl_4(g)$

$$\begin{aligned}
\Delta S^\circ_{sys} &= 1\times S^\circ(SiCl_4,g) - [1\times S^\circ(Si,s) + 2\times S^\circ(Cl_2,g)] \\
&= 330.73\,J\,K^{-1}\,mol^{-1} - [18.83\,J\,K^{-1}\,mol^{-1} + 2\times223.066\,J\,K^{-1}\,mol^{-1}] \\
&= -134.23\ J\,K^{-1}\,mol^{-1}
\end{aligned}$$

$$\begin{aligned}
\Delta H^\circ_{sys} &= 1\times\Delta_f H^\circ(SiCl_4,g) - [1\times\Delta_f H^\circ(Si,s) + 2\times\Delta_f H^\circ(Cl_2,g)] \\
&= -657.01\,kJ\,mol^{-1} - [0+0] \\
&= -657.01\,kJ\,mol^{-1}
\end{aligned}$$

i.e., this is just the formation reaction for $SiCl_4(g)$.

$$\Delta S^\circ_{surr} = -\frac{\Delta H^\circ_{sys}}{T} = -\frac{-657.01\times10^3\,J\,mol^{-1}}{298\,K} = +2205\ J\,K^{-1}\,mol^{-1}$$

$$\begin{aligned}
\Delta S^\circ_{univ} &= \Delta S^\circ_{sys} + \Delta S^\circ_{surr} = -134.23\ J\,K^{-1}\,mol^{-1} + 2205\ J\,K^{-1}\,mol^{-1} \\
&= +2071\ J\,K^{-1}\,mol^{-1}
\end{aligned}$$

Since the change in entropy for the universe upon reaction in the direction indicated by the equation is positive, the reaction is spontaneous.

Exercise 17.12—Contributions of $\Delta_r S°$ and $\Delta_r H°$ to spontaneity of reaction

Reaction	$\Delta_r H°$ kJ mol^{-1}	$\Delta_r S°$ J K^{-1} mol^{-1}	Reaction Type
(a) $CH_4(g) + 2O_2(g) \longrightarrow 2H_2O(l) + CO_2(g)$	−890.6	−242.8	Type 2: spontaneous at low T NOT spontaneous at high T $T_{cross} = \Delta_r H°/\Delta_r S°$ = 890600/242.8 K = 3668 K
(b) $2Fe_2O_3(s) + 3C(graphite) \longrightarrow 4Fe(s) + 3CO_2(g)$	+467.9	+560.7	Type 3: NOT spontaneous at low T spontaneous at high T
(c) $C(graphite) + O_2(g) \longrightarrow CO_2(g)$	−393.5	+3.1	Type 1: spontaneous at all T
(d) $N_2(g) + 3F_2(g) \longrightarrow 2NF_3(g)$	−264.2	−277.8	Type 2: spontaneous at low T NOT spontaneous at high T

Exercise 17.14—Calculating $\Delta_r G°$ from $\Delta_r H°$ and $\Delta_r S°$

$$N_2(g) + 3H_2(g) \longrightarrow 2NH_3(g)$$

$$\Delta_r S° = 2 \times S°(NH_3, g) - [1 \times S°(N_2, g) + 3 \times S°(H_2, g)]$$
$$= 2 \times 192.45\,J\,K^{-1}\,mol^{-1} - [191.61\,J\,K^{-1}\,mol^{-1} + 3 \times 130.684\,J\,K^{-1}\,mol^{-1}]$$
$$= -198.76\,J\,K^{-1}\,mol^{-1}$$

$$\Delta_r H° = 2 \times \Delta_f H°(NH_3, g) - [\Delta_f H°(N_2, g) + 3 \times \Delta_f H°(H_2, g)]$$
$$= -2 \times (-45.9\,kJ\,mol^{-1}) = -91.8\,kJ\,mol^{-1}$$

$$\Delta_r G° = \Delta_r H° - T\Delta_r S°$$
$$= -91.8\,kJ\,mol^{-1} - 298\,K \times (-198.76\,J\,K^{-1}mol^{-1}) / 1000\,J\,/kJ$$
$$= -32.6\,kJ\,mol^{-1}$$

Exercise 17.15—Standard molar free energy change of formation, $\Delta_f G°$

(a) $\frac{1}{2} N_2(g) + \frac{3}{2} H_2(g) \rightarrow NH_3(g)$

(b) $2 Fe(s) + \frac{3}{2} O_2(g) \rightarrow Fe_2O_3(s)$

(c) $3 C(s) + 3 H_2(g) + O_2(g) \rightarrow CH_3CH_2COOH(l)$

(d) $Ni(s) \rightarrow Ni(s)$ ($Ni(s)$ is the element in its standard state. Its $\Delta_f G° = 0$)

Exercise 17.17—$\Delta_r G°$ from $\Delta_f G°$ values

$SO_2(g) + \frac{1}{2} O_2(g) \rightarrow SO_3(g)$

$$\Delta_r G° = 1 \times \Delta_f G°(SO_3, g) - [1 \times \Delta_f G°(SO_2, g) + \tfrac{1}{2} \times \Delta_f G°(O_2, g)]$$
$$= -371.06 \text{ kJ mol}^{-1} - [-300.194 \text{ kJ mol}^{-1} + \tfrac{1}{2} \times 0 \text{ kJ mol}^{-1}]$$
$$= -70.87 \text{ kJ mol}^{-1}$$

Exercise 17.18—$\Delta_r G°$ and spontaneity

$3 H_2(g) + CO(g) \longrightarrow CH_4(g) + H_2O(g)$

$$\Delta_r G° = 1 \times \Delta_f G°(CH_4, g) + 1 \times \Delta_f G°(H_2O, g) - [3 \times \Delta_f G°(H_2, g) + 1 \times \Delta_f G°(CO, g)]$$
$$= -50.72 \text{ kJ mol}^{-1} + (-228.572 \text{ kJ mol}^{-1}) - [3 \times 0 \text{ kJ mol}^{-1} - 137.168 \text{ kJ mol}^{-1}]$$
$$= -142.12 \text{ kJ mol}^{-1}$$

This reaction is spontaneous in the forward direction (at 25 °C, and if all of the reagents are simultaneously at 1 bar pressure)

Exercise 17.19—Finding the temperature at which $\Delta_r G° = 0$

$2 HgO(s) \longrightarrow 2 Hg(l) + O_2(g)$

$$\Delta_r S° = 2 \times S°(Hg, l) + 1 \times S°(O_2, g) - 2 \times S°(HgO, s)$$
$$= 2 \times 76.02 \text{ J K}^{-1} \text{mol}^{-1} + 205.138 \text{ J K}^{-1} \text{mol}^{-1} - 2 \times 70.29 \text{ J K}^{-1} \text{mol}^{-1}$$
$$= 216.598 \text{ J K}^{-1} \text{mol}^{-1}$$

$$\Delta_r H° = 2 \times \Delta_f H°(Hg, l) + 1 \times \Delta_f H°(O_2, g) - 2 \times \Delta_f H°(HgO, s)$$
$$= 0 + 0 - 2 \times (-90.83 \text{ kJ mol}^{-1}) = +181.66 \text{ kJ mol}^{-1}$$

$$\Delta_r G° = \Delta_r H° - T\Delta_r S° = 0$$

when

$$T = \frac{\Delta_r H°}{\Delta_r S°} = \frac{181.66 \times 1000 \text{ J mol}^{-1}}{216.598 \text{ J K}^{-1} \text{mol}^{-1}} = 838.7 \text{ K}$$

Exercise 17.21—Spontaneity and temperature

(a) $N_2(g) + 2O_2(g) \longrightarrow 2NO_2(g)$

We can deduce that $\Delta_r S° < 0$ ($\Delta_r S°$ is negative) because three moles of gas are converted into two moles of gas. The value can be calculated using $S°$ values from Appendix C.
Increasing temperature makes the $-T\Delta_r S°$ component of $\Delta_r G°$ larger (and positive). Consequently, $\Delta_r G°$ increases with increasing T.

(b) $2C(s) + O_2(g) \longrightarrow 2CO(g)$

$\Delta_r S° > 0$ because two moles of gas are formed from one mole of gas (plus a solid) – the gases account for most of the entropy change. The value can be calculated using $S°$ values from Appendix C.

$\Delta_r G°$ decreases with increasing T.

(c) $CaO(s) + CO_2(g) \longrightarrow CaCO_3(s)$

$\Delta_r S° < 0$ because one mole of gas is consumed by the process. The value can be calculated using $S°$ values from Appendix C.

$\Delta_r G°$ increases with increasing T.

(d) $2NaCl(s) \longrightarrow 2Na(s) + Cl_2(g)$

$\Delta_r S° > 0$ because one mole of gas is produced by the process. The value can be calculated using $S°$ values from Appendix C.

$\Delta_r G°$ decreases with increasing T.

Exercise 17.22—Calculation of $\Delta_r G$ for a non-standard reaction mixture

$$H_2O(l) \rightleftharpoons H_2O(g)$$

$$\Delta_r G = \Delta_r G° + RT \ln Q$$

$$= +8.56 \times 10^3 \, J \, mol^{-1} + (8.314 \, J \, K^{-1} mol^{-1})(298 \, K) \times \ln p(H_2O)$$

$$= +8.56 \times 10^3 \, J \, mol^{-1} + (8.314 \, J \, K^{-1} mol^{-1})(298 \, K) \times \ln(0.020)$$

$$= -1.13 \, kJ \, mol^{-1}$$

Exercise 17.25—The relationship between $\Delta_r G°$ and K

$$C(s) + CO_2(g) \longrightarrow 2CO(g)$$

$$\Delta_r G° = 2 \times \Delta_f G°(CO,g) - [1 \times \Delta_f G°(C,s) + 1 \times \Delta_f G°(CO_2,g)]$$

$$= 2 \times (-137.168 \, kJ \, mol^{-1}) - [0 + (-394.359 \, kJ \, mol^{-1})]$$

$$= +120.023 \, kJ \, mol^{-1}$$

$$\ln K = -\frac{\Delta_r G°}{RT} = -\frac{120.023 \times 10^3 \, J \, mol^{-1}}{8.314 \, J \, K^{-1} \, mol^{-1} \times 298 \, K}$$

$$= -48.44$$

$$K = e^{-48.44} = 9.18 \times 10^{-22}$$

Exercise 17.26—The relationship between $\Delta_r G°$ and K

$$\Delta_r G° = -RT \ln K = = -8.314 \, \text{J} \, \cancel{\text{K}}^{-1} \, \text{mol}^{-1} \times 298 \, \cancel{\text{K}} \times \ln(1.6 \times 10^7)$$
$$= -41 \, \text{kJ} \, \text{mol}^{-1}$$

Exercise 17.27—$\Delta_r G°$ and $E°_{cell}$ at the same temperature

$$H_2(g) + Zn^{2+}(aq) \longrightarrow Zn(s) + 2\,H^+(aq)$$

$n = 2$ here

$$\Delta_r G° = -nFE°_{cell} = -(2)(96\,450 \, \text{C} \, \text{mol}^{-1})(-0.76 \, \text{V}) = 1.47 \times 10^5 \, \text{CV} \, \text{mol}^{-1} = 147 \, \text{kJ} \, \text{mol}^{-1}$$

Because $\Delta_r G° > 0$, the reverse reaction is spontaneous in a standard cell at 25 °C.

Exercise 17.28—Interrelationships among $\Delta_r G°$, K, and $E°_{cell}$

(a) $\quad 2\,Fe^{3+}(aq) + 2\,I^-(aq) \longrightarrow 2\,Fe^{2+}(aq) + I_2(s)$

$n = 2$ here

$$E° = E°[Fe^{3+}|Fe^{2+}] - E°[I_2|I^-] = 0.771 \, \text{V} - 0.535 \, \text{V} = 0.236 \, \text{V}$$
$$\Delta_r G° = -nFE°_{cell} = -(2)(96\,450 \, \text{C} \, \text{mol}^{-1})(0.236 \, \text{V}) = -4.55 \times 10^4 \, \text{CV} \, \text{mol}^{-1} = -45.5 \, \text{kJ} \, \text{mol}^{-1}$$
$$K = \exp(-\Delta_r G°/RT) = \exp\left(\frac{4.55 \times 10^4 \, \text{J} \, \text{mol}^{-1}}{8.314 \, \text{J} \, \text{K}^{-1} \, \text{mol}^{-1} \times 298 \, \text{K}}\right)$$
$$= 9.46 \times 10^7$$

(b) $\quad I_2(aq) + 2\,Br^-(aq) \longrightarrow 2\,I^-(aq) + Br_2(aq)$

$n = 2$ here

$$E° = E°[I_2|I^-] - E°[Br_2|Br^-] = 0.535 \, \text{V} - 1.066 \, \text{V} = -0.531 \, \text{V}$$
$$\Delta_r G° = -nFE°_{cell} = -(2)(96\,450 \, \text{C} \, \text{mol}^{-1})(-0.531 \, \text{V}) = 1.02 \times 10^5 \, \text{CV} \, \text{mol}^{-1} = -102 \, \text{kJ} \, \text{mol}^{-1}$$
$$K = \exp(-\Delta_r G°/RT) = \exp\left(\frac{102 \times 10^3 \, \text{J} \, \text{mol}^{-1}}{8.314 \, \text{J} \, \text{K}^{-1} \, \text{mol}^{-1} \times 298 \, \text{K}}\right)$$
$$= 7.58 \times 10^{17}$$

Exercise 17.30—Dependence of K on temperature

$$2SO_3(g) \rightleftharpoons 2SO_2(g) + O_2(g)$$

(a) $\Delta_r G°$ at 298 K

$$\Delta_r G° = 2 \times \Delta_f G°(SO_2,g) + 1 \times \Delta_f G°(O_2,g) - 2 \times \Delta_f G°(SO_3,g)$$
$$= 2 \times (-300.194 \, \text{kJ mol}^{-1}) + 0 \, \text{kJ mol}^{-1} - 2 \times (-371.06 \, \text{kJ mol}^{-1})$$
$$= 141.73 \, \text{kJ mol}^{-1}$$

(b) K at 298 K

$$\ln K = -\frac{\Delta_r G°}{RT} = -\frac{141.73 \times 10^3 \, \text{J mol}^{-1}}{8.314 \, \text{J K}^{-1} \, \text{mol}^{-1} \times 298 \, \text{K}}$$
$$= -57.21$$

$$K = e^{-57.21} = 1.43 \times 10^{-25}$$

(c) K at 1500 °C

$$\ln K = -\frac{\Delta_r G°}{RT} = -\frac{141.73 \times 10^3 \, \text{J mol}^{-1}}{8.314 \, \text{J K}^{-1} \, \text{mol}^{-1} \times 1773 \, \text{K}}$$
$$= -9.61$$

$$K = e^{-9.61} = 6.67 \times 10^{-5}$$

Exercise 17.32—Dependence of equilibrium vapour pressure on temperature

T (K)	$1/T$ (K^{-1})	p (bar)	$\ln p$	$\Delta(\ln p)/\Delta(1/T)$ / K
273.15	0.0036610	4.00×10^{-5}	−10.13	
288.15	0.0034704	1.75×10^{-4}	−8.65	−1.48 / 0.0001906 = −7766
303.15	0.0032987	6.25×10^{-4}	−7.38	−1.27 / 0.0001717 = −7396
323.15	0.0030945	2.87×10^{-3}	−5.85	−1.53 / 0.0002042 = −7494
363.15	0.0027537	3.57×10^{-2}	−3.33	−2.52 / 0.0003408 = −7394

The average of the $\Delta(\ln p)/\Delta(1/T)$ / K provides the best estimate of the slope of a $\ln p$ vs. $1/T$ plot.
Average = −7512 K
Alternatively feed the data sets of $\ln p$ and $1/T$ into a plotting program to obtain an equation for the straight line of best fit, with the slope indicated.

$$\text{Slope} = -7512 \, \text{K} = -\frac{\Delta_{sub} H°}{R} = -\frac{\Delta_{sub} H°}{8.314 \, \text{J K}^{-1} \, \text{mol}^{-1}}$$

(c) So $\Delta_{sub} H° = 8.314 \times 7512 \, \text{J mol}^{-1} = 62.5 \, \text{kJ mol}^{-1}$

(a) The equilibrium vapour pressure at 40 °C

We have the equilibrium vapour pressure at 30 °C. We can estimate the vapour pressure at 40 °C using a linear approximation.

$$\ln p_{40°C} = \ln p_{30°C} + \text{slope} \times (1 / 313.15 - 1 / 303.15)$$
$$= -7.38 + (-7512 \text{ K}) \times (-0.0001053 \text{ K}^{-1})$$
$$= -7.38 + 0.79 = -6.59$$
$$p_{40°C} = 1.37 \times 10^{-3} \text{ bar}$$

(b) The equilibrium vapour pressure at 100 °C

Estimate using the linear approximation about 90 °C (the nearest temperature from the data set).

$$\ln p_{100°C} = \ln p_{90°C} + \text{slope} \times (1 / 373.15 - 1 / 363.15)$$
$$= -3.33 + (-7512 \text{ K}) \times (-7.380 \times 10^{-5} \text{ K}^{-1})$$
$$= -3.33 + 0.55 = -2.78$$
$$p_{40°C} = 6.20 \times 10^{-2} \text{ bar}$$

REVIEW QUESTIONS

17.4—Entropy and the second law of thermodynamics

17.33
$\Delta S = q / T = 6.01 \times 10^3 \text{ J mol}^{-1} / 273.15 \text{ K} = 22.0 \text{ J K}^{-1} \text{ mol}^{-1}$

17.35
(a) NaCl(g) has more entropy than NaCl(aq), which has more entropy than NaCl(s).
The greatest disorder is in the gas state. NaCl(s) is a highly constrained arrangement of the ions.
Check: $S^\circ[\text{NaCl(g)}] = 229.79 > S^\circ[\text{NaCl(aq)}] = 115.5 > S^\circ[\text{NaCl(s)}] = 72.11$ (all in J K^{-1})
(b) $H_2S(g)$ has more entropy than $H_2O(g)$ because S is heavier than O. Heavier molecules have more entropy (everything else equal). This is because energy distributed among heavier gas molecules corresponds to a larger range of momenta—which means more ways of distributing the energy. At the same temperature, molecules with less mass travel faster (on average).
Check: $S^\circ[\text{H}_2\text{S(g)}] = 205.79 > S^\circ[\text{H}_2\text{O(g)}] = 188.84$
(c) $C_2H_4(g)$ has more entropy than $N_2(g)$ (with the same molar mass).
More complex molecules have more entropy than less complex molecules with the same mass. They have more internal motions which mean more ways of distributing energy.
Check: $S^\circ[\text{C}_2\text{H}_4\text{(g)}] = 219.36 > S^\circ[\text{N}_2\text{(g)}] = 191.56$
(d) $H_2SO_4(l)$ has more entropy than $H_2SO_4(aq)$.
On the surface, it would appear that since $H_2SO_4(aq)$ explores a larger volume—the volume of its aqueous environment—it should have a greater entropy than the same amount of $H_2SO_4(l)$. (Note that standard entropies correspond to 1 M concentration in case of aqueous species.) However, the aqueous sulfuric acid is strongly associated with water via the acid-base reaction. The stronger forces in the ionized solution reduce its entropy, more than compensating for the increase in entropy due to the greater freedom of the aqueous species.
Check: $S^\circ[\text{H}_2\text{SO}_4\text{(l)}] = 156.9 > S^\circ[\text{H}_2\text{SO}_4\text{(aq)}] = 20.1$

17.37
$2\,C(\text{graphite}) + 3\,H_2(g) \longrightarrow C_2H_6(g)$
$\Delta_r S^\circ = 1 \times S^\circ(C_2H_6, g) - [2 \times S^\circ(C, \text{graphite}) + 3 \times S^\circ(H_2, g)]$
$= 229.60 \text{ J K}^{-1} \text{mol}^{-1} - [2 \times 5.740 \text{ J K}^{-1} \text{mol}^{-1} + 3 \times 130.684 \text{ J K}^{-1} \text{mol}^{-1}]$
$= -173.93 \text{ J K}^{-1} \text{mol}^{-1}$

17.41
1. $C(s) + 2\,H_2(g) \longrightarrow CH_4(g)$
$\quad C(s) = C(\text{graphite})$
$\quad \Delta_r S^\circ_1 = S^\circ[\text{CH}_4\text{(g)}] - (S^\circ[\text{C(s)}] + 2\,S^\circ[\text{H}_2\text{(g)}])$
$\quad\quad = 186.3 \text{ J K}^{-1} \text{mol}^{-1} - (5.74 \text{ J K}^{-1} \text{mol}^{-1} + 2 \times 130.7 \text{ J K}^{-1} \text{mol}^{-1}) = -80.8 \text{ J K}^{-1} \text{mol}^{-1}$

2. $CH_4(g) + \dfrac{1}{2}O_2(g) \longrightarrow CH_3OH(l)$
$\quad \Delta_r S^\circ_2 = S^\circ[\text{CH}_3\text{OH(l)}] - (S^\circ[\text{CH}_4\text{(g)}] + \frac{1}{2}\,S^\circ[\text{O}_2\text{(g)}])$
$\quad\quad = 126.8 \text{ J K}^{-1} \text{mol}^{-1} - (186.3 \text{ J K}^{-1} \text{mol}^{-1} + \frac{1}{2} \times 205.1 \text{ J K}^{-1} \text{mol}^{-1}) = -162.1 \text{ J K}^{-1}$
$\quad\quad \text{mol}^{-1}$

3. $C(s) + 2 H_2(g) + \dfrac{1}{2}O_2(g) \longrightarrow CH_3OH(l)$

$$\Delta_r S^\circ_3 = S^\circ[CH_3OH(l)] - (S^\circ[C(s)] + 2\,S^\circ[H_2(g)] + \tfrac{1}{2}\,S^\circ[O_2(g)])$$
$$= 126.8\ \text{J K}^{-1}\,\text{mol}^{-1} - (5.74\ \text{J K}^{-1}\,\text{mol}^{-1} + 2 \times 130.7\ \text{J K}^{-1}\,\text{mol}^{-1} + \tfrac{1}{2} \times 205.1\ \text{J K}^{-1}$$
$\text{mol}^{-1})$
$$= -242.9\ \text{J K}^{-1}\,\text{mol}^{-1}$$

Just as reactions 1 & 2 add to give reaction 3, the $\Delta_r S^\circ$ values add in the same fashion.

$$\Delta_r S^\circ_1 + \Delta_r S^\circ_2 = -80.8\ \text{J K}^{-1}\,\text{mol}^{-1} - 162.1\ \text{J K}^{-1}\,\text{mol}^{-1} = -242.9\ \text{J K}^{-1}\,\text{mol}^{-1} = \Delta_r S^\circ_3$$

This is an example of Hess's Law applied to entropy, which can only be the case if S is a state function.

17.5—Entropy changes and spontaneity

17.43

$H_2(g) + Cl_2(g) \longrightarrow 2\,HCl(g)$

$$\Delta S^\circ_{sys} = 2\,S^\circ[HCl(g)] - (S^\circ[H_2(g)] + S^\circ[Cl_2(g)])$$
$$= 2 \times 186.9\ \text{J K}^{-1}\,\text{mol}^{-1} - (130.7\ \text{J K}^{-1}\,\text{mol}^{-1} + 223.1\ \text{J K}^{-1}\,\text{mol}^{-1}) = 20.0\ \text{J K}^{-1}\,\text{mol}^{-1}$$

$$\Delta H^\circ_{sys} = 2\,\Delta_f H^\circ[HCl(g)] - (\Delta_f H^\circ[H_2(g)] + \Delta_f H^\circ[Cl_2(g)])$$
$$= 2 \times (-92.31)\ \text{kJ mol}^{-1} - 0 - 0 = -184.62\ \text{kJ mol}^{-1}$$

$$\Delta S^\circ_{surr} = -\frac{\Delta H^\circ_{sys}}{T} = -\frac{-184.62 \times 10^3\ \text{J mol}^{-1}}{298\,\text{K}} = +619.5\ \text{J K}^{-1}\,\text{mol}^{-1}$$

$$\Delta S^\circ_{univ} = \Delta S^\circ_{sys} + \Delta S^\circ_{surr} = 20.0\ \text{J K}^{-1}\,\text{mol}^{-1} + 619.5\ \text{J K}^{-1}\,\text{mol}^{-1}$$
$$= +639.5\ \text{J K}^{-1}\,\text{mol}^{-1}$$

This reaction is spontaneous in the forward direction, under the reaction mixture conditions specified.

17.45

$H_2O(l) \longrightarrow H_2(g) + \tfrac{1}{2}O_2(g)$

$$\Delta S^\circ_{sys} = (S^\circ[H_2(g)] + \tfrac{1}{2}\,S^\circ[O_2(g)]) - S^\circ[H_2O(l)]$$
$$= (130.7\ \text{J K}^{-1}\,\text{mol}^{-1} + \tfrac{1}{2} \times 205.1\ \text{J K}^{-1}\,\text{mol}^{-1}) - 69.91\ \text{J K}^{-1}\,\text{mol}^{-1} = 163.3\ \text{J K}^{-1}\,\text{mol}^{-1}$$

$$\Delta H^\circ_{sys} = (\Delta_f H^\circ[H_2(g)] + \tfrac{1}{2}\,\Delta_f H^\circ[O_2(g)]) - \Delta_f H^\circ[H_2O(l)]$$
$$= -285.8\ \text{kJ mol}^{-1} - 0 - 0 = -285.8\ \text{kJ mol}^{-1}$$

$$\Delta S^\circ_{surr} = -\frac{\Delta H^\circ_{sys}}{T} = -\frac{-285.8 \times 10^3\ \text{J mol}^{-1}}{298\,\text{K}} = -959.1\ \text{J K}^{-1}\,\text{mol}^{-1}$$

$$\Delta S^\circ_{univ} = \Delta S^\circ_{sys} + \Delta S^\circ_{surr} = 163.3\ \text{J K}^{-1}\,\text{mol}^{-1} - 959.1\ \text{J K}^{-1}\,\text{mol}^{-1}$$
$$= -795.8\ \text{J K}^{-1}\,\text{mol}^{-1}$$

This reaction is NOT spontaneous. There is no danger of water spontaneously decomposing into its elements. On the contrary, reaction in the opposite direction is spontaneous (but very slow) if in the reaction mixture at 25 °C, $p(H_2) = p(O_2) = 1$ bar.

17.47

$$COCl_2(g) \longrightarrow CO(g) + Cl_2(g)$$

Because one mole of gas is converted to two moles of gas, we know that the change in entropy is positive for this reaction. Since the reaction is NOT spontaneous at 25 °C, it must be endothermic — i.e., causing a positive change in the entropy of the surroundings. Increasing the temperature decreases the impact of the heat flow from the surroundings — i.e., the decrease in the entropy of the surroundings is not as great at higher T. At sufficiently high T, the system change in entropy will be the dominant term and the reaction will be spontaneous.

17.6—Gibbs free energy

17.49

(a) $2\,Al(s) + 3\,Cl_2(g) \longrightarrow 2\,AlCl_3(s)$

$\Delta_r S^\circ = 2 \times S^\circ[AlCl_3(s)] - (2 \times S^\circ[Al(s)] + 3 \times S^\circ[Cl_2(g)])$
$\quad = 2 \times 110.67\ J\ K^{-1}\ mol^{-1} - (2 \times 28.3\ J\ K^{-1}\ mol^{-1} + 3 \times 223.066\ J\ K^{-1}\ mol^{-1}) =$
$-504.5\ J\ K^{-1}\ mol^{-1}$
$\Delta_r H^\circ = 2\ \Delta_f H^\circ[AlCl_3(s)] - [0 + 0] = 2 \times (-704.2\ kJ\ mol^{-1}) = -1408.4\ kJ\ mol^{-1}$
$\Delta_r G^\circ = \Delta_r H^\circ - T\,\Delta_r S^\circ = -1408.4\ kJ\ mol^{-1} - (298\ K) \times (-504.5\ J\ K^{-1}\ mol^{-1})\ /\ (1000\ J\ /\ kJ)$
$\quad = -1258.1\ kJ\ mol^{-1}$

This reaction is spontaneous at 25 °C in a reaction mixture with all reagent species in their standard states. It is enthalpy driven—the negative enthalpy change more than compensates for the negative entropy change (positive $-T\,\Delta_r S^\circ$) at 25 °C.

(b) $6\,C(graphite) + 3H_2(g) \longrightarrow C_6H_6(l)$

$\Delta_r S^\circ = 1 \times S^\circ[C_6H_6(l)] - (6 \times S^\circ[C(graphite)] + 3 \times S^\circ[H_2(g)])$
$\quad = 172.8\ J\ K^{-1}\ mol^{-1} - (6 \times 5.740\ J\ K^{-1}\ mol^{-1} + 3 \times 130.684\ J\ K^{-1}\ mol^{-1}) =$
$-253.7\ J\ K^{-1}\ mol^{-1}$
$\Delta_r H^\circ = \Delta_f H^\circ[C_6H_6(l)] = 49.03\ kJ\ mol^{-1}$
$\Delta_r G^\circ = \Delta_r H^\circ - T\,\Delta_r S^\circ = 49.03\ kJ\ mol^{-1} - (298\ K) \times (-253.7\ J\ K^{-1}\ mol^{-1})\ /\ (1000\ J\ /\ kJ)$
$\quad = 124.6\ kJ\ mol^{-1}$

This reaction is NOT spontaneous at 25 °C in a reaction mixture with all reagent species in their standard states. Both enthalpy and entropy terms disfavour this process. It is NOT spontaneous at any temperature.

17.51

This problem is answered using thermochemical data from the NIST webbook (webbook.nist.gov).

$$2\,C(graphite) + 2\,H_2(g) + O_2(g) \longrightarrow CH_3COOH(l)$$

$\Delta_r S^\circ = S^\circ[CH_3COOH(l)] - (2 \times S^\circ[C(graphite)] + 2 \times S^\circ[H_2(g)] + 1 \times S^\circ[O_2(g)])$
$= 158.0\ J\ K^{-1}\ mol^{-1} - (2 \times 5.6\ J\ K^{-1}\ mol^{-1} + 2 \times 130.680\ J\ K^{-1}\ mol^{-1} + 205.152\ J\ K^{-1}$
$mol^{-1}) = -319.7\ J\ K^{-1}\ mol^{-1}$

$\Delta_r H^\circ = \Delta_f H^\circ[CH_3COOH(l)] = -483.52\ kJ\ mol^{-1}$

$\Delta_r G^\circ = \Delta_r H^\circ - T\,\Delta_r S^\circ = -483.52\ kJ\ mol^{-1} - (298.15\ K) \times (-319.7\ J\ K^{-1}\ mol^{-1})\ /\ (1000\ J\ /\ kJ)$
$= -388.2\ kJ\ mol^{-1}$

This value agrees with the tabulated free energy of formation ($\Delta_f G^\circ$) for acetic acid, as long as the same database is used—to be consistent (these data are not always accurately determined, or tabulated). This should not be surprising because the reaction of interest is the one for which $\Delta_r G^\circ = \Delta_f G^\circ$, by definition.

17.53

$$BaCO_3(s) \longrightarrow BaO(s) + CO_2(g)$$

$\Delta_r G^\circ = \Delta_f G^\circ[BaO(s)] + \Delta_f G^\circ[CO_2(g)] - \Delta_f G^\circ[BaCO_3(s)] = 219.7\ kJ\ mol^{-1}$
$= -525.1\ kJ\ mol^{-1} + (-394.359\ kJ\ mol^{-1}) - \Delta_f G^\circ[BaCO_3(s)] = 219.7\ kJ\ mol^{-1}$

So,

$\Delta_f G^\circ[BaCO_3(s)] = = -525.1\ kJ\ mol^{-1} + (-394.359\ kJ\ mol^{-1}) - 219.7\ kJ\ mol^{-1} = -1139.2\ kJ\ mol^{-1}$

17.55

$$Ni(CO)_4(l) \longrightarrow Ni(s) + 4\,CO(g) \qquad \Delta_r G^\circ\ at\ 25\,^\circ C = 40\ kJ\ mol^{-1}$$

$\Delta_r S^\circ = S^\circ[Ni(s)] + 4\,S^\circ[CO(g)] - S^\circ[Ni(CO)_4(l)]$
$= 29.87\ J\ K^{-1}\ mol^{-1} + 4 \times 197.674\ J\ K^{-1}\ mol^{-1} - 320\ J\ K^{-1}\ mol^{-1} = 501\ J\ K^{-1}\ mol^{-1}$

$\Delta_r H^\circ = \Delta_f H^\circ[Ni(s)] + 4\,\Delta_f H^\circ[CO(g)] - \Delta_f H^\circ[Ni(CO)_4(l)]$
$= 0 + 4 \times (-110.525\ kJ\ mol^{-1}) - (-632\ kJ\ mol^{-1}) = 190\ kJ\ mol^{-1}$

Assuming $\Delta_r H^\circ$ and $\Delta_r S^\circ$ are temperature independent, we get the temperature where $Ni(CO)_4(l)$ is in equilibrium with 1 bar CO(g) via

$\Delta_r G^\circ = \Delta_r H^\circ - T_{eq}\,\Delta_r S^\circ = 0$

i.e.,

$T_{eq} = \Delta_r H^\circ\ /\ \Delta_r S^\circ = 190 \times 1000\ J\ mol^{-1}\ /\ (501\ J\ K^{-1}\ mol^{-1}) = 379\ K = 106\,^\circ C.$

17.57

$$C_2H_6(g) + \tfrac{7}{2}O_2(g) \longrightarrow 2\,CO_2(g) + 3\,H_2O(g)$$

$\Delta_r G^\circ = 2\,\Delta_f G^\circ[CO_2(g)] + 3\,\Delta_f G^\circ[H_2O(g)] - (\Delta_f G^\circ[C_2H_6(g)] + \tfrac{7}{2}\,\Delta_f G^\circ[O_2(g)])$
$= 2 \times (-394.359\ kJ\ mol^{-1}) + 3 \times (-228.572\ kJ\ mol^{-1}) - (-32.82\ kJ\ mol^{-1} + \tfrac{7}{2} \times 0) = -1441.61\ kJ\ mol^{-1}$

This is in accord with experience. Ethane is a component of natural gas—it is a fuel. Its combustion is highly favoured—it releases a lot of free energy.

17.7—The relationship between $\Delta_r G°$ and K

17.59

$\frac{1}{2} N_2(g) + \frac{1}{2} O_2(g) \longrightarrow NO(g)$ $\Delta_r G°$ at 25 °C = +86.58 kJ mol^{-1}

$\ln K = -\dfrac{\Delta_r G°}{RT} = -\dfrac{86.58 \times 10^3 \, J \, mol^{-1}}{8.314 \, J \, K^{-1} \, mol^{-1} \times 298 \, K}$

$\qquad = -34.95$

$K = e^{-34.95} = 6.63 \times 10^{-16}$

K is very small, in accord with a positive standard free energy change—the forward reaction is NOT favoured in a reaction mixture at 25 °C and all reactants and products in their standard states (i.e., the conditions that apply for specification of the <u>standard</u> free energy change of reaction).

17.61

$N_2O_4(g) \rightleftharpoons 2 NO_2(g)$

$\Delta_r G° = -RT \ln K = -8.314 \, J \, K^{-1} \, mol^{-1} \times 298 \, K \, \ln(0.14)$

$\qquad = 4871 \, J \, mol^{-1} = 4.871 \, kJ \, mol^{-1}$

$\Delta_r G° = 2 \, \Delta_f G°[NO_2(g)] - \Delta_f G°[N_2O_4(g)]$
$\qquad = 2 \times 51.31 \, kJ \, mol^{-1} - (97.89 \, kJ \, mol^{-1}) = 4.73 \, kJ \, mol^{-1}$

This is pretty close to the above value, based upon the equilibrium constant (with only two significant digits). Also, note that tabulated data are not always as accurate as the number of digits indicates.

17.9—Dependence of equilibrium constants on temperature

17.63

$C(graphite) + \frac{1}{2} O_2(g) + 2 H_2(g) \longrightarrow CH_3OH(l)$

$\Delta_r G° = \Delta_f G°[CH_3OH(g)] = -166.27 \, kJ \, mol^{-1}$

This is the formation reaction for methanol.

$\ln K = -\dfrac{\Delta_r G°}{RT} = -\dfrac{-166.27 \times 10^3 \, J \, mol^{-1}}{8.314 \, J \, K^{-1} \, mol^{-1} \times 298 \, K}$

$\qquad = 67.11$

$K = e^{67.11} = 1.398 \times 10^{29}$

This reaction is decidedly product-favoured at 25 °C (when all reagents are in their standard states). Another temperature is not needed to better favour the product.

17.65

For water to boil in Denver, its vapour pressure need only be raised to 0.829 atm or 0.840 bar ($= Q$ for the vaporization process when the units are converted to bar).

$\Delta_{vap}H^{\circ} = \Delta_fH^{\circ}[H_2O(g)] - \Delta_fH^{\circ}[H_2O(l)]$
$\quad = -241.818 \text{ kJ mol}^{-1} - (-285.830 \text{ kJ mol}^{-1}) = 44.012 \text{ kJ mol}^{-1}$

$\Delta_{vap}S^{\circ} = S^{\circ}[H_2O(g)] - S^{\circ}[H_2O(l)]$
$\quad = 188.825 \text{ J K}^{-1} \text{ mol}^{-1} - (69.91 \text{ J K}^{-1} \text{ mol}^{-1}) = 118.92 \text{ J K}^{-1} \text{ mol}^{-1}$

With water vapour at 0.829 atm $=$ 0.840 bar, and $T = T_{bp}$ in Denver,

$\Delta_{vap}G = \Delta_{vap}G^{\circ} + RT_{bp} \ln Q = \Delta_{vap}H^{\circ} - T_{bp} \Delta_{vap}S^{\circ} + 8.314 \text{ J K}^{-1} \text{ mol}^{-1} \times T_{bp} \times \ln(0.840) = 0$

$44.012 \text{ kJ mol}^{-1} - T_{bp} (118.92 \text{ J K}^{-1} \text{ mol}^{-1} - 8.314 \text{ J K}^{-1} \text{ mol}^{-1} \times \ln(0.840)) = 0$

$T_{bp} = 44.012 \times 10^3 \text{ J mol}^{-1} / (118.92 \text{ J K}^{-1} \text{ mol}^{-1} - 8.314 \text{ J K}^{-1} \text{ mol}^{-1} \times (-0.174)) = 365.6 \text{ K} = 92.4 \,^{\circ}\text{C}$

SUMMARY AND CONCEPTUAL QUESTIONS

17.67

$$2\,HgO(s) \longrightarrow 2\,Hg(l) + O_2(g)$$

(a) $\Delta S°_{sys}$ is clearly positive—a gas is produced from a solid. $\Delta_r H°$ is surely positive—it is the reverse of a redox reaction between a metal and a non-metal. $\Delta S°_{surr}$ is correspondingly negative. $\Delta_r G°$ is positive at ordinary temperatures—otherwise, mercuric oxide would decompose in nature. At the temperatures Priestly studied, $\Delta_r G°$ is less than zero. The forward reaction proceeded, allowing Priestly to produce pure oxygen. Note that the mercury impurity is incredibly small because the vapour pressure of mercury is very small.

$\Delta_r S° = 2\,S°[Hg(l)] + S°[O_2(g)] - 2\,S°[HgO(s)]$
$= 2 \times 76.02\ J\ K^{-1}\ mol^{-1} + 205.138\ J\ K^{-1}\ mol^{-1} - 2 \times 70.29\ J\ K^{-1}\ mol^{-1} =$
$216.60\ J\ K^{-1}\ mol^{-1}$ √
$\Delta_r H° = 0 + 0 - 2 \times \Delta_f H°[HgO(s)] = -2 \times (-90.83\ kJ\ mol^{-1}) = 181.66\ kJ\ mol^{-1}$ √
A positive $\Delta_r H°$ corresponds to a negative $\Delta H°_{surr}$ and a negative $\Delta S°_{surr}$. √
$\Delta_r G° = \Delta_{vap}H° - T\,\Delta_{vap}S°$
$= 181.66\ kJ\ mol^{-1} - 298.15\ K \times 216.60\ J\ K^{-1}\ mol^{-1} / (1000\ J\ /kJ) = 117\ kJ\ mol^{-1}$ √
√

We have $\Delta_r G°$ at 25 °C, so we can calculate K at 25 °C (298 K) from
$\Delta_r G° - RT\ lnK$
$117\ 000\ J\ mol^{-1} = (-8.314\ J\ K^{-1}\ mol^{-1})(298\ K)\ lnK$
$lnK = (117\ 000\ J\ mol^{-1})/(-8.314\ J\ K^{-1}\ mol^{-1})(298\ K) = -47.22$
K at 298 K $= 3.1 \times 10^{-21}$

17.69

$$CH_3OH(l) \longrightarrow CH_4(g) + \tfrac{1}{2}O_2(g)$$

(a) $\Delta_r S° = S°[CH_4(g)] + \tfrac{1}{2}\,S°[O_2(g)] - S°[CH_3OH(l)]$
$= 186.264\ J\ K^{-1}\ mol^{-1} + \tfrac{1}{2} \times 205.138\ J\ K^{-1}\ mol^{-1} - 160.7\ J\ K^{-1}\ mol^{-1} =$
$128.133\ J\ K^{-1}\ mol^{-1}$
A large positive entropy change is consistent with the production of 1.5 mol gas from 1 mol liquid.
(b) $\Delta_r G° = \Delta_f G°[CH_4(g)] - \Delta_f G°[CH_3OH(l)]$
$= -50.72\ kJ\ mol^{-1} - (-166.27\ kJ\ mol^{-1}) = 115.55\ kJ\ mol^{-1}$
The reaction is NOT spontaneous at 25 °C and under standard conditions.
(c) $\Delta_r H° = \Delta_r G° + T\,\Delta_r S°$
$= 115.55\ kJ\ mol^{-1} - 298\ K \times 128.133\ J\ K^{-1}\ mol^{-1} / (1000\ J\ /kJ) = 77.37\ kJ\ mol^{-1}$
The minimum temperature at which the reaction becomes spontaneous is such that
$\Delta_r G° = \Delta_r H° - T_{spon}\,\Delta_r S° = 0$ (above T_{spon}, $\Delta_r G° < 0$)
$T_{spon} = \Delta_r H° / \Delta_r S° = 77.37 \times 10^3\ J\ mol^{-1} / 115.55\ J\ K^{-1}\ mol^{-1} = 669.5\ K$

17.71

(a)

S_8 (s, rhombic) $\longrightarrow$ S_8 (s, monoclinic)

$\Delta_r H^\circ = 3.213\,\text{kJ mol}^{-1}$, $\Delta_r S^\circ = 8.7\,\text{J K}^{-1}\,\text{mol}^{-1}$

$\Delta_r G^\circ = \Delta_r H^\circ - T\Delta_r S^\circ = 3.213\,\text{kJ mol}^{-1} - 353.15\,\text{K} \times (8.7\,\text{J K}^{-1}\,\text{mol}^{-1})\,/\,(1000\,\text{J /kJ}) = 0.141$ kJ mol^{-1}

The rhombic form of S_8 solid is more stable than the monoclinic form at 80 °C.

$\Delta_r G^\circ = \Delta_r H^\circ - T\Delta_r S^\circ = 3.213\,\text{kJ mol}^{-1} - 383.15\,\text{K} \times (8.7\,\text{J K}^{-1}\,\text{mol}^{-1})\,/\,(1000\,\text{J /kJ}) = -0.120$ kJ mol^{-1}

The monoclinic form of S_8 solid is more stable than the rhombic form at 110 °C.

(b)

$\Delta_r G^\circ = \Delta_r H^\circ - T_{\text{trans}}\Delta_r S^\circ = 0$

at temperature, T_{trans}, the temperature at which rhombic sulfur transforms to monoclinic sulfur.

$T_{\text{trans}} = \Delta_r H^\circ\,/\,\Delta_r S^\circ = 3.213 \times 10^3\,\text{J mol}^{-1}/(8.7\,\text{J K}^{-1}\,\text{mol}^{-1}) = 369\,\text{K} = 96\,°\text{C}$.

17.73

$4\,\text{Ag(s)} + \text{O}_2\text{(g)} \longrightarrow 2\,\text{Ag}_2\text{O(s)}$

(a)

$\Delta_r H^\circ = 2\,\Delta_f H^\circ[\text{Ag}_2\text{O(s)}] = 2 \times (-31.05\,\text{kJ mol}^{-1}) = -62.10\,\text{kJ mol}^{-1}$

$\Delta_r S^\circ = 2\,S^\circ[\text{Ag}_2\text{O(s)}] - (4\,S^\circ[\text{Ag(s)}] + S^\circ[\text{O}_2\,\text{(g)}])$

$\qquad = 2 \times 121.3\,\text{J K}^{-1}\,\text{mol}^{-1} - (4 \times 42.55\,\text{J K}^{-1}\,\text{mol}^{-1} + 205.138\,\text{J K}^{-1}\,\text{mol}^{-1})$

$\qquad = -132.738\,\text{J K}^{-1}\,\text{mol}^{-1}$

$\Delta_r G^\circ = \Delta_r H^\circ - T\Delta_r S^\circ = -62.10\,\text{kJ mol}^{-1} - 298.15\,\text{K} \times (-132.738\,\text{J K}^{-1}\,\text{mol}^{-1})\,/\,(1000\,\text{J /kJ}) =$ $-22.524\,\text{kJ mol}^{-1}$

(b)

$$\ln K = -\frac{\Delta_r G^\circ}{RT} = -\frac{-22.524 \times 10^3\,\text{J mol}^{-1}}{8.314\,\text{J K}^{-1}\,\text{mol}^{-1} \times 298.15\,\text{K}}$$

$$= 9.087$$

$$K = e^{9.087} = 8.835 \times 10^3 = \frac{1}{p(\text{O}_2)}$$

So, $p(\text{O}_2) = 1\,/\,(8.835 \times 10^3) = 1.13 \times 10^{-4}$ bar

(c)

$\Delta_r G^\circ = \Delta_r H^\circ - T\Delta_r S^\circ = 0$

at $T = \Delta_r H^\circ\,/\,\Delta_r S^\circ = -62.10 \times 10^3\,\text{J mol}^{-1}/\,(-132.738\,\text{J K}^{-1}\,\text{mol}^{-1}) = 468\,\text{K}$

At this temperature, the vapour pressure of oxygen is 1.00 bar.

17.75

$Hg(l) \rightarrow Hg(g)$

$\Delta_r H° = \Delta_f H°[Hg(g)] = 61.38 \text{ kJ mol}^{-1}$

$\Delta_r S° = S°[Hg(g)] - S°[Hg(l)] = 174.97 \text{ J K}^{-1} \text{ mol}^{-1} - 76.02 \text{ J K}^{-1} \text{ mol}^{-1}$
  $= 98.95 \text{ J K}^{-1} \text{ mol}^{-1}$

$\Delta_r G° = \Delta_r H° - T \Delta_r S° = 61.38 \text{ kJ mol}^{-1} - 298.15 \text{ K} \times (98.95 \text{ J K}^{-1} \text{ mol}^{-1}) / (1000 \text{ J /kJ}) =$
31.88 kJ mol^{-1}

(a) $p(Hg) = 1.00 \text{ bar}$
  $K = 1.00$

$\Delta_r G° = -RT \ln(K) = 0$
  $= \Delta_r H° - T \Delta_r S°$

So,

$T = \Delta_r H° / \Delta_r S° = 61.38 \times 10^3 \text{ J mol}^{-1} / 98.95 \text{ J K}^{-1} \text{ mol}^{-1} = 620 \text{ K}$

(b) $p(Hg) = 1.33 \times 10^{-3} \text{ bar}$
  $K = 1.33 \times 10^{-3}$

$\Delta_r G° = -RT \ln(K) = \Delta_r H° - T \Delta_r S°$

Solve for T.

$T = \Delta_r H° / (\Delta_r S° - R \ln(K))$
  $= 61.38 \times 10^3 \text{ J mol}^{-1} / (98.95 \text{ J K}^{-1} \text{ mol}^{-1} - 8.314 \text{ J K}^{-1} \text{ mol}^{-1} \ln(1.33 \times 10^{-3})) = 399 \text{ K}$

17.77

(a) Entropy *of the system* decreases in some spontaneous reactions. These reactions occur because of the increase in the entropy of the surroundings caused by the heat liberated—they are enthalpy driven reactions. However it is correct that the entropy *of the universe* increases with every spontaneous reaction.

(b) Reactions with a negative free energy change ($\Delta_r G° < 0$) are product-favoured. However, these reactions sometimes occur very slowly. $\Delta_r G°$ tells us how far a reaction will go to attain a condition of equilibrium, but does not tell us the speed of the reaction.

(c) Some spontaneous processes are endothermic. These reactions occur because the system entropy increases sufficiently to exceed the entropy lost by surroundings due to heat absorbed by the system—they are entropy driven reactions.

(d) Endothermic processes are sometimes spontaneous—see answer to (c).

17.79

$\Delta_r S°$ is usually positive for the dissolution of condensed substances. Dissolved species have more freedom as a solute than as a pure condensed phase. There is a positive component to the entropy of dissolution simply due to the disorder of having a mixture of two species. In this typical case of positive entropy of dissolution, if $\Delta_r H° = 0$

$\Delta_r G° = \Delta_r H° - T \Delta_r S° < 0$

and the dissolution is spontaneous. Such dissolutions are driven by entropy.
Some condensed substances—e.g., liquid hydrocarbons—actually have negative entropy of dissolution in water. The hydrophobic effect is actually attributable to the negative change in entropy associated with the water molecules solvating the hydrophobic molecule.

17.81

$C_6H_6(s) \rightarrow C_6H_6(l)$

(a) $\Delta_r H° > 0$

It takes energy (enthalpy) to break the intermolecular bonds holding the benzene molecules rigidly in their lattice positions.

(b) $\Delta_r S° > 0$

A liquid has more disorder—more entropy—than the corresponding solid.

(c) $\Delta_r G° = 0$ at 5.5 °C, the melting point, where the solid and liquid co-exist.

(d) $\Delta_r G° > 0$ at 0.0 °C $<$ the melting point, where the solid is the more stable phase.

(e) $\Delta_r G° < 0$ at 25.0°C $>$ the melting point, where the liquid is the more stable phase.

17.83

$I_2(s) \longrightarrow I_2 \text{ (in CCl}_4 \text{ solution)}$

$\Delta_r G° < 0$

This process is spontaneous—iodine dissolves readily in CCl₄.

$\Delta_r S° = (\Delta_r H° - \Delta_r G°) / T > 0$ since $\Delta_r H° = 0$

This dissolution is an entropy-driven process.

17.85

(a) $N_2H_4(l) + O_2(g) \rightarrow N_2(g) + 2\ H_2O(l)$

This is an oxidation-reduction reaction—oxygen is the oxidizing agent and hydrazine is the reducing agent.

(b)

$\Delta_r H° = 2\ \Delta_f H°[H_2O(l)] - \Delta_f H°[N_2H_4(l)]$

$\quad\quad = 2 \times (-285.830\ \text{kJ mol}^{-1}) - (50.63\ \text{kJ mol}^{-1}) = -622.29\ \text{kJ mol}^{-1}$

$\Delta_r S° = 2\ S°[H_2O(l)] + S°[N_2(g)] - (S°[N_2H_4(l)] + S°[O_2(g)])$

$\quad\quad = 2 \times (69.91\ \text{J K}^{-1}\ \text{mol}^{-1}) + 191.61\ \text{J K}^{-1}\ \text{mol}^{-1} - (121.21\ \text{J K}^{-1}\ \text{mol}^{-1} + 205.138\ \text{J K}^{-1}$ mol^{-1})

$\quad\quad = 5.08\ \text{J K}^{-1}\ \text{mol}^{-1}$

$\Delta_r G° = \Delta_r H° - T\ \Delta_r S° = -622.29\ \text{kJ mol}^{-1} - 298\ \text{K} \times 5.08\ \text{J K}^{-1}\ \text{mol}^{-1}\ /(1000\ \text{J/kJ}) =$ $-623.80\ \text{kJ mol}^{-1}$

Chapter 18
Spontaneous Change: How Fast?

IN-CHAPTER EXERCISES

Exercise 18.2—Relative rates of [reactant] decrease and [product] increase

$$2\,NOCl(g) \longrightarrow 2\,NO(g) + Cl_2(g)$$

For every 2 mol of NOCl(g) that react, 2 mol of NO(g) and 1 mol of Cl_2(g) are produced. The relative rate of appearance of Cl_2 is ½ the rate of appearance of NO which equals the rate of disappearance of NOCl.

$$-\frac{\Delta[NOCl]}{\Delta t} = \frac{\Delta[NO]}{\Delta t} = 2 \times \frac{\Delta[Cl_2]}{\Delta t}$$

Exercise 18.3—Estimation of rate of reaction

In the first two hours, the concentration of sucrose decreases from 0.050 to 0.033 mol L^{-1}. The average rate of decrease of sucrose concentration
$$= -((0.033 - 0.050)\ mol\ L^{-1})\ /\ 2\ h\ =\ 0.0085\ mol\ L^{-1}\ h^{-1}$$
In the last two hours, the concentration of sucrose decreases from 0.014 to 0.010 mol L^{-1}. The average rate of decrease of sucrose concentration
$$= -((0.010 - 0.014)\ mol\ L^{-1})\ /\ 2\ h\ =\ 0.002\ mol\ L^{-1}\ h^{-1}$$
At $t = 4.0$ h, we draw a tangent to the concentration vs. time curve. The straight tangent line intersects (you may get slightly different values)
$t = 0$ hr at 0.039 mol L^{-1} and
$t = 8$ hr at 0.005 mol L^{-1}.
The instantaneous rate of decrease of [sucrose] at $t = 4.0$ is minus the slope of this line
$$= -((0.005 - 0.039)\ mol\ L^{-1})\ /\ 8\ h\ =\ 0.0044\ mol\ L^{-1}\ h^{-1}$$

Exercise 18.4—Rate equations and order of reaction

(a) The rate equation for the reaction is
   Rate $= k\,[NO_2][O_3]$
(b) If the concentration of NO_2(g) is tripled, the rate triples.
(c) If the concentration of O_3(g) is halved, the rate is halved.

Exercise 18.7—Determining a rate equation from initial rates

| Experiment | Initial concentrations (mol L^{-1}) | | Initial rate (mol L^{-1} s^{-1}) |
	[NO] (mol L^{-1})	[O$_2$] (mol L^{-1})	
1	0.020	0.010	0.028
2	0.020	0.020	0.057
3	0.020	0.040	0.114
4	0.040	0.020	0.227
5	0.010	0.020	0.014

In Experiments 1, 2, and 3, with [NO] the same, the relative [O$_2$] values are 1: 2: 4. The corresponding relative rates are 0.028: 0.057: 0.114 = 1: 2: 4 showing that the reaction is first order with respect to the O$_2$(g) concentration. In Experiments 5, 2, and 4, with [O$_2$] the same, the relative [NO] values are 1, 2, and 4. The corresponding relative rates are 0.014: 0.057: 0.227 $\approx$ 1: 4: 16 = $(1)^2$: $(2)^2$: $(4)^2$ showing that the reaction is second order with respect to the NO(g) concentration.

$$\text{Rate} = k\,[NO]^2[O_2]$$

Using the data from Experiment 4, we have

$$k = \frac{\text{Rate}}{[NO_2]^2[O_2]} = \frac{0.227\,\text{mol}\,L^{-1}\,s^{-1}}{(0.040\,\text{mol}\,L^{-1})(0.040\,\text{mol}\,L^{-1})(0.020\,\text{mol}\,L^{-1})} = 7.1 \times 10^3\,L^2\,\text{mol}^{-2}\,s^{-1}$$

Exercise 18.9—Using the integrated rate equation for a first-order reaction

$$\ln \frac{[\text{sucrose}]_t}{[\text{sucrose}]_0} = -k\,t = -(0.21\,h^{-1})(5.0\,h) = -1.05$$

$$\text{Fraction remaining} = \frac{[\text{sucrose}]_t}{[\text{sucrose}]_0} = 0.35$$

$$[\text{sucrose}]_t = 0.010\,\text{mol}\,L^{-1} \times 0.35 = 0.0035\,\text{mol}\,L^{-1}$$

Exercise 18.11—Using the integrated rate equation for a second-order reaction

$$\frac{1}{[HI]_t} - \frac{1}{[HI]_0} = \frac{1}{[HI]_{12\ min}} - \frac{1}{0.010\ mol\,L^{-1}}$$

$$= (30\,L\,mol^{-1}\,min^{-1})\,t = (30\,L\,mol^{-1}\,min^{-1}) \times (12\ min)$$

$$= 360\ L\,mol^{-1}$$

So

$$\frac{1}{[HI]_{12\ min}} = 360\ L\,mol^{-1} + \frac{1}{0.010\ mol\,L^{-1}}$$

$$= 360\ L\,mol^{-1} + 100\ L\,mol^{-1} = 460\ L\,mol^{-1}$$

$$[HI]_{12\ min} = \frac{1}{460\ L\,mol^{-1}} = 0.00217\ mol\,L^{-1}$$

Exercise 18.13—Determining reaction order by graphical methods

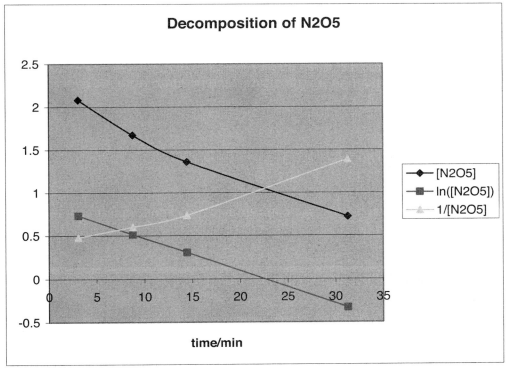

The plot of $\ln[N_2O_5]()$ most closely resembles a straight line. The reaction is first order.

$k = -\text{slope} = (\ln([N_2O_5]_{3.07\ min}) - \ln([N_2O_5]_{31.28\ min})) / (31.28 - 3.07\ min)$
$= (0.732368 - (-0.3285)) / (31.28 - 3.07\ min) = 0.0143\ min^{-1}$

Exercise 18.15—Half-life of a first-order process

(a) The half-lives of ^{241}Am and ^{125}I are given by

$$t_{1/2} = \frac{0.693}{k} = \frac{0.693}{0.0016\ \text{y}^{-1}} = 430\ \text{years}$$

and

$$t_{1/2} = \frac{0.693}{k} = \frac{0.693}{0.011\ \text{d}^{-1}} = 63\ \text{days}$$

respectively.

(b) ^{125}I decays much faster than ^{241}Am.

(c)

$$\ln\frac{[^{125}\text{I}]_t}{[^{125}\text{I}]_0} = \ln\frac{[^{125}\text{I}]_{2.0\,d}}{1.6\times10^{15}\ \text{atoms}} = -kt = -(0.011\ \text{d}^{-1})(2\ \text{d}) = -0.022$$

$$[^{125}\text{I}]_{2.0\,d} = 1.6\times10^{15}\ \text{atoms} \times 0.978 = 1.57\times10^{15}\ \text{atoms}$$

Note that we can use number of atoms in place of concentration for first order reactions. For first order reactions, the rate depends only on the amount of the substance—the rate is unaffected by a change in volume (the concentrations always appear as ratios).

Exercise 18.17—Calculating E_a from the temperature dependence of k

$$\ln\frac{1.00\times10^4\ \text{s}^{-1}}{4.5\times10^3\ \text{s}^{-1}} = \frac{E_a}{8.314\ \text{J K}^{-1}\text{mol}^{-1}}\left(\frac{1}{274\,\text{K}} - \frac{1}{283\,\text{K}}\right)$$

$$E_a = 8.314\ \text{J K}^{-1}\text{mol}^{-1} \times 0.799 \times \left(1.16\times10^{-4}\right)^{-1}$$

$$= 57.2\times10^3\ \text{J mol}^{-1} = 57.2\ \text{kJ mol}^{-1}$$

Exercise 18.19—Elementary steps of a reaction

$$2\ \text{NO} \longrightarrow \text{N}_2\text{O}_2$$
$$\text{N}_2\text{O}_2 + \text{H}_2 \longrightarrow \text{N}_2\text{O} + \text{H}_2\text{O}$$
$$\underline{\text{N}_2\text{O} + \text{H}_2 \longrightarrow \text{N}_2 + \text{H}_2\text{O}}$$
$$2\ \text{NO} + 2\,\text{H}_2 \longrightarrow \text{N}_2 + 2\,\text{H}_2\text{O}$$

All three of these processes are bimolecular. The rate equation for the third step is
$$\text{Rate} = k\,[\text{N}_2\text{O}]\,[\text{H}_2]$$

Exercise 18.21—Elementary steps and reaction mechanisms

(a)

Step 1 (Fast): $NH_3(aq) + OCl^-(aq) \longrightarrow \cancel{NH_2Cl(aq)} + \cancel{OH^-(aq)}$

Step 2 (Slow): $\cancel{NH_2Cl(aq)} + NH_3(aq) \longrightarrow N_2H_5^+(aq) + Cl^-(aq)$

Step 3 (Fast): $\cancel{N_2H_5^+(aq)} + \cancel{OH^-(aq)} \longrightarrow N_2H_4(aq) + H_2O(l)$

$\overline{\qquad 2\,NH_3(aq) + OCl^-(aq) \longrightarrow N_2H_4(aq) + H_2O(l) + Cl^-(aq) \qquad}$

(b) Step 2, the slow step is the rate-determining step.

(c) The rate equation for elementary step 2 is

$$\text{Rate} = k\,[NH_2Cl]\,[NH_3]$$

(d) NH_2Cl, $N_2H_5^+$ and OH^- are intermediates in this reaction mechanism.

Exercise 18.22—Nucleophilic substitution reactions

(a) $\text{Rate} = k\,[CH_3(CH_2)_5CHBrCH_3]\,[OH^-]$

(b) If the concentration of (S)-2-bromooctane is doubled, keeping the concentration of hydroxide ions constant, the rate of reaction will double.

(c) If the concentration of hydroxide ions is doubled, keeping the concentration of (S)-2-bromooctane constant, the rate of reaction will double.

(d)

(e) (R)-octan-2-ol

Exercise 18.23—Nucleophilic substitution reactions

(a) The rate of substitution will double if the concentration of *tert*-butyl chloride is doubled, keeping the concentration of iodide ions constant.

(b) The rate of substitution will NOT change if the concentration of iodide ions is doubled, keeping the concentration of *tert*-butyl chloride constant. The rate is independent of [I$^-$].

(c) Carbocation intermediate.

(d) Whether or not inversion takes place is irrelevant here because the product is the same in both cases—due to symmetry.

(e) The proportion of this reaction that proceeds by the S_N2 mechanism is negligible because the rate of the S_N1 mechanism is enhanced by the relatively stable carbocation intermediate, and because the rate of the S_N2 mechanism is reduced by the presence of the three methyl groups about the carbon center. The portion of collisions between iodide and *tert*-butyl chloride which can lead to reaction is much smaller than the value for methyl chloride (for example). The three methyl groups present an obstacle limiting the rate of the S_N2 reaction. The S_N1 mechanism presents no such obstacle—the iodide reacts with a reactive planar intermediate.

REVIEW QUESTIONS

18.2—The concept of reaction rate

18.25

$N_2(g) + 3H_2(g) \longrightarrow 2\,NH_3(g)$

If $-\Delta[H_2]/\Delta t = 4.5 \times 10^{-4}$ mol L^{-1} min^{-1}, $\Delta[NH_3]/\Delta t = \frac{2}{3} \times 4.5 \times 10^{-4}$ mol L^{-1} min^{-1}

$$= 3.0 \times 10^{-4} \text{ mol } L^{-1} \text{ min}^{-1}$$

18.3—Conditions that affect the rate of a reaction

18.27

Rate $= k[A]^2[B]$

The reaction is second order in [A] and first order in [B]. The total order of the reaction is three.

18.4—Dependence of rate on reactant concentration

18.29

$Pt(NH_3)_2Cl_2(aq) + H_2O(l) \longrightarrow [Pt(NH_3)_2(H_2O)Cl^+(aq) + Cl^-(aq)$

Rate $= k\,[Pt(NH_3)_2Cl_2]$

$\qquad = 0.090$ h$^{-1} \times 0.020$ mol $L^{-1} = 1.8 \times 10^{-3}$ mol L^{-1} h^{-1}

In this reaction, the rate of reaction is the rate of change of [Cl$^-$]—i.e., 1.8×10^{-3} mol L^{-1} h^{-1}

18.31
(a) Doubling [NO], keeping [O$_2$] fixed, (second vs. first experiment) increases the rate of reaction by the factor, $4 = 2^2$. The reaction is second order in NO concentration.
Doubling [O$_2$], keeping [NO] fixed, (third vs. first experiment) increases the rate of reaction by the factor, $2 = 2^1$. The reaction is first order in oxygen concentration.
(b)

$\qquad$ Rate $= k\,[NO]^2[O_2]$

(c) Using the middle data set, we have
$\qquad k = 1.0 \times 10^{-4}$ mol L^{-1} s^{-1} / $(0.020$ mol $L^{-1})^2(0.010$ mol $L^{-1}) = 25$ L^2 mol^{-2} s^{-1}
(d) If [NO] $= 0.015$ mol L^{-1} and [O$_2$] $= 0.0050$ mol L^{-1},
$\qquad$ Rate $= 25$ L^2 mol^{-2} s$^{-1} \times (0.015$ mol $L^{-1})^2(0.0050$ mol $L^{-1}) = 2.8 \times 10^{-5}$ mol L^{-1} s^{-1}
(e) If NO(g) reacts at rate $= 1.0 \times 10^{-4}$ mol L^{-1} s^{-1}, O$_2$ (g) reacts at rate $= 5.0 \times 10^{-5}$ mol L^{-1} s^{-1};
i.e., one half the rate of reaction of NO(g). NO$_2$(g) forms at the same rate as NO(g) reaction, 1.0×10^{-4} mol L^{-1} s^{-1}. The relative rates are given by the stoichiometric coefficients of the reaction.

18.33

$$2\,NO(g) + 2\,H_2(g) \longrightarrow N_2(g) + 2\,H_2O(g)$$

(a) When [NO] is doubled with [H₂] fixed, the rate changes by the factor

$$0.136 \,/\, 0.0339 \approx 4$$

We can conclude that the reaction is second order with respect to [NO].

When [H₂] is doubled with [NO] fixed, the rate changes by the factor

$$0.0678 \,/\, 0.0339 \approx 2$$

The reaction is first order with respect to [H₂].

(b)

$$\text{Rate} = k\,[NO]^2[H_2]$$

(c) Using the first data set,

$$k = 0.136 \text{ mol L}^{-1}\text{ s}^{-1} \,/\, (0.420 \text{ mol L}^{-1})^2 (0.122 \text{ mol L}^{-1}) = 6.32 \text{ L}^2\text{ mol}^{-2}\text{ s}^{-1}$$

(d) The rate of formation of N₂(g) if [NO] = 0.350 mol L⁻¹ and [H₂] = 0.205 mol L⁻¹ is

$$\text{Rate} = 6.32 \text{ L}^2\text{ mol}^{-2}\text{ s}^{-1} \times (0.350 \text{ mol L}^{-1})^2 (0.205 \text{ mol L}^{-1}) = 0.159 \text{ mol L}^{-1}\text{ s}^{-1}$$

18.35 Carbon monoxide reacts with O₂ to form CO₂

$$2\,CO(g) + O_2(g) \longrightarrow 2\,CO_2(g)$$

(a) + (b) When [CO] is doubled, with [O₂] fixed, the rate increases by the factor $1.47 \times 10^{-4} \,/\, 3.68 \times 10^{-5} \approx 4 = 2^2$.

The reaction is second order in [CO].

When [O₂] is doubled with [CO] fixed, the rate increases by the factor $7.36 \times 10^{-5} \,/\, 3.68 \times 10^{-5} = 2$.

The reaction is first order in [O₂].

$$\text{Rate} = k\,[CO]^2[O_2]$$

The overall order of the reaction is three.

(c) Using the second data set,

$$k = 1.47 \times 10^{-4} \text{ mol L}^{-1}\text{ min}^{-1} \,/\, (0.04 \text{ mol L}^{-1})^2 (0.02 \text{ mol L}^{-1}) = 4.6 \text{ L}^2\text{ mol}^{-2}\text{ min}^{-1}$$

18.36

(a) Doubling [NO], keeping [O₂] fixed (Exp 2 vs. 1), increases the rate of reaction by the factor, $4 = 2^2$. The reaction is second order with respect to NO(g) concentration.

Doubling [O₂], keeping [NO] fixed, increases the rate of reaction by the factor, $2 = 2^1$. The reaction is first order with respect to O₂(g) concentration.

$$\text{Rate} = k\,[NO]^2[O_2]$$

(b) Using the third data set,

$$k = 1.7 \times 10^{-8} \text{ mol L}^{-1}\text{ s}^{-1} \,/\, (1.8 \times 10^{-4} \text{ mol L}^{-1})^2 (1.04 \times 10^{-2} \text{ mol L}^{-1}) = 50 \text{ L}^2\text{ mol}^{-2}\text{ s}^{-1}$$

(c) For the fourth experiment,

$$\text{Rate} = 50 \text{ L}^2\text{ mol}^{-2}\text{ s}^{-1} \times (1.8 \times 10^{-4} \text{ mol L}^{-1})^2 (5.2 \times 10^{-2} \text{ mol L}^{-1}) = 8.4 \times 10^{-8} \text{ mol L}^{-1}\text{ s}^{-1}$$

We could get this answer directly by noting that experiment 4 is like experiment 3 except that the oxygen concentration is increased by a factor of 5. So, we just have to scale the experiment 3 rate by 5, since the reaction is first order with respect to [O₂].

18.5—Concentration-time relationships: Integrated rate equations

18.37

(a) $2\,NO_2(g) \longrightarrow 2\,NO(g) + O_2(g)$

For first order kinetics,

$$\ln\frac{[NO_2]_{150\,s}}{[NO_2]_0} = -kt = -(3.6 \times 10^{-3}\ s^{-1})(150\ s) = -0.54$$

$$\frac{[NO_2]_{150\,s}}{[NO_2]_0} = 0.58$$

(b)

If 99% of the NO_2 is decomposed, then $[NO_2]_t/[NO_2]_0 = 0.01$.

$$\ln\frac{[NO_2]_t}{[NO_2]_0} = \ln 0.01 = -4.6 = -kt$$

$$t = 4.6/k = 4.6/3.6 \times 10^{-3}\ s^{-1} = 1280\ s$$

18.39

$$SO_2Cl_2(g) \longrightarrow SO_2(g) + Cl_2(g)$$

$$\ln\frac{[SO_2Cl_2]_t}{[SO_2Cl_2]_0} = -kt = -(2.8 \times 10^{-3}\ min^{-1})\,t$$

$$t = -\frac{1}{2.8 \times 10^{-3}\ min^{-1}}\ln\frac{0.31 \times 10^{-3}}{1.24 \times 10^{-3}} = 495\ min$$

18.43

The integrated rate equation for a second order reaction gives

$$kt = \frac{1}{[HI]_t} - \frac{1}{[HI]_0} = \frac{1}{0.180\ mol\ L^{-1}} - \frac{1}{0.229\ mol\ L^{-1}}$$

$$t = \frac{1.19\,L\,mol^{-1}}{0.0113\,L\,mol^{-1}\,min^{-1}} = 105\ min$$

18.45

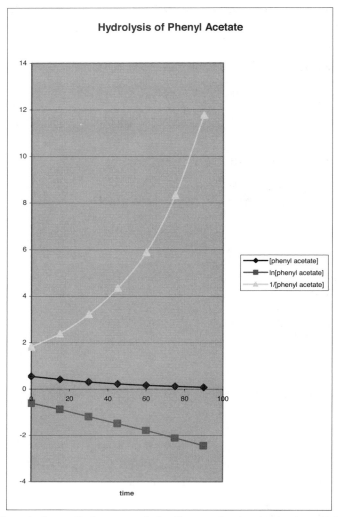

The ln [phenyl acetate] data is best fit by a straight line. The hydrolysis of phenyl acetate is a first reaction.

$$\ln \frac{[\text{phenyl acetate}]_{90.0\,s}}{[\text{phenyl acetate}]_0} = -kt = -k\,(90.0\ s)$$

$$k = -\frac{1}{90.0\ s}\ln \frac{0.085}{0.55} = 0.021\ s^{-1}$$

18.47

This reaction is second order in $[NO_2]$. The rate equation is

$$\text{Rate} = -\Delta[NO_2]\,/\,\Delta t = k\,[NO_2]^2$$

The slope of the $1/[NO_2]$ vs. time plot is the rate constant for a second order reaction.

$$k = 1.1\ \text{L mol}^{-1}\ s^{-1}$$

18.49

(a) This is a first order reaction. There is therefore a fixed half-life associated with $N_2O_5(g)$.

$$t_{1/2} = \frac{0.693}{k} = \frac{0.693}{5.0 \times 10^{-4}\,s^{-1}} = 1400\,s = 23\,min$$

(b) $[N_2O_5]$ decreases to one tenth of its original value after

$$t_{1/10} = \frac{\ln 10}{k} = \frac{2.303}{5.0 \times 10^{-4}\,s^{-1}} = 4600\,s = 77\,min$$

18.51

If 75% of the sample has decomposed, then 25% remains.

$$\ln\frac{[HCOOH]_{72\,s}}{[HCOOH]_0} = \ln 0.25 = -kt = -k\,(72\,s)$$

$$k = \frac{-\ln 0.25}{72\,s}$$

$$t_{1/2} = \frac{0.693}{k} = \frac{0.693 \times 72\,s}{-\ln 0.25} = 36\,s$$

Of course we could have deduced this answer by noting that as ¼ of the sample remains, two half-lives must have passed after 72 s.

18.6—A microscopic view of reaction rates: Collision theory

18.53

$$\ln\frac{k_{310K}}{k_{300K}} = \frac{E_a}{8.314\,J\,K^{-1}\,mol^{-1}}\left(\frac{1}{300\,K} - \frac{1}{310\,K}\right)$$

$$E_a = 8.314\,J\,K^{-1}\,mol^{-1} \times \ln(3) \times \left(1.08 \times 10^{-4}\right)^{-1}$$

$$= 8.49 \times 10^4\,J\,mol^{-1} = 84.9\,kJ\,mol^{-1}$$

18.55

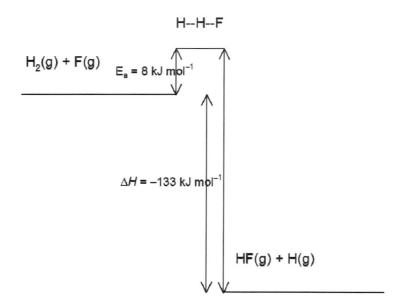

18.7—Reaction mechanisms

18.57
(a) $Cl(g) + ICl(g) \rightarrow I(g) \ \square + Cl_2(g)$
  Rate $= k\,[Cl][ICl]$
(b) $O(g) + O_3(g) \rightarrow 2\,O_2(g)$
  Rate $= k\,[O][O_3]$
(c) $2\,NO_2(g) \rightarrow N_2O_4(g)$
  Rate $= k\,[NO_2]^2$

18.59

$$2\,O_3 \longrightarrow 3\,O_2$$

Step 1   Fast, reversible $O_3 \rightleftharpoons O_2 + O$
Step 2   Slow $O_3 + O \longrightarrow 2\,O_2$

(a) Step 2, the slow step, is the rate-determining step.
(b) Rate $= k_2\,[O_3]\,[O]$
The concentration of O atoms is determined by the equilibrium represented in the equation for step 1.

$$[O] = K_1\,[O_3]\,/\,[O_2]$$

So,

$$Rate = k_2\,K_1\,[O_3]^2\,/\,[O_2]$$

The reaction is second order in $[O_3]$.

18.61

(a)

Step 1	Slow	$H_2O_2(aq) + I^-(aq) \longrightarrow H_2O(l) + OI^-(aq)$
Step 2	Fast	$H^+(aq) + OI^-(aq) \longrightarrow HOI(aq)$
Step 3	Fast	$HOI(aq) + H^+(aq) + I^-(aq) \longrightarrow I_2(aq) + H_2O(l)$

$$H_2O_2(aq) + 2\,I^-(aq) + 2\,H^+(aq) \longrightarrow 2\,H_2O(l) + I_2(aq)$$

(b) The first two steps are bimolecular, while the last is ternary.

(c) The rate is determined by the slow reaction.

Rate $= k_2\,[H_2O_2]\,[I^-]$

18.63

(a)

Step 1	Slow, endothermic	$2\,NO_2(g) \longrightarrow NO(g) + NO_3(g)$
Step 2	Fast, exothermic	$NO_3(g) + CO(g) \longrightarrow NO_2(g) + CO_2(g)$
Overall Reaction	Exothermic	$NO_2(g) + CO(g) \longrightarrow NO(g) + CO_2(g)$

$NO_2(g)$ and $CO(g)$ are reactants.

$NO_3(g)$ is an intermediate.

$CO_2(g)$ and $NO(g)$ are products.

(b)

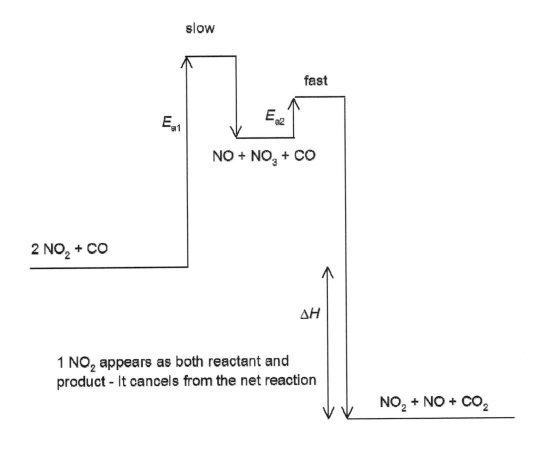

SUMMARY AND CONCEPTUAL QUESTIONS

18.65
No answers provided. They will be derived from your exploration of this activity.

18.67
$$H_2(g) + I_2(g) \longrightarrow 2\,HI(g)$$
Rate = $k\,[H_2][I_2]$
(a) The reaction must occur in a single step. **FALSE**
While the observed rate equation is consistent with a single step process, there may be other multi-step mechanisms also consistent with the observed rate equation.
(b) This is a second-order reaction overall. **TRUE**
(c) Raising the temperature will cause the value of k to decrease. **FALSE**
Increasing temperature means more frequent and more energetic collisions (that are more frequently successful)—i.e., the rate increases with temperature.
(d) Raising the temperature lowers the activation energy for this reaction. **FALSE**
The activation energy is fixed by the reaction. Its dependence upon temperature is negligible.
(e) If the concentrations of both reactants are doubled, the rate will double. **FALSE**
The rate will increase fourfold if both concentrations are doubled.
(f) Adding a catalyst in the reaction will cause the initial rate to increase. **TRUE**

18.69
(a) The rate-determining elementary step in a reaction is the slowest step in a mechanism. **TRUE**
(b) It is possible to change the rate constant by changing the temperature. **TRUE**
(c) As a reaction proceeds at constant temperature, the rate remains constant. **FALSE**
The rate generally varies with time—e.g., in first and second order reactions. Only for zero order reactions does the rate remain constant. Zero order reactions arise when there is a limited amount of a catalyst and the reactants are present in abundance. In any case, reactions are only zero order for a time. Eventually, reactants deplete sufficiently that the amount of catalyst is no longer rate limiting, and the reaction is no longer zero order.
(d) A reaction that is third order overall must involve more than one step. **FALSE**
It is possible that the reaction proceeds via a single ternary process. Ternary processes are infrequent, but they can play important roles.

Chapter 19
Alkenes and Alkynes

IN-CHAPTER EXERCISES

Exercise 19.2—IUPAC names of alkenes

(a) 2-Methylhex-1-ene

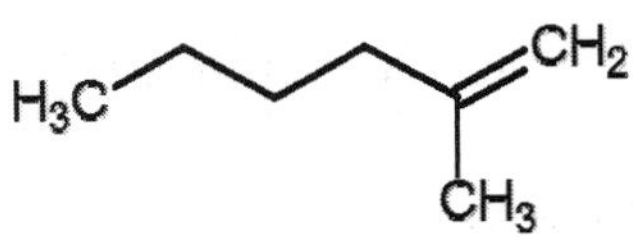

(b) 4,4-Dimethylpent-2-ene

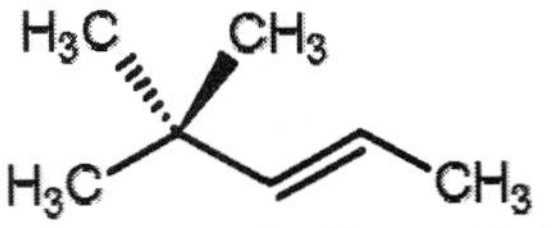

(c) 2-Methylhexa-1,5-diene

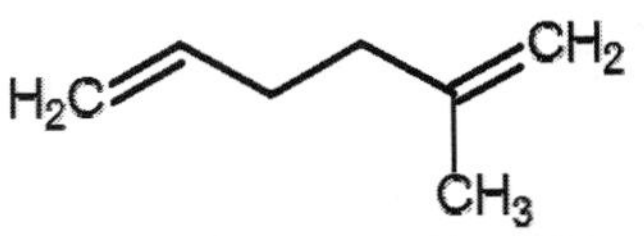

(d) 3-Ethyl-2,2-dimethylhept-3-ene

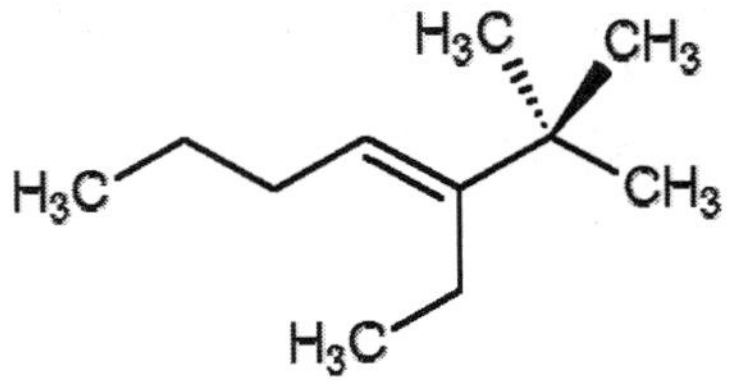

Exercise 19.4—*Cis-trans* isomers

(a) $CH_3CH=CH_2$
No *cis-trans* isomers.
(b) $(CH_3)_2C=CHCH_3$
No *cis-trans* isomers.

(c) ClCH=CHCl

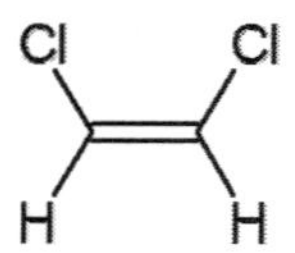

cis 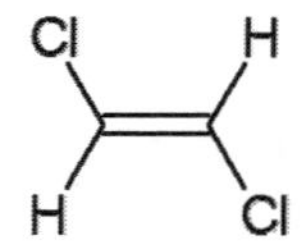   trans

(d) $CH_3CH_2CH=CHCH_3$

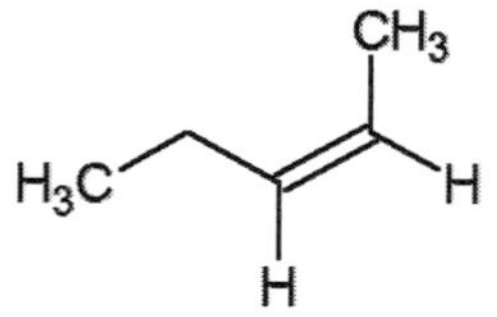

cis 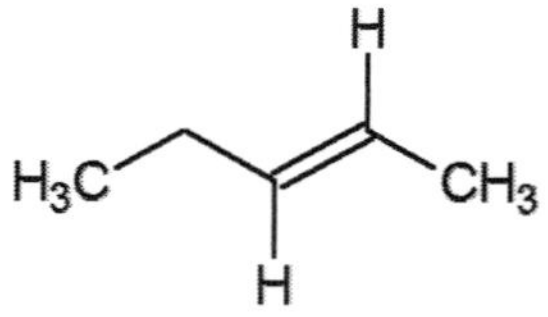   trans

(e) $CH_3CH_2CH=C(Br)CH_3$

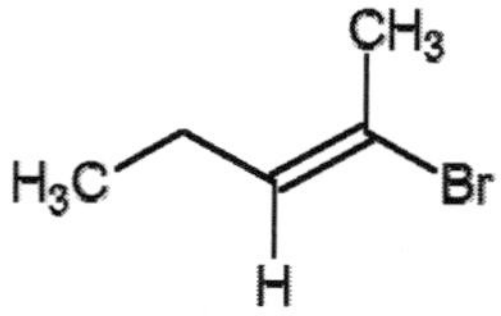

cis 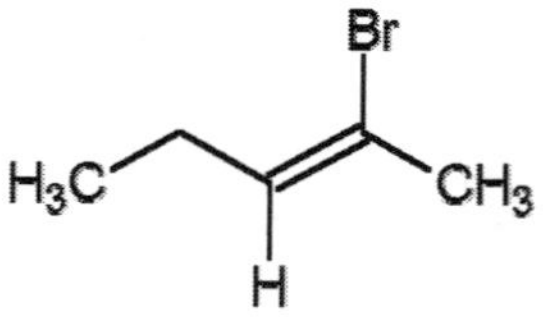   trans

Note that the *cis-trans* designations refer to the relative positions of the two alkyl groups (similar groups). However, this practice is, in general, ambiguous and the *E-Z* designations would normally be used for a trisubstituted alkene such as this.

(f) 3-Methylhept-3-ene

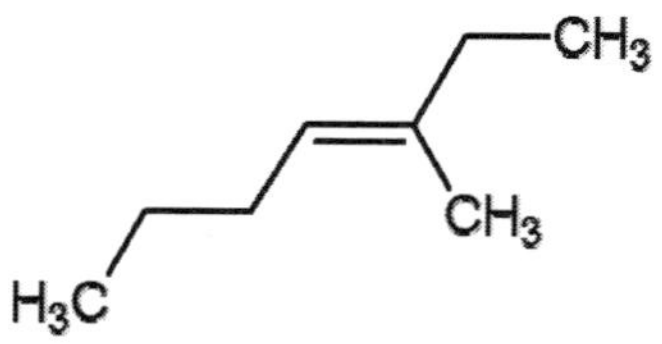

cis 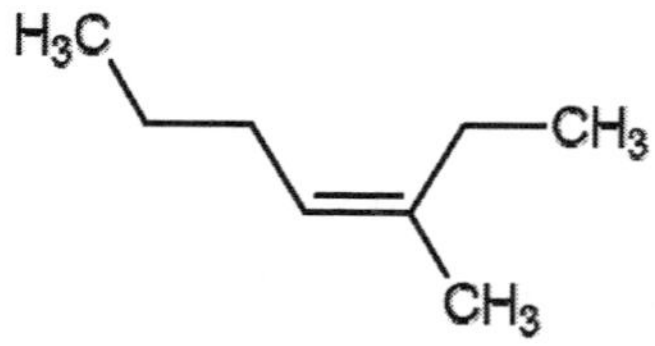   trans

Exercise 19.6—Assigning *E* / *Z* configurations

(a) –H or –Br
–Br has higher priority
(b) –Cl or –Br
–Br has higher priority
(c) –CH₃ or –CH₂CH₃
–CH₂CH₃ has higher priority
(d) –NH₂ or –OH
–OH has higher priority
(e) –CH₂OH or –CH₃
–CH₂OH has higher priority
(f) –CH₂OH or –CH=O
–CH=O has higher priority

Exercise 19.7—Assigning *E* / *Z* configurations

(a)

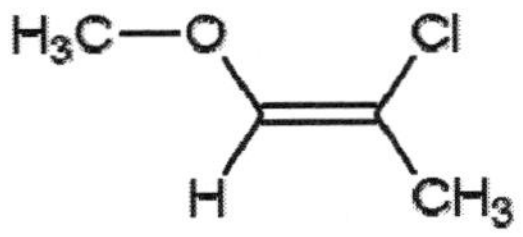

Z

methoxy and chloro are the higher priority substituents on the two alkene carbons.

(b)

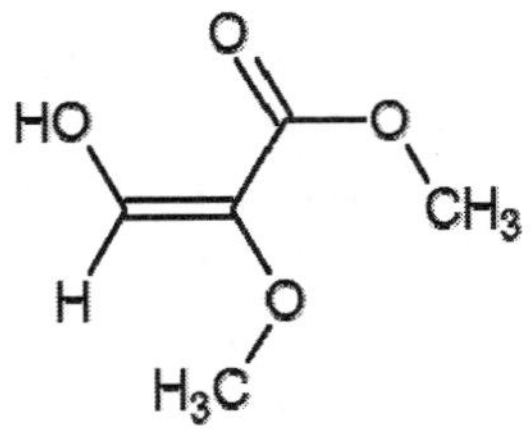

E

methyl and methoxy are the higher priority substituents. Note that O has higher priority than the carbonyl carbon atom.

Exercise 19.8—Asssigning *E / Z* configurations

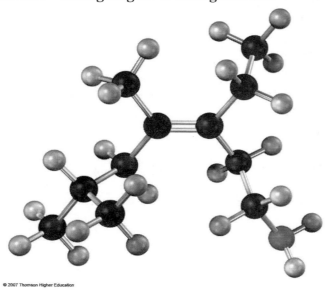

© 2007 Thomson Higher Education

This is the *Z* isomer.

Exercise 19.9—^{13}C NMR spectroscopy of alkenes

The two downfield lines, at about 139 and 113 ppm, are the two alkene carbon atom peaks.

Exercise 19.10—Using IR and NMR to determine the structure of alkenes

The ^{13}C NMR spectrum shows three C atoms, only one of them a downfield alkene C. Because we know there are 6 C atoms, the molecule must be symmetrical, with three pairs of indistinguishable carbons, including a pair of equivalent alkene carbon atoms. The IR spectrum shows alkyl C-H stretches, a C=C stretch and C-H stretch just above 3000 cm^{-1}, consistent with an alkene. Note that the formula, C_6H_{10}, corresponds to two degrees of unsaturation—i.e., a total of two double bonds plus rings. The ^{1}H NMR spectrum shows three types of H atoms in the ratio 1:2:2. This corresponds to 2, 4, and 4 H atoms of each type. The 2 equivalent (one cluster of peaks) downfield H atoms have chemical shift around 5.8 ppm, consistent with two equivalent vinyl H atoms. The equivalence of 4 H atoms corresponds to two pairs of CH_2 type H atoms equivalent by symmetry. These observations are consistent with cyclohexene.

Other structures involving two C=C double bonds are ruled out because they would produce at least two inequivalent alkene carbon atoms.

Exercise 19.11—Classifying organic reactions by type

(a) $CH_3Br + KOH \rightarrow CH_3OH + KBr$
is a substitution reaction – OH^- substitutes for Br^-.
(b) $CH_3CH_2OH \rightarrow H_2C=CH_2 + H_2O$
is an elimination reaction – H_2O is eliminated from ethanol to give ethene.
(c) $H_2C=CH_2 \square + H_2 \rightarrow CH_3CH_3$
is an addition reaction —H_2 is added to ethene to give ethane.

Exercise 19.13—Bond polarity of functional groups

(a) Aldehyde

The carbonyl C=O bond is the most polar, due to the electronegativity difference between C and O atoms.

(b) Ether

(c) Ester

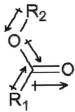

(d) Alkylmagnesium bromide, R—MgBr

$$H_3C-Mg-Br$$

In this functional group, the carbon (the first carbon in R) has the excess negative charge.

Exercise 19.14—Recognizing electrophiles and nucleophiles

(a) NH_4^+
Ammonium is an electrophile. It is attracted to concentrations of negative charge, and will donate a proton to proton acceptors—i.e., nucleophiles.
(b) CN^-
Cyanide is a nucleophile. It is attracted to concentrations of positive charge (i.e., regions of depleted electron density) such as acidic protons and carbonyl carbon atoms.
(c) Br^-
Bromide is a nucleophile. It is attracted to concentrations of positive charge (i.e., regions of depleted electron density).
(d) CH_3NH_2
The amine N atom is a nucleophile. It is attracted to concentrations of positive charge (i.e., regions of depleted electron density).
(e) $H–C≡C–H$
Ethyne is a nucleophile. It has a relatively loosely held cloud of π electrons that electrophiles are attracted to—and vice versa.

Exercise 19.15—Electrostatic potential maps

BF_3 is electron deficient at the low-electronegativity boron atom. In the Lewis structure of BF_3, the boron atom has only 6 electrons in its valence shell. BF_3 is electrophilic at the boron centre.

Exercise 19.16—Addition of HX to alkenes

Exercise 19.17—Addition of HX to alkenes

In accord with Markovnokov's rule, the halogen attaches to most substituted carbon atom.
(a) $CH_3CH_2CHClCH_3$
(b) $(CH_3)_2ClCH_2CH_2CH_3$
(c) chlorocyclohexane

Exercise 19.18—Addition of HX to alkenes

(a) Bromocyclopentane can be prepared from cyclopentene and HBr.
(b) 3-Bromohexane can be prepared from hex-2-ene and HBr.
(c) 1-Iodo-1-isopropylcyclohexane can be prepared from:

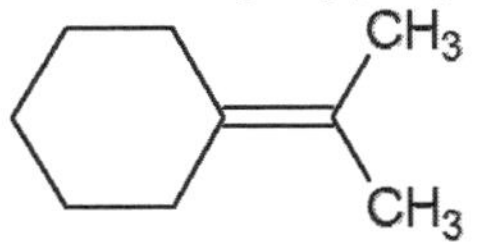

(d)

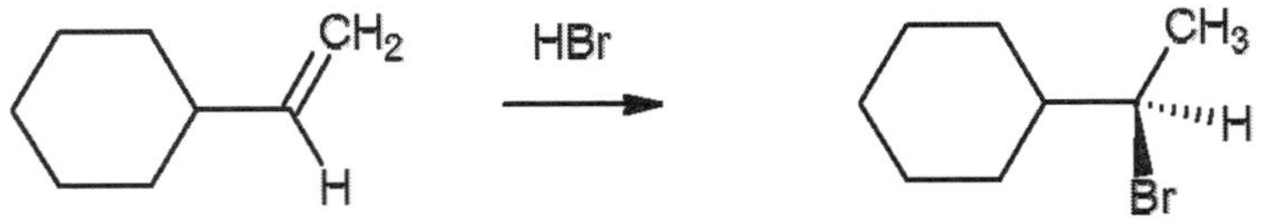

Exercise 19.19—Addition of HX to alkenes

In accord with Markovnikov's rule, the carbocation will be formed on the more substituted carbon.

(a)

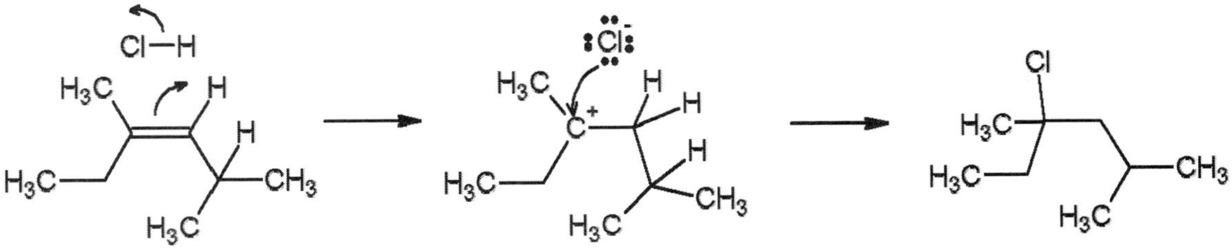

(b)

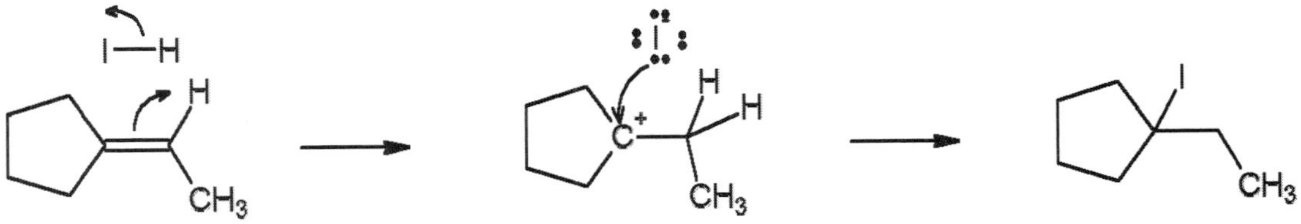

Exercise 19.20—Addition of Br₂ to dienes

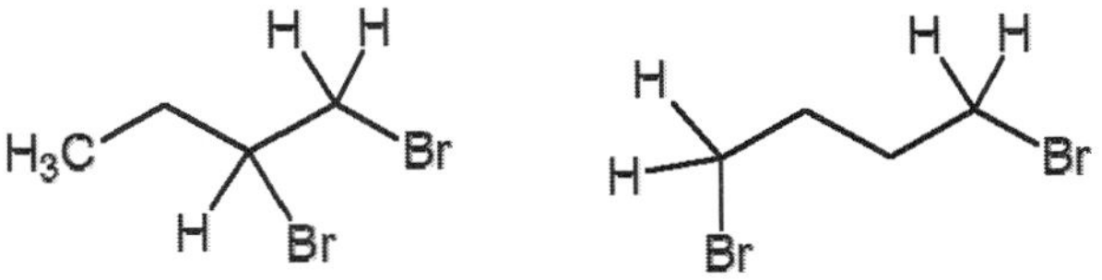

1,2-dibromobutane   1,4-dibromobutane

Exercise 19.21—Addition of H₂O to alkenes

In accord with Markovnikov's rule, the hydroxyl group attaches to more substituted carbon.

(a) Addition of water to 3-methylhex-3-ene gives 3-methylhexan-3-ol,

(b) Addition of water to 1-methylcyclopentene gives 1-methylcyclopentanol

(c) Addition of water to 2,5-dimethylhept-2-ene gives 2,5-dimethylheptan-2-ol

Exercise 19.22—Addition of H₂O to alkenes

(a) Butan-2-ol can be made by addition of water to but-1-ene or but-2-ene.
(b) 3-Methylpentan-3-ol can be made by addition of water to 3-methylpent-2-ene or 2-ethylbut-1-ene.
(c) 1,2-Dimethylcyclohexanol can be made by addition of water to 1,2-dimethylcyclohexene or 1,6-dimethylcyclohexene.

Exercise 19.23—Addition of Br₂ to alkenes

If the second bromine atom added to the other carbon atom—the one on the left, then the enantiomeric product would be formed. By symmetry, this pathway is just as likely as the one shown in Figure 19.29 of the text. The two enantiomeric products would be formed in equal amounts.

Exercise 19.24—Addition of Br₂ to alkenes

The two enantiomers of 1,2-dibromo-1,2-dimethylcyclohexane are expected to be formed in equal amounts (a racemate) upon bromination of 1,2-dimethylcyclohexene.

Exercise 19.25—Addition of Br₂ to alkenes

Intermediate bromonium ion.

Exercise 19.26—Addition of H₂ to alkenes

(a) Catalytic hydrogenation of $(CH_3)_2C=CHCH_2CH_3$ produces $(CH_3)_2CHCH_2CH_2CH_3$.
(b) Catalytic hydrogenation of 3,3-dimethylcyclopentene gives 1,1-dimethylcyclopentane.

Exercise 19.27—Oxidation of alkenes

(a) Reaction of 1,2-dimethylcyclohexene with $KMnO_4$, H_3O^+ produces hexane-2,7-dione. This is the stronger, more oxidizing formulation of $KMnO_4$. It oxidizes the double bond, cleaving the ring to produce a diketone.

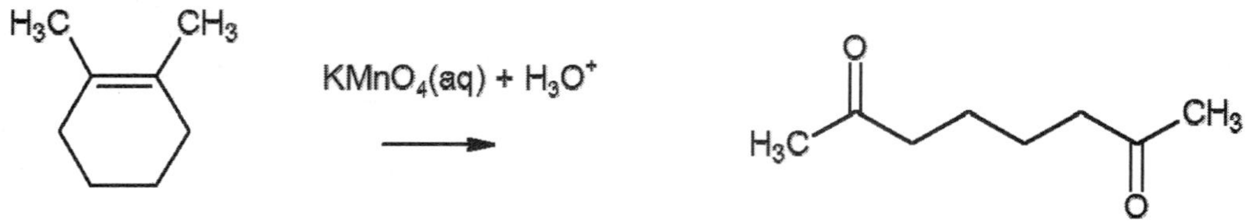

(b) Reaction of 1,2-dimethylcyclohexene with $KMnO_4$, OH^-, H_2O produces *cis*-1,2-dimethylcyclohexan-1,2-diol. In base, $KMnO_4$ is a milder oxidizing agent. producing the diol (glycol) from the alkene. The C=C bond does not get cleaved.

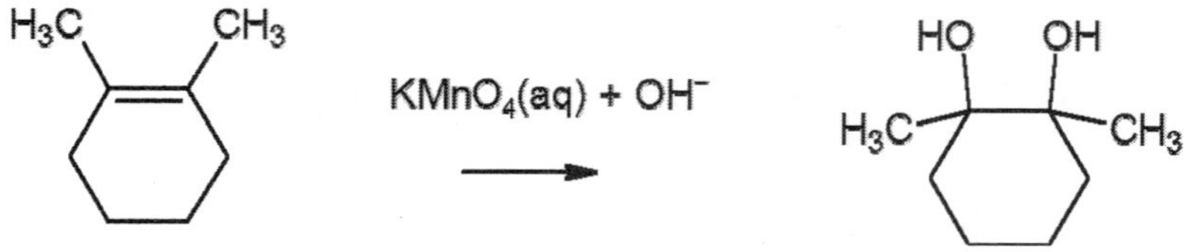

Exercise 19.28—Oxidation of alkenes

(a) $(CH_3)_2C=O + CO_2$ is the product upon treating $(CH_3)_2C=CH_2$ with acidic $KMnO_4$.
(b) 2 mol $CH_3CH_2CO_2H$ is the product upon treating one mol of $CH_3CH_2CH=CHCH_2CH_3$ with acidic $KMnO_4$.

Exercise 19.29—IUPAC nomenclature

(a) 6-methylhept-3-yne
(b) 3,3-dimethylbut-1-yne
(c) 5-methylhex-2-yne
(d) hept-2-en-5-yne

Exercise 19.30—IR spectra of alkynes

The peak just above 3300 cm^{-1} looks like an alkyne C–H stretch, while the peak just above 2100 cm^{-1} looks like an alkyne C≡C stretch. There also appear to be alkyl C–H stretch peaks just below 3000 cm^{-1}. The compound likely has alkyl H atoms in addition to the alkynyl H atom.

Exercise 19.31—NMR spectra of alkynes

The compound has 6 peaks, or 6 unique kinds of C atoms. The two downfield peaks, ~68 and ~85 ppm, are characteristic of an alkyne. Alkene ^{13}C chemical shifts occur further downfield.

Exercise 19.32—Identification of unknown alkyne from IR, MS, and NMR

We are looking for a primary alkyne – there is an alkynyl C−H peak in the IR spectrum. Six carbon atoms and 10 hydrogen atoms fit the observed molecular ion peak at m/z = 82 in the mass spectrum. There are six distinct peaks in the ^{13}C NMR spectrum, ruling out molecules with some symmetry. The unknown compound is likely either $CH_3CH_2CH_2CH_2C{\equiv}CH$ or $(CH_3)_2CHCH_2C{\equiv}CH$.

Exercise 19.33—Addition of H_2O to alkynes

Exercise 19.34—Addition of H_2O to alkynes

(a) To make pentan-2-one you could hydrate pent-1-yne, $CH_3CH_2CH_2C{\equiv}CH$.
(b) To make 3-hexan-2-one you could hydrate hex-2-yne, $CH_3CH_2CH_2C{\equiv}CCH_3$.

Exercise 19.35—Synthesis of alkynes

(a) To make 5-methylhex-1-yne, $(CH_3)_2CHCH_2CH_2C\equiv CH$ you could start with
ethyne (acetylene) , $HC\equiv CH$, treat it with $NaNH_2$ to make sodium acetylide
add to 1-bromo-3-methylbutane, $(CH_3)_2CHCH_2CH_2Br$

(b) To make hex-2-yne, $CH_3CH_2CH_2C\equiv CCH_3$ you could start with
propyne , $HC\equiv CCH_3$, treat it with $NaNH_2$ to make its acetylide ion conjugate base
add to bromopropane, $CH_3CH_2CH_2Br$

Alternatively, you could treat
Pent-1-yne, $CH_3CH_2CH_2C\equiv CH$, with $NaNH_2$ to make its acetylide ion conjugate base
add to methylbromide, CH_3Br

(c) To make 4-methylpent-2-yne, $(CH_3)_2CHC\equiv CCH_3$ you could start with
propyne, $HC\equiv CCH_3$, treat it with $NaNH_2$ to make its acetylide ion conjugate base
add to isopropylbromide, $(CH_3)_2CHBr$

Alternatively, you could treat
3-methylbut-1-yne, $(CH_3)_2CHC\equiv CH$, with $NaNH_2$ to make its acetylide ion conjugate base
add to methylbromide, CH_3Br

REVIEW QUESTIONS

19.1 Pheasant wattles, photoprotection, and photonics

19.37
β-carotene is added to commercial food for ornamental fish to make the colour of the fish (specifically the red-orange component of the colour) brighter and richer. β-carotene is a red-orange pigment, used in nature along with other pigments to make colours.

19.2 Reactions of alkenes: Patterns and explanations

19.39
(a) Alkyl and nitrile (or cyano) groups are present
(b) Alkyl and ether groups are present.
(c) Alkyl, ketone, and ester groups are present
(d) Alkene and ketone groups are present,

(e) Alkene and amide groups are present,

(f) Phenyl and carboxyl groups are present,

19.3 Electronic structure of alkenes

19.41
The double bonds within the rings at the ends of β-carotene are between carbon 1 (the carbon connected to the chain) and carbon 2, whereas the ring double bonds in β-carotene are between carbon 2 and carbon 3. Otherwise, the two molecules are identical.
The HOMO-LUMO gap for α-carotene is larger than that for lycopene. Whereas α-carotene absorbs blue light and appears orange, lycopene absorbs green light (lower frequency and lower energy than blue light) and appears red.

19.4 Naming alkenes

19.43

(a) *cis*-4,5-Dimethylcyclohexene

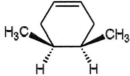

(b) 3,3,4,4-Tetramethylcyclobutene

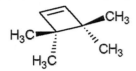

19.45
(a) 4-methylpent-1-ene
(b) hept-3-ene
(c) hepta-1,5-diene
(d) 2-methylhex-3-ene

19.47
Molecules (b) and (c) have *cis-trans* isomers.

19.5 *E* and *Z* isomers of alkenes

19.49
Methylcarboxylate, $-COOCH_3$, has higher prioroty than carboxyl, $-COOH$.

19.51
(a) Z. Although the methyl groups are on opposite sides, the higher priority group on the left and on the right are on the same side.
(b) Z.

19.7 Classifying organic reactions by type

19.53
(a) A substitution reaction—CN^- substitutes for Br^-.
(b) An elimination reaction—H_2O is eliminated to form a double bond.
(c) An addition reaction—a C=C double bond adds to a cyclic diene to form a bicyclic alkene.
(d) A substitution reaction—NO_2 substitutes for H on cyclohexane.

19.9 Electrophilic addition of HX to alkenes—Hydrohalogenation

19.55

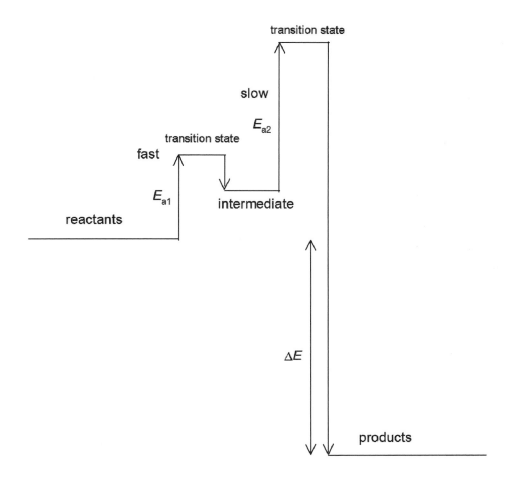

19.57

$CH_3CH_2CH=CHCH_3$ + HCl $\rightarrow$ $CH_3CH_2CHClCH_2CH_3$ or $CH_3CH_2CH_2CHClCH_3$

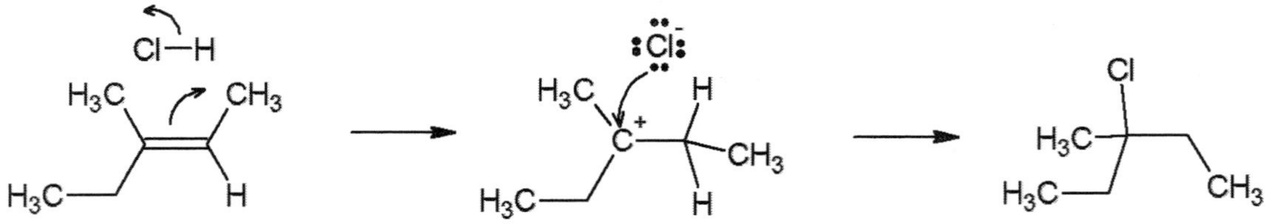

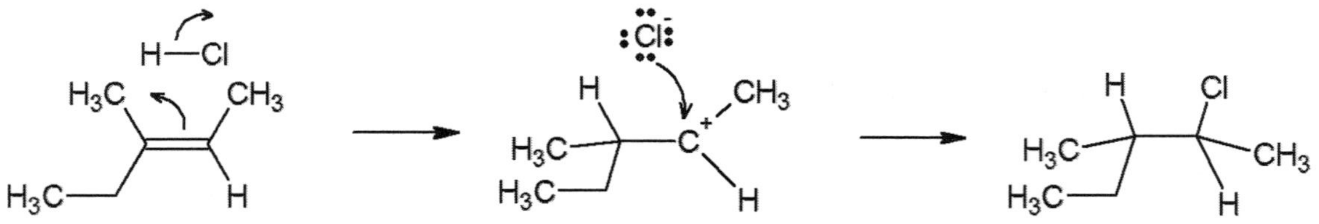

19.9–19.14 Reactions of alkenes

19.59

The carbocation on the left is the more stable, having two alkyl substituents rather than one.

19.61

(a)

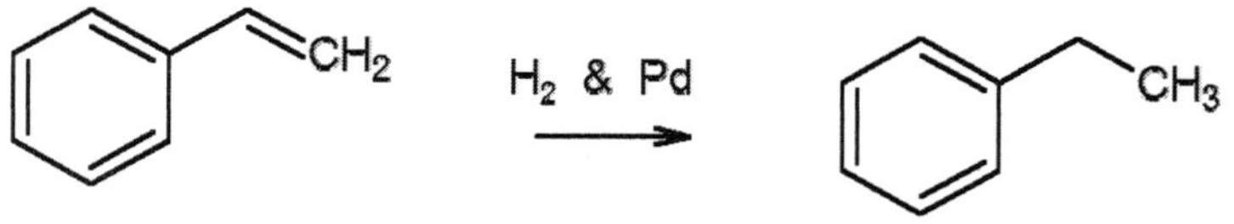

(b)

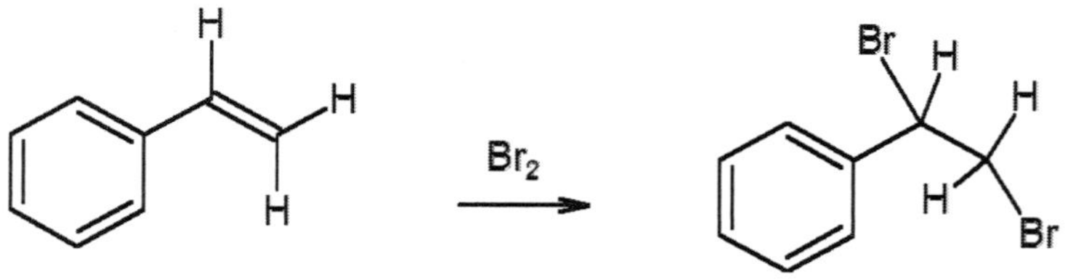

(c)

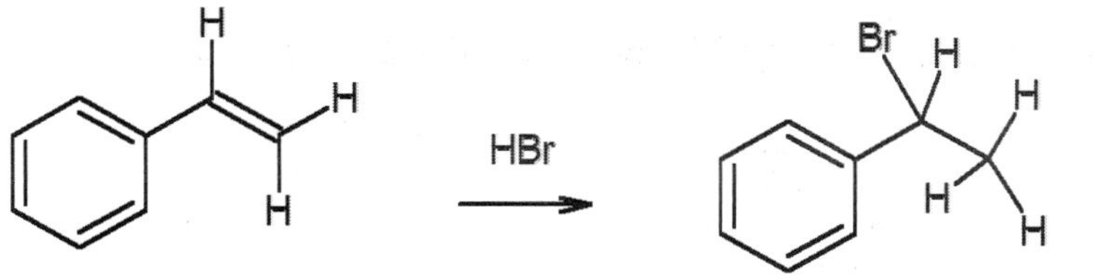

(d)

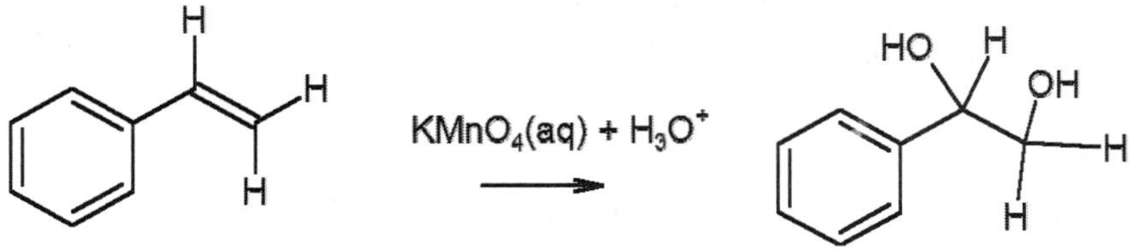

19.63
Oxidative cleavage with acidic KMnO₄ of

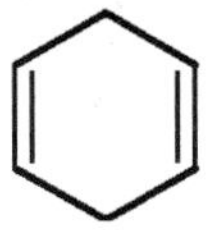

yields a single product,

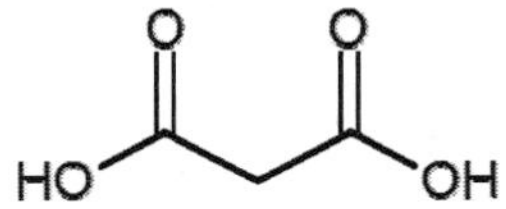

because of symmetry.
Oxidative cleavage of

gives two distinct products,

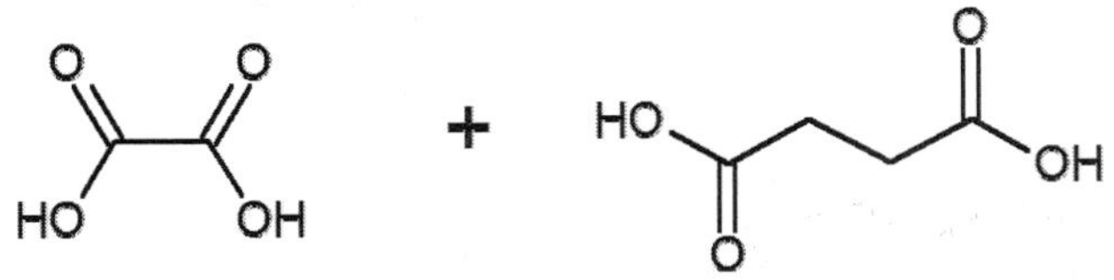

19.65

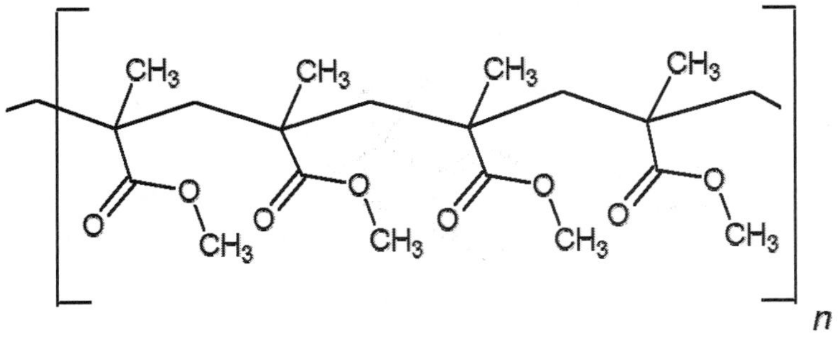

19.15 Alkynes—Electronic structure, spectroscopy, and reactions

19.67
(a) Hex-1-yne has 6 distinct peaks (6 distinct carbon atoms) in its ^{13}C NMR spectrum.

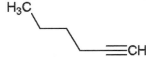

(b) Hex-2-yne has 6 distinct peaks (6 distinct carbon atoms) in its ^{13}C NMR spectrum.

(c) Hex-3-yne has 3 distinct peaks (3 distinct carbon atoms) in its ^{13}C NMR spectrum. This molecule has symmetry so the carbon atoms are equivalent in pairs.

19.69

19.71

$H_3C\!-\!\!\equiv\!\!CH$ $\xrightarrow{\text{NaNH}_2}$ $H_3C\!-\!\!\equiv\!\!C^-\ Na^+$

$\xrightarrow{\text{CH}_3\text{Br}}$ $H_3C\!-\!\!\equiv\!\!-CH_3$ $\xrightarrow{\text{H}_2 + \text{Lindlar catalyst}}$

SUMMARY AND CONCEPTUAL QUESTIONS

19.73
(a) An alkene, C_6H_{12}, that does not have *cis and trans* isomers has one degree of saturation—i.e., one carbon-carbon double bond—and must have the same two substituents on at least one of the alkene carbons. Three possibilities are

1-hexene,   2-methylpent-2-ene   &   2,3-dimethylbut-2-ene

(b) The *E* isomer of a trisubstituted alkene, C_6H_{12}. To have an *E* isomer, the alkene must have two different substituents on each alkene carbon atom.

(c) A cycloalkene, C_7H_{12}, with a tetrasubstituted double bond.

19.75

An O−H bond is formed, a C−H bond is broken, a C−C bond is formed (to make a double bond) and a C−Br bond is broken.

19.77

Because of these three additional resonance structures, the + charge of the carbocation is spread out —delocalized—over four carbon atoms. This makes the benzyl carbocation especially stable—for a carbocation.

19.79

19.81

The observation of a cyclohexanone product upon oxidation reveals a cyclohexane ring with a double bond outside the ring in A. This leaves 3 C atom and 2 H atoms. One of these C atoms is the alkene carbon doubly bonded to the cyclohexane ring. This double bond accounts for 1 equivalent of H_2 absorbed on reduction of A. Because the other two equivalents are consumed by only two carbon atoms—i.e., the 2 C atoms have 2 degrees of unsaturation—they must be associated with an alkyne group.

19.83

Treat acetylene with $NaNH_2$ to get acetylide. $HC{\equiv}C^-$.
Add to $CH_3(CH_2)_{11}CH_2Br$ to get
$\qquad CH_3(CH_2)_{12}C{\equiv}CH$
Treat with $NaNH_2$ to get its conjugate base.
Add to 1-bromooctane, $CH_3(CH_2)_6CH_2Br$ to get
$\qquad CH_3(CH_2)_{12}C{\equiv}C(CH_2)_7CH_3$
Treatment of this alkyne with H_2 and a Lindlar catalyst produces Muscalure, which has been shown to be a *cis*-alkene

19.85

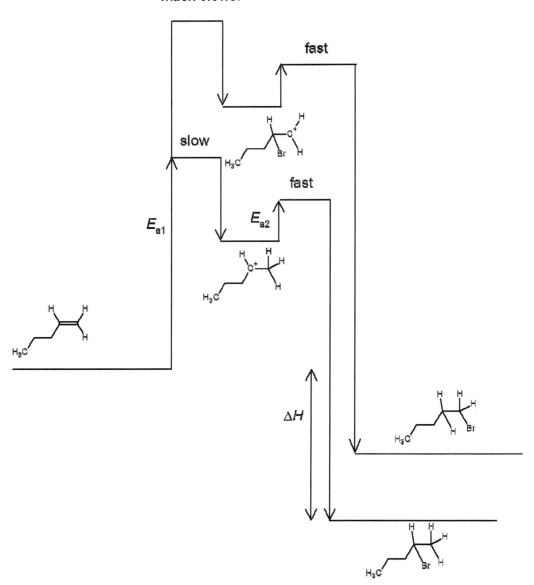

19.87

Chapter 20
Aromatic Compounds

IN-CHAPTER EXERCISES

Exercise 20.1—Relative stability of dienes

The enthalpy change of hydrogenation of cyclohexa-1,3-diene is given in Figure 20.4 of the textbook as -230 kJmol^{-1}. The enthalpy change of hydrogenation of cyclohexa-1,4-diene should be something close to twice the enthalpy change of hydrogenation of cyclohexene. The two double bonds in cyclohexa-1,4-diene are not conjugated—i.e., they are not next to each other allowing an additional resonance structure (such as that shown below for cyclohexa-1,3-diene). Figure 20.4 of the text gives the value -118 kJ mol^{-1} for the enthalpy change of hydrogenation of cyclohexene. We therefore expect the enthalpy change of hydrogenation of cyclohexa-1,4-diene to be about -236 kJ mol^{-1}. Cyclohexa-1,3-diene is stabilized relative to cyclohexa-1,4-diene due to conjugation of the two double bonds—the four π electrons are delocalized over four carbon atoms. This delocalization is shown with orbital views in Figure 19.23.

Exercise 20.2—Hückel's rule

(a) Cyclobutadiene is monocyclic, planar and fully conjugated. However, there are $4 = 4 \times 1 \neq 4n + 2$ π electrons. Cyclobutadiene is anti-aromatic.

(b) Cyclohepta-1,3,5-triene is monocyclic and planar, but not fully conjugated—there is a methylene group separating the two outer double bonds.

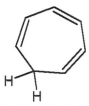

(c) Cyclopenta-1,3-diene is monocyclic and planar, but not fully conjugated—there is a methylene group separating the double bonds on one side.

(d) Cyclooctatetraene is monocyclic, planar and fully conjugated. However, there are $8 = 4 \times 2 \neq 4n + 2$ π electrons. Cyclooctatetraene is anti-aromatic.

Exercise 20.3—^{13}C NMR spectra of aromatic compounds

(a) Benzene has just one ^{13}C NMR absorption peak – all the C atoms are equivalent.

(b) Chlorobenzene has four ^{13}C NMR absorption peaks. The C atom bonded to Cl and the C atom para to the Cl each give one peak. The two ortho carbon atoms and two meta carbon atoms each contribute an additional peak—the two ortho C atoms are equivalent, as are the meta C atoms.

(c) Naphthalene has three ^{13}C NMR absorption peaks. Carbon atoms related via reflection through the long axis of naphthalene are equivalent. Also, the pairs of such C atoms to the left of centre are equivalent to the pairs to the right of centre—by reflection (through the short axis of the molecule) symmetry.

(d) 1,3-dichlorobenzene has four ^{13}C NMR absorption peaks. The two carbon atoms bonded to chlorine are equivalent, as are the two C atoms below the Cl-substituted atoms. The other two C atoms produce the remaining two peaks.

Exercise 20.4—IR spectra of aromatic compounds

(a) There is an aryl C-H stretching band with three peaks at 3100, 3070, and 3030 cm^{-1}. There is an alkyl C-H stretching band with principal peaks at 2925 and 2850 cm^{-1}.
(b) The out-of-plane C-H bending vibrations give rise to two peaks at 740 and 700 cm^{-1}.
(c) The in-plane C-H bending vibrations produce peaks at 1090 and 1030 cm^{-1}. The aryl C-C stretches produce a band with principal peaks at 1610, 1510, and 1470 cm^{-1}.

Exercise 20.5—Aromaticity of heterocyclic compounds

Furan has $6 = 4 \times 1 + 2$ electrons in its π shell (4 electrons from each of four C atoms and 2 electrons from the O atom—note that the other 2 lone-pair electrons on O are in an sp^2 orbital in the plane of the ring). Therefore, furan is aromatic. Its ^1H NMR shifts are at 6.4 and 7.4 ppm—they are downfield from where alkenyl H atoms normally absorb. This is because the delocalized electrons readily respond to an applied magnetic field—they circulate in such a way as to generate a magnetic field countering the applied field inside the ring (see Figure 20.7). Since the magnetic field lines curl back the other way on the outside of the ring, the magnetic field is reinforced in the neighbourhood of the attached hydrogens. The net effect is that these H atoms are deshielded by the induced field caused by the ring currents.

Exercise 20.6—Aromaticity of cyclic ions

(a) $C_3H_3^+$ has $2 = 4 \times 0 + 2$ π electrons. It is aromatic.
(b) $C_5H_5^+$ has $4 = 4 \times 1$ π electrons. It is anti-aromatic.
(c) $C_7H_7^+$ has $6 = 4 \times 1 + 2$ π electrons. It is aromatic.

Exercise 20.8—Nomenclature of aromatic compounds

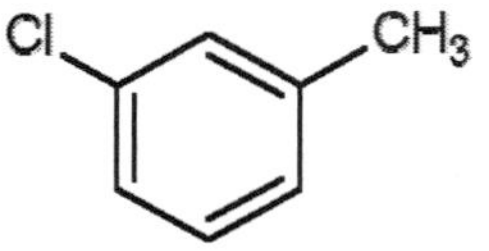

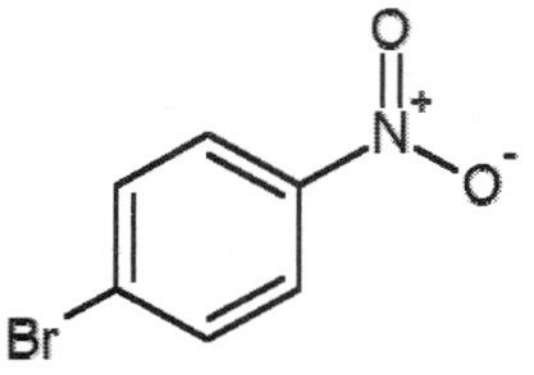

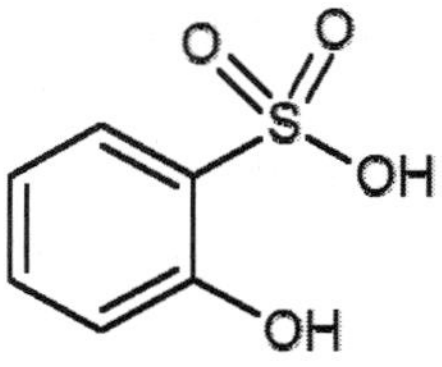

(a)   (b)   (c)

(a) 1,3 is meta
(b) 1,4 is para
(c) 1,2 is ortho

Exercise 20.9—Nomenclature of aromatic compounds

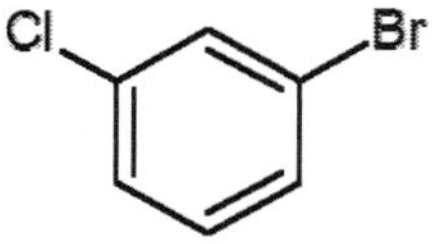

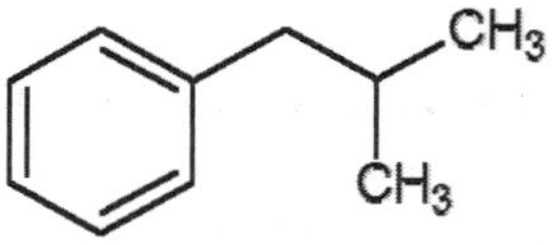

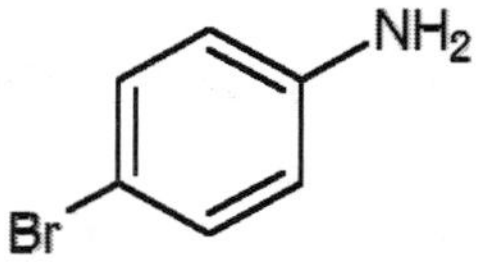

(a)   (b)   (c)

(a) 1-bromo-3-chlorobenzene, or *m*-bromochlorobenzene
(b) (2-methylpropyl)benzene, or 2-methyl-1-phenylpropane
(c) 1-amino-4-bromobenzene, or *p*-aminobromobenzene or *p*-bromoaniline

Exercise 20.10—Electrophilic aromatic bromination

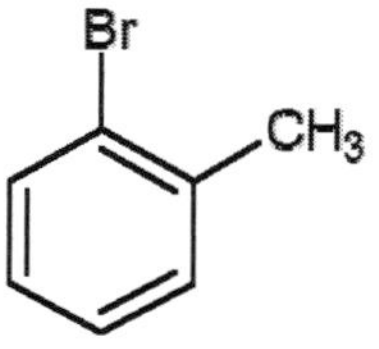

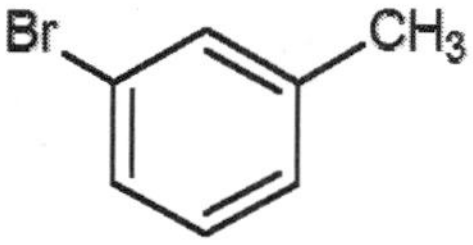

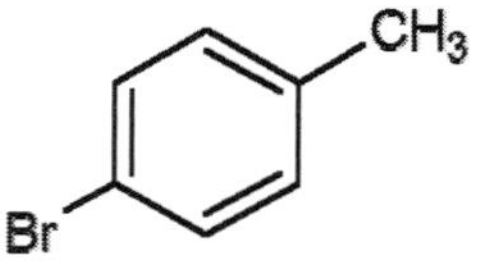

o-bromotoluene, *m*-bromotoluene, and *p*-bromotoluene, respectively

Exercise 20.11—Mechanism for sulfonation

Deuterium (^{2}H) atoms replace all six hydrogen atoms in the aromatic ring when benzene is treated with D_2SO_4 because the D_2SO_4 sulfonates the benzene ring reversibly. On sulfonation, an H^+ is lost to the solvent (D_2SO_4). On de-sulfonation, D^+ adds to the ring. Since the solvent far outnumbers the benzene, the H atoms are lost (as ions) in a sea of D atoms on the solvent. The benzene ends up with all H atoms replaced by D atoms.

Exercise 20.12—Substituent effects in electrophilic aromatic substitution

(a) Nitrobenzene < toluene < phenol (hydroxybenzene)
nitro is electron withdrawing—an N with a +1 formal charge is attached to the ring
methyl (in toluene) is electron donating
hydroxyl (in phenol) is electron donating—via a resonance structure wherein O accepts the +1 formal charge
e.g.,

(b) Benzoic acid < chlorobenzene < benzene < phenol
(c) Benzaldehyde < bromobenzene < benzene < aniline (aminobenzene)

Exercise 20.13—Substituent effects in electrophilic aromatic substitution

The products of electrophilic bromination of nitrobenzene, benzoic acid, benzaldehyde, bromobenzene, chlorobenzene, benzene, toluene, phenol, and aniline, respectively, are

Exercise 20.14—Orientation of electrophilic aromatic substitution

(a) Sulfonation of nitrobenzene gives

(b) Sulfonation of bromobenzene gives

(c) Sulfonation of toluene gives

(d) Sulfonation of benzoic acid gives

(e) Sulfonation of benzonitrile gives

Exercise 20.15—Oxidation of aromatic compounds

Oxidizing *m*-chloroethylbenzene with $KMnO_4$ produces *m*-chlorobenzoic acid,

Oxidizing tetralin with $KMnO_4$ produces 1,2-bezenedicarboxylic acid a.k.a., phthalic acid,

Exercise 20.16—Multi-step synthesis

(a) To make 2-bromo-4-nitrotoluene, we can start with toluene.
Nitrate using nitric and sulfuric acids
Because $-CH_3$ (in toluene) is electron donating, it is ortho para directing.
Separate the para product, 4-nitro-toluene
Brominate 4-nitro-toluene using $Br_2 + FeBr_3$
Now, $-CH_3$ (in 4-nitro-toluene) is ortho-para directing, while $-NO_2$ is meta directing. The Br ends up in the desired 2 position:

Separation steps can be achieved via distillation or chromatography, or maybe even a molecular sieve. There are "molecular sieve" membranes that will pass the narrower para molecules but not the wider ortho molecules.

(b) To make 1,3,5-trinitrobenzene, simply nitrate benzene with 3 equivalents of HNO_3 in H_2SO_4. Because nitro is meta directing, the 1,3,5-trinitro compound will be formed.

(c) To make 2,4,6-tribromoaniline, we first make aniline from benzene.
Nitrate benzene using 1 equivalent of nitric acid in sulfuric acids
Reduce the nitrobenzene product using hydrogen to get aniline
Brominate the aniline with 3 equivalents of Br_2 to get the fully brominated product, 2,4,6-tribromoaniline

REVIEW QUESTIONS

20.1 Aromatic roots of chemistry: Curing, death, and dyeing

20.17
Many skilled Huguenot silk weavers and dyers emigrated from France to Germany following the revocation of the Edict of Nantes. Coal tar was available, at the time, as a source of starting materials. The Rhine River provided the water needed to do the chemistry.

20.19
Nitrates are needed to make trinitrotoluene (TNT), an important explosive, from toluene. They are also used to make other explosives (e.g., nitroglycerin) and as fertilizers. During the lead up to World War II, the British controlled the world's supply of naturally occurring nitrates—in massive guano deposits in Chile. The Germans needed a synthetic supply of nitrates. Fritz Haber and Carl Bosch developed an effective means of producing ammonia from nitrogen (from the air) and hydrogen. The ammonia could then be oxidized to produce the desired nitrate.

20.2 Benzene and aromaticity—Reactivity, structure, spectroscopy

20.21
The planar arrangement of 1,3,5,7-tetramethylcyclooctatraene produces an anti-aromatic π electron configuration. It is destabilized with respect to simply having four localized double bonds, as it does in the tub conformer. The tub conformer, therefore, provides the observed lowest energy conformation.

20.23
The four resonance structures of anthracene

20.25
There are three non-equivalent H's in anthracene—the central vertical pair, and two pairs (related via reflection through the vertical axis) of vertical pairs of H's. The chemical shifts are expected to be in the 6.5 to 8 ppm range typical of aromatics.

20.27

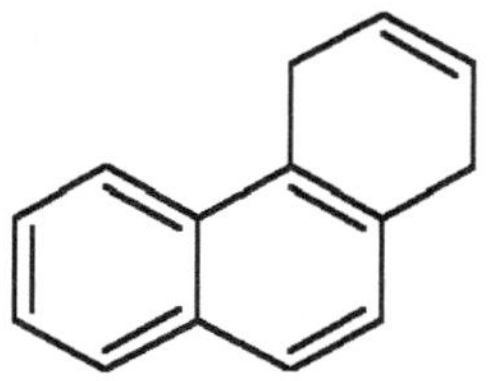

Phenanthrene has $14 = 4 \times 3 + 2$ π electrons. Although it is not monocyclic, it certainly qualifies as aromatic. The outer ring of atoms provides the principal path over which $4 \times 3 + 2$ electrons are delocalized in accord with Hückel's rule. Consequently, the term aromatic applies equally to polycyclic systems.

20.29

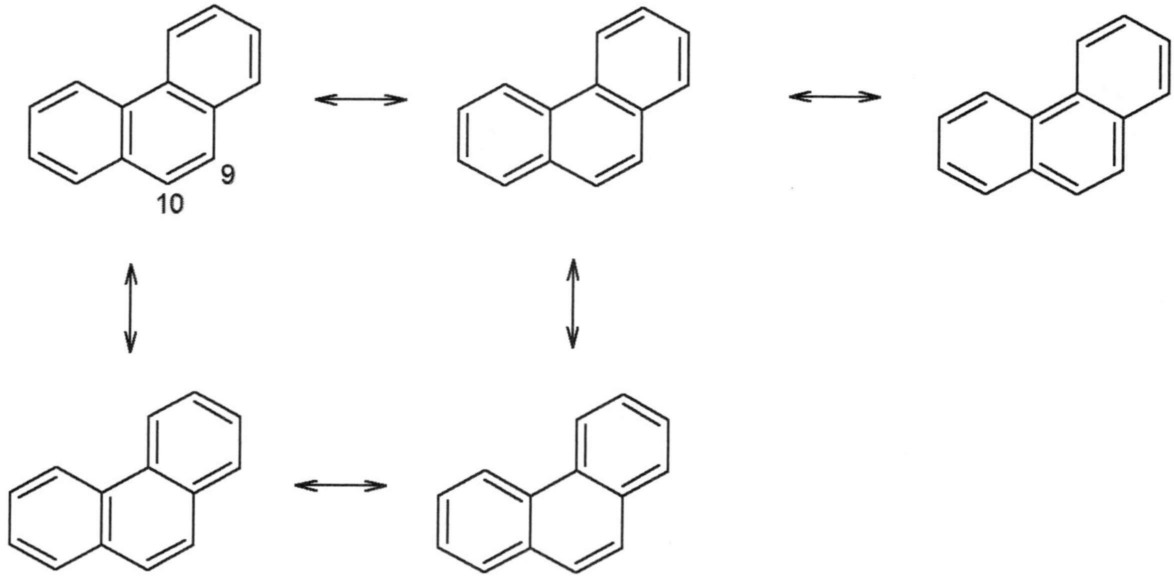

The shortest bonds have the highest order—look for the bonds which are double more often than the others. The shortest C–C bond is the C9-C10 bond (labelled in the first structure). It is a double bond in 4 of the 5 structures. Its bond order is $1\frac{4}{5}$.

20.31
C_8H_9Br has 4 degrees of unsaturation, consistent with a single phenyl ring (a ring and 3 double bonds). The ^{1}H NMR spectrum shows two downfield peaks (7.3 and 7.0 ppm) associated with $2 + 2 = 4$ H atoms. These are consistent with two pairs of equivalent aromatic H atoms. We are thus looking for a compound with a doubly substituted benzene ring—either ortho or para (the meta configuration would rise to three distinct peaks integrating to 1, 2, and 1). The other two peaks in the ^{1}H NMR spectrum are consistent with a $-CH_2CH_3$ substituent. Bromine provides the second substituent. The compound is *o*- or *p*- bromoethylbenzene,

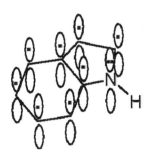

20.3 Aromatic heterocycles and ions

20.33
The 1,3,5,7-tetramethylcyclooctatetraene dianion should be planar, as Hückel's rule is satisfied and the planar structure should have a special stability.

benzene, there is a filled two-orbital shell—i.e., the electrons are optimally packed for interaction

20.35

Indole has $10 = 4 \times 2 + 2$ π electrons. It is aromatic. Note the N atom is sp^2 hybridized. Its lone pair goes in a *p* orbital that forms part of the π system.

20.37
1,3,5,7-cyclononatetraene is expected to be unusually acidic because of the stabilization of the anion. It can be converted to a salt by treatment with a strong base such as $NaNH_2$. The sodium salt of the aromatic cation would be formed.

20.4 Naming aromatic compounds

20.39

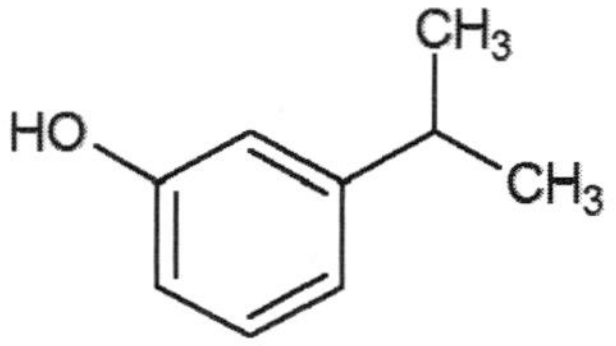

m-isopropylphenol

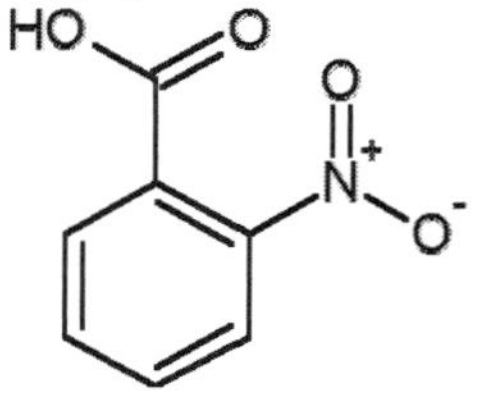

20.41
(a) *o*-dinitrobenzene, *m*-dinitrobenzene and *p*-dinitrobenzene

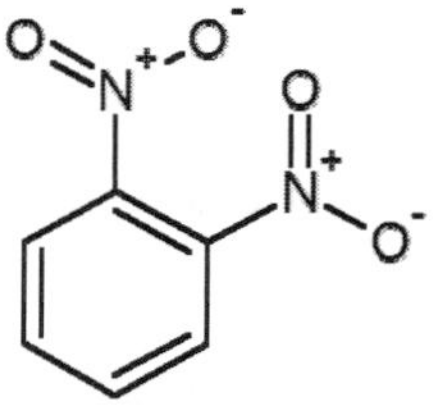

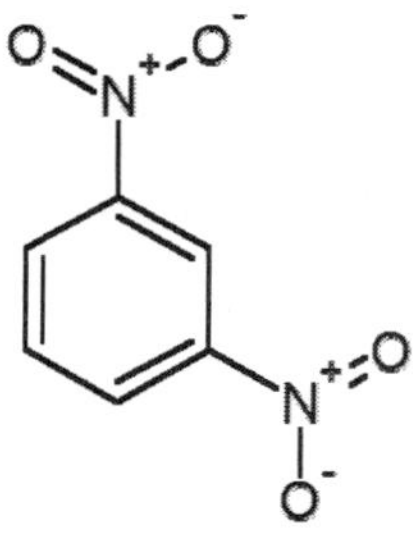

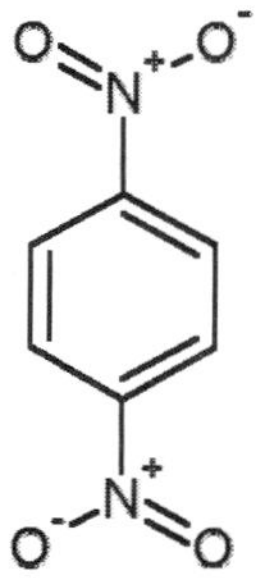

(b) 1-bromo-2,3-dimethylbenzene, 1-bromo-2,4-dimethylbenzene, 2-bromo-1,4-dimethylbenzene, 2-bromo-1,3-dimethylbenzene, 4-bromo-1,2-dimethylbenzene, and 1-bromo-3,5-dimethylbenzene

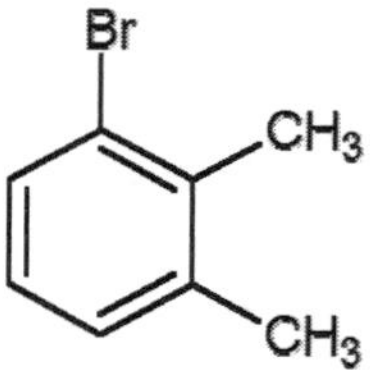

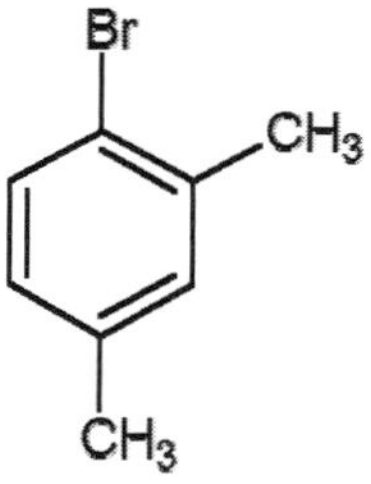

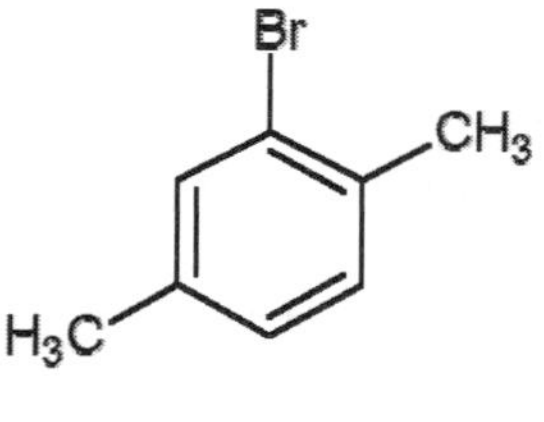

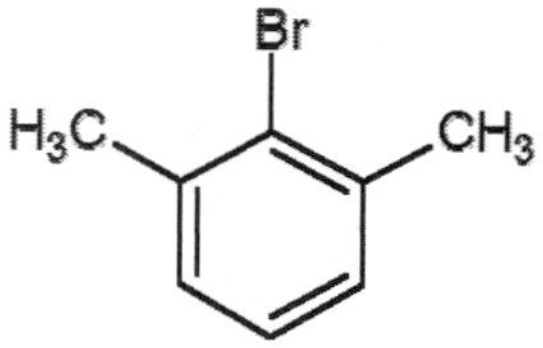

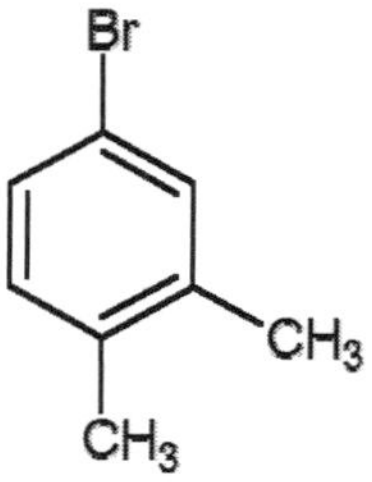

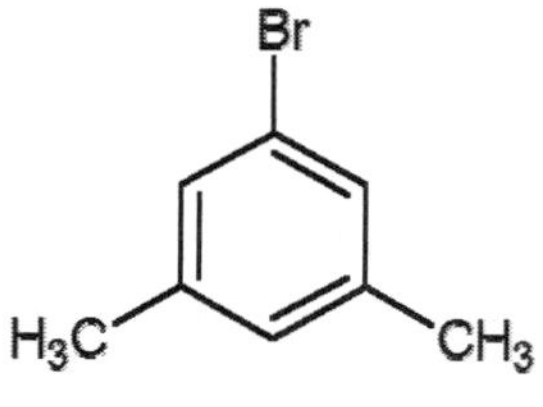

20.5–20.7 Electrophilic aromatic substitution

20.43

20.8 Substituent effects in electrophilic aromatic substitution

20.47

(a)

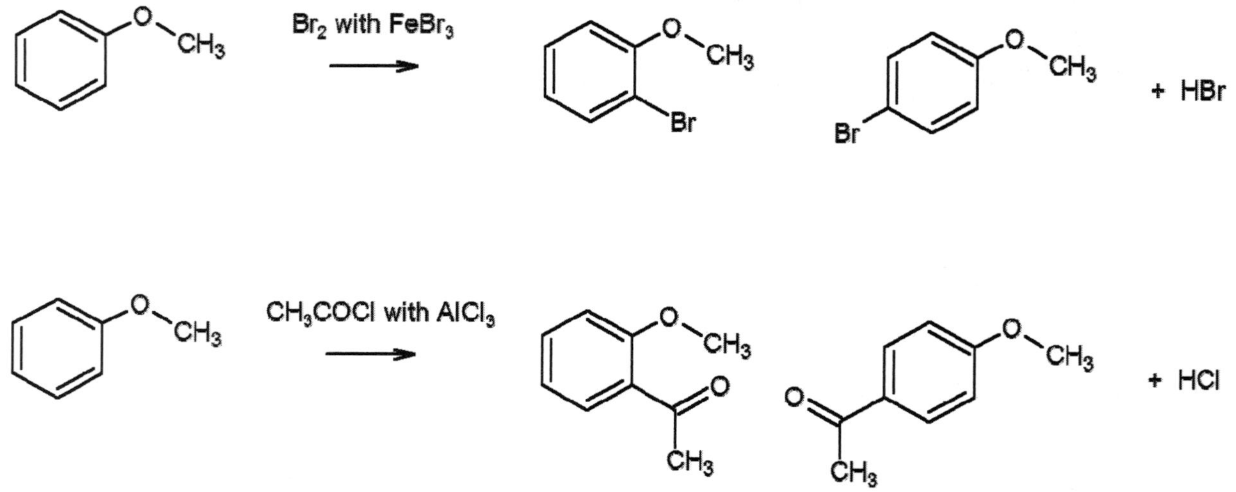

(b)

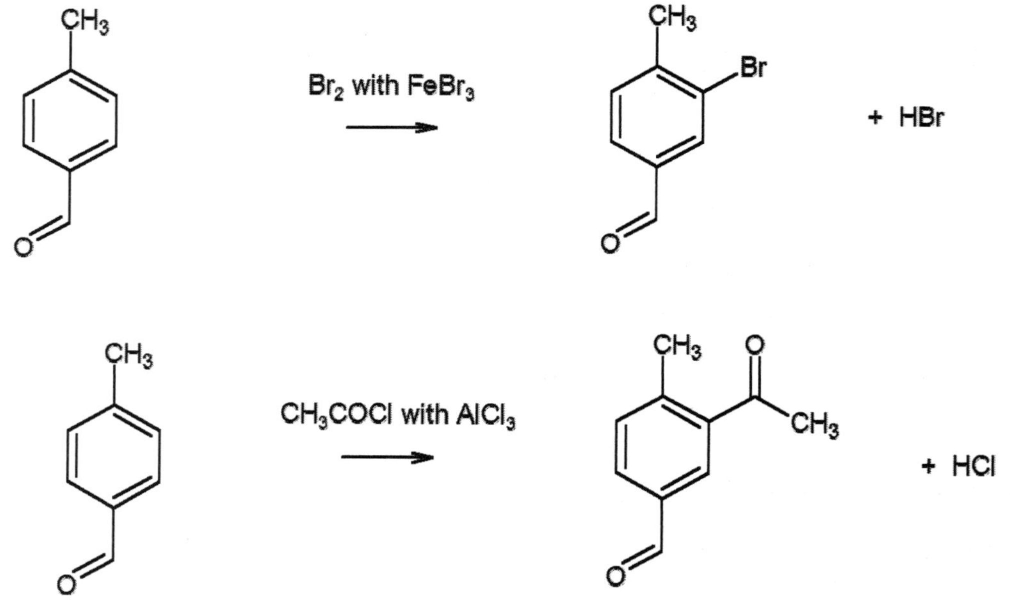

20.49

ortho addition

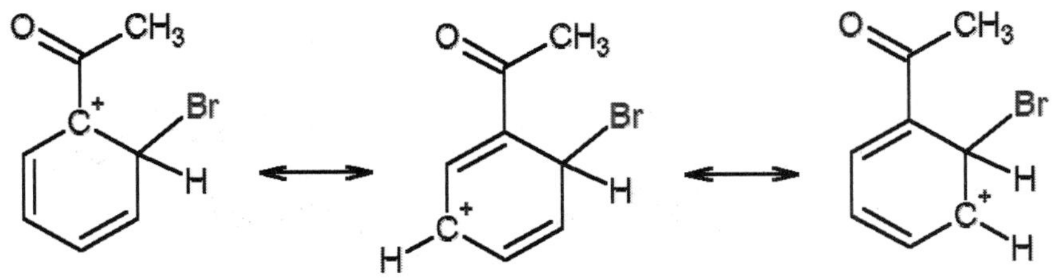

meta addition

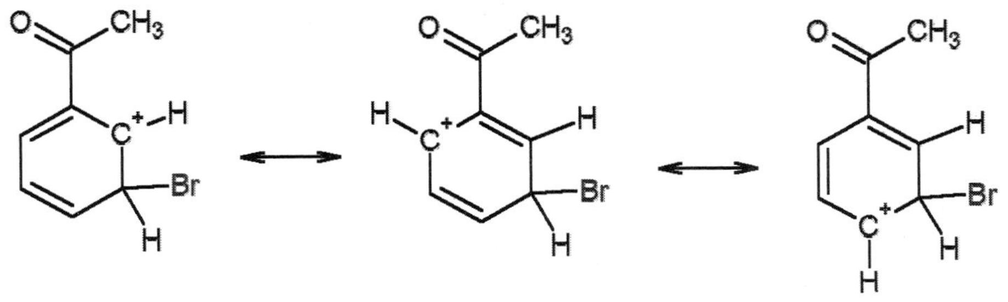

para addition

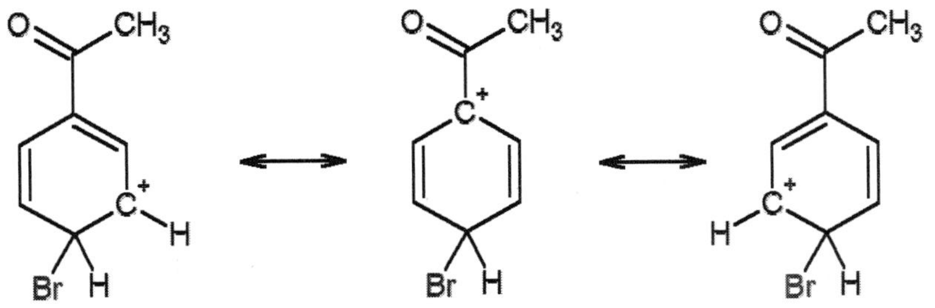

In this case, the three carbocation intermediates all have the same number of resonance structures. However, in the case of the ortho and para addition intermediates, one of the resonance structures has the positive charge on a carbon atom bonded to the electron-withdrawing acetyl group. This resonance structure is disfavoured, making the carbocation intermediates less stable in the case of ortho and para addition. Meta addition is the favoured pathway, by default.

20.51

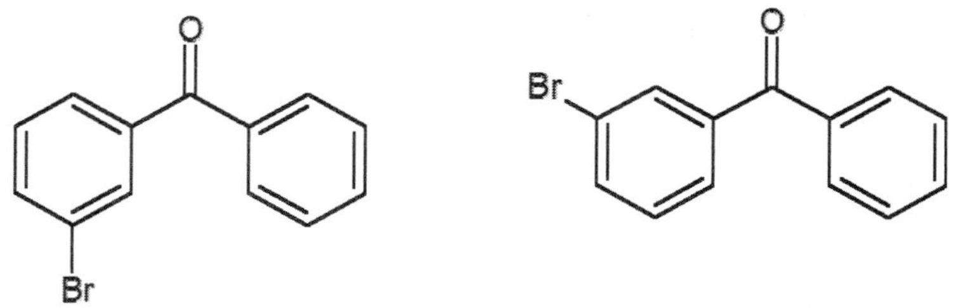

The carbonyl is electron withdrawing. It leads to the meta bromination which corresponds to two products in this case. However, these compounds readily interconvert at room temperature. There is, in effect, only one product—a mixture of these two species.

20.53

(a)

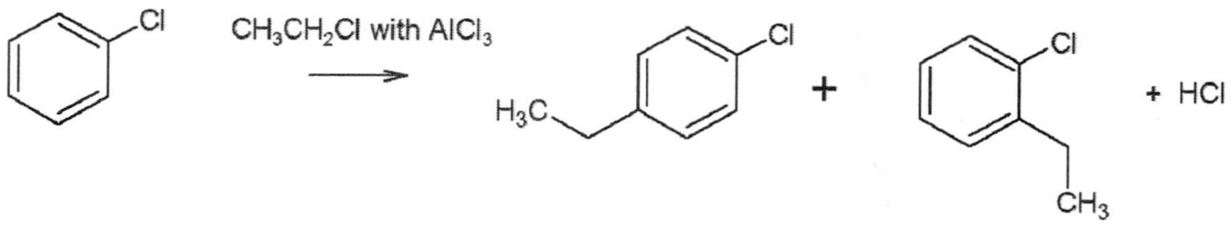

(b)

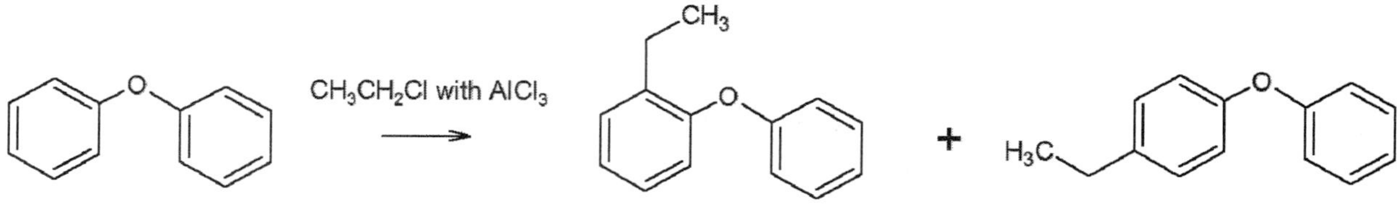

(c)

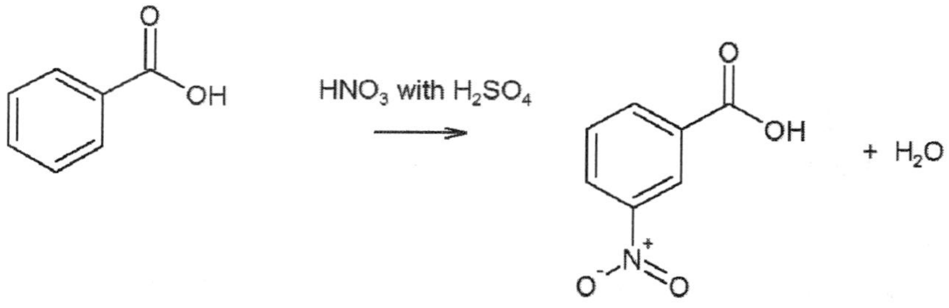

(d)

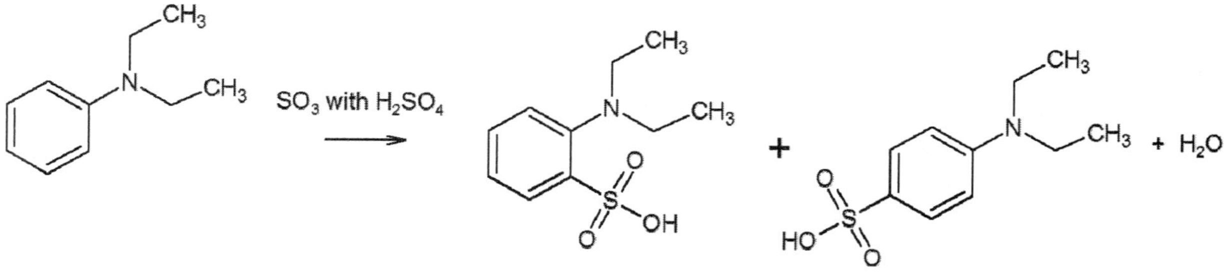

20.55
Diphenyl ether, phenol, and diethylphenylamine react faster than benzene, while chlorobenzene, benzonitrile, and benzoic acid react slower.

20.10 Organic synthesis: Thinking backwards

20.57
(a) *o*-Bromotoluene can be synthesized from benzene via a Friedel-Crafts methylation (i.e., treat with 1 equivalent CH_3Cl and $AlCl_3$) to make toluene, followed by bromination (treat toluene with 1 equivalent Br_2 and $FeBr_3$) and separation of the ortho and para products. This separation might be achieved by distillation. Alternatively, there are "molecular sieves" that can be used as membrane filters, passing only molecules smaller than a certain size. There are some such devices that can distinguish between ortho and para disubstituted benzenes—the ortho is slightly wider.
(b) 2-Bromo-1,4-dimethylbenzene is obtained if we apply another Friedel-Crafts methylation to the product, *o*-bromotoluene, from part (a). Because the first methyl is an activating ortho-para directing, whereas bromine is a deactivating ortho-para directing, the 1,4-dimethyl and 1-6-dimethyl products are obtained. A final separation step is required to obtain the desired compound. This synthesis requires three steps and several separations. A much better synthesis is to dimethylate benzene via Friedel-Crafts methylation. Separate the para product and then brominate, to get a single product.

20.59
Methylate benzene using CH_3Cl and $AlCl_3$. Nitrate using HNO_3 and H_2SO_4 to get *o*- and *p*-nitrobenzene. Separate the para product. Oxidize it using $KMnO_4$ to obtain the desired *p*-nitrobenzoic acid.

20.61
(a) *o*-Nitrobenzoic acid synthesis:
Methylate benzene using CH_3Cl and $AlCl_3$.
Nitrate the toluene product using HNO_3 and H_2SO_4. Separate the ortho product.
Oxidize using $KMnO_4$ to convert the methyl group to a carboxylic acid.
(b) *p-tert*-Butylbenzoic acid synthesis:
Alkylate benzene using $(CH_3)_3CCl$ and $AlCl_3$. Then methylate *t*-butyl benzene with CH_3Cl and $AlCl_3$. Due to the size of the *t*-butyl group, the para product will predominate. Then oxidize using $KMnO_4$ to convert the methyl group to a carboxylic acid. The *t*-butyl group will not be oxidized, since there are no H atoms on the benzylic carbon atom.

SUMMARY AND CONCEPTUAL QUESTIONS

20.63
Trimethylammonium is a deactivating substituent. Unlike the amine substituent, ammonium has no lone pair of electrons to participate in the π system, and provide an additional resonance structure stabilizing the carbocation intermediates associated with ortho and para substitution. Nitrogen, being more electronegative than carbon, is otherwise somewhat electron withdrawing. More importantly, the positive charge on the ammonium substituent destabilizes the carbocation intermediate.

20.65
(a) Sulfonation of *o*-Chlorotoluene yields

(b) Sulfonation of *m*-Bromophenol yields

(c) Sulfonation of *p*-Nitrotoluene yields

20.67

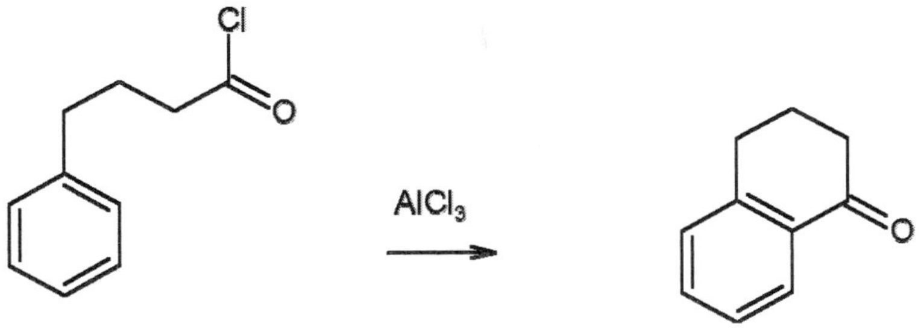

20.69

Nitration of 4-bromobiphenyl produces 4-bromophenyl-2-nitrobenzene and 4-bromophenyl-4-nitrobenzene.

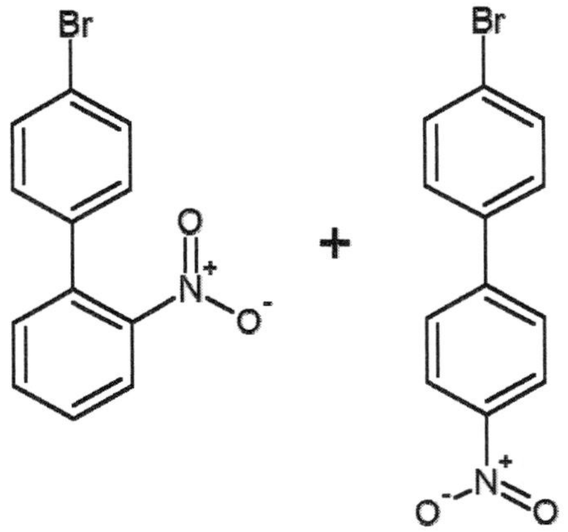

Note that substitution occurs on the phenyl without the bromo substituent. A bromine atom deactivates the brominated phenyl toward substitution. So the preferred pathway of nitration is for the unbrominated ring, and the aryl group attached to it directs attack ortho and para.

20.71

Here is the mechanism for addition of HBr to 1-phenylpropene with Br adding to the 2 position on the propyl group.

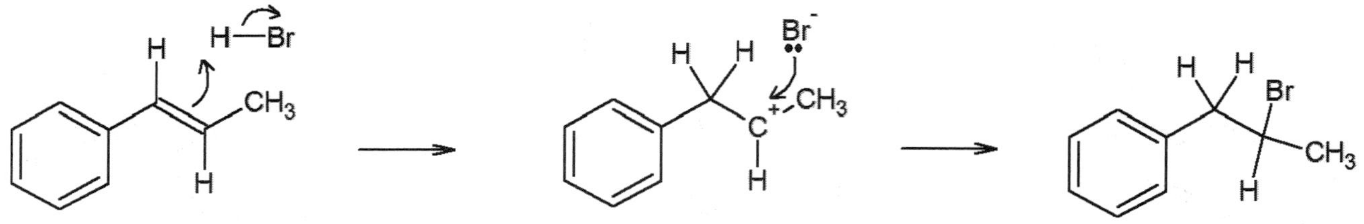

Here is the mechanism for addition of HBr to 1-phenylpropene with Br adding to the 1 position on the propyl group.

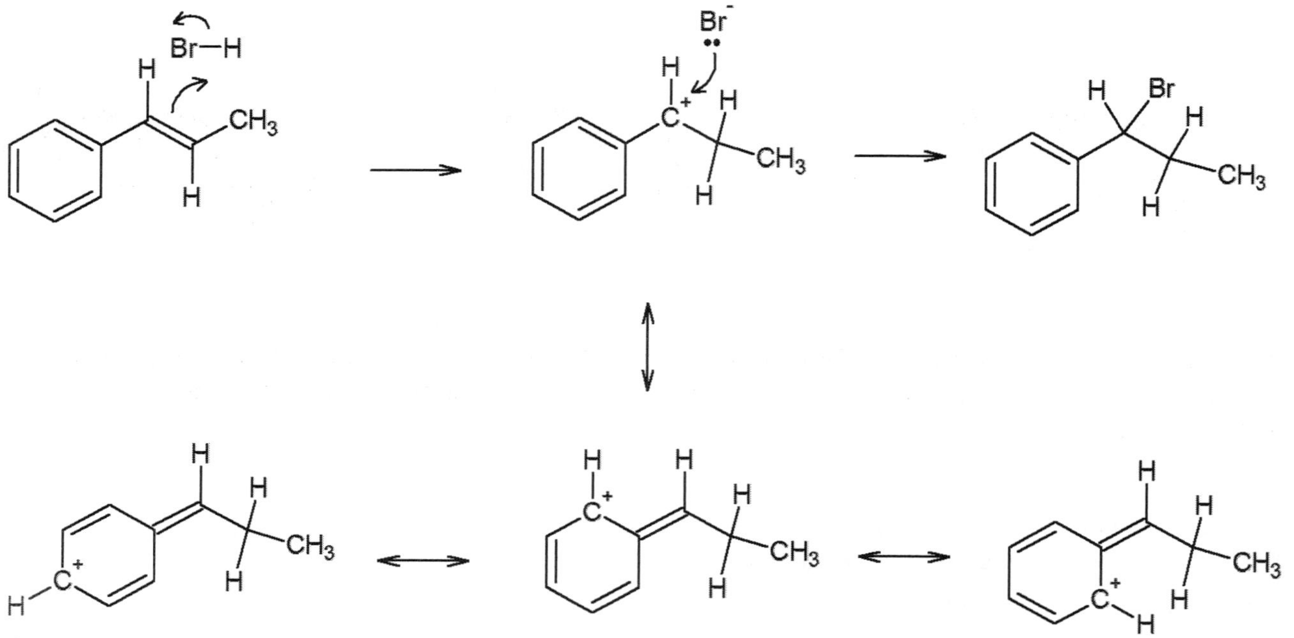

The extra resonance structures associated with the carbocation intermediate, in the case of addition of Br to the 1 position, stabilize the intermediate, enhancing the rate of the associated pathway, making the 1-bromopropyl product the exclusive product.

20.73

(a) Toluene is chlorinated at the meta position. Toluene chlorinates at ortho and para positions.

(b) *tert*-Butyl is oxidized to make carboxylic acid. However, *tert*-butyl is inert to oxidation. Oxidation of alkyl benzenes requires at least one benzylic H atom.

20.75 The mechanism for addition of HBr to the nitro compound is

Here we see that the carbocation intermediate is destabilized by the electron-withdrawing nitro group. The resonance structure with the + charge next to the formally + charged nitro N atom is especially destabilized.

The mechanism for addition of HBr to the methoxy compound is

Here we see that the carbocation intermediate is stabilized by an extra resonance structure. Addition of HBr to the methoxy compound is much faster than addition to the nitro compound.

20.77
The acetylation of isobutylbenzene at the para position.

The carbocation intermediate here is stabilized by the resonance structure with the positive charge next to the isobutyl substituent—isobutyl is electron-donating. In the case of meta substitution, none of the resonance structures have the positive charge next to the alkyl substituent. The meta product is not formed. Ortho substitution is possible, though, because the associated carbocation intermediate is stabilized by the isobutyl substituent. However, the para product might come in higher yield due to steric hindrance associated with the isobutyl and acetyl substituents next to each other in the ortho configuration.

Chapter 21
Alkyl Halides

IN-CHAPTER EXERCISES

Exercise 21.1—Nomenclature of alkyl halides

(a) 2-bromobutane
(b) 3-chloro-2-methylpentane
(c) 1-choro-3-methylbutane
(d) 1,3-dichloro-3-methylbutane
(e) 1-bromo-4-chlorobutane
(f) 4-bromo-1-chloropentane

Exercise 21.3—Nomenclature of alkyl halides

(a) 2-Chloro-3,3-dimethylhexane

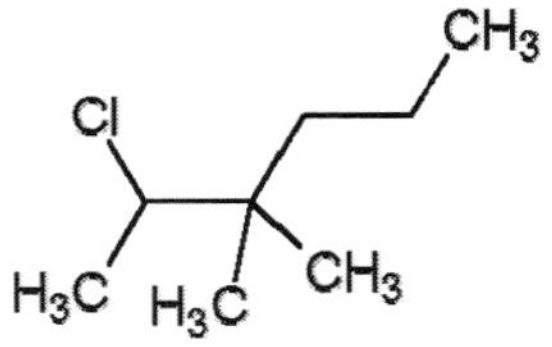

(b) 3,3-Dichloro-2-methylhexane

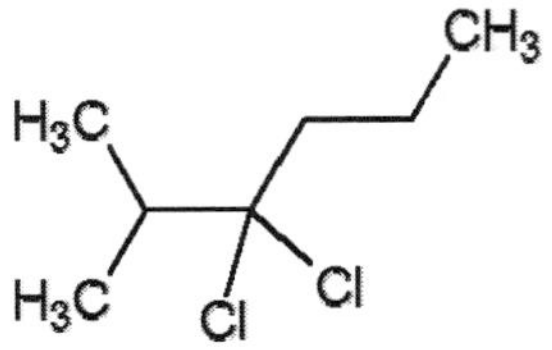

(c) 3-Bromo-3-ethylpentane

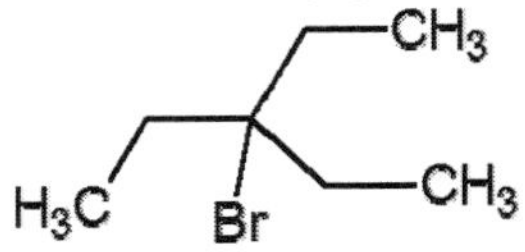

(d) 2-Bromo-5-chloro-3-methylhexane

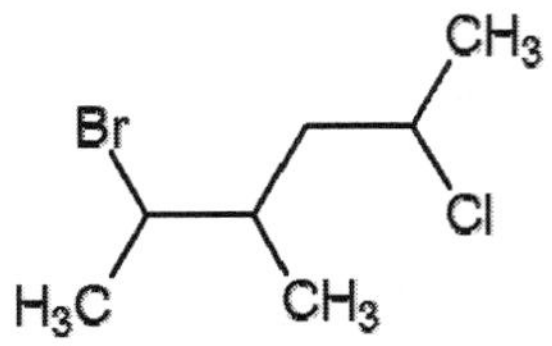

Exercise 21.5—Synthesis of alkyl halides

(a)

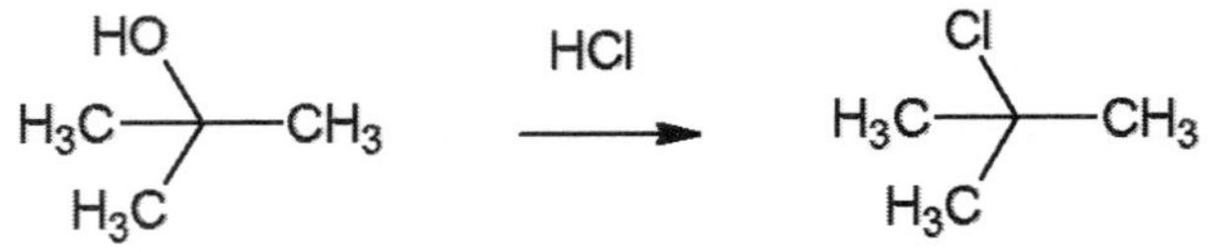

(b)

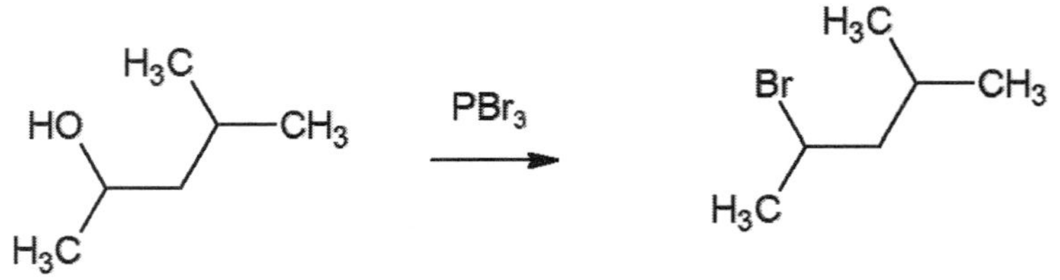

(c)

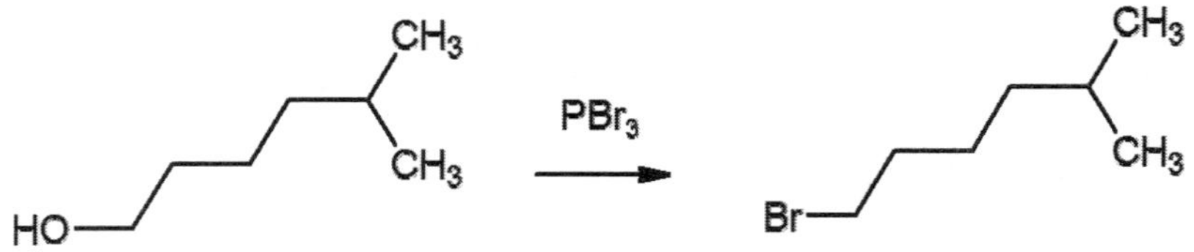

(d)

Exercise 21.6—Synthesis of alkyl halides

(a)

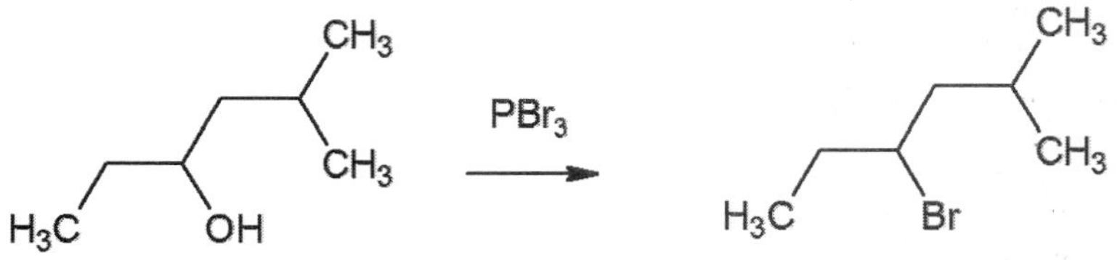

(b)

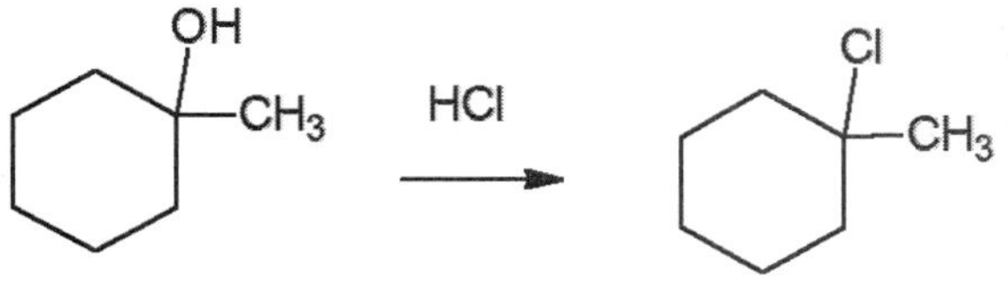

(c)

Exercise 21.8—Nucleophilic substitution reactions

(a)

(b)

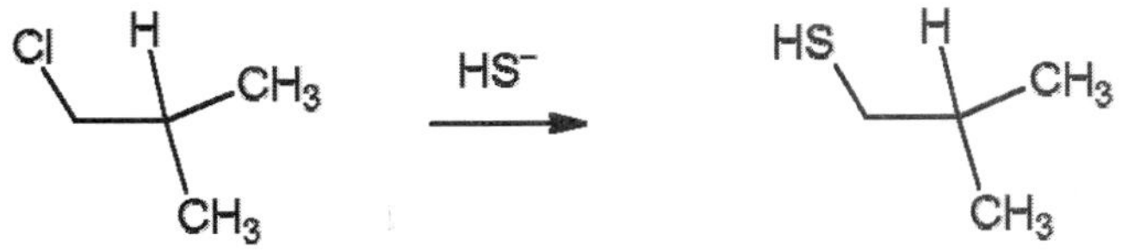

(c)

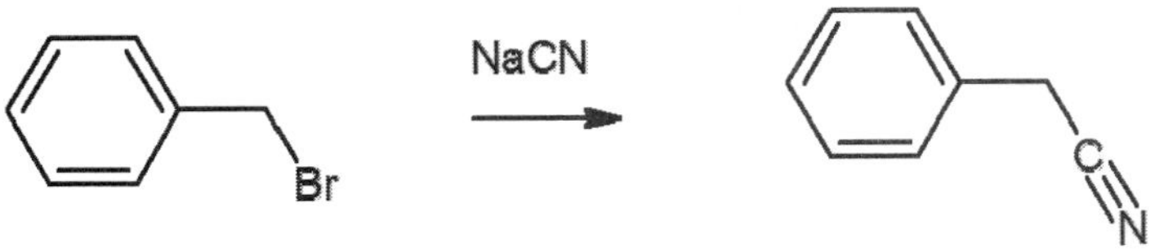

Exercise 21.9—Nucleophilic substitution reactions

(a)

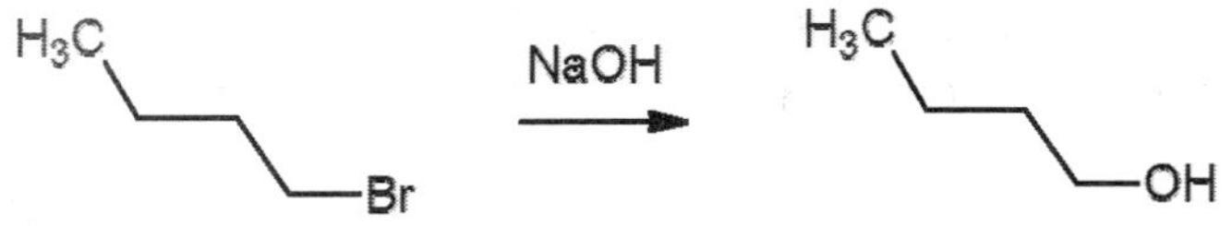

(b)

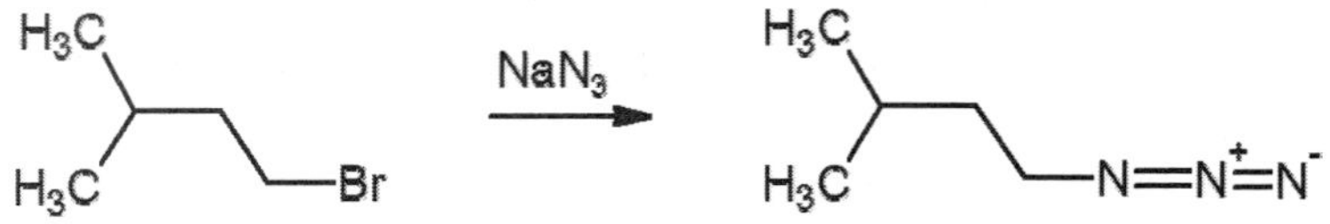

Exercise 21.10—Kinetics of S$_N$2 reaction

(a) S$_N$2 reactions are bimolecular. The rate of reaction is proportional to the product of the concentrations of the two reactants. In particular, it is proportional to the concentration of CH$_3$I. Therefore, if the methyl iodide concentration is tripled, the rate is tripled.
(b) If the concentrations of methyl iodide and sodium acetate are doubled, the rate increases by the factor $2 \times 2 = 4$.

Exercise 21.11—S$_N$2 reactions

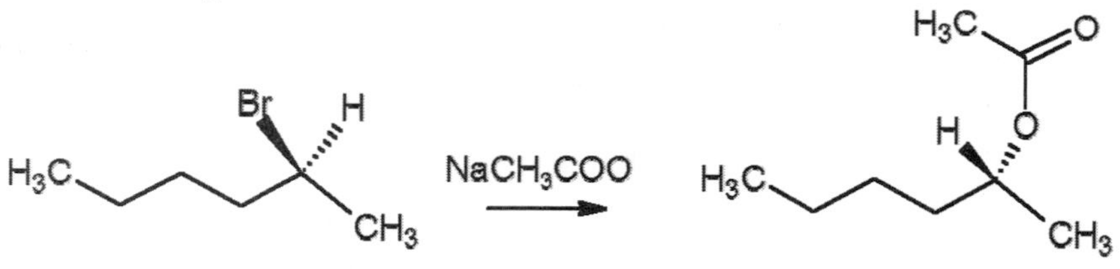

(S)-2-bromohexane

(R)-2-hexyl acetate

Exercise 21.12—S$_N$2 reactions

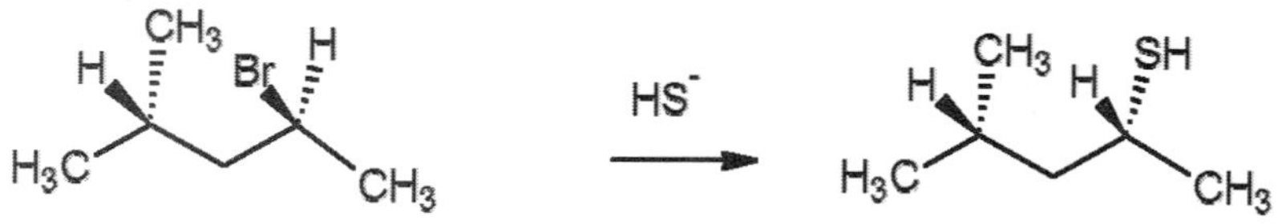

(S)-2-bromo-4-methylpentane

4-methylpentane-(R)-2-thiol

Exercise 21.14—Kinetics of S$_N$2 reactions

(a) CN$^{\square}$ (cyanide ion) reacts faster with CH$_3$CH$_2$CH$_2$Br than CH$_3$CH(Br)CH$_3$. The back of the bromine-bearing carbon atom (where the cyanide reacts) is more accessible in a primary bromide than a secondary bromide. A smaller fraction of cyanide bromide collisions will produce a reaction in the case of the secondary bromide.
(b) Reaction of I$^{\square}$ with (CH$_3$)$_2$CHCH$_2$Cl is faster. In fact, H$_2$C=CHCl is unreactive with iodide. Vinylic halides do not undergo S$_N$2 substitution reactions.

Exercise 21.15—Leaving groups in S$_N$2 reactions

According to reactivity with respect to S$_N$2 substitution, the three compounds are ranked as follows: CH$_3$I > CH$_3$Br > CH$_3$F. The halides are the best leaving groups (i.e., they leave most readily via S$_N$2 substitution), with I$^-$ > Br$^-$ > Cl$^-$ > F$^-$.

Exercise 21.16—Kinetics of S$_N$1 mechanism

(a) The rate determining step of S$_N$1 reactions is unimolecular. The rate of S$_N$1 substitution of *tert*-butyl alcohol and HBr is proportional only to the concentration of *tert*-butyl alcohol. Therefore, tripling the concentration of HBr has no effect on the rate of reaction.
(b) Halving the HBr concentration has no effect. However, doubling the *tert*-butyl alcohol concentration doubles the rate of reaction.

Exercise 21.17—Stereochemistry of S$_N$1 reaction

(S)-3-methyloctan-3-ol

(S)-3-bromo-3-methyloctane

(R)-3-bromo-3-methyloctane

We get a 50/50 mixture (racemate) of the 3-bromo-3-methyloctanes, because of the S$_N$1 mechanism.

Exercise 21.18—Stereochemistry of S$_N$1 reaction

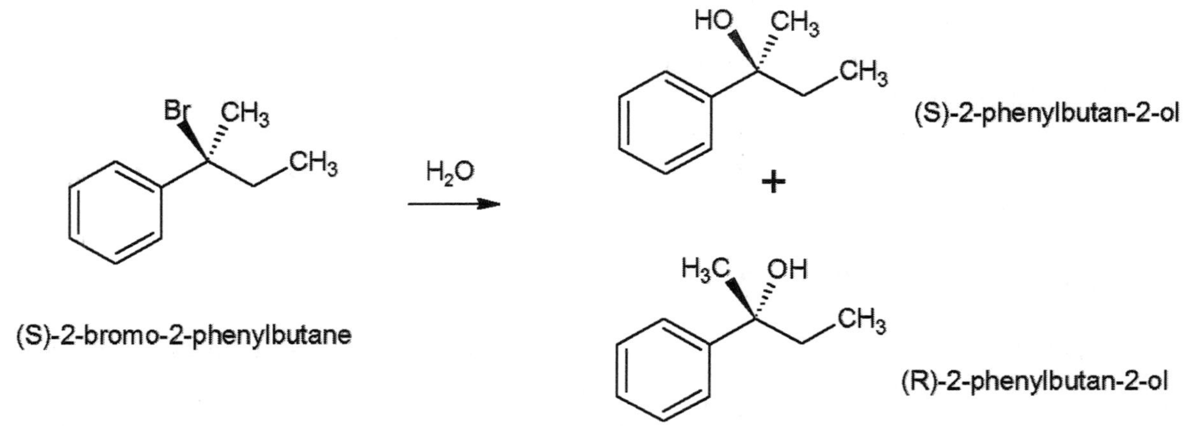

(S)-2-phenylbutan-2-ol

(S)-2-bromo-2-phenylbutane

(R)-2-phenylbutan-2-ol

The product is a 50/50 mixture (racemate) of the enantiomeric 2-phenylbutan-2-ols.

Exercise 21.19—E2 reaction

(a)

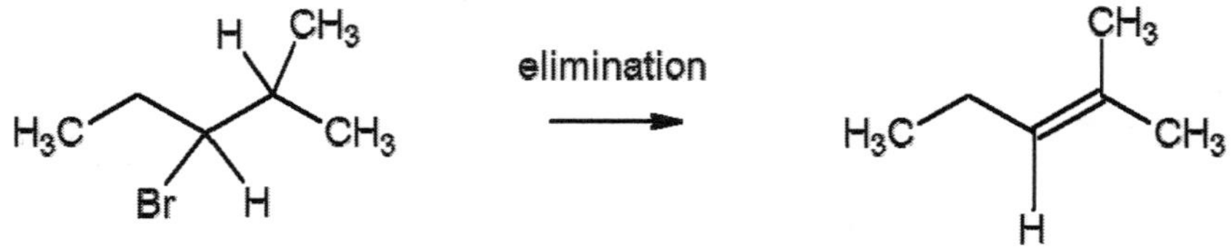

Note that the trisubstituted alkene is formed, as opposed to the disubstituted product resulting from elimination at the other C–CBr bond.

(b)

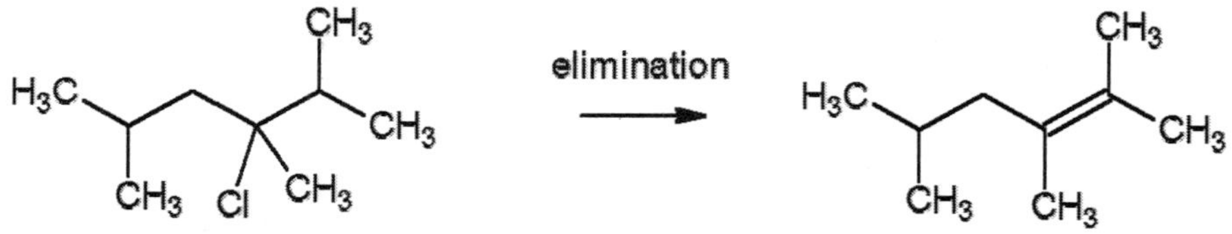

Here, we get a tetrasubstituted alkene, as opposed to the trisubstituted alkene.

(c)

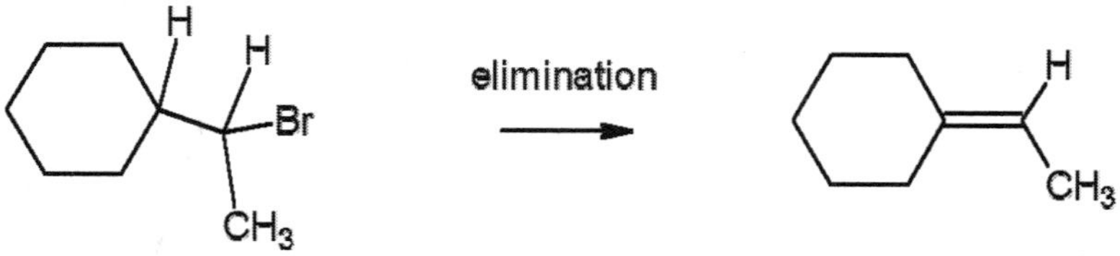

Here, we get a trisubstituted alkene, as opposed to the disubstituted alkene.

Exercise 21.20—E2 reaction

(a) For example,

Note that we must start with a primary halide. Otherwise, the 2-heptene will be formed (it is trisubstituted, as opposed to disubstituted).

(b) For example,

We must start with the 1-chloro-3,4-dimethylcyclopentane, as opposed to 1-chloro-2,3-dimethylcyclopentane. The latter reactant would produce 1,5-dimethylcyclopentene (trisubstituted), as opposed to the desired 4,5-dimethylcyclopentene.

Exercise 21.21—E1 reaction mechanism

The rate determining step of E1 reactions is unimolecular. The rate of this E1 elimination is proportional to the concentration of 2-chloro-2-methylpropane. Therefore, tripling the concentration of this alkyl halide triples the rate of the elimination reaction.

Exercise 21.22—Predicting the mechanism

(a) Because 1-bromobutane is a primary halide, it will not form a carbocation intermediate. Instead, the reaction takes place in a single, concerted bimolecular step—S_N2 or E2. This reaction is a substitution reaction—so it is S_N2. Note that substitutions are favoured for primary alkyl halides.
(b) Here, if a carbocation intermediate could be formed (S_N1 and E1), it would be secondary. However, the same strong base with high concentration conditions that favour elimination over substitution, favour E2 over E1. Recall that the rate of E2 reactions is proportional to the nucleophile concentration in addition to alkyl halide concentration. A higher concentration of hydroxide increases the rate of the E2 reaction, but not the E1 reaction. We expect the E2 mechanism to predominate.
(c) In this case, the rates of bimolecular processes (S_N2 and E2) are significantly reduced by steric hindrance. Furthermore, the carbocation intermediate is tertiary—much more stable than secondary or primary. Since this is a substitution reaction, the mechanism is S_N1.

Exercise 21.23—Synthesis using Grignard reagents

Grignard reagent

Exercise 21.24—Synthesis using Grignard reagents

REVIEW QUESTIONS

21.1 Chlorofluorocarbons and stratospheric ozone depletion

21.25
Methyl bromide is manufactured from bromide salts recovered from seawater and methane—the principal component of natural gas. It is also naturally occurring—in the oceans, plants and soil. Methyl bromide has been used as a pesticide, or fumigant, in the agricultural industry. Because it is a potent ozone depleting substance, its use has been phased out. A large number of alternative pesticides are in use, or are being developed. For example, phosphine is used to fumigate food storage, transportation, and processing facilities. Sulfuryl fluoride and 1,3-dichloropropene are two other alternative fumigants used to treat foods and forest seedlings.

21.2 Alkyl halides—Synthetic and natural

21.27

2-bromo-2-chloro-1,1,1-trifluoroethane

Halothane has alkyl and alkyl fluoride, chloride and bromide functionality.

21.29
In G.W. Gribble, *The diversity of naturally produced organohalogens*, Chemosphere **52**, 289 (2003), the author states that there are more than 3800 different organohalogen compounds produced naturally by living organisms or by forest fires, volcanoes, or other geothermal processes. The single largest biogenic source of these compounds is the oceans.

21.5–21.7 Nucleophilic substitution reactions

21.31
(a) Substrate structure plays an important role in determining whether an S_N1 or S_N2 substitution reaction will occur. S_N2 reactions are inhibited by bulky substituents—completely so in the case of attack at a tertiary carbon atom. This is because of the steric hindrance caused by the substituents. The S_N2 pathway is most favoured in the case of primary carbons. S_N1 reactions are favoured by substituents (unless they are electron-withdrawing) because they stabilize the carbocation intermediate of this mechanism.
(b) A better leaving group increases the rate of both S_N1 and S_N2 reactions.

21.33
(a) Br^- is a better leaving group than F^-
(b) Cl^- is a better leaving group than NH_2^-
(c) I^- is a better leaving group than OH^-

21.35

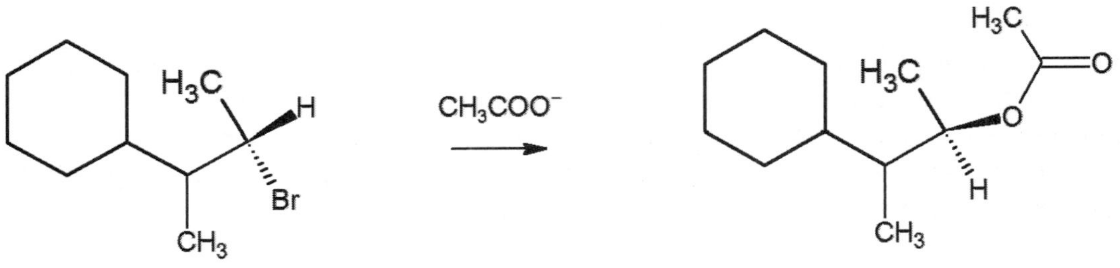

21.37
S_N2 reaction rates are proportional to the concentration of both the alkyl halide (i.e., in the case of S_N2 reactions of alkyl halides) and the nucleophile.
(a) If the concentration of CH_3Br is tripled and that of CN^- is halved, the rate increases by the factor
  $3 \times 1/2 = 3/2$.
(b) If the concentration of CH_3Br is halved and that of CN^- is tripled, the rate increases by the factor
  3/2.
(c) If the concentration of CH_3Br is tripled and that of CN^- is doubled, the rate increases by the
  factor $3 \times 2 = 6$.
(d) If the reaction temperature is raised, the rate increases—collisions are more energetic, and a
larger fraction of collisions are effective in causing a reaction.
(e) If the volume of the reacting solution is doubled by addition of more solvent, the concentration
of all species is halved. The rate of reaction decreases $2 \times 2 = 4$ fold.

21.39
According to decreasing S_N2 reactivity, the reagents are ranked as follows:
(a)

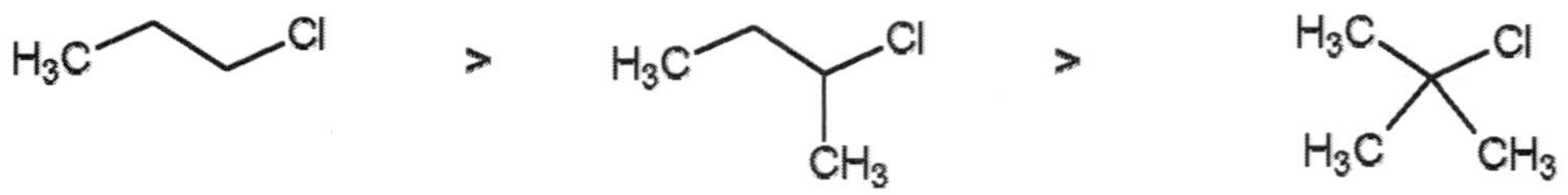

(b)

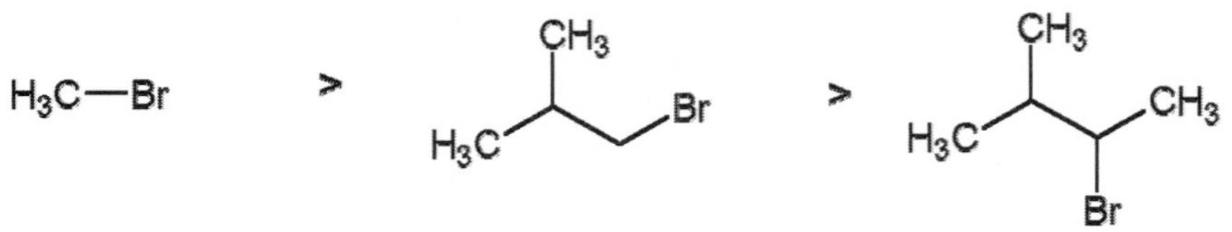

The rule for S_N2 reactions is the less hindered alkyl halides have the greater reactivity (e.g., primary
> secondary > tertiary—for the same leaving group)

21.41

(a) This is an S_N2 reaction at a tertiary carbon atom. It is strongly hindered and would simply be too slow.

(b) This is an S_N2 reaction at a primary carbon—this is good. The only trouble is that hydroxide is a bad leaving group. This reaction will not go. Hydroxide can only be eliminated under acid conditions where it is protonated, and water is the leaving group.

(c) This is an elimination reaction at a tertiary carbon—this is good. However, S_N1 reaction to give a product where the OH group has been substituted by a Br atom also occurs and separation will be required. Under many reaction conditions, the substitution product is favoured. To separate the products, we might use a distillation—the dehydration product should have a lower boiling point than the substitution product (higher molecular weight and a little more polar)—or perhaps some form of chromatography—the differences in molecular weight and polarity should allow the dehydration product to elute through a chromatographic column faster than the substitution product.

21.9–21.11 Elimination reactions and the elimination/substitution continuum

21.43

21.45

This is an S_N2 reaction. The rate is proportional to both the concentration of the alkyl iodide and cyanide.

(a) If the CN^- concentration is halved and 1-iodo-2-methylbutane concentration is doubled, the rate stays the same.

(b) If both CN^- and 1-iodo-2-methylbutane concentrations are tripled, the rate increases $3 \times 3 = 9$ fold.

21.47

(a) 1-bromoethylbenzene + KOH $\rightarrow$ phenylethene is an E2 elimination reaction (strong base present)

(b) 1-bromoethylbenzene + CH_3OH $\rightarrow$ (1-methoxyethyl)benzene is an S_N1 substitution (benzylic carbocation formed)

21.49

2-chlorobutane yields 3 isomeric alkenes upon treatment with a strong base; (*E*)-but-2-ene, (*Z*)-but-2-ene and but-1-ene The *n*-butyl and *tert*-butyl chlorides each form only one alkene.

SUMMARY AND CONCEPTUAL QUESTIONS

21.51
The oceans have high concentrations of halide ions. The oceans are also the likely origin of all life, and are still now brimming with life. It is natural that organohalogen compounds would be formed there, and formed there first. Moreover—and this is likely why the oceans have been so crucial to life—the oceans provide a solvent supporting vastly more chemistry than is possible of the land surface.

21.53

(a)

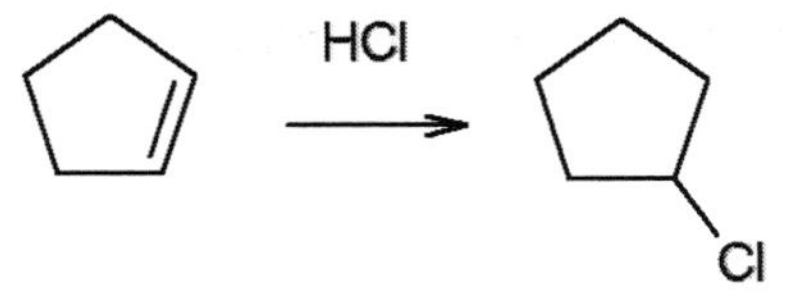

(b)

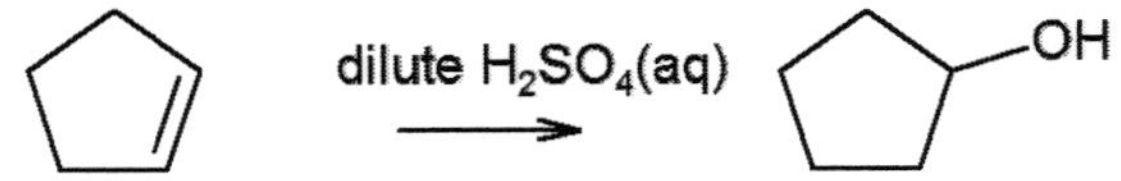

(c)

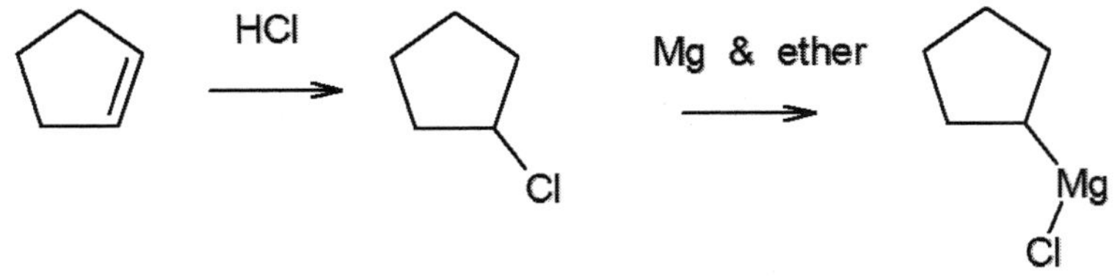

(d)

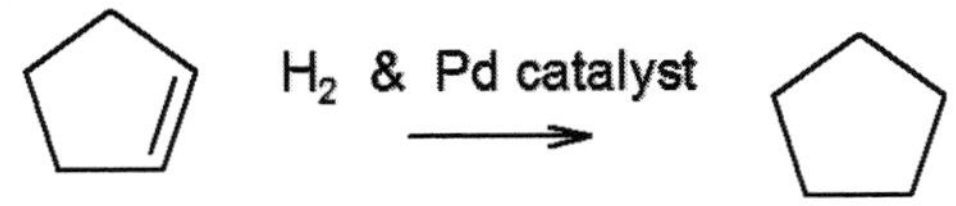

21.55
(a) 1-iodopropane **(b)** butanenitrile **(c)** 1-propanol
(d) Upon reaction with Mg in ether, we get propylmagnesium bromide. Subsequent reaction with water produces propane.
(e) methyl propyl ether

21.57
Arranged in order of decreasing reactivity in the S$_N$2 reaction, the isomers of C$_4$H$_9$Br are

1-bromobutane > 1-bromo-2-methylpropane > 2-bromobutane > 2-bromo-2-methylpropane

21.59

21.61
E2 elimination reactions yield non-Zaitsev products because elimination follows a concerted mechanism requiring a base to abstract a proton from an adjacent carbon atom that is anti to the bromide that becomes the leaving group. This stereochemical relationship is not available to the proton that could otherwise be lost to give the more highly substituted alkene.

21.63
Reaction of HBr with (*R*)-3-methylhexan-3-ol yields (±)-3-bromo-3-methylhexane, because it is an S$_N$1 substitution. S$_N$2 is not possible because of the significant hindrance around the tertiary carbon atom. The same alkyl substituents that hinder S$_N$2 stabilize the carbocation intermediate, allowing the S$_N$1 pathway to be used. Once the carbocation is formed, the stereochemistry of the starting material is lost. Since the carbocation intermediate is planar, reaction with bromide ion can occur with equal probability on either face of the cation, and the (*R*) and (*S*) products are equally likely.

21.65

A is an alkyl dibromide that forms diene, B, under elimination reaction conditions. We know that B is a diene, because it absorbs two equivalents of H_2 when reduced. It cannot be an alkyne because A would need to be a geminal dibromide, and geminal dibromides can only under a single elimination step—to make the vinyl bromide. Vinyl bromides are relatively unreactive. The formula, $C_{16}H_{16}Br_2$, shows 8 degrees of unsaturation—consistent with two phenyl rings. Oxidation of B produces oxalic acid and only one other acid, $C_7H_6O_2$. To account for B having two double bonds, and the total of 16 carbon atoms in B, we look for a symmetric structure about the two carbon atoms that become oxalic acid. Note that the formula for C, $C_7H_6O_2$ is consistent with benzoic acid.

Chapter 22
Alcohols, Phenols, and Ethers

IN-CHAPTER EXERCISES

Exercise 22.2—Naming alcohols

(a) 2-Methylhexan-2-ol

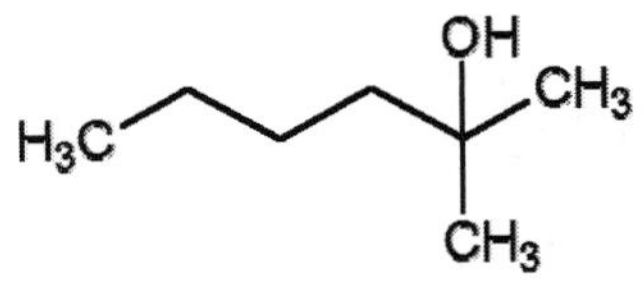

(b) Hexane-1,5-diol

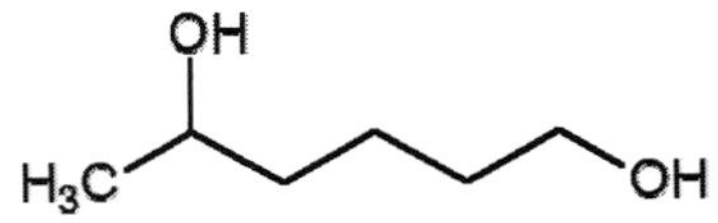

(c) 2-Ethylbut-2-en-1-ol

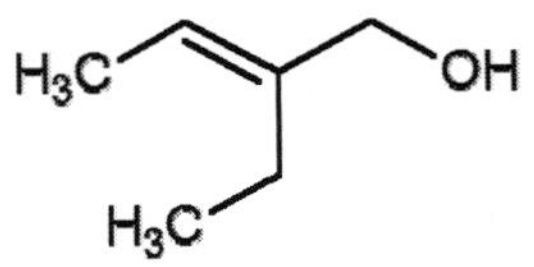

(d) Cyclohex-3-en-1-ol

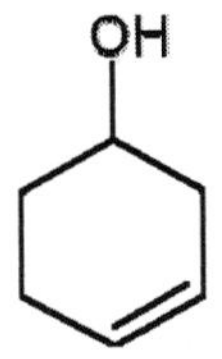

(e) 2-Bromophenol

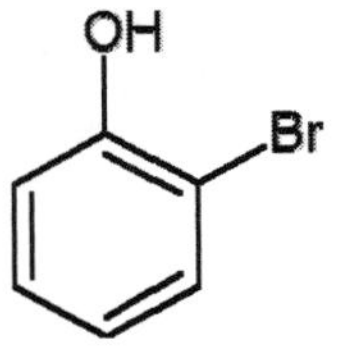

(f) 2,4,6-Trinitrophenol

Exercise 22.3—Identifying alcohols

(a) 2-Methylhexan-2-ol is a tertiary alcohol.
(b) Hexane-1,5-diol has both primary and secondary alcohol functional groups.
(c) 2-Ethylbut-2-en-1-ol is a primary alcohol.
(d) Cyclohex-3-en-1-ol is a secondary alcohol.
(e) 2-Bromophenol is a phenol—a distinct functionality.
(f) 2,4,6-Trinitrophenol is a phenol.

Exercise 22.4—Naming ethers

(a) 2-isopropoxypropane or diisopropyl ether
(b) propoxycyclopentane or cyclopentyl propyl ether
(c) *p*-bromoanisole or 4-bromomethoxybenzene
(d) 1-ethoxy-2-methylpropane or ethyl isobutyl ether

Exercise 22.5—IR spectroscopy of alcohols

The O–H stretch of benzyl alcohol appears in its IR spectrum as the strong, broad peak centred at about 3320 cm^{-1}. The aromatic C–H stretches give rise to the three sharp peaks at 3090, 3070, and 3010 cm^{-1}. Alkyl C–H stretches correspond to the broader peaks at 2950 and 2890 cm^{-1}. The C–O stretches corresponds to the strong, somewhat broad peak at 1020 cm^{-1}. The other peaks are associated with the C–C stretches of the benzene ring—peaks (three sharp ones) from 1500 to 1370 cm^{-1}—and the benzene C–H in-plane and out-of-plane bends—peaks at 1080 and 1040 cm^{-1}, and 730 and 700 cm^{-1}, respectively.

Exercise 22.6—Spectroscopy of alcohols

An IR spectrum of the product would test whether a synthesis of 5-cholestene-3-one from cholesterol was successful or not. The IR spectrum of the product should not show the strong, broad alcohol O–H peak around 3300 cm^{-1}, or the strong C–O peak near 1000 cm^{-1}. Instead, it should show a strong C=O stretching peak near 1720 cm^{-1}.

Exercise 22.7—Reduction of carbonyl compounds

Exercise 22.8—Williamson synthesis of ethers

Could you make cyclohexyl ethyl ether by first forming sodium ethoxide (product of treatment of ethanol with sodium metal—hydrogen gas is the other product), then reacting it with cyclohexyl iodide? No. Since the alkyl iodide substrate in this case is much more sterically hindered, an E2 reaction to produce cyclohexene will compete with an S_N2 reaction, giving a much poorer yield of the substitution product.

Exercise 22.9—Williamson synthesis of ethers

According to decreasing reactivity toward alkoxide substitution, the species are ranked as follows:

bromoethane > chloroethane > 2-bromopropane > 2-chloro-2-methylpropane

bromoethane > chloroethane because bromide is a better leaving group than chloride. The other two species have successively greater steric hindrance at the electrophilic carbon (i.e. that which the nucleophile attacks).

Exercise 22.10—Dehydration of alcohols

(a)

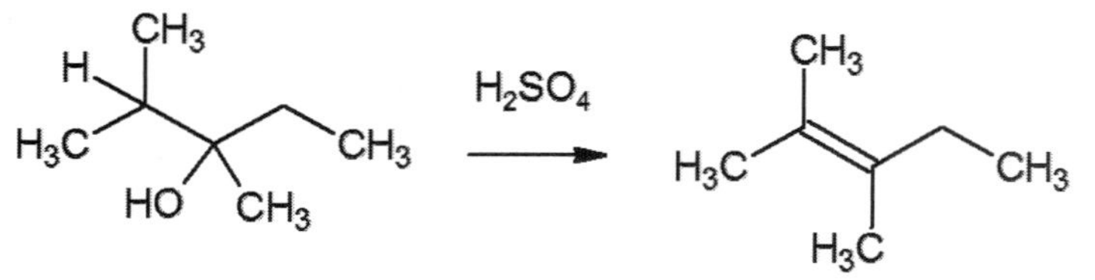

(b)

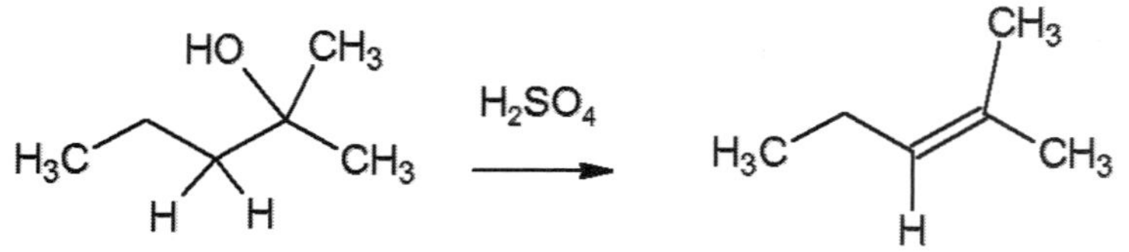

Note that the most substituted alkene is formed in each case. Concentrated sulfuric acid can achieve the reverse of the hydration of an alkene because the water eliminated on dehydration is quickly taken up by the solvent.

Exercise 22.11—Oxidation of alcohols

(a)

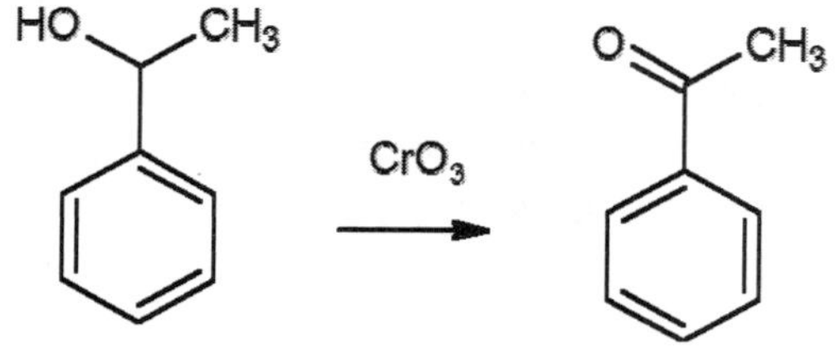

(b)

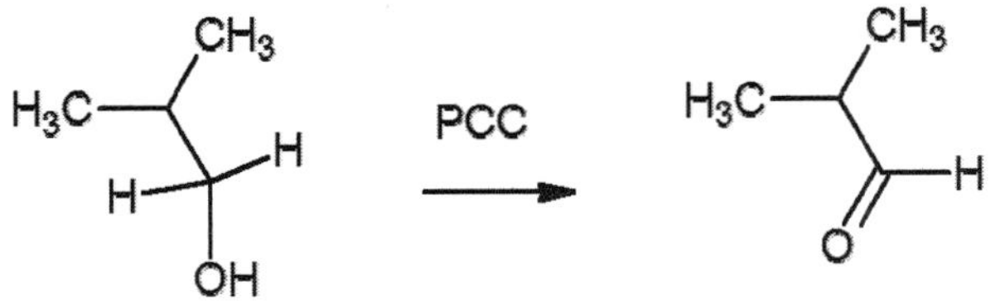

(c)

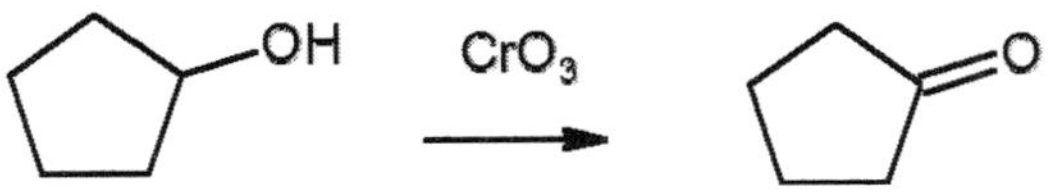

Note that milder conditions are needed to get the aldehyde in part (b).

Exercise 22.12—Oxidation of alcohols

(a)

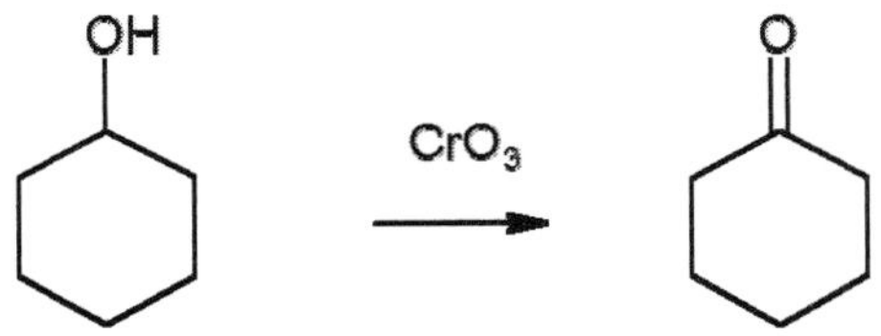

(b)

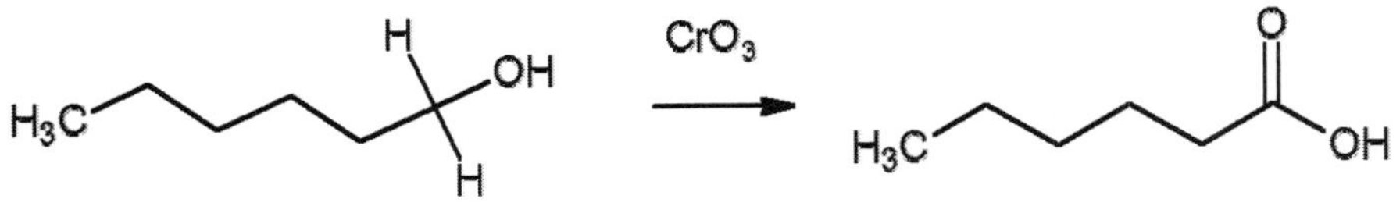

(c)

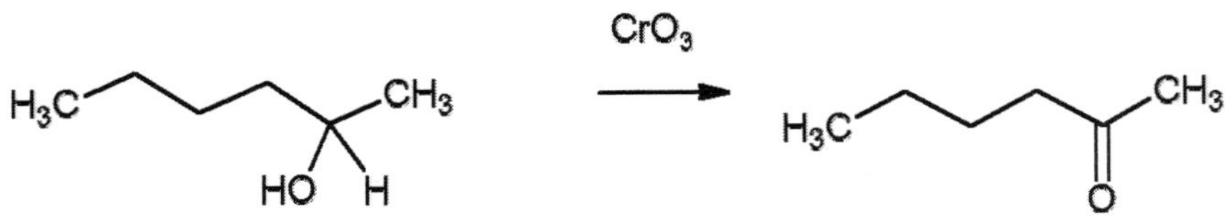

Exercise 22.13—Oxidation of alcohols

Oxidation of the alcohols in Exercise 22.12 with pyridinium chlorochromate (PCC) gives the same products in parts (a) and (b), and the aldehyde in part (b).

(a)

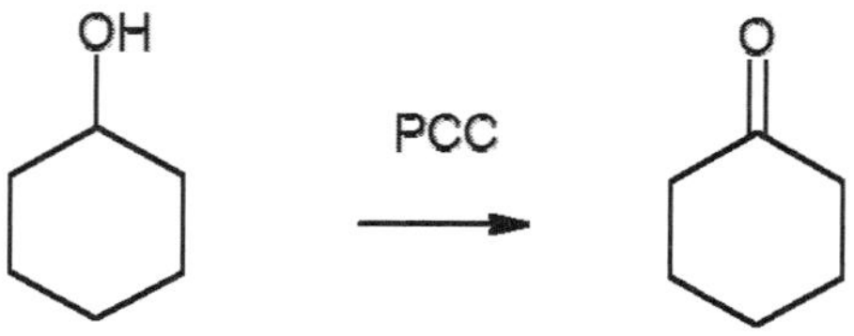

(b)

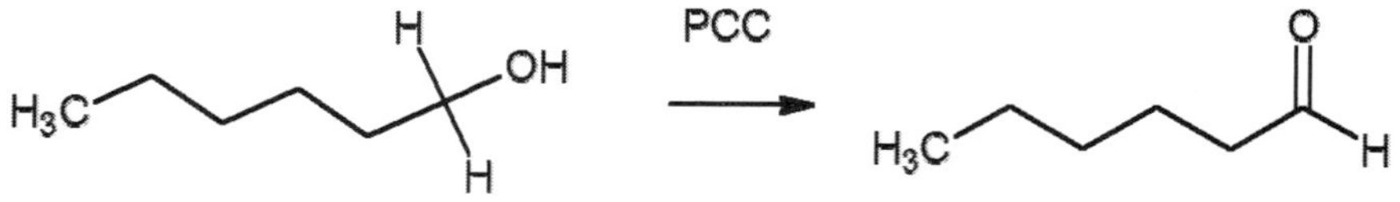

(c)

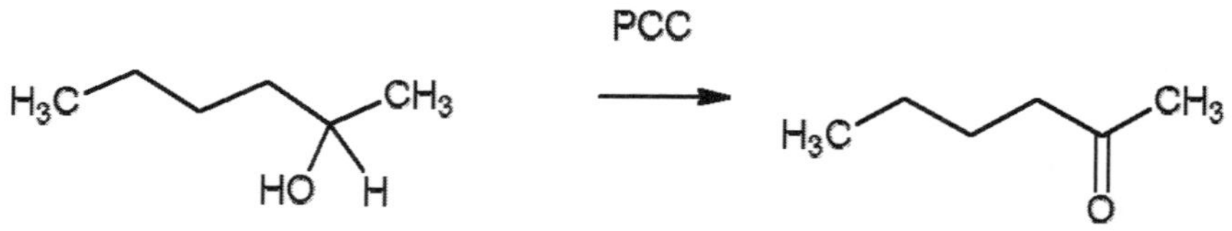

Exercise 22.14—Synthesis of phenols

Here, we add the hydroxyl group before the methyl. Since both substituents are activating and *ortho-para* directing, it would be possible to do it the other way around—add methyl, then hydroxyl. However, because the Friedel-Crafts alkylation is the slowest step, the above synthesis takes advantage of the strong activating power of the hydroxyl group.

Exercise 22.15—Mechanism for reactions of phenols

Exercise 22.16—Reactions of ethers

(a)

$$H_3C \diagdown O \diagdown CH_3 \quad \xrightarrow{HI} \quad H_3C \diagdown I \quad + \quad H_3C \diagdown OH$$

(b)

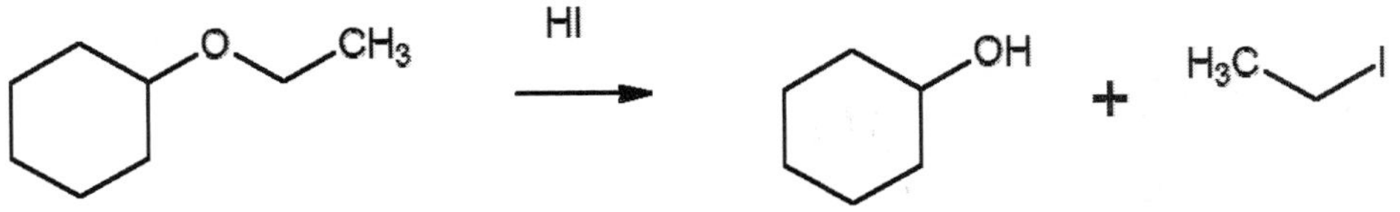

(c)

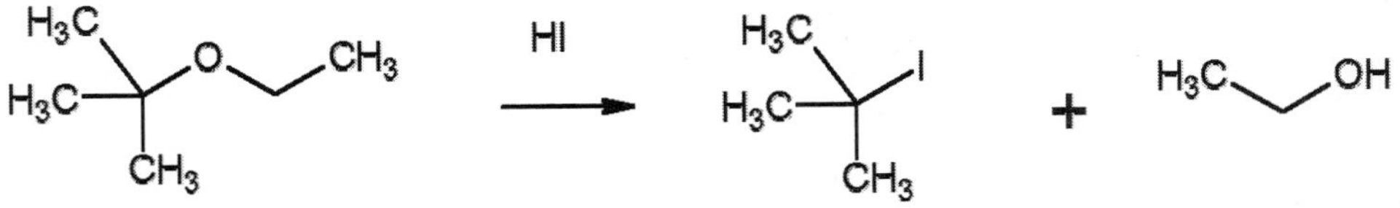

Exercise 22.17—Reactions of epoxides

The stereochemistry of the product diol has the two –OH groups trans to each other, since the ring opening involved backside reaction by an S_N2 mechanism.

Exercise 22.18—Naming thiols

(a) 2-butanethiol
(b) 2,2,6-trimethylheptane-4-thiol
(c) cyclopent-3-ene-1-thiol

Exercise 22.19—Naming thiols and sulfides

(a) ethyl methyl sulfide
(b) *tert*-butyl ethyl sulfide
(c) *o*-(dimethylthio)benzene

Exercise 22.20—Synthesis of thiols

To make but-2-ene-1-thiol from but-2-ene-1-ol:
You could treat but-2-en-1-ol with PBr_3 to make 1-bromobut-2-ene.
1-bromobut-2-ene is then treated with NaSH to make but-2-ene-1-thiol.
To make but-2-ene-1-thiol from methyl but-2-enoate:
First reduce methyl but-2-enoate with $LiAlH_4$ to get methanol and but-2-ene-1-ol.
but-2-ene-1-ol can then be processed as above.

REVIEW QUESTIONS

22.1 Cyclodextrins

22.21
The hydroxyl groups on the outside of cyclodextrins make them soluble in water, allowing the drug encapsulated by the cyclodextrin to be delivered throughout the body.

22.23
The article describes the use of cyclodextrins to form aggregates and give rise to dispersed systems (i.e., not solutions) with more complex structures for dispersed drug delivery. Such dispersed structures include emulsions, micro and nano capsules and spheres, as well as liposomes and niosomes.
Sublingual literally means "under the tongue," referring to the administration of a drug through dissolution and absorption of a tablet held under the tongue. In the context of this case study, sublingual might be used as a metaphor meaning "under the ability-to-detect (i.e., taste) by the tongue."

22.25
Some types of wounds—e.g., exudating (oozing) wounds, venous leg ulcers, cancerous lesions, and wounds with necrotic (dead) tissue—produce unpleasant odours. Cyclodextrins have been used to treat this problem by encapsulating odour molecules. This is in addition to their use in increasing bioavailability of antibiotics to the wound.

22.2 Naming alcohols, ethers, and phenols

22.27
(a)

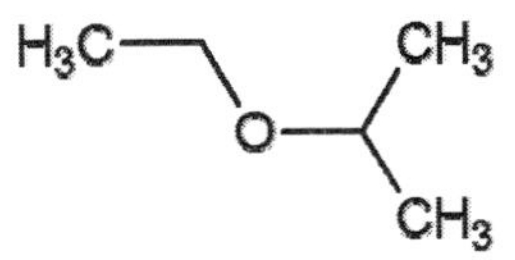

(b)

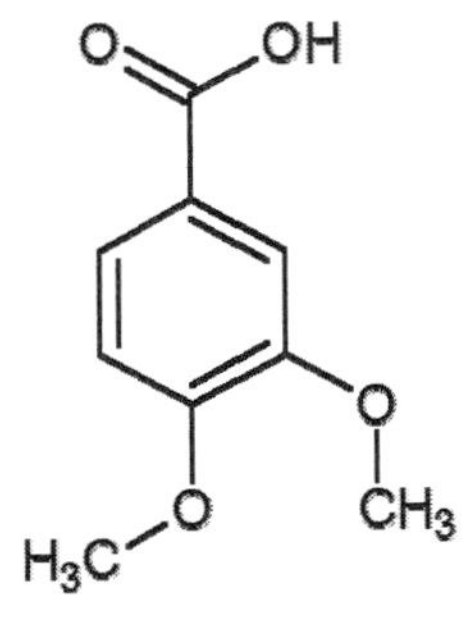

(c)

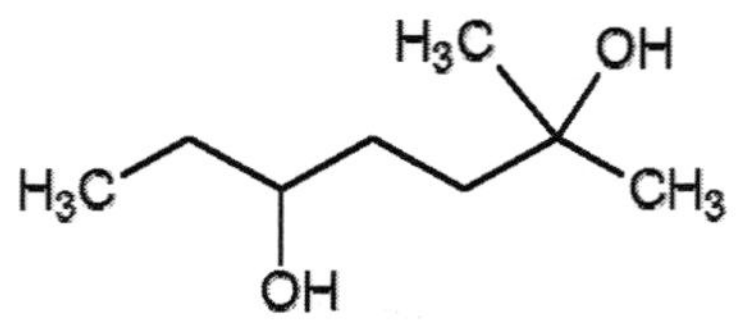

(d)

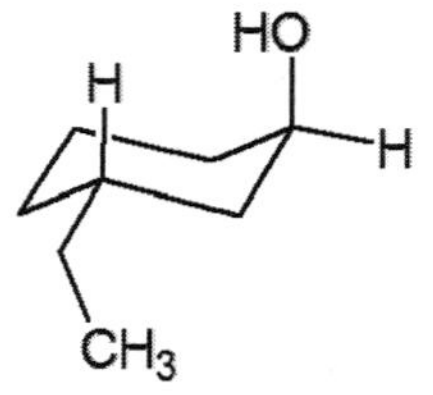

(e)

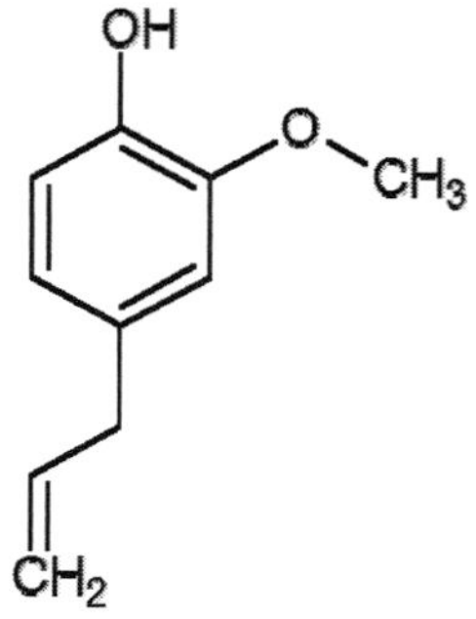

22.29

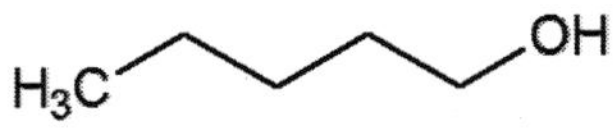

1-pentanol

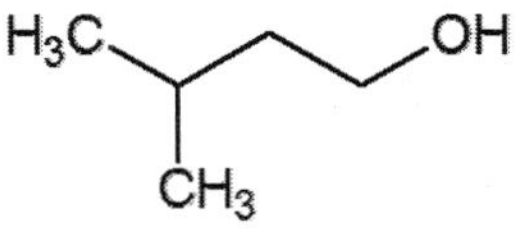

3-methylbutan-1-ol

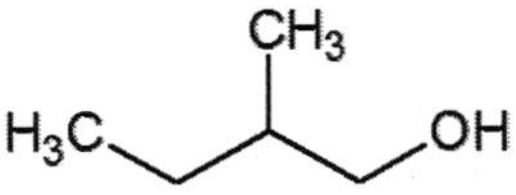

2-methylbutan-1-ol

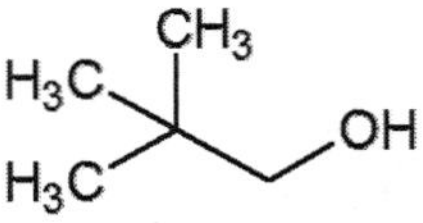

2,2-dimethylpropan-1-ol

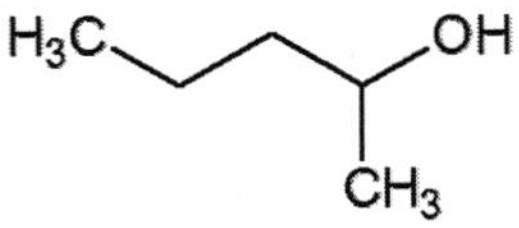

2-pentanol

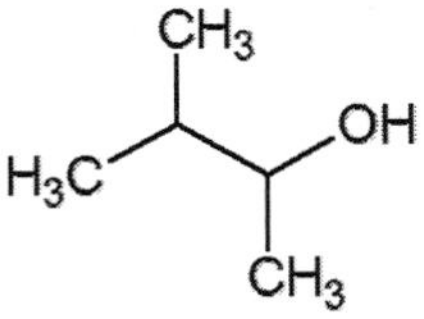

3-methylbutan-2-ol

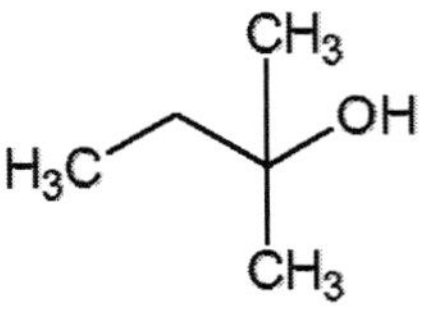

2-methylbutan-2-ol

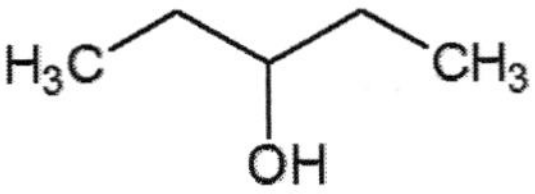

3-pentanol

22.31

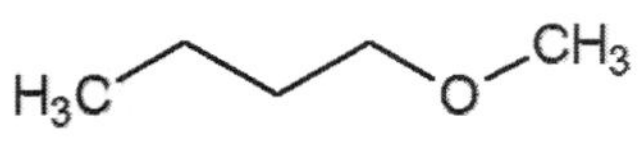

1-methoxybutane

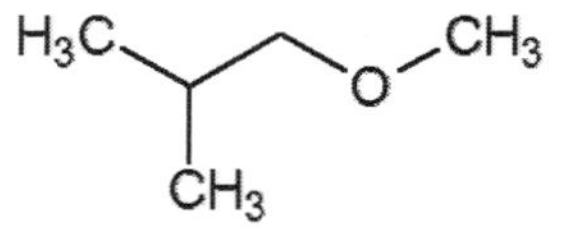

1-methoxy-2-methylpropane

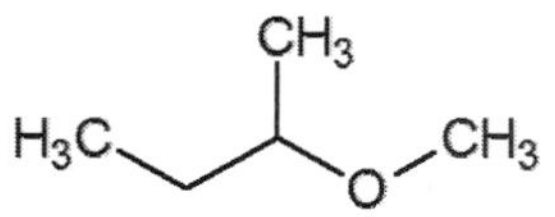

2-methoxybutane

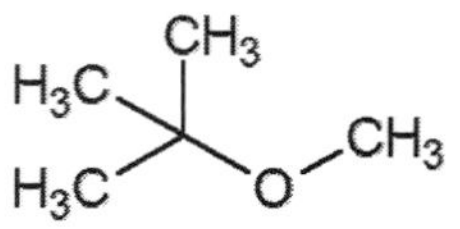

2-methoxy-2-methylpropane

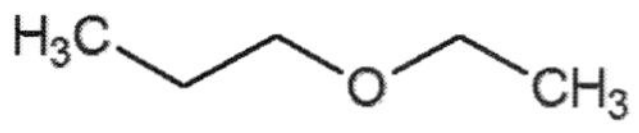

1-ethoxypropane

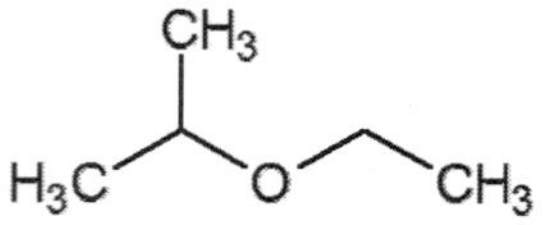

2-ethoxypropane

22.35

In the ^{1}H NMR spectrum of an alcohol in deuterated dimethyl sulfoxide, the hydroxyl H peak is split into a triplet if the alcohol has two equivalent neighbouring C–H protons , a doublet if the alcohol has one neighbouring C–H proton, and it is not split if the alcohol is tertiary with no neighbouring protons—a singlet is seen in this case. The expected splitting patterns are

(a) a singlet for 2-methylpropan-2-ol
(b) a doublet for cyclohexanol
(c) a triplet for ethanol
(d) a doublet for propan-2-ol
(e) a doublet for cholesterol—cholesterol is a secondary alcohol
(f) a singlet for 1-methylcyclohexanol

22.37

^{1}H NMR: 1.24 δ (12 H, singlet); 1.56 δ (4 H, singlet); 1.95 δ (2 H, singlet)
The formula, $C_8H_{18}O_2$, is that of a fully saturated compound. The IR absorption peak at 3350 cm^{-1} is consistent with the O–H functionality. The ^{1}H NMR spectrum shows 12 equivalent alkyl H atoms (e.g., four equivalent methyl groups), another 4 equivalent H atoms a little downfield, and another two equivalent H atoms yet further downfield. The molecule must be symmetrical. The following structure is consistent with these data.

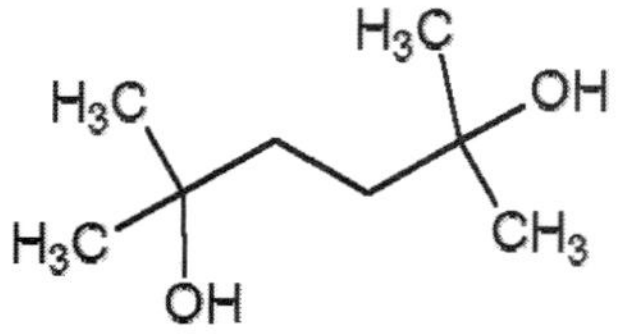

22.39

(a) $C_5H_{12}O$

The integrals in the 1H NMR spectrum correspond to the pattern of 1,1,4 and 6 protons of each distinct type. The 4 and 6 protons are consistent with two equivalent ethyl groups, leading to the structure,

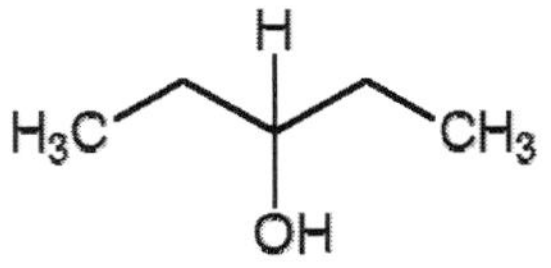

The broader singlet at around 3.4 ppm corresponds to the hydroxyl proton. The other singlet at 1.9 ppm, downfield from the other alkyl peaks, corresponds with the remaining methine proton.

(b) $C_8H_{10}O$

Here we have four degrees of unsaturation and phenyl peaks, around 7.3 ppm, integrating to a total intensity of 5 relative to 1, 1 and 3 for the other three multiplets. This is consistent with a mono-substituted phenyl group, C_6H_5-. The hydroxyl proton absorbs at 2.4, appearing as a singlet. The proton geminal to the hydroxyl appears as a quartet at 4.85 ppm with a relative integration of 1. The quartet is due to coupling to three methyl protons. The methyl protons give the doublet at 2.4 ppm with a relative integration of 3. The doublet is due to coupling with the single proton geminal to the O atom. The structure must be

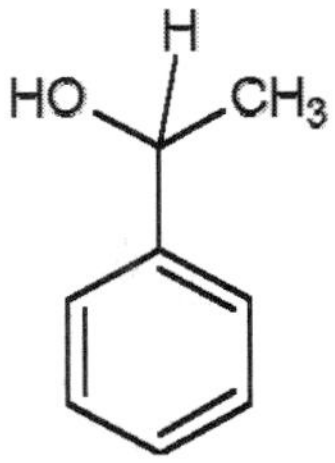

22.5-22.8 Synthesis and reactions

22.41

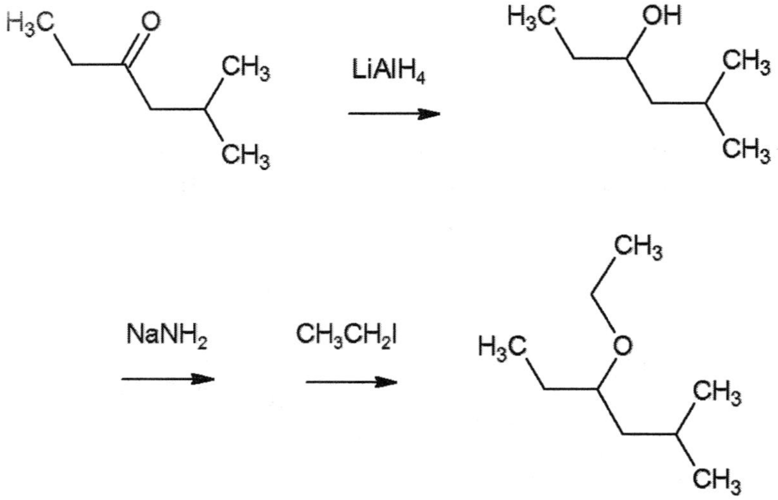

22.43

(a)

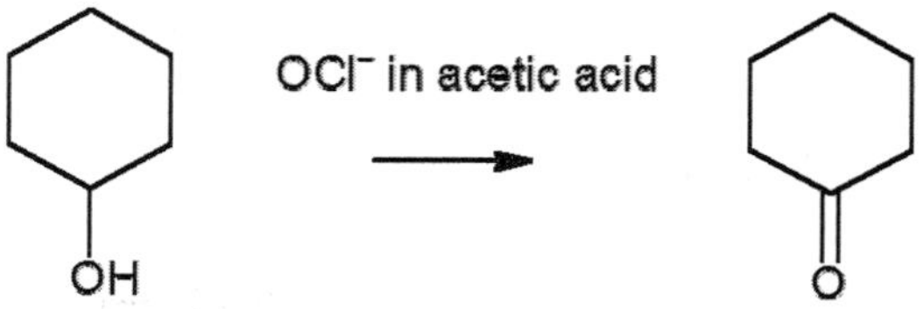

(b)

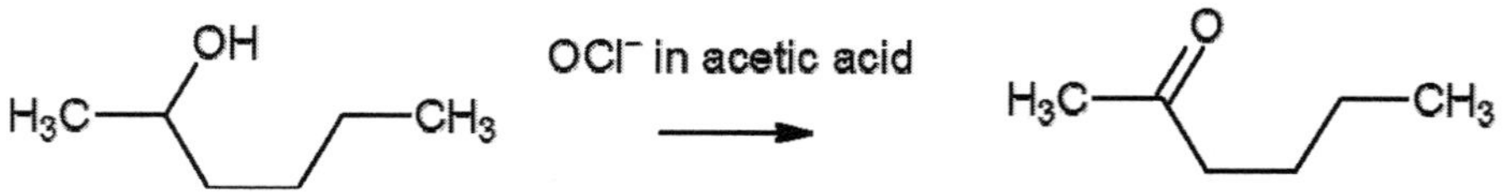

(c)

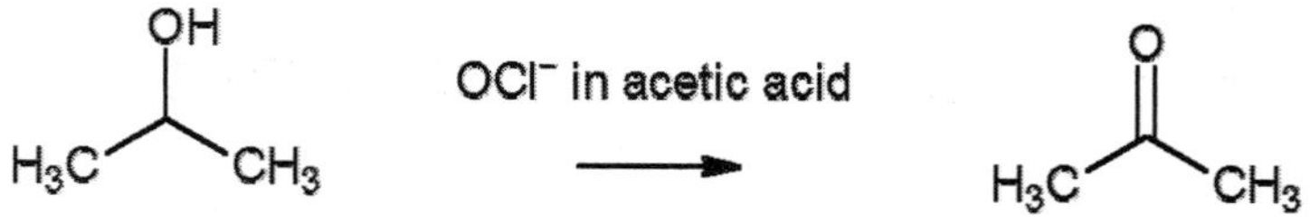

22.45

S$_N$2 substitution is much faster at the least substituted carbon. Only for the 5th ether do both potential sites of attack have the same degree of substitution—they are both secondary carbons. One side might be slightly more hindered—the carbon with the ethyl substituent. This favours the top two products. However, both sets of products are likely. In the case of the 3rd ether, the S$_N$1 mechanism is active—it is a tertiary ether. The product arises from the relatively stable tertiary carbocation intermediate.

22.47

(a)

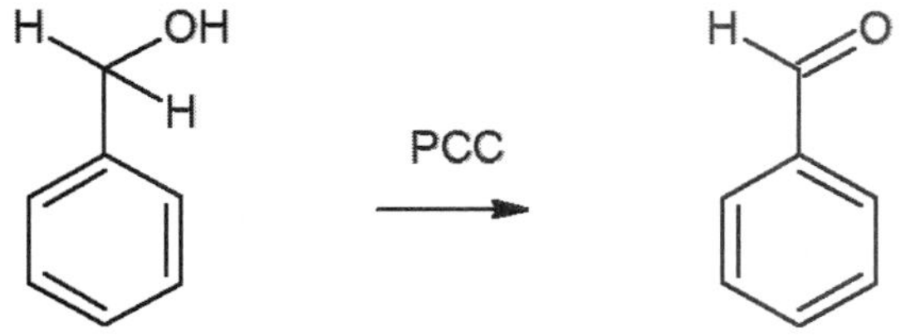
H
OH
H
PCC
H
O

(b)

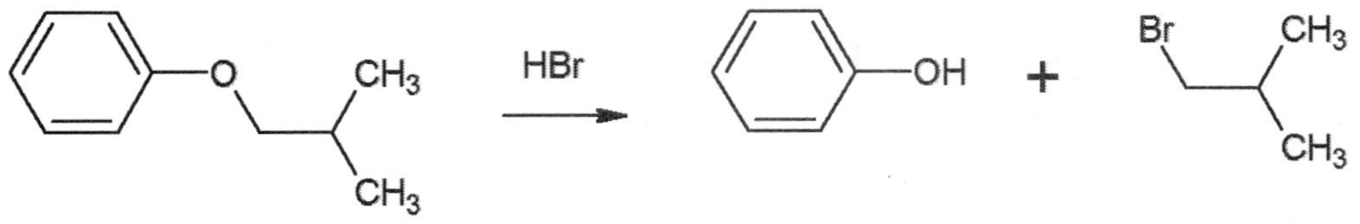
O
CH3
CH3
HBr
OH
+
Br
CH3
CH3

(c)

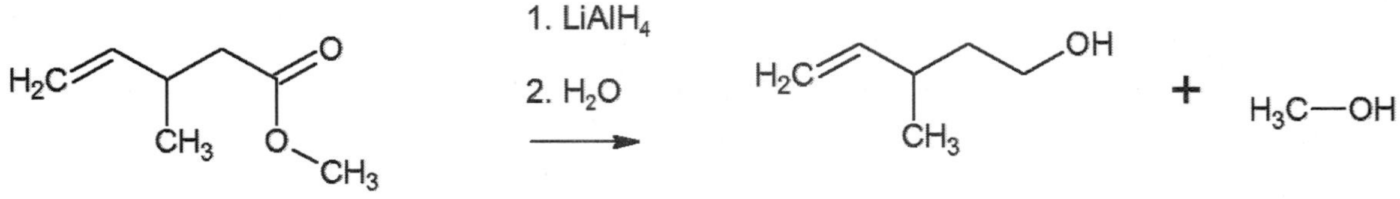
H2C
O
CH3
O
CH3
1. LiAlH4
2. H2O
H2C
OH
CH3
+
H3C—OH

(d)

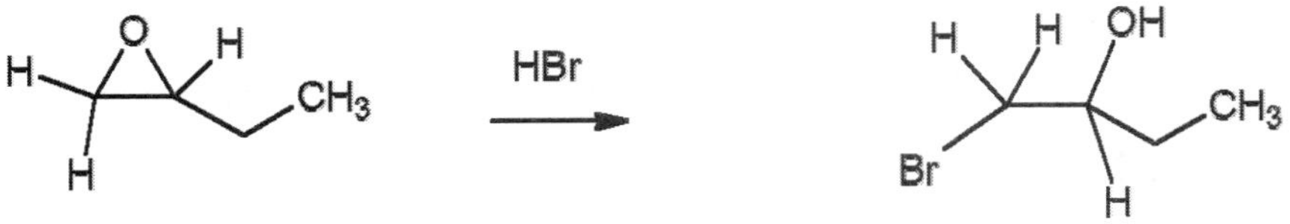
O
H
H
CH3
H
HBr
H
H
OH
CH3
Br
H

22.49

(a)

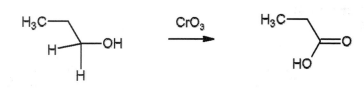

(b)

(c)

(d)

22.51

(a)

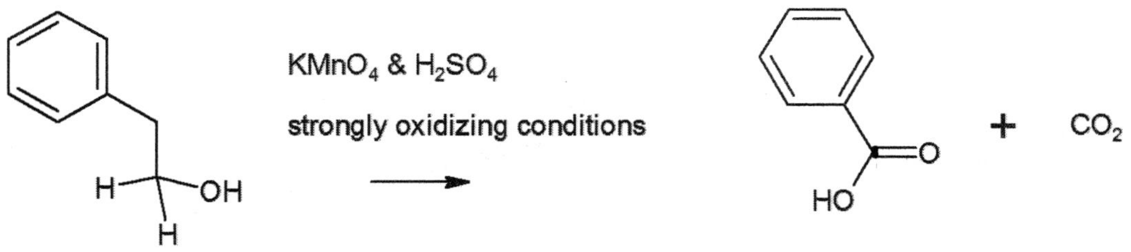

KMnO$_4$ & H$_2$SO$_4$

strongly oxidizing conditions

+ CO$_2$

oxidation takes place at the benzyl
position - in addition to the alcohol

intermediates are stabilized by the
neighboring phenyl ring

(b)

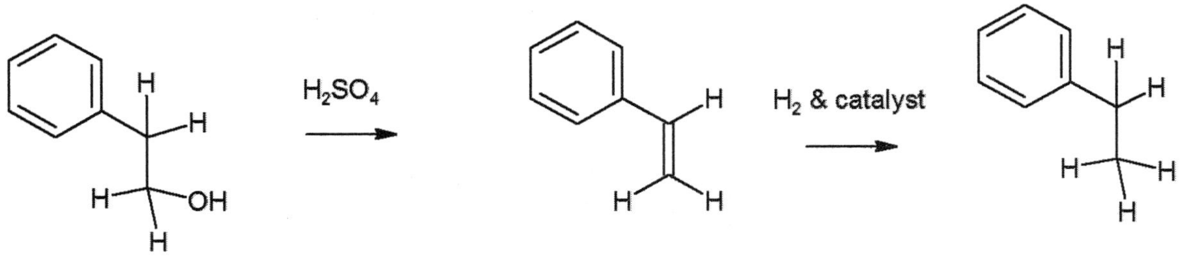

H$_2$SO$_4$

H$_2$ & catalyst

(c)

PBr$_3$

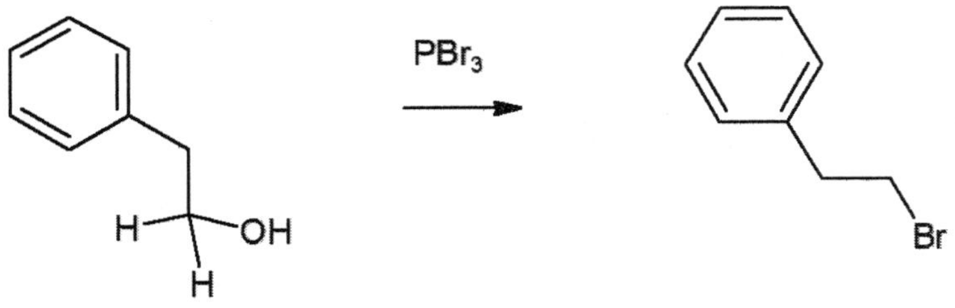

(d)

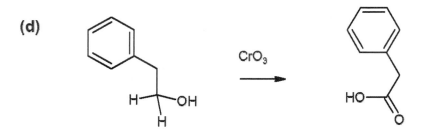

(e)

22.52

(a)

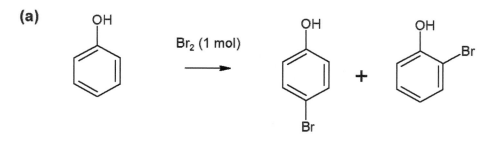

OH → Br$_2$ (1 mol) → OH ... Br + OH ... Br

(b)

OH → Br$_2$ (3 mol) → Br ... OH ... Br ... Br

(c)

OH → NaOH, then CH$_3$I → O—CH$_3$

(d)

OH → Na$_2$Cr$_2$O$_7$ & H$_3$O$^+$ → O ... O

22.53

(a)

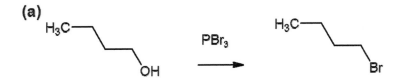

(b)

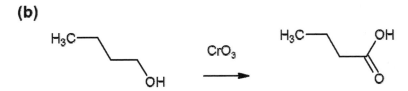

(c)

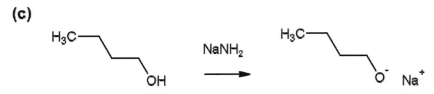

(d)

22.55

(a)

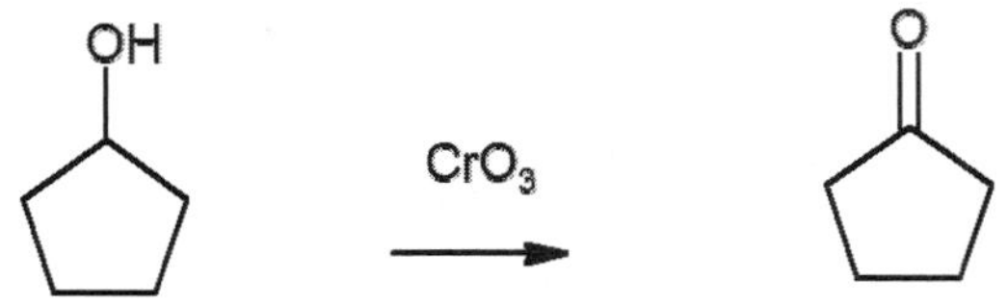

(b)

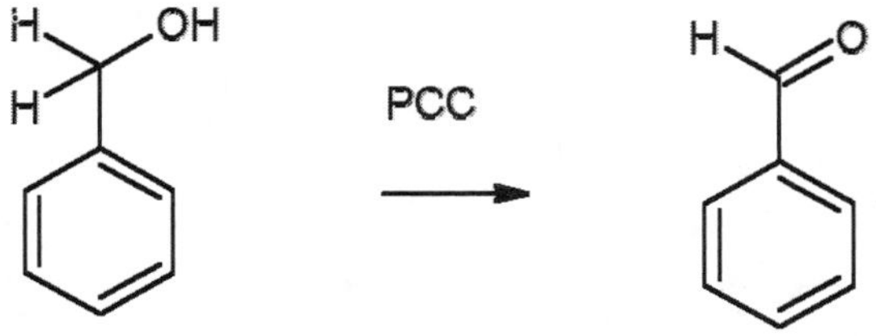

(c)

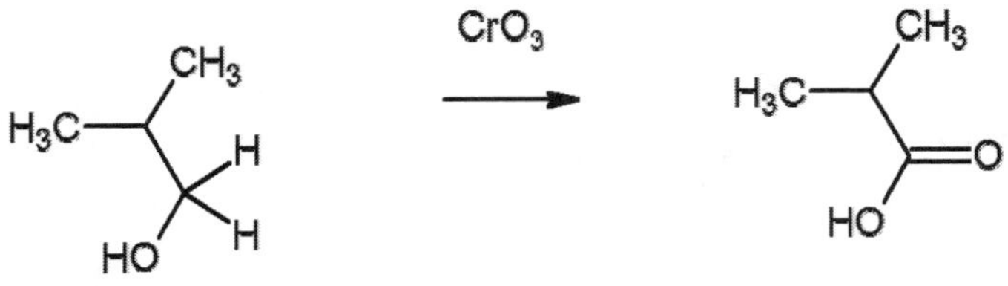

22.57

(a)

(b)

(c)

22.59

Acetic acid, pentane-2,4-dione and phenol will all react, essentially to completion, with NaOH. These three species have lower pK_a than water.

22.61

Only acetic acid will react with NaHCO$_3$. Of the substances in Exercise 22.58, only acetic acid has a pK_a lower than that of carbonic acid.

22.9 Cyclic ethers

22.63

The O atoms of the crown ether will coordinate a cation, holding it at one end of the molecule (the crown). To bind Cs$^+$, one could prepare a larger crown ether by adding $-OCH_2CH_2O-$ units. Figure 22.12 (c) shows that one more unit (i.e., making crown-7) is sufficient to bind cesium.

22.64

Note the two arrows in succession represent the propagation phase, where the polymer grows by successively adding to epoxide monomers. The portion—between parentheses—of the polymer molecules in the last line repeats many times.

SUMMARY AND CONCEPTUAL QUESTIONS

22.65
Bombykol, the sex pheromone secreted by the female silkworm moth.

22.67
Williamson ether synthesis cannot be used to prepare diphenyl ether because one of the two reagents in Williamson ether synthesis is a halide which undergoes S_N2 substitution by (in this case) phenoxide. The trouble is that phenyl halides are not amenable to S_N2 substitution. Nucleophilic attack will not occur at an aromatic carbon atom.

22.69

The *cis* isomer will oxidize faster with CrO_3—the hydroxyl group is axial in this isomer. Upon oxidation, the alcohol is turned into a ketone, which reduces the steric strain between that functional group and the cyclohexane ring.

22.71

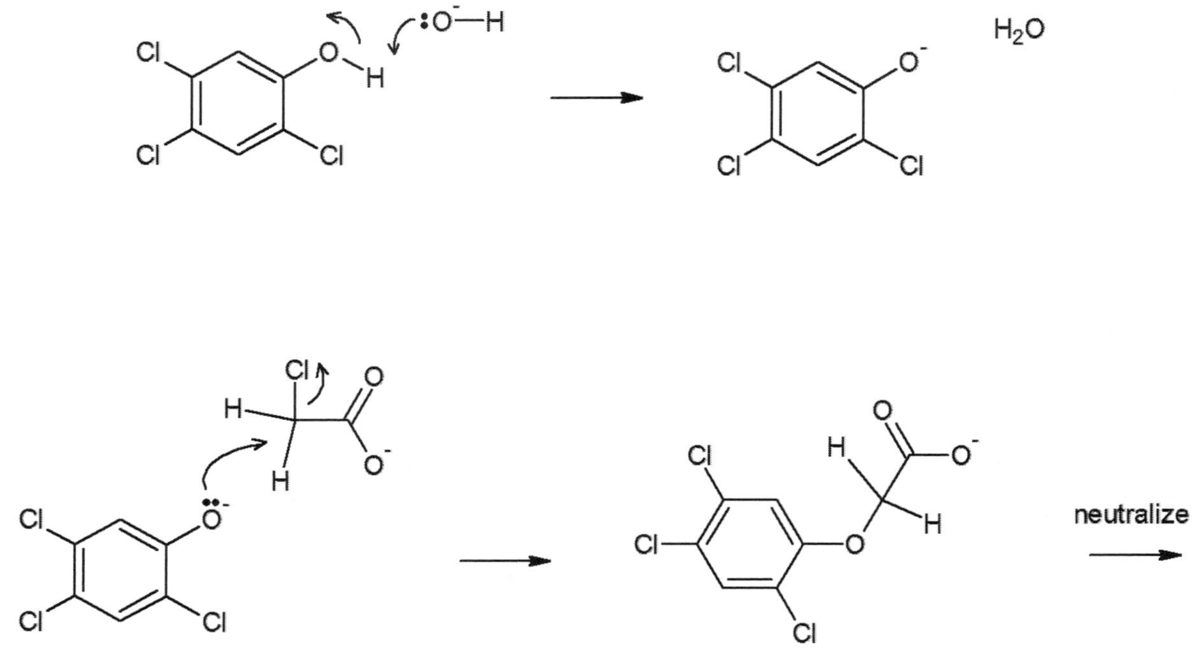

22.73

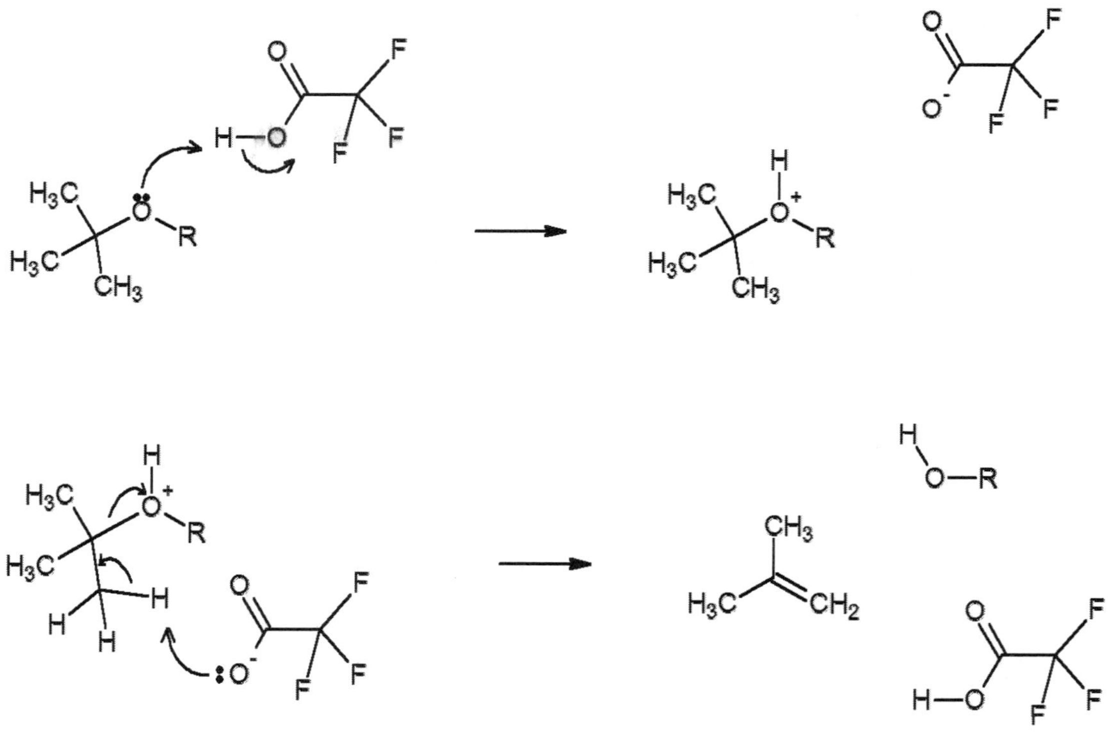

This is a dehydration reaction, proceeding by an E1 elimination reaction mechanism.

22.75

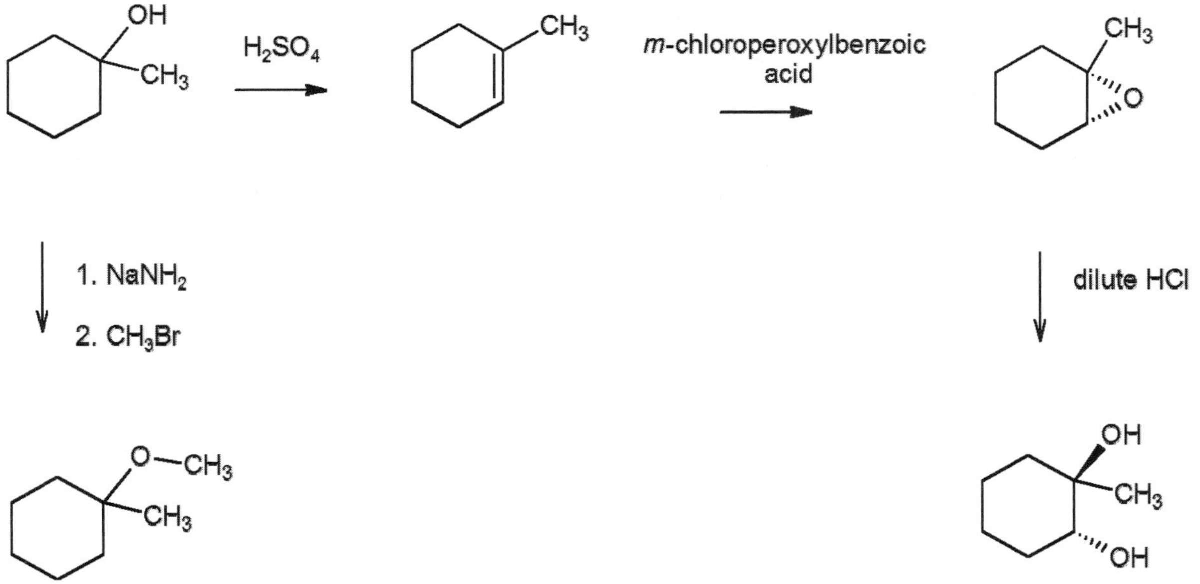
OH
CH3
H2SO4
CH3
m-chloroperoxylbenzoic acid
CH3
O
1. NaNH2
2. CH3Br
dilute HCl
O—CH3
CH3
OH
CH3
OH

22.77

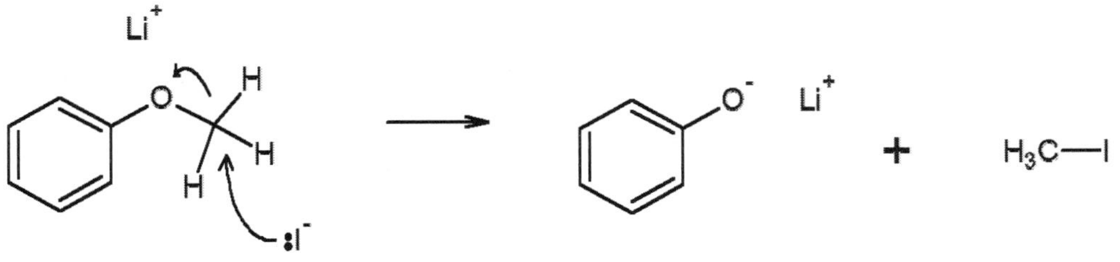
Li+
O
H
H
H
:I-
O-
Li+
+
H3C—I

22.79

Compound A, $C_8H_{10}O$, has an O—H stretch peak, and C=C stretching peaks characteristic of a phenyl ring, in its IR spectrum. A phenyl ring would account for all four degrees of unsaturation. The ^{1}H NMR spectrum shows 2 downfield (7.0 and 6.7 ppm) doublets with a relative integration of 2/2 characteristic of two pairs of equivalent protons attached to a phenyl ring in the 2,6 and 3,5 positions. The other peaks—a 2-proton quartet at 2.6 ppm and a 3-proton triplet at 1.15 ppm—are consistent with a $-CH_2CH_3$ substituent. The 2 benzyl protons are downfield because of the induced magnetic dipole of the phenyl ring—it enhances the field just outside the ring. Disappearance of the broad peak with an integration of 1 at 5.5 ppm, on adding D_2O, confirms its assignment as the hydroxyl H.

Chapter 23
Carbonyl Compounds: Part I

IN-CHAPTER EXERCISES

Exercise 23.2—Naming aldehydes and ketones

(a) 2-methylpentan-3-one
(b) 3-phenylpropanal
(c) octane-2,6-dione
(d) *trans*-2-methylcyclohexanecarbaldehyde
(e) pentanedial
(f) *cis*-2,5-dimethylcyclohexanone

Exercise 23.3—Naming aldehydes and ketones

(a)

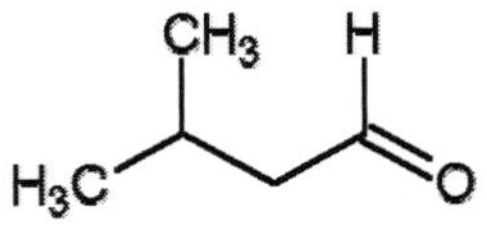

(b)

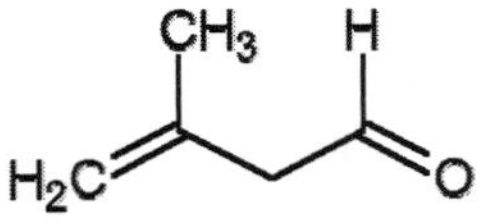

(c)

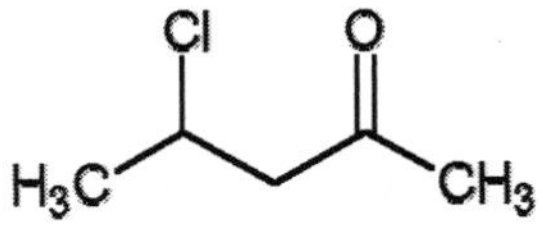

(d)

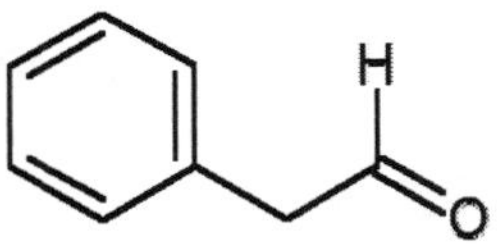

(e)

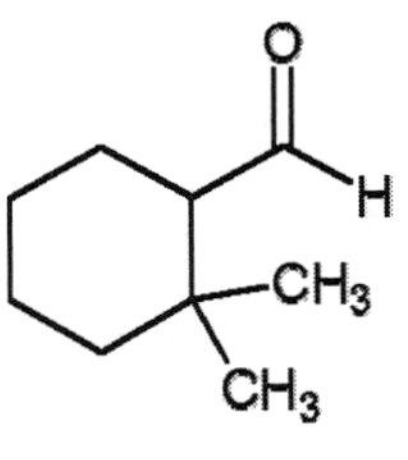

(f)

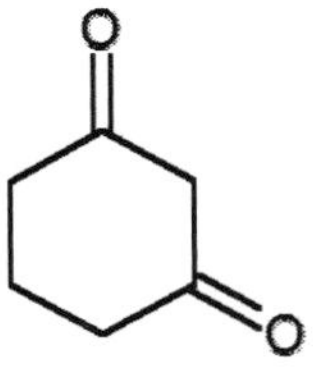

Exercise 23.4—Synthesis of aldehydes

(a)

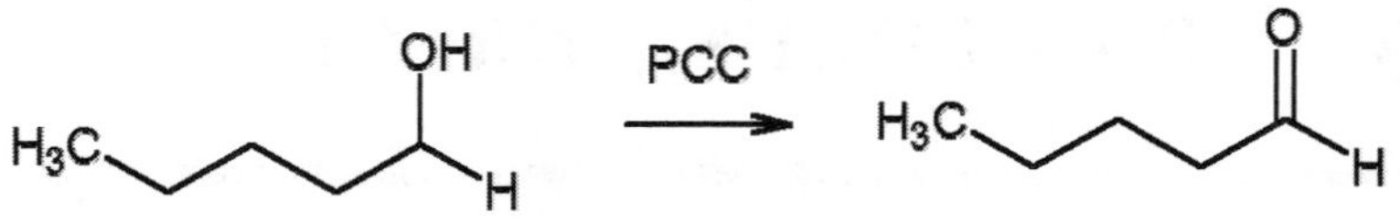

(b)

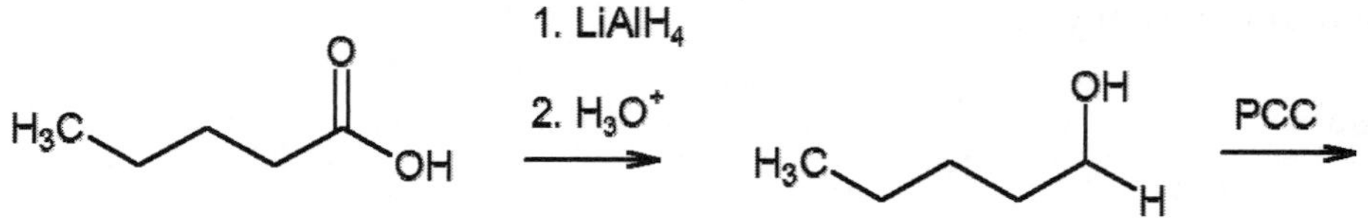

(c)

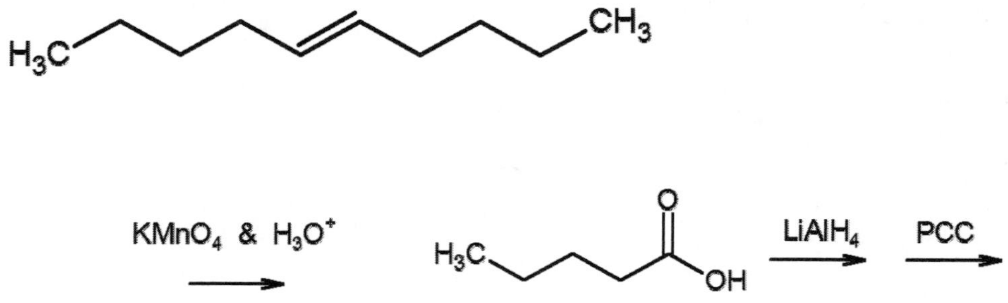

Exercise 23.5—Synthesis of ketones

(a)

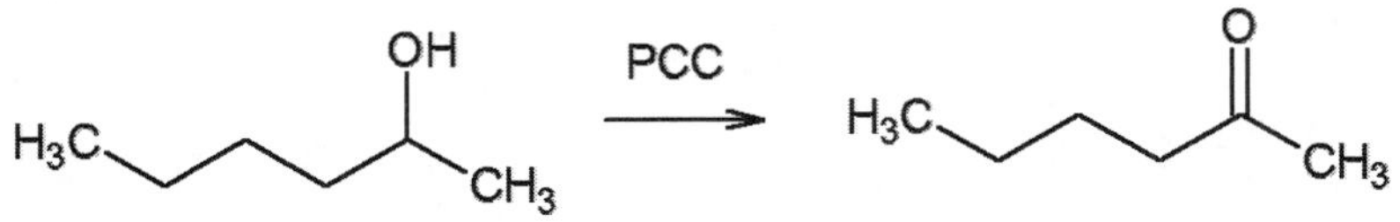

PCC

(b)

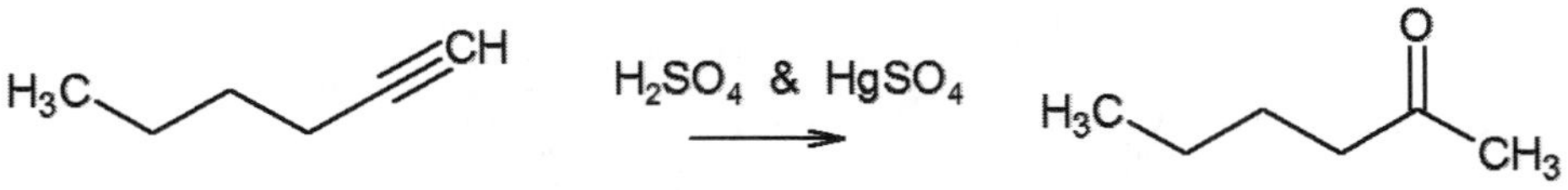

H₂SO₄ & HgSO₄

(c)

KMnO₄ & H₃O⁺

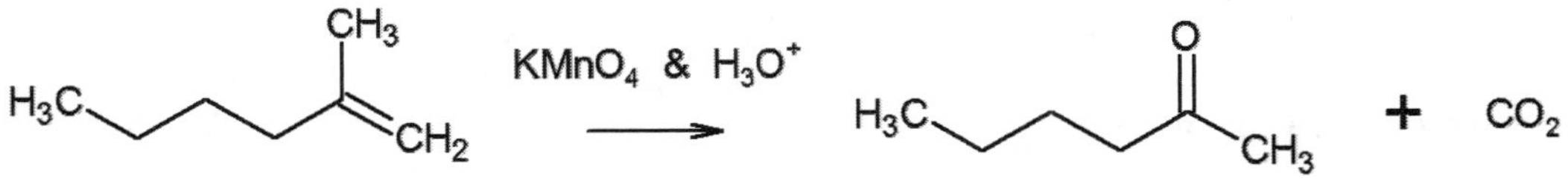

Exercise 23.6—Nucleophilic addition reactions

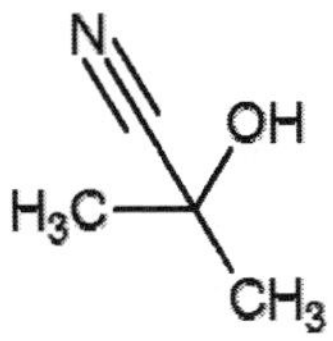

2-hydroxy-2methylpropanenitrile

Exercise 23.7—Nucleophilic addition reactions

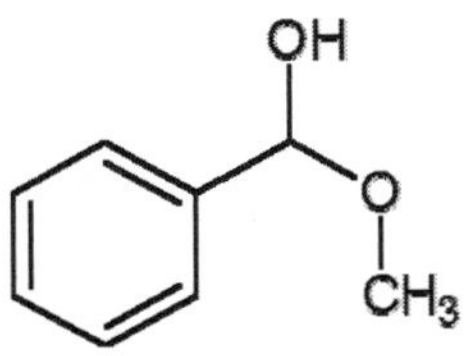

Nucleophilic addition of an alkoxide ion to an aldehyde, followed by protonation, produces a hemiacetal.

Exercise 23.8—Addition of water

choral hydrate

Exercise 23.9—Addition of water

Aldehydes and ketones undergo reversible addition and elimination of water. At the stage of the geminal diol, the two oxygen atoms (the original carbonyl oxygen and the oxygen from water) are indistinguishable. As such, upon reversal of the addition process, half of the time the resulting ketone or aldehyde will have the oxygen atom that came from the water—i.e., the two O atoms exchange.

Exercise 23.10—Addition of alcohols

1 mol of ethanol:

2 mol of ethanol:

Exercise 23.11—Addition of alcohols

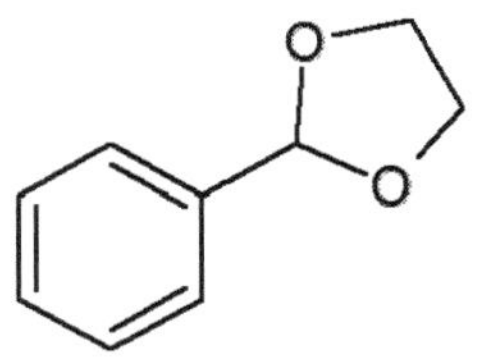

Exercise 23.12—Addition of alcohols

protection step

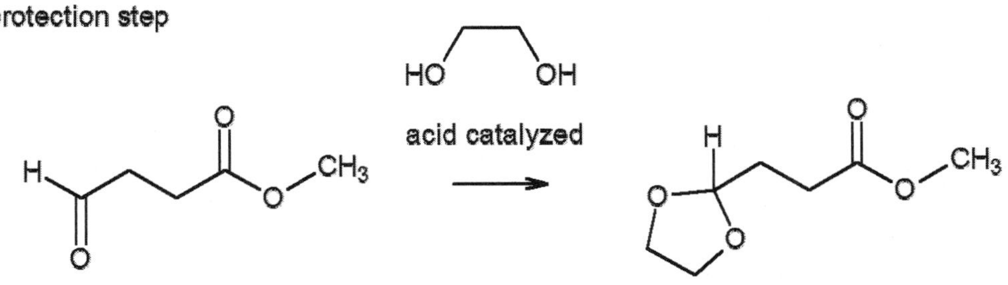
HO
OH
acid catalyzed
H
CH3

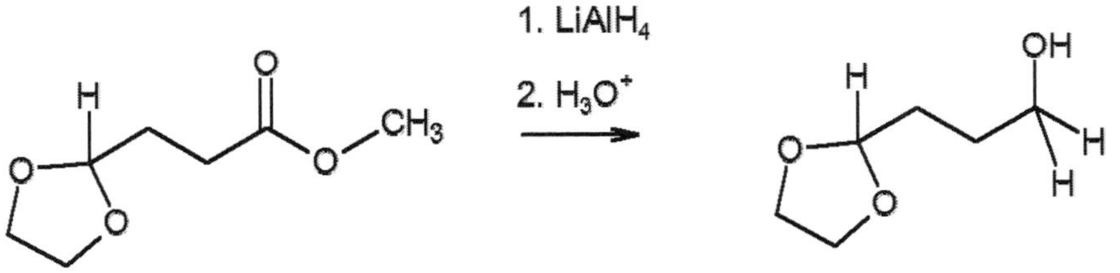
1. LiAlH4
2. H3O+
H
CH3
OH

deprotection step

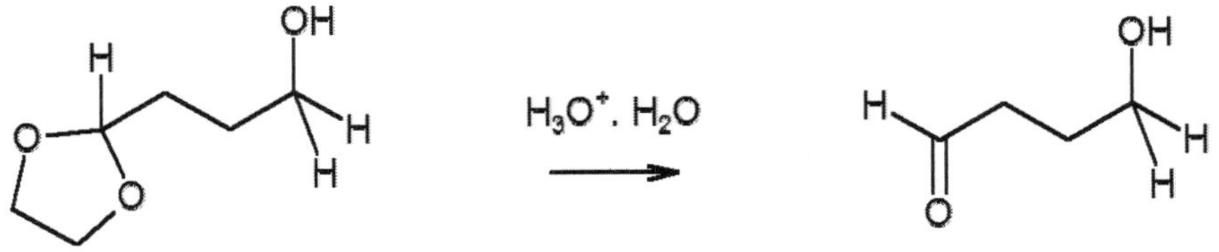
H3O+, H2O
OH
H

Exercise 23.13—Reactions of ketones

(a)

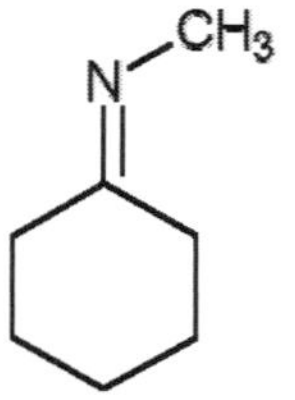

(b)

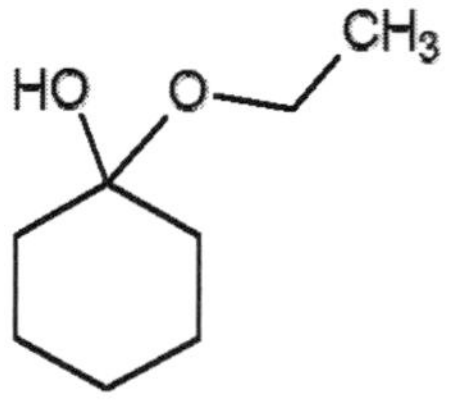

(c)

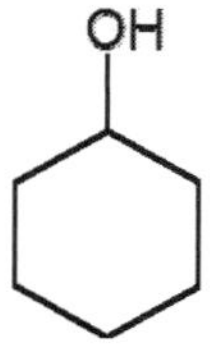

Exercise 23.14—Reaction with amines

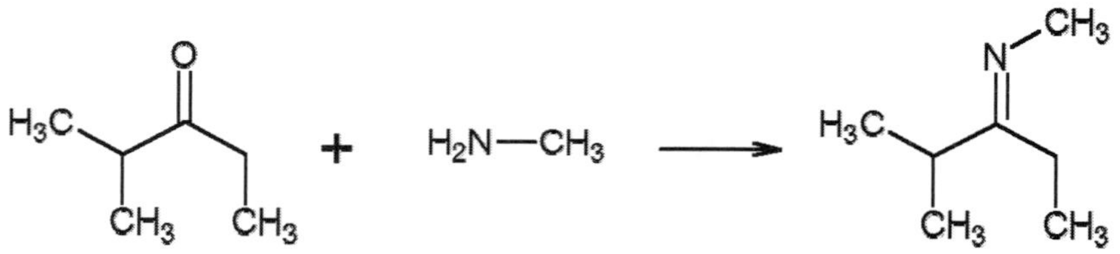

Exercise 23.15—Using Grignard reagents

(a)

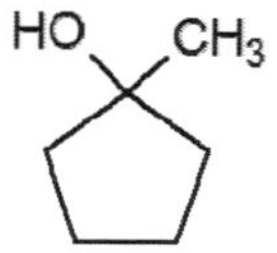

(b)

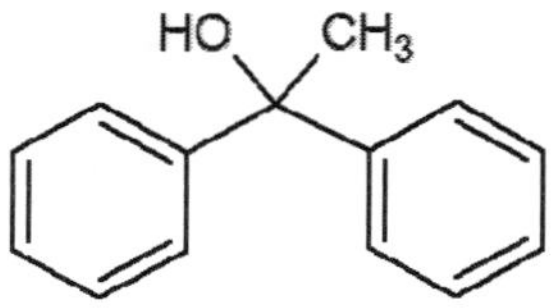

(c)

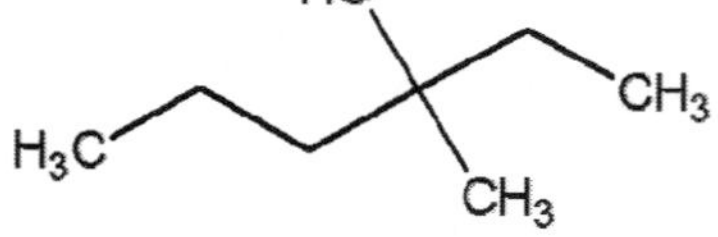

Exercise 23.16—Conjugate addition

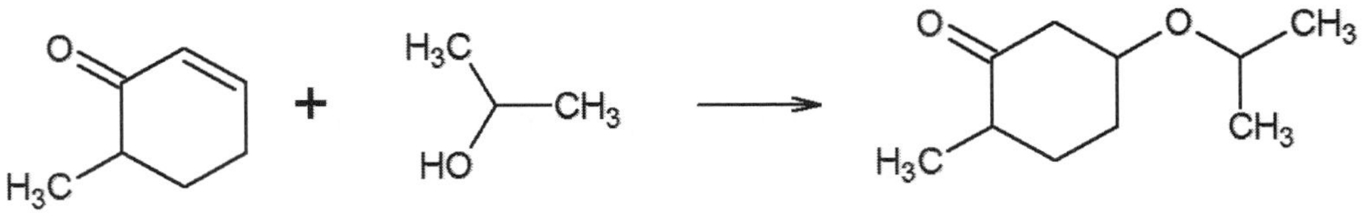

Exercise 23.17—Oxidation reactions

(a)

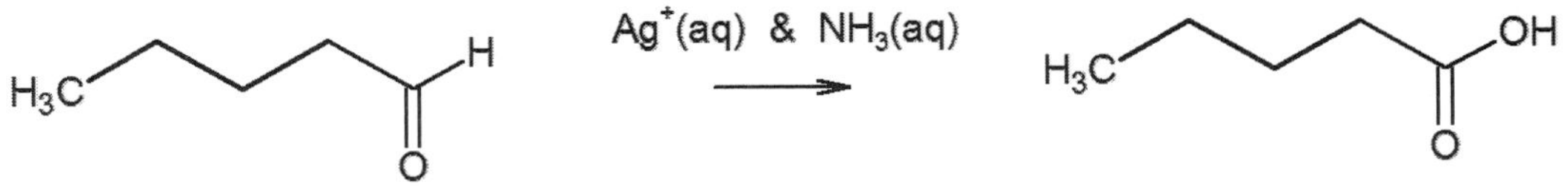

(b)

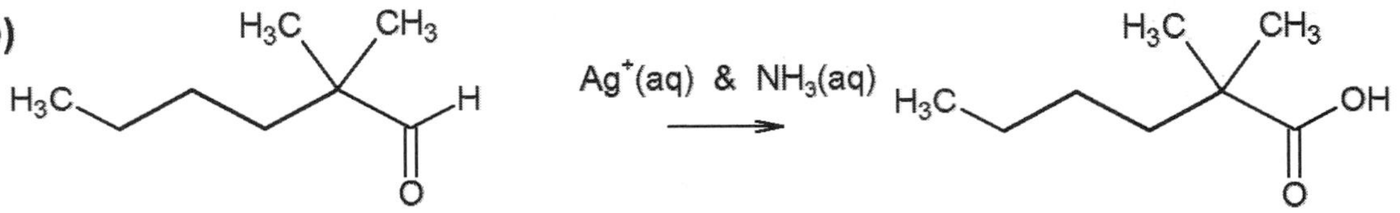

(c)

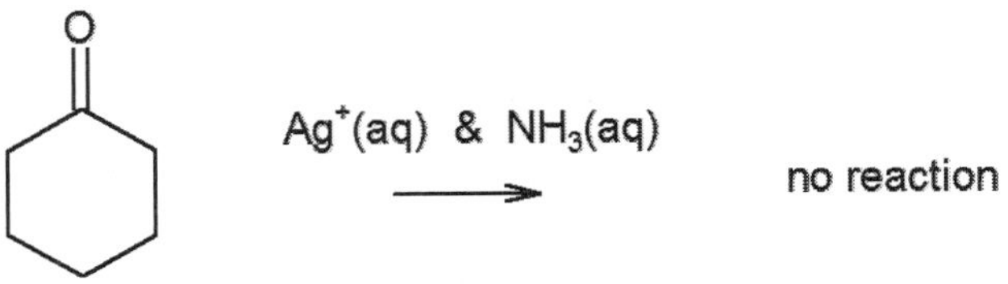

REVIEW QUESTIONS

23.1 Making scents of the mountain pine beetle

23.19
Functionalities exhibited by pheromones include alkyl chains and rings, alcohol, aldehyde, ester—especially acetate, alkene, ketone, and ether—especially epoxide—groups.

23.21
Carbonyl functional groups shown:
(a) amide (specifically, acetamide)
(b) ester (specifically, acetate)
(c) aldehyde & carboxylic acid

23.23

(a) 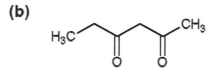 **(b)**

(c) **(d)**

23.2 Naming aldehydes and ketones

23.25

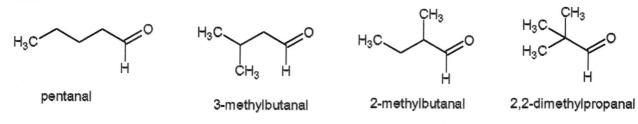

pentanal

3-methylbutanal

2-methylbutanal

2,2-dimethylpropanal

pentan-2-one

3-methylbutan-2-one

pentan-3-one

23.27

(a) 3-methylcyclohex-3-enone
(b) (R)-2,3-dihydroxypropanal
(c) 5-isopropyl-2-methylcyclohex-2-enone
(d) 2-methylpentan-3-one
(e) 3-hydroxybutanal
(f) 1,4-benzenedicarbaldehyde

23.3 Synthesis and spectroscopy

23.29

(a)

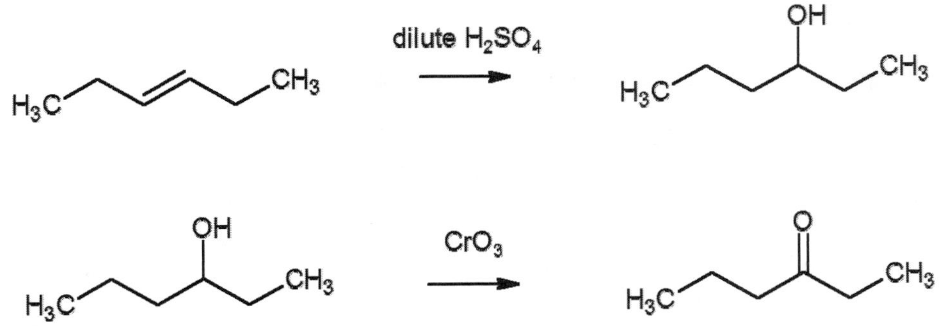

(b)

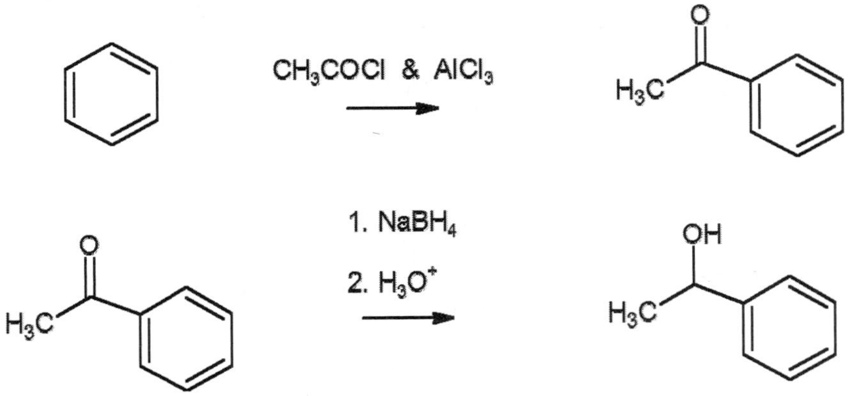

23.31

Here, the IR absorption for the C=O stretching vibration is a little lower, typical of a ketone. The ^{1}H NMR spectrum shows no aldehyde H atom peak—suggesting a ketone. The two downfield alkyl H peaks (2.4 and 2.1 ppm) are characteristic of a methine (integrates to 1) H and three methyl H's on either side of the carbonyl group. The doublet at 1.2 ppm integrating to 6 looks like two equivalent methyl groups split by the downfield methine H (it appears as a septet—i.e., split by 6 equivalent H's).

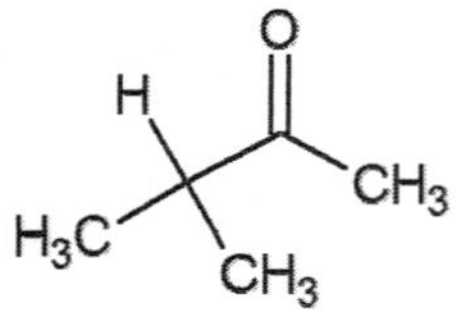

B

23.4–23.10 Reactions of aldehydes and ketones

23.33

(a)

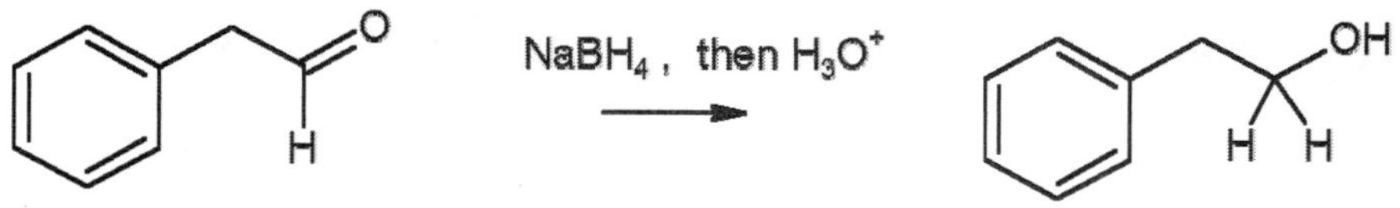

NaBH$_4$, then H$_3$O$^+$

(b)

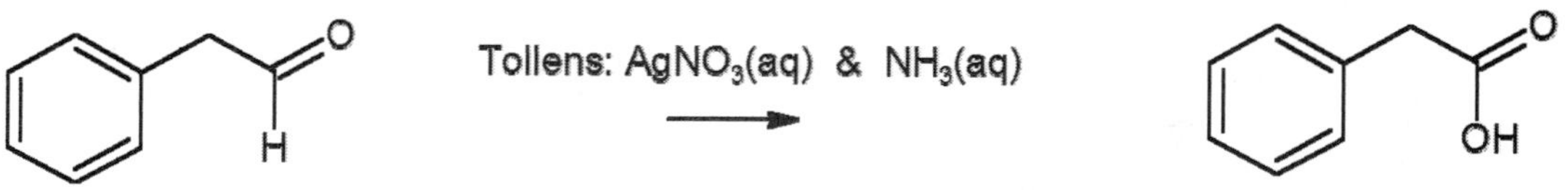

Tollens: AgNO$_3$(aq) & NH$_3$(aq)

(c)

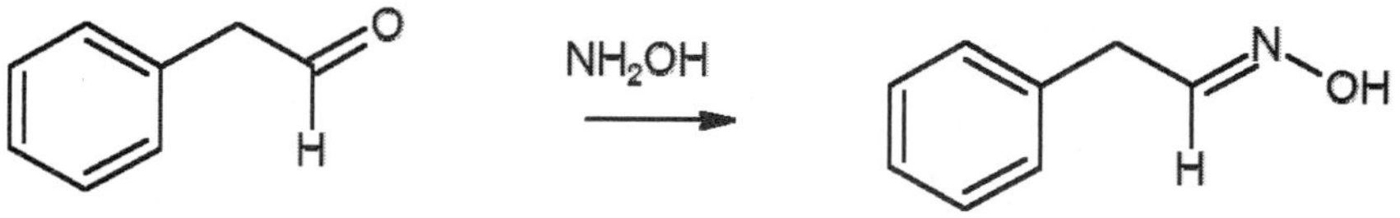

NH$_2$OH

(d)

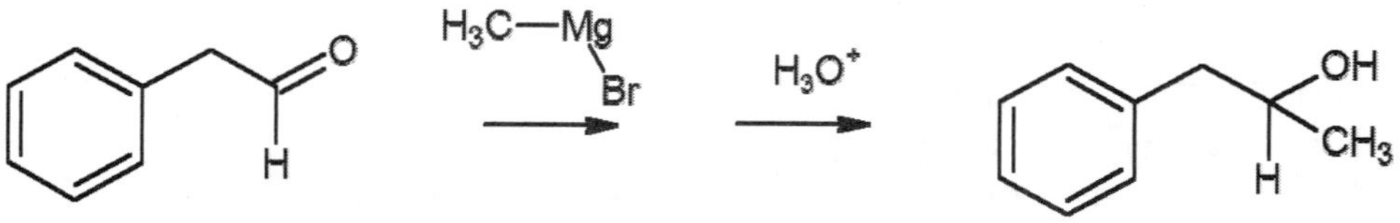

H$_3$C—Mg–Br   H$_3$O$^+$

(e)

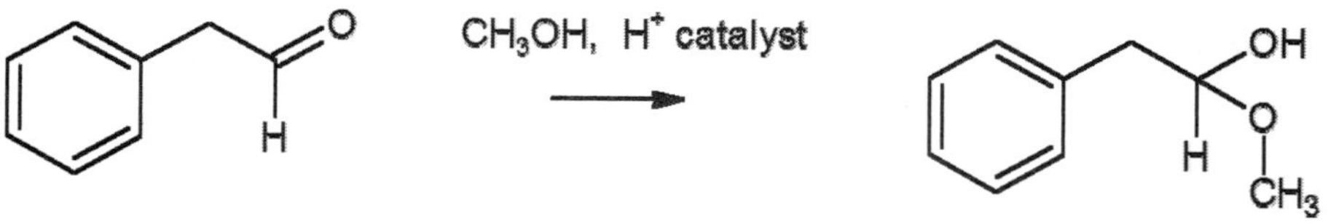

CH$_3$OH, H$^+$ catalyst

23.35

(a)

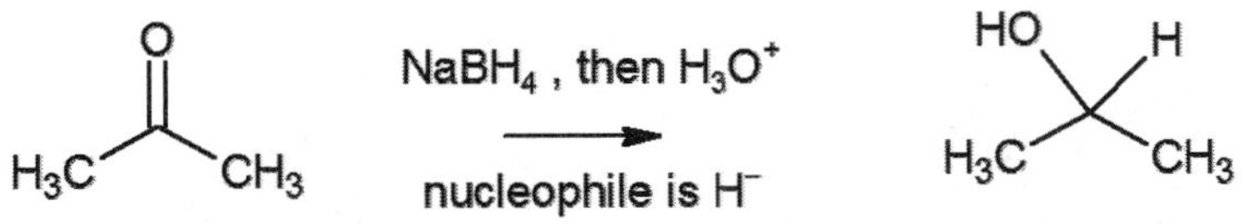
O
H3C
CH3
NaBH4 , then H3O+
nucleophile is H−
HO
H
H3C
CH3

(b)

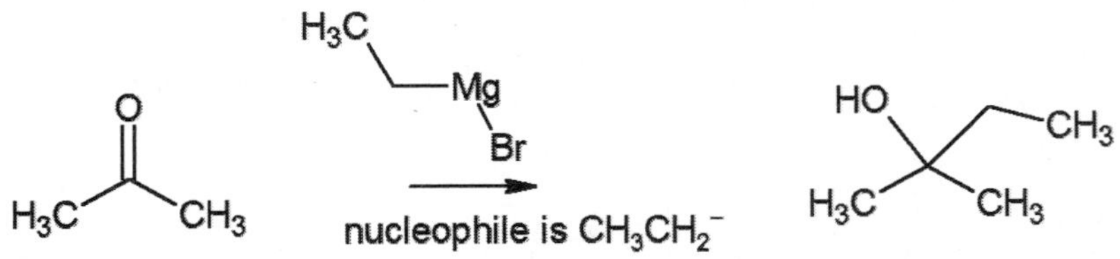
O
H3C
CH3
H3C
Mg
Br
nucleophile is CH3CH2−
HO
CH3
H3C
CH3

(c)

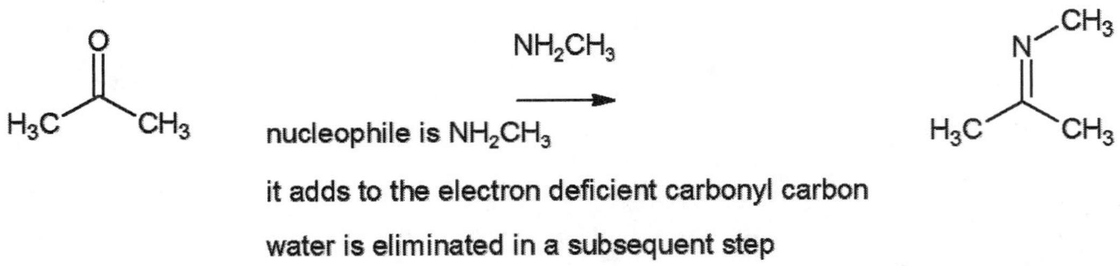
O
H3C
CH3
NH2CH3
nucleophile is NH2CH3
it adds to the electron deficient carbonyl carbon
water is eliminated in a subsequent step
CH3
N
H3C
CH3

(d)

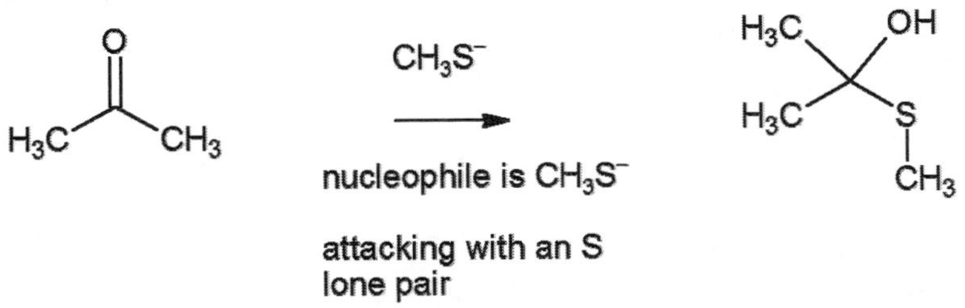
O
H3C
CH3
CH3S−
nucleophile is CH3S−
attacking with an S
lone pair
H3C
OH
H3C
S
CH3

23.37

The reaction of phenylmagnesium bromide with butan-2-one produces a racemate of (*R*) and (*S*) 2-phenylbutan-2-ol.

23.7 Addition of alcohols in nature and the laboratory

23.39

(a)

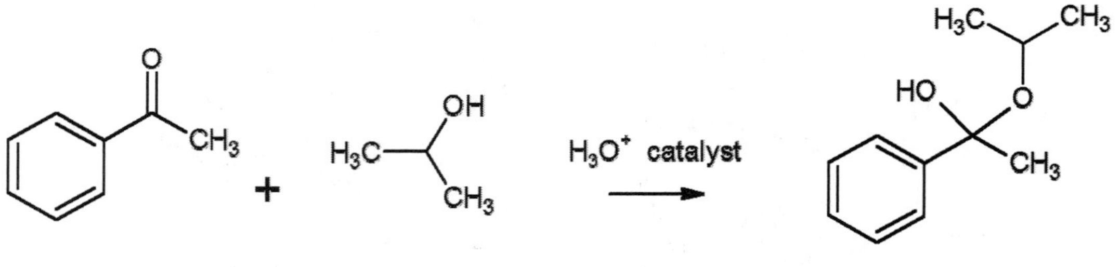

a hemiacetal

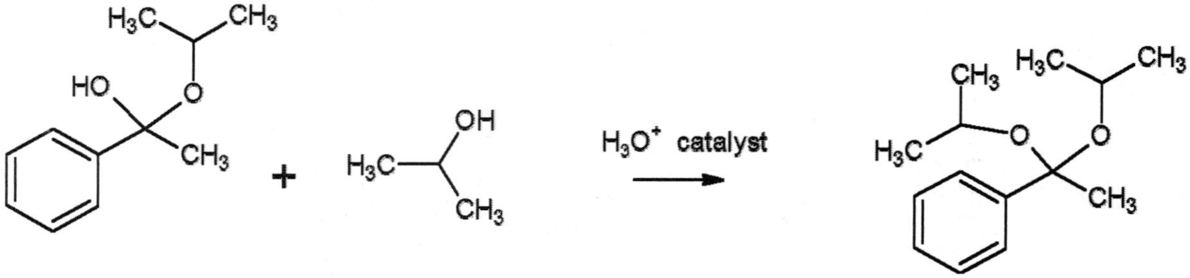

(b)

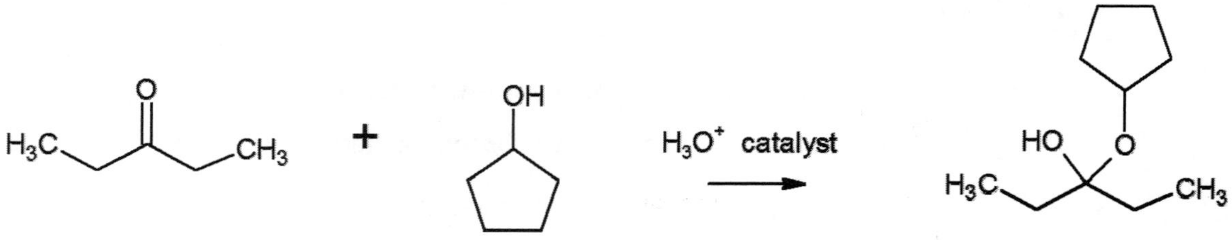

a hemiacetal

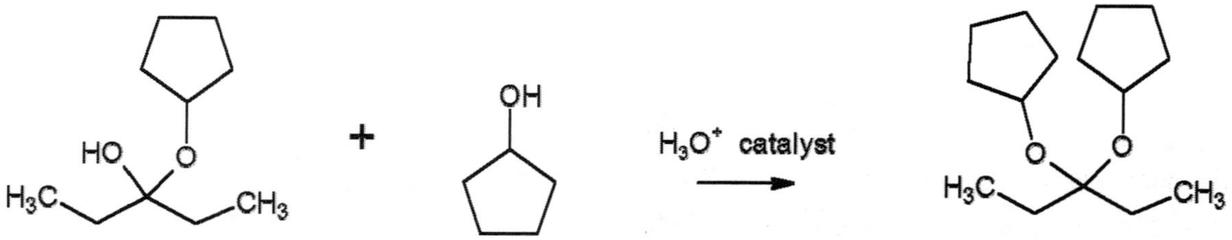

an acetal

23.8 Addition of amines to form imines

23.41

add another ketone with
the same steps

23.10 Addition of Grignard reagents

23.43

(a)

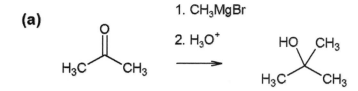

H3C⟍(C=O)⟋CH3

1. CH3MgBr

2. H3O+

⟶

HO CH3
H3C⟍(C)⟋CH3

(b)

1. CH3MgBr

2. H3O+

⟶

HO CH3

(c)

1. CH3MgBr

2. H3O+

H3C⟍CH2⟋(C=O)⟍CH3

⟶

HO CH3
H3C⟍(C)⟋CH3

23.45

(a)

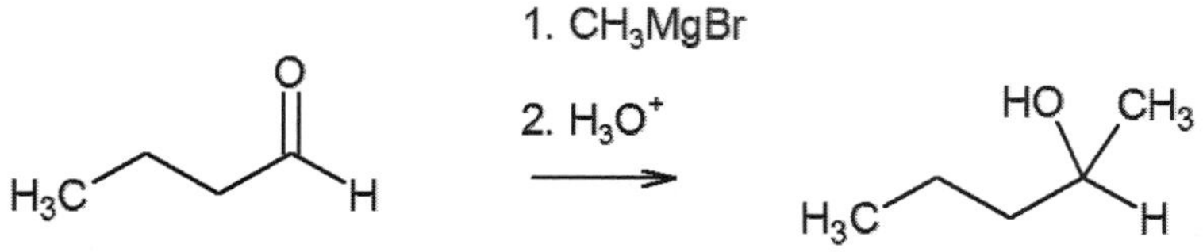

or

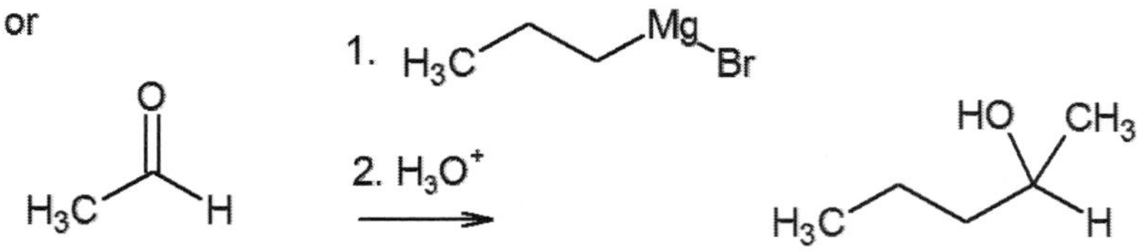

(b)

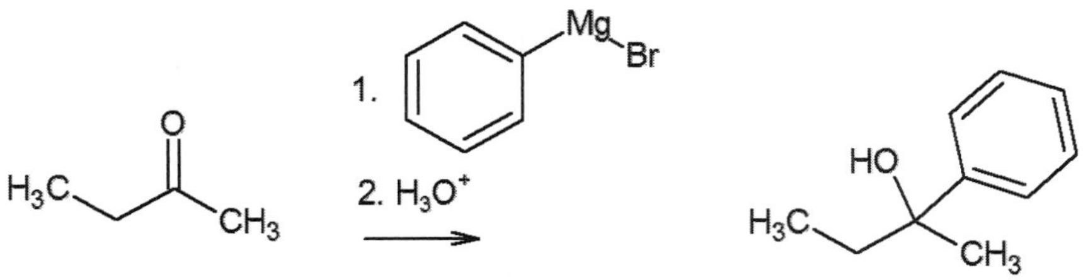

(c)

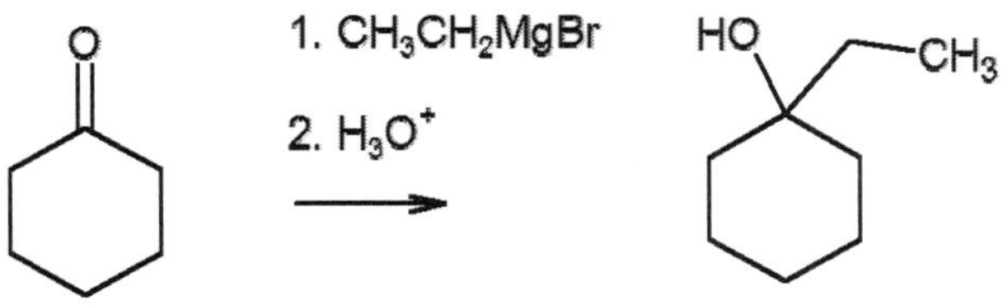

(d)

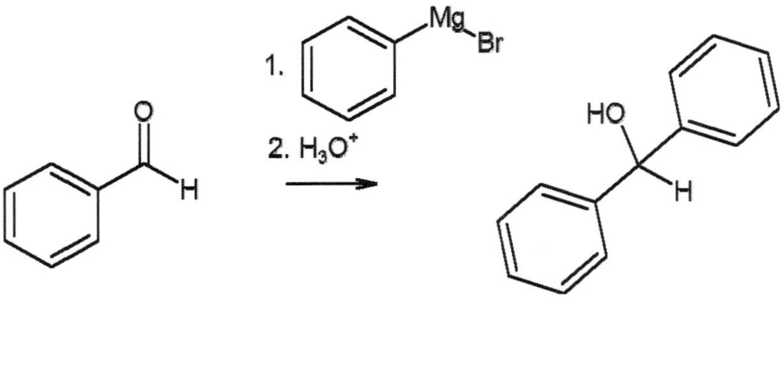

23.47

(a)

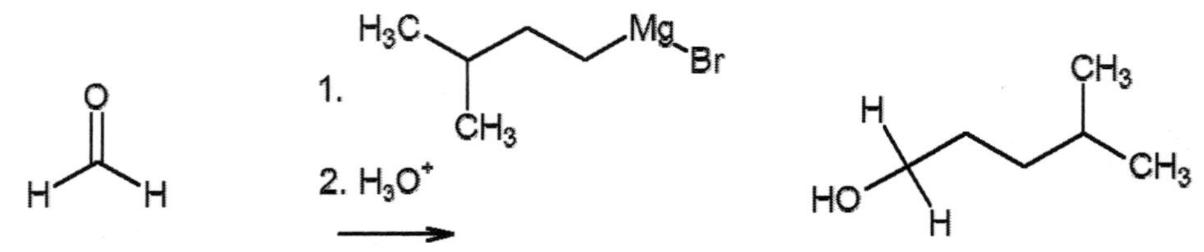
O
H
H
1.
H3C
Mg
Br
CH3
2. H3O+
CH3
H
CH3
HO
H

(b)

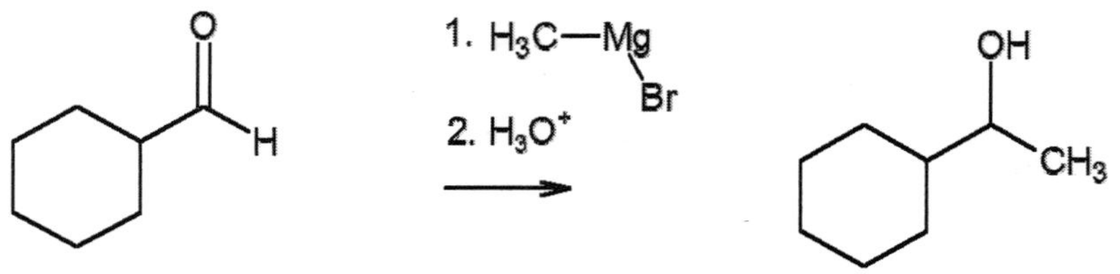
O
H
1. H3C—Mg
Br
2. H3O+
OH
CH3

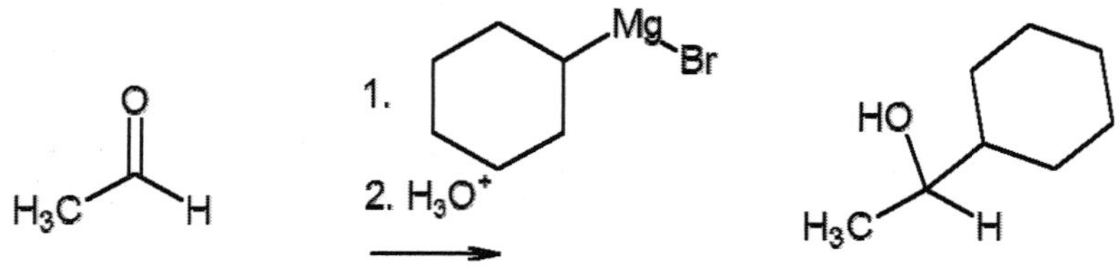
O
H3C
H
1.
Mg
Br
2. H3O+
HO
H3C
H

(c)

this choice leads to an additional aldehyde product
which could then react with the Grignard reagent to
form yet additional products

23.11 Conjugate addition reactions

23.49

(a)

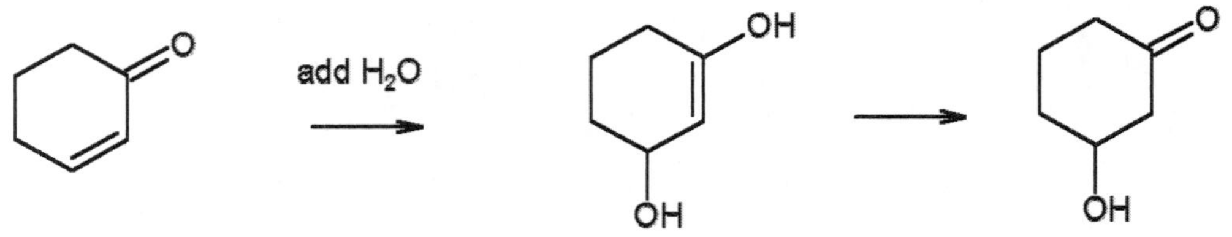

(b)

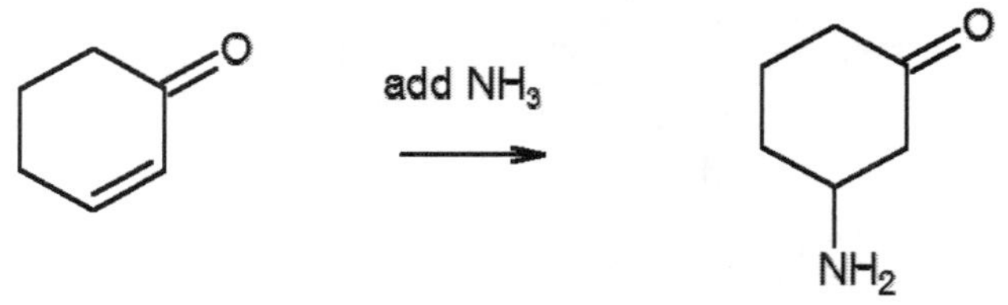

(c)

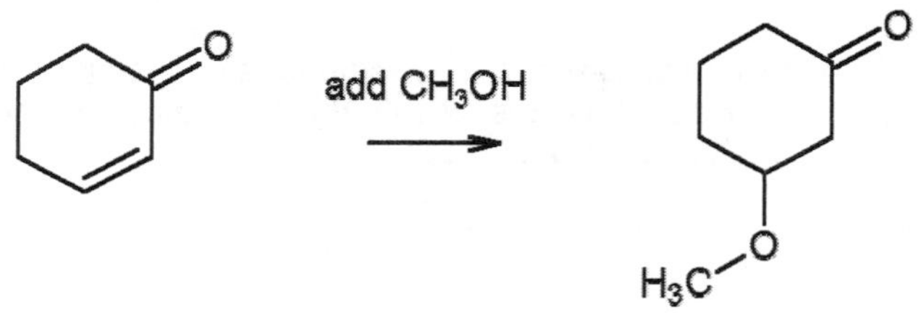

(d)

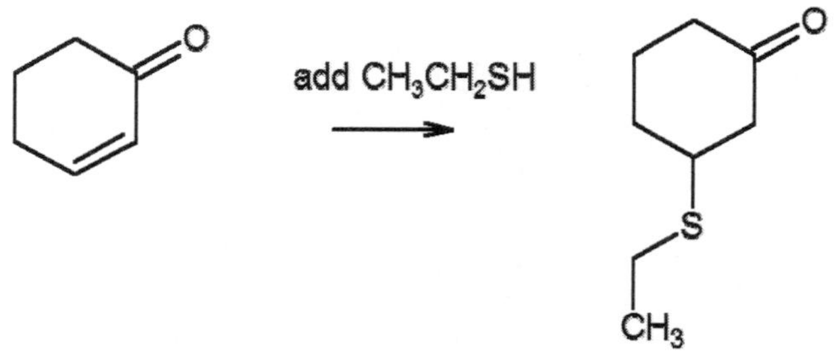

SUMMARY AND CONCEPTUAL QUESTIONS

23.51

(a)

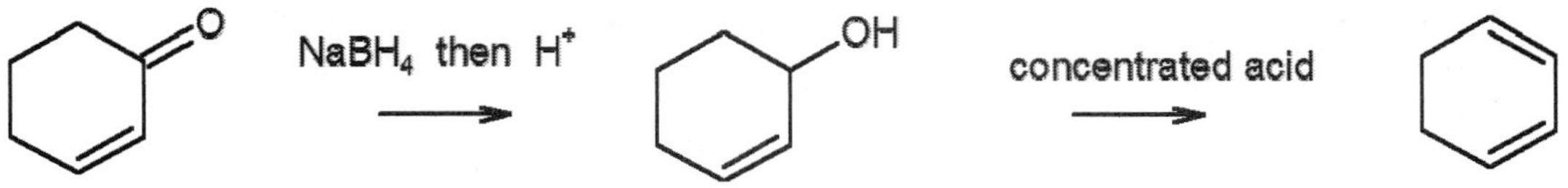

(b)

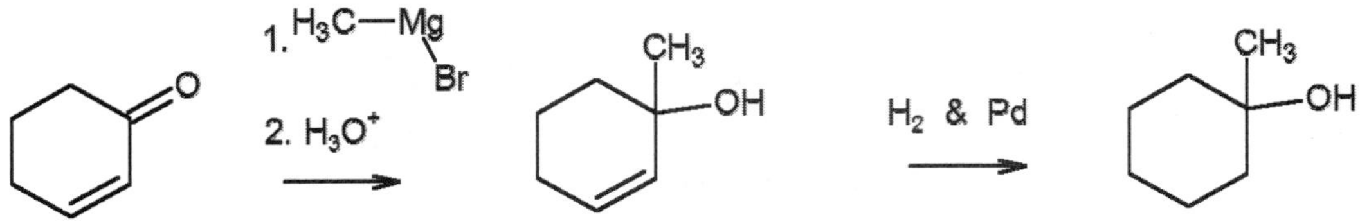

(c)

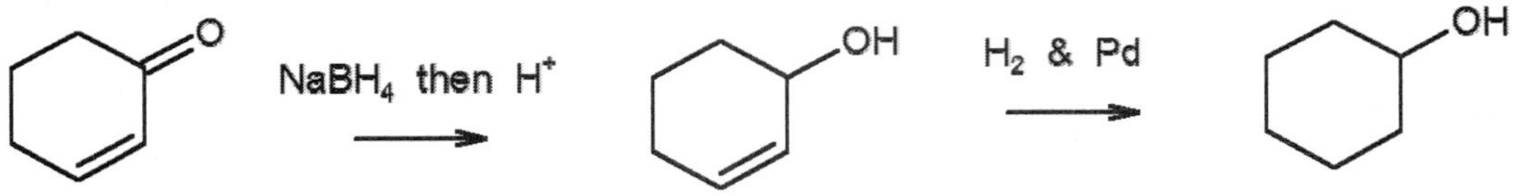

(d)

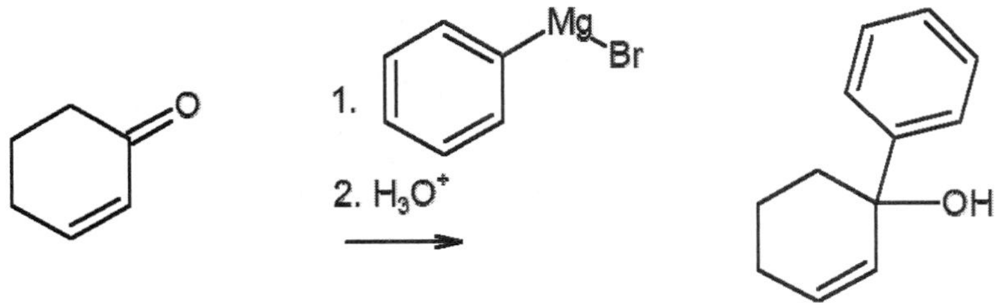

23.53

(a)

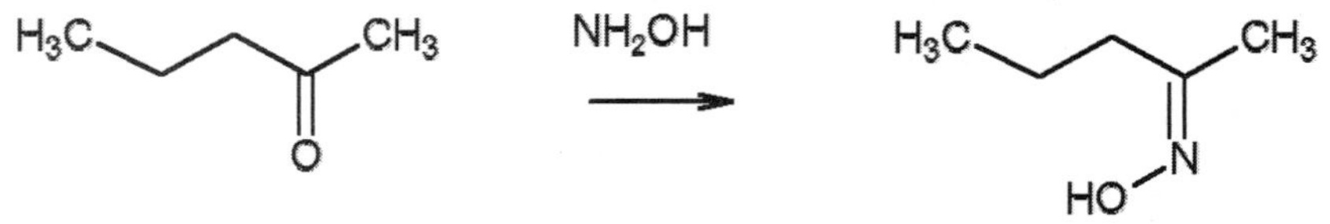

(b)

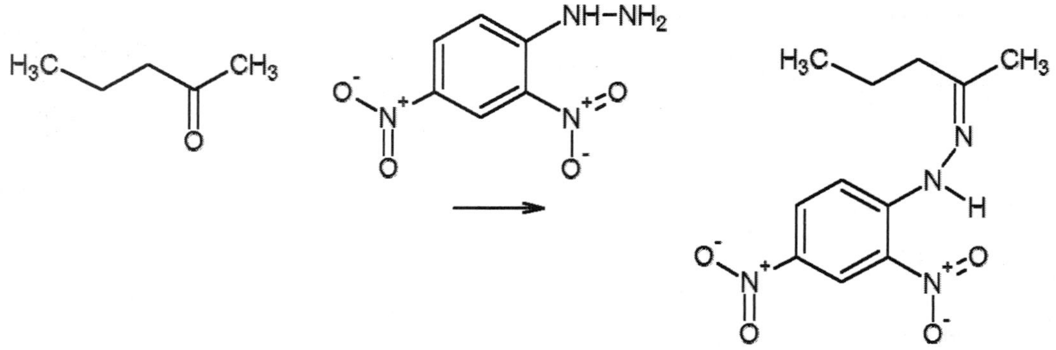

(c)

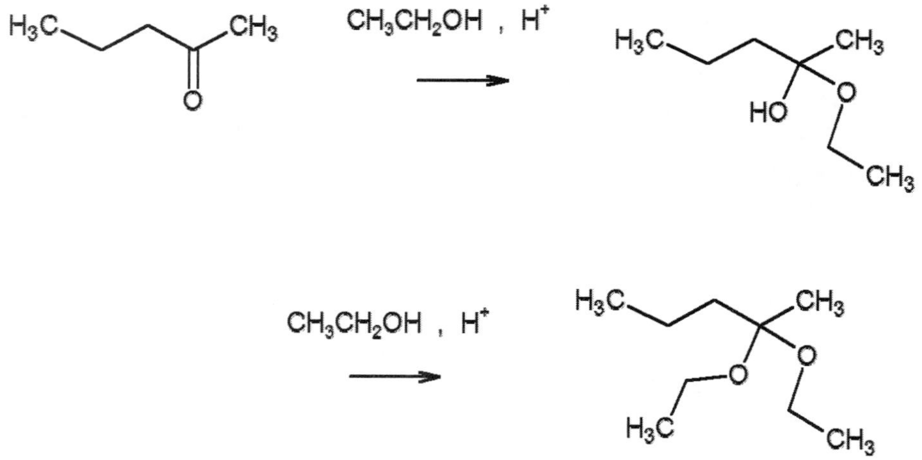

23.55
Thioacetal formation is analogous to acetal formation. In this case, it is thiols adding to a ketone rather than alcohols adding to a ketone.

23.57

(a)

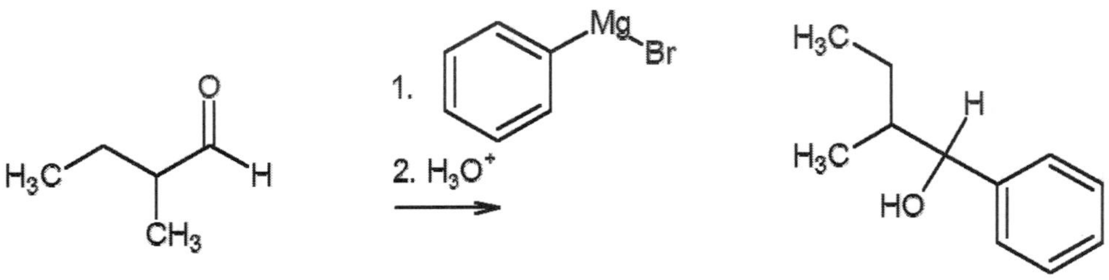
H3C
O
H
CH3
Mg
Br
1.
2. H3O+
H3C
H
H3C
HO

(b)

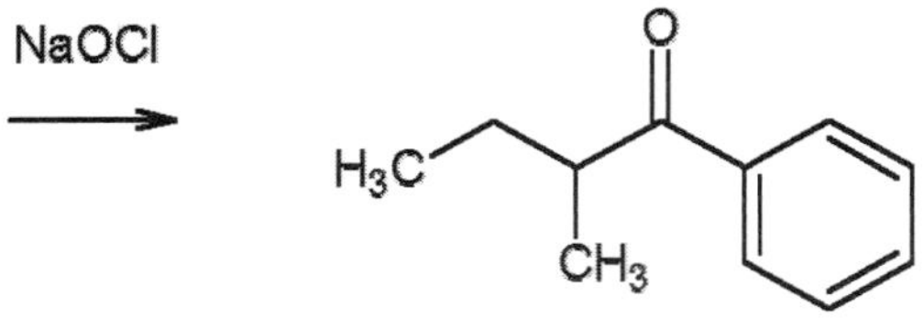
NaOCl
O
H3C
CH3

(c)

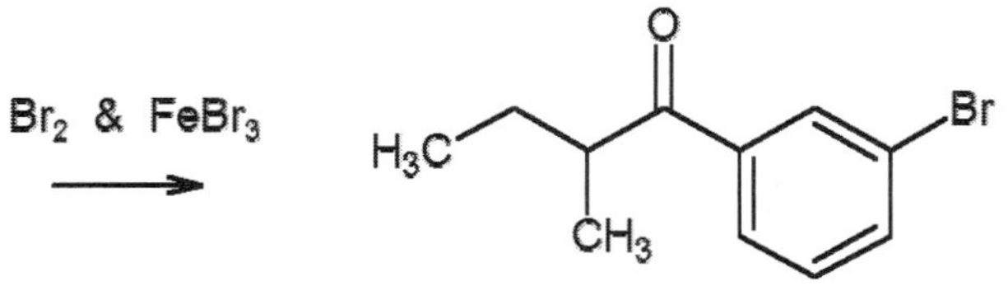
Br2 & FeBr3
O
Br
H3C
CH3

(d)

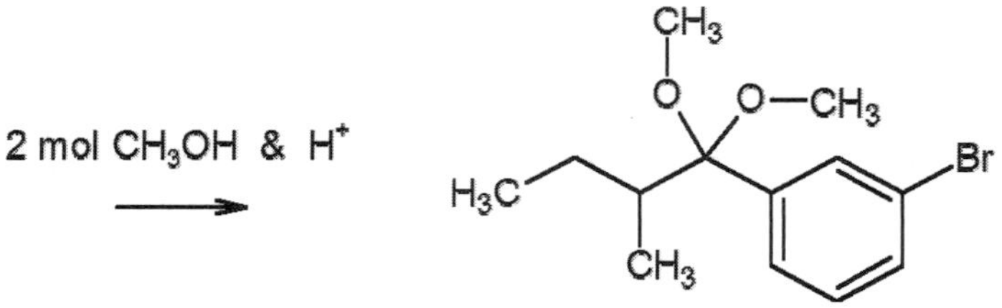
2 mol CH3OH & H+
CH3
O
O—CH3
Br
H3C
CH3

23.59

(a)

1,4 addition products

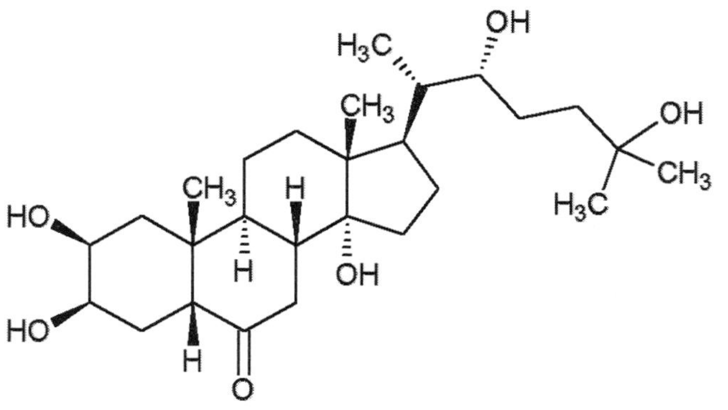

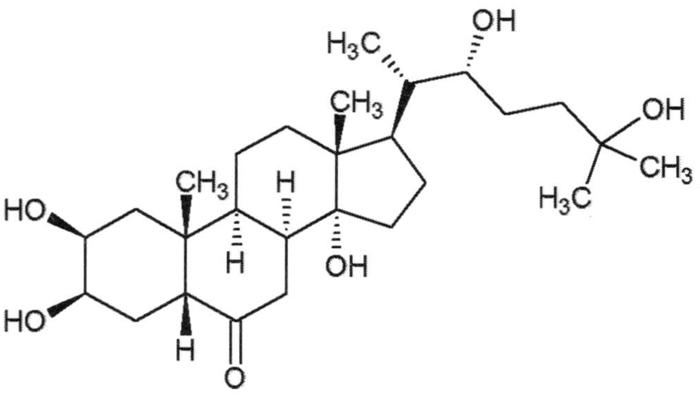

1,2 addition products

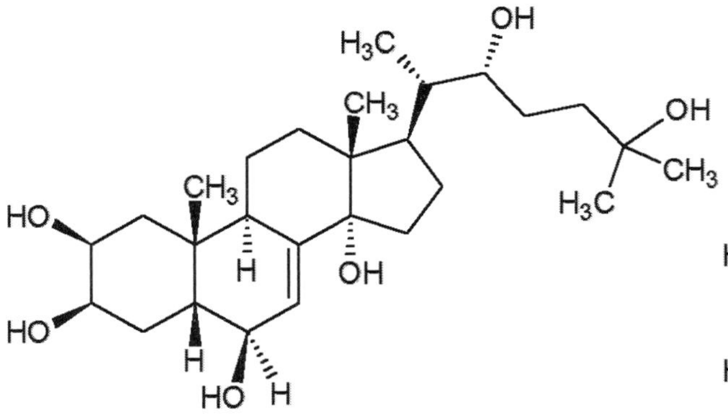

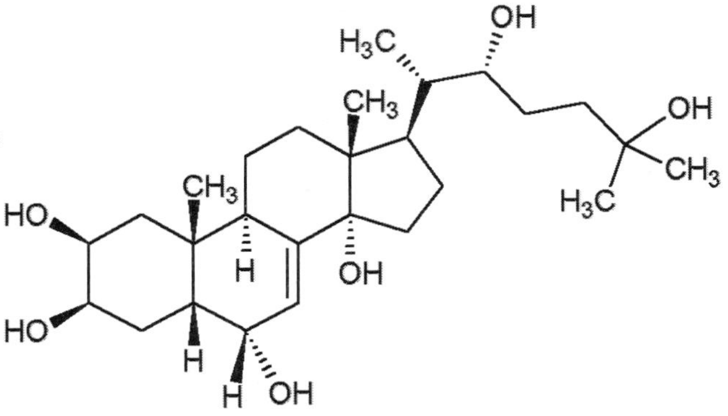

(b) Each of the pairs of stereoisomers produced in 1,2 and 1,4 addition are a pair of diastereomers. This is because a new stereocentre is created, but the other stereocentres on the molecule are unchanged. So the two 1,2 and the two 1,4 addition products do not have a mirror image relationship to each other.

(c) Since diastereomers are formed, we would not expect the 1,4 or 1,2 addition products to yield in equal amounts. Although the addition can take place from below or above the ring structure, the stereochemistry of the substrate at existing stereocentres can affect the rate of these two pathways.

Chapter 24
The Carbonyl Group: Part II

IN-CHAPTER EXERCISES

Exercise 24.2—Naming carboxylic acid derivatives

(a)

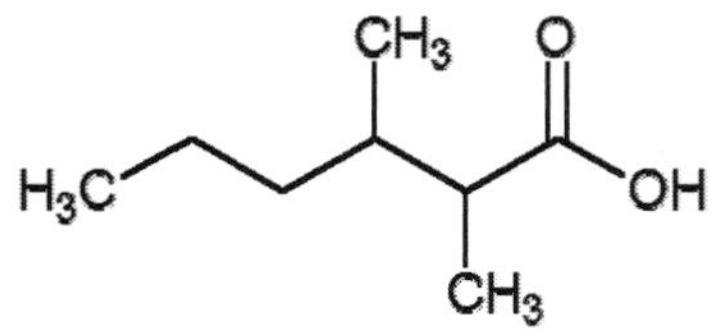

(b)

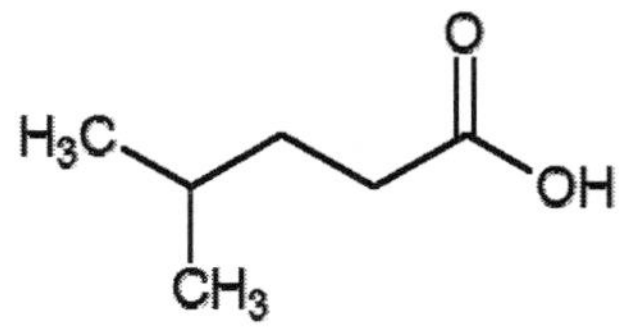

(c)

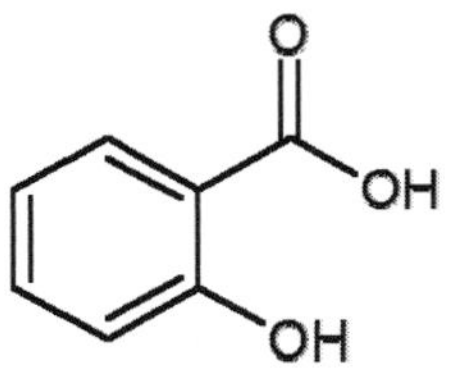

(d)

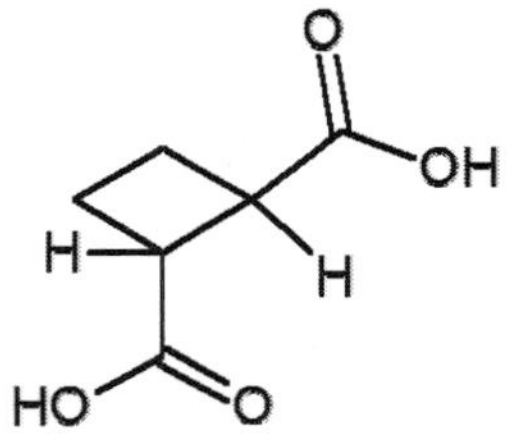

Exercise 24.3—Naming carboxylic acid derivatives

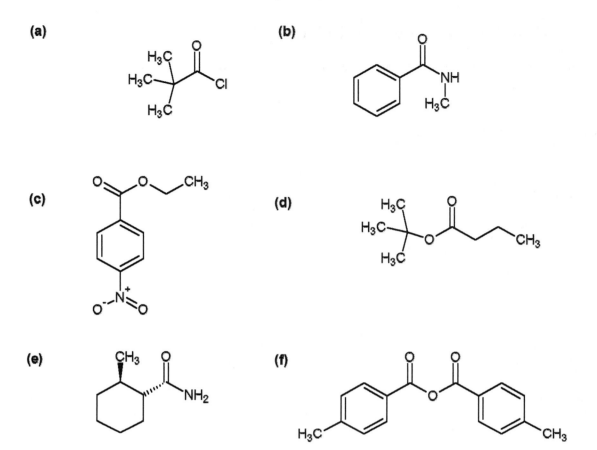

(a)

(b)

(c)

(d)

(e)

(f)

Exercise 24.4—IR spectroscopy

(a) An IR absorption at 1735 cm^{-1} indicates a C=O stretch—most likely a saturated ester.

(b) An absorption at 1810 cm^{-1} is likely the C=O stretching band of an acyl halide.

(c) Absorptions at 2400-3300 cm^{-1} and 1710 cm^{-1} are associated with carboxylic acids. The O–H stretch has a very wide range.

(d) Absorption at 1715 cm^{-1} indicates a ketone C=O stretch (unless a carboxylic acid is suggested by an additional very broad O–H stretch—2400-3300 cm^{-1}).

Exercise 24.5—IR spectroscopy

(a)

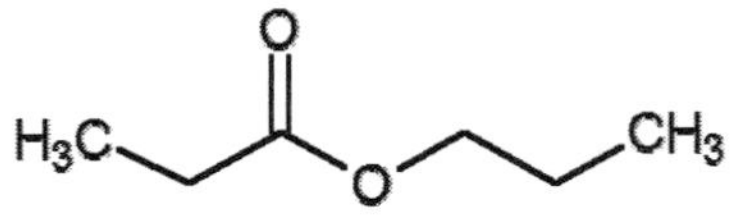

1735 cm^{-1} - an ester C=O stretch

(b)

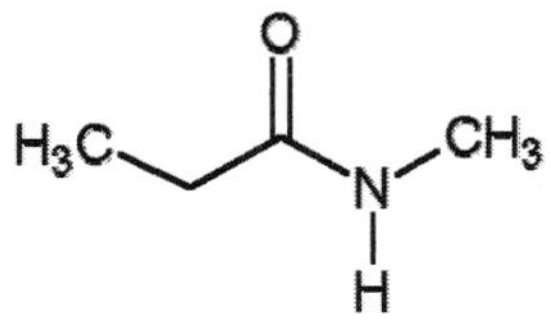

1650 cm^{-1} - an amide C=O stretch

(c)

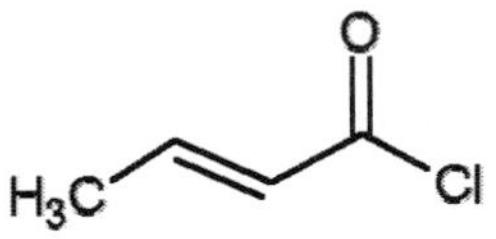

1735 cm^{-1} - an acid chloride C=O stretch

Exercise 24.6—NMR spectroscopy

Isomers cyclopentanecarboxylic acid and 4-hydroxycyclohexanone are possibly distinguished by their ^{13}C NMR spectra via the position of the carbonyl carbon absorption. The carboxylic acid carbonyl carbon will absorb around 180-185 ppm. The ketone carbonyl range is from 180-220 ppm. Thus, an absorption well downfield from 185 ppm indicates the ketone. Also, 4-hydroxycyclohexanone has a hydroxyl functionality and will have a C−O carbon absorption in the 50-90 ppm range. The other alkyl carbons will mostly absorb in the 10-30 ppm range—though the carbon next to the carbonyl (in both molecules) can be as far as 50 ppm downfield. In the ^{1}H NMR spectra, the molecules are easily distinguished by the single peak between 10-12 ppm for the proton of the carboxylic acid. A broad singlet is found for the alcohol O-H of 4-hydroxycyclohexanone, but it typically comes between 2-5 ppm.

Exercise 24.8—Reactions of carboxylic acids

(a)

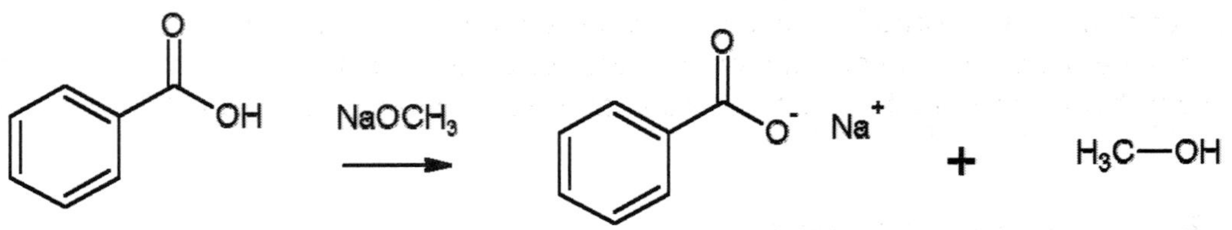

(b)

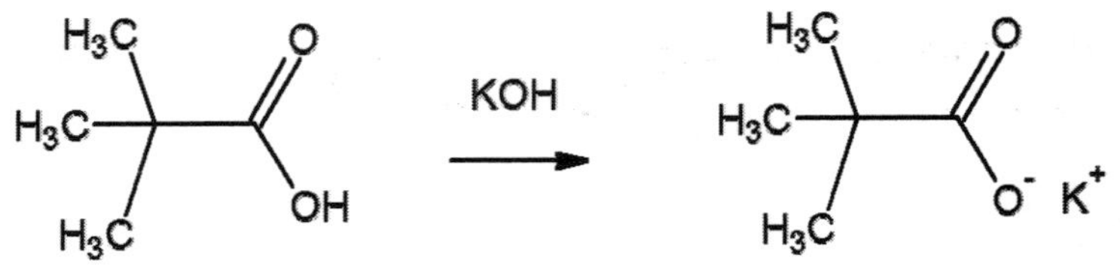

Exercise 24.9—Relative acidity

According to increasing acidity, we have the following ranking:
methanol < phenol < *p*-nitrophenol < acetic acid < sulfuric acid

Exercise 24.10—Relative acidity

According to increasing acidity, we have the following rankings:
(a) $CH_3CH_2CO_2H$ < $BrCH_2CH_2CO_2H$ < $BrCH_2CO_2H$
(b) ethanol < benzoic acid < *p*-cyanobenzoic acid

Example 24.11—Reactivity of carboxylic acid derivatives

With respect to nucleophilic acyl substitution reactivity we have
(a) CH_3COCl > $CH_3CO_2CH_3$ **(b)** $CH_3CH_2CO_2CH_3$ > $(CH_3)_2CHCONH_2$
(c) $CH_3CO_2COCH_3$ > $CH_3CO_2CH_3$ **(d)** $CH_3CO_2CH_3$ > CH_3CHO
Note that acid chlorides and then anhydrides are the most reactive. The carbonyl carbon atom is highly positively polarized in these compounds, and they have good leaving groups—chloride and carboxylate. Esters are more reactive than amides—the amide leaving group is a stronger base than the alkoxide leaving group of an ester. Aldehydes are unreactive with respect to acyl substitution (but not addition) reactions—hydride is a very poor leaving group (it is a very strong base).

Example 24.12—Reactivity of carboxylic acid derivatives

Methyl trifluoroacetate, $CF_3CO_2CH_3$, is more reactive than methyl acetate, $CH_3CO_2CH_3$, with respect to nucleophilic acyl substitution reactions because the fluorine substituents are electron withdrawing, making the carbonyl carbon atom more electron deficient (attracting nucleophiles) and stabilizing the tetrahedral intermediate, which has additional electron density due to the bonded nucleophile.

Exercise 24.13—Reactions of acyl chlorides

(a)

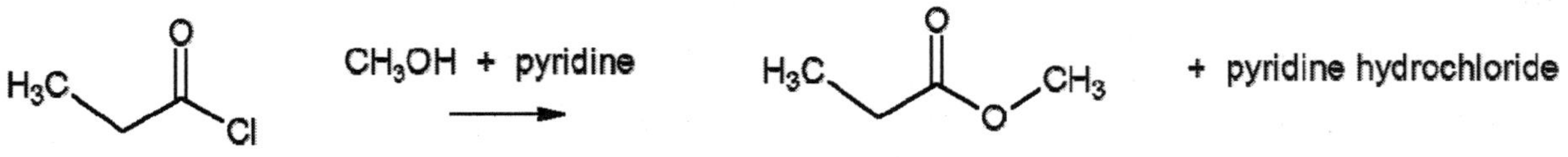

(b)

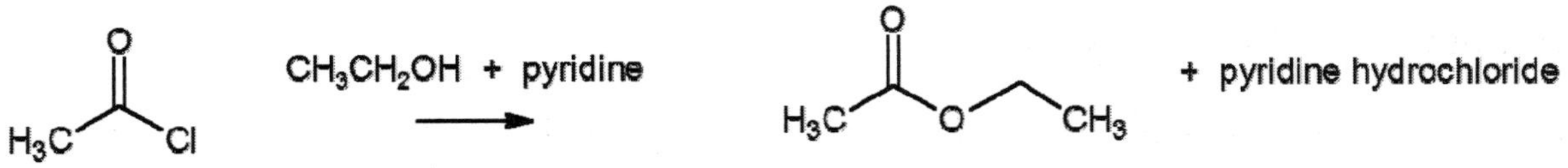

(c)

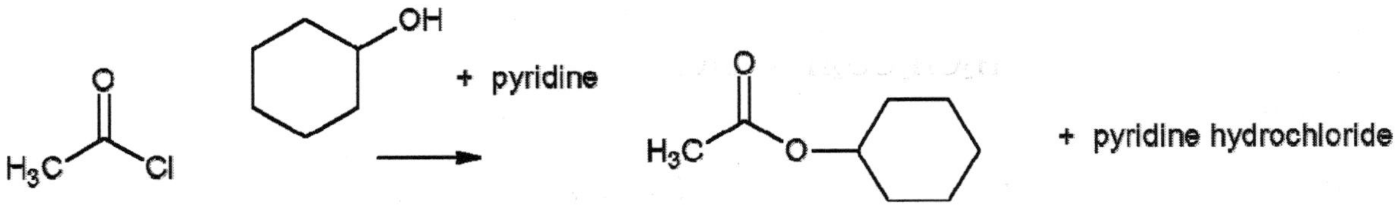

Exercise 24.14—Mechanism for reactions of acyl chlorides

Exercise 24.15—Reactions of acyl chlorides

(a)

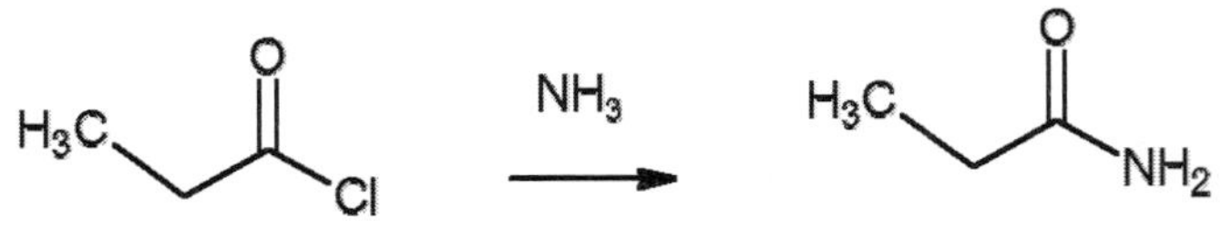

(b)

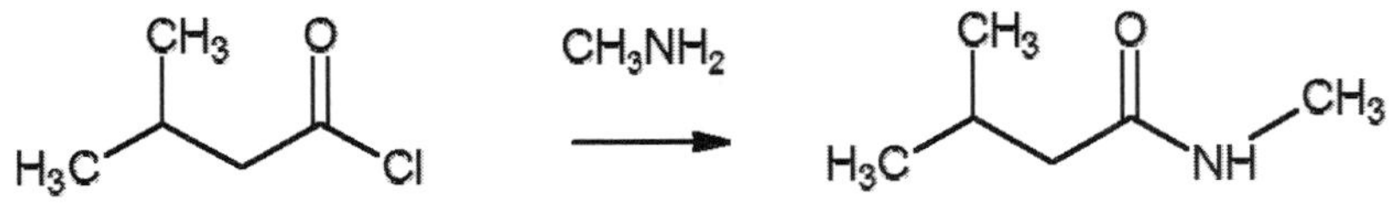

(c)

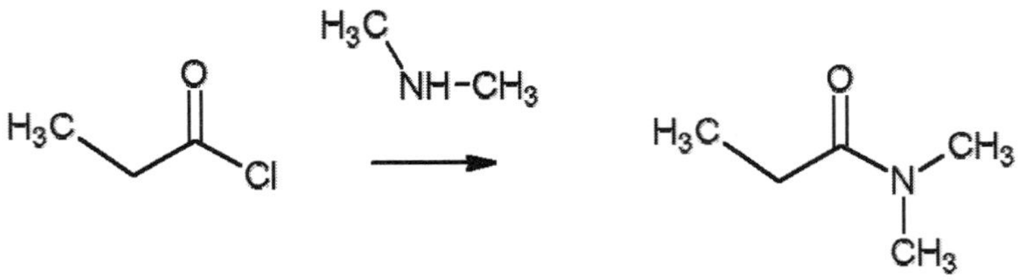

(d)

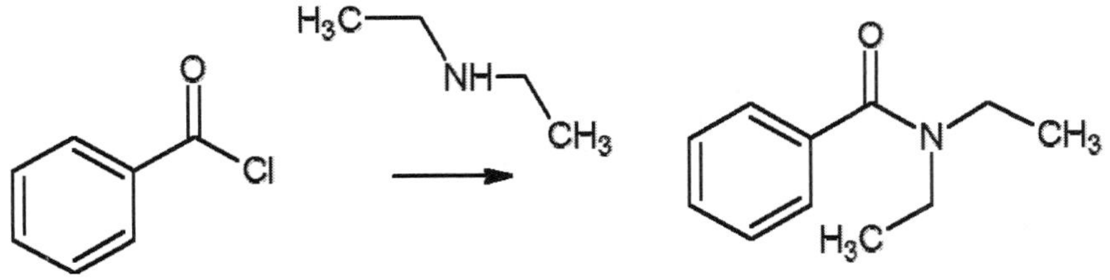

Exercise 24.16—Mechanism for acid anhydride reactions

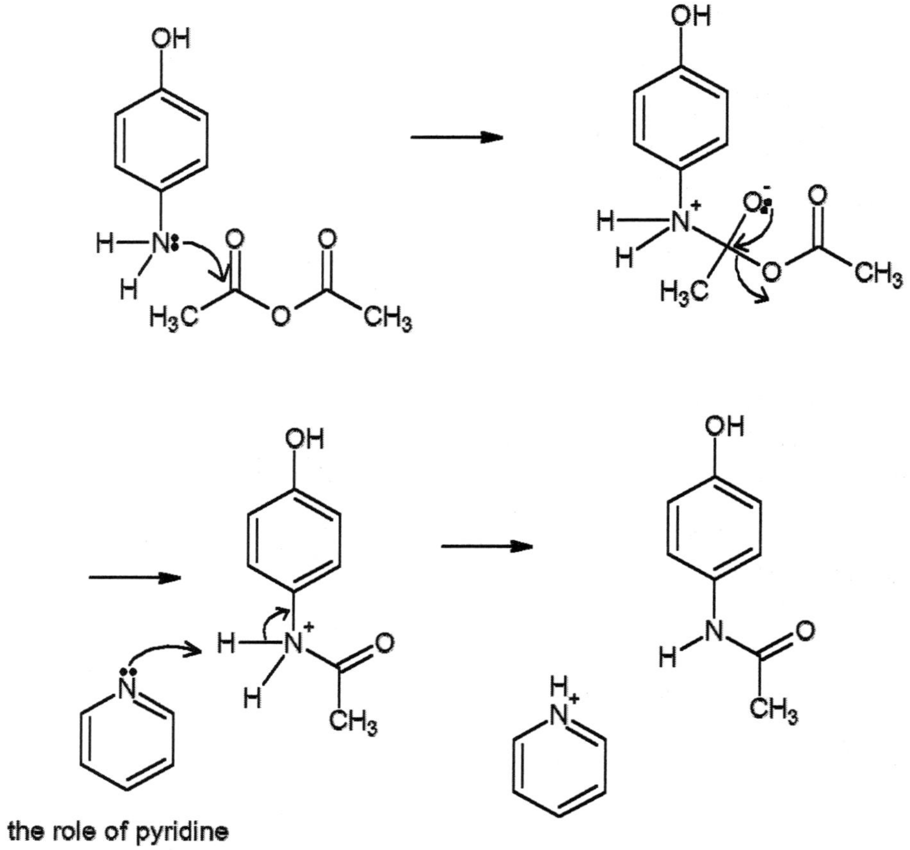

the role of pyridine

Exercise 24.17—Reactions of acid anhydrides

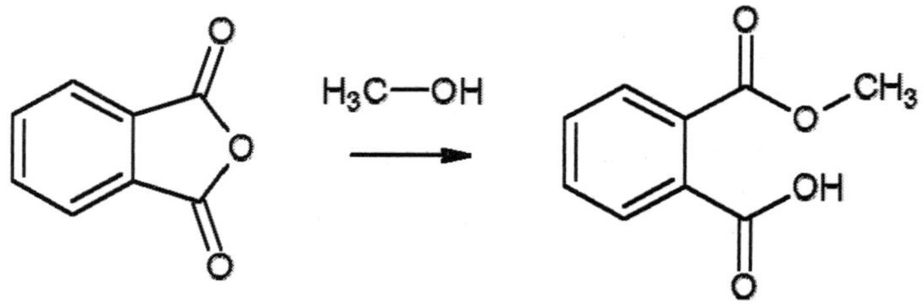

H₃C—OH →

Exercise 24.18—Reactions of esters

(a)

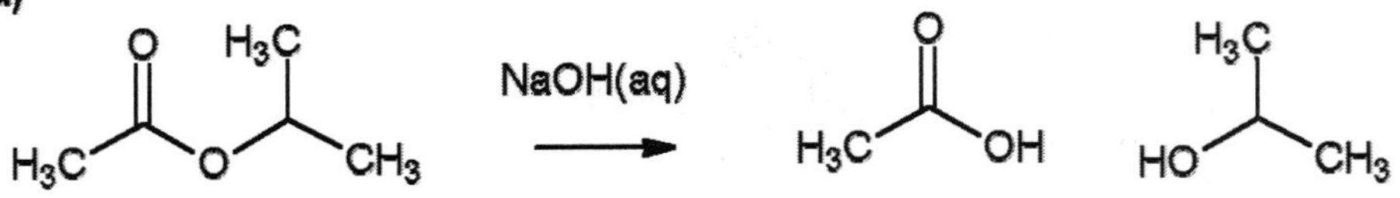

(b)

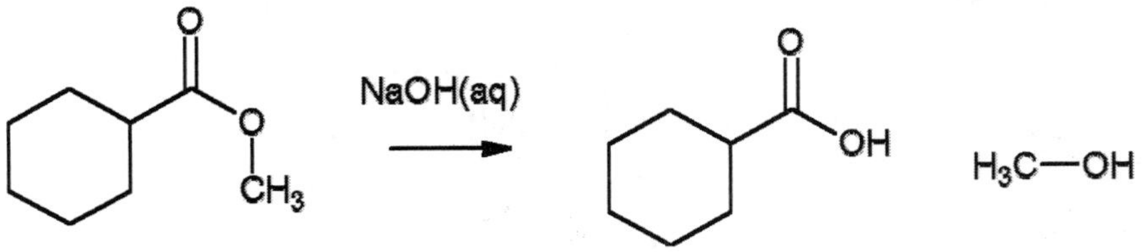

Exercise 24.19—Reactions of esters

(a)

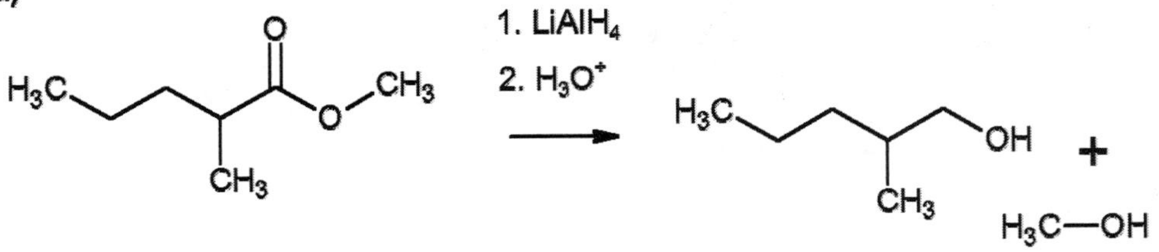

(b)

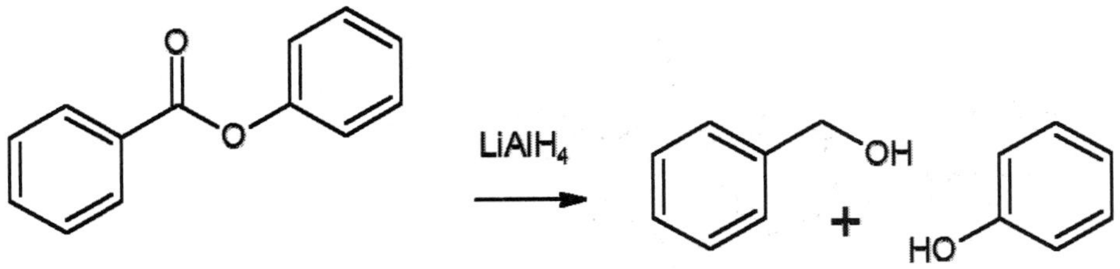

Exercise 24.20—Grignard reactions of esters

(a)

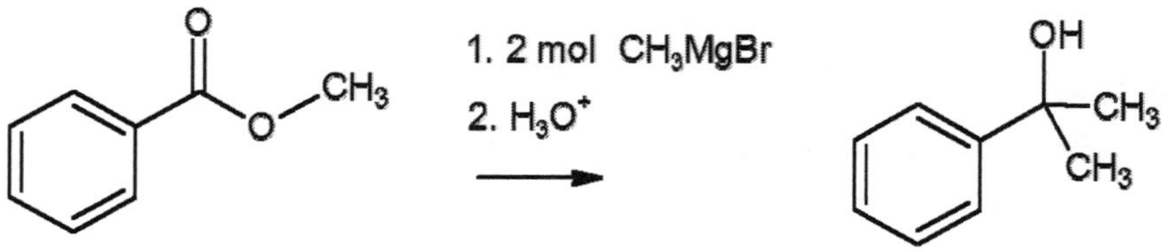

1. 2 mol CH$_3$MgBr
2. H$_3$O$^+$

(b)

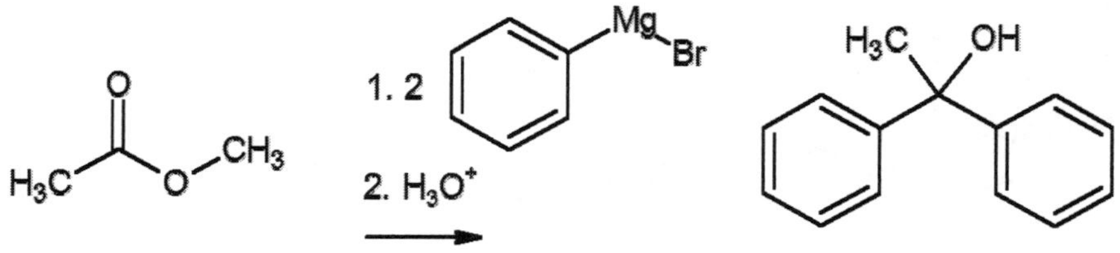

1. 2
2. H$_3$O$^+$

(c)

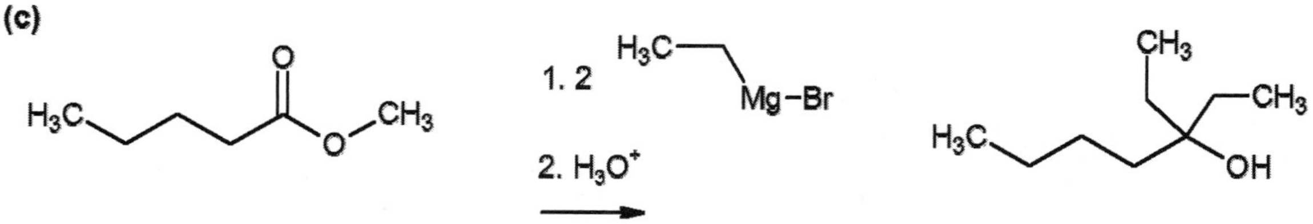

1. 2
2. H$_3$O$^+$

Exercise 24.21—Reactions of amides

(a)

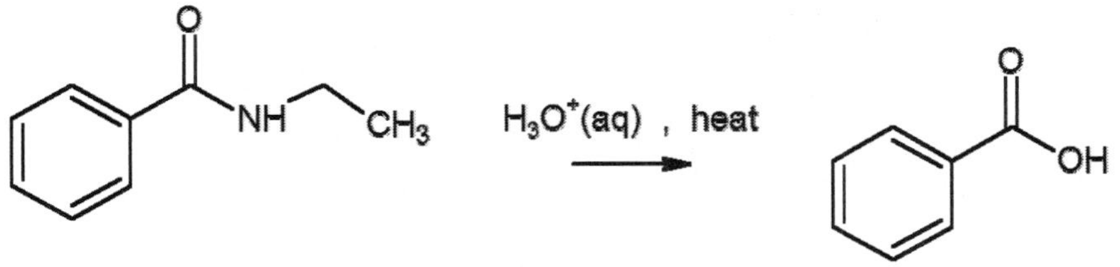
O
NH
CH3
H3O+(aq) , heat
O
OH

(b)

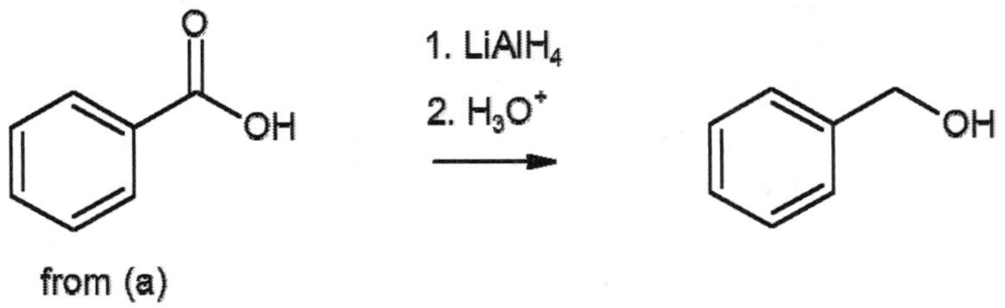
O
OH
1. LiAlH4
2. H3O+
OH
from (a)

(c)

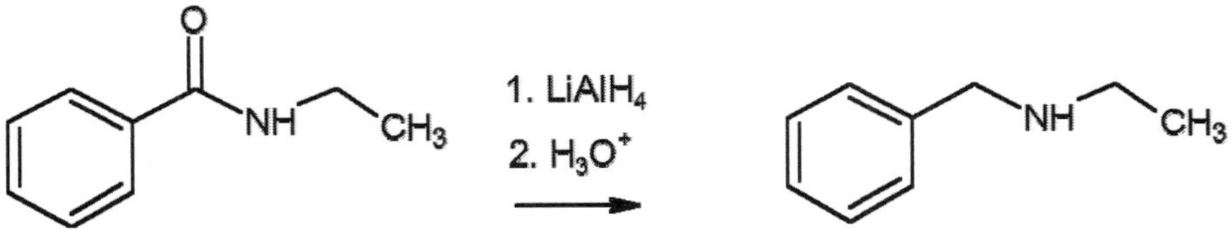
O
NH
CH3
1. LiAlH4
2. H3O+
NH
CH3

Exercise 24.22—Polyamides

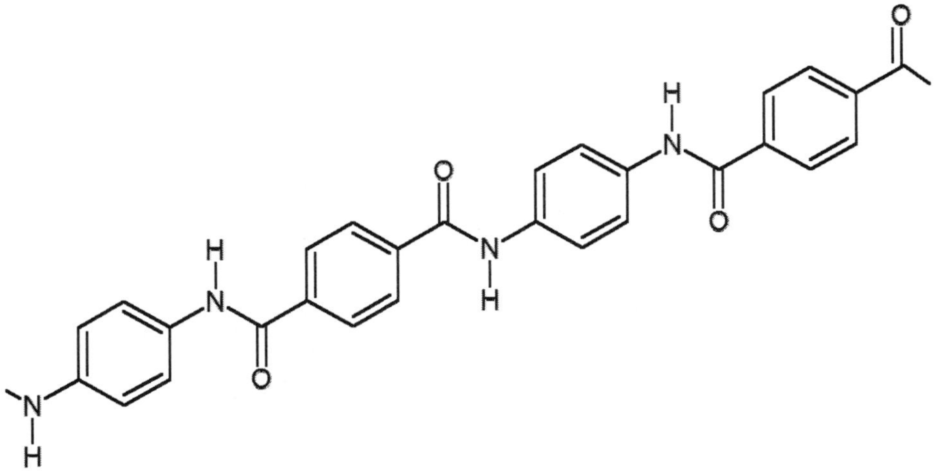
O
H
N
O
N
H
H
N
O
N
H

Kevlar

Exercise 24.23—Acidic hydrogen atoms

H atoms α to one carbonyl group are weakly acidic, because the electron pair on the carbon atom of the enolate ion that is formed can be delocalized onto the adjacent carbonyl oxygen atom. Those that are α to two carbonyl groups are much more acidic. This can be shown by drawing resonance structures.

(a) CH_3CH_2CHO has two weakly acidic hydrogen atoms on the α carbon atom (next to the carbonyl carbon atom).
(b) The hydrogen atoms on the methyl group α to the carbonyl carbon of $(CH_3)_3CCOCH_3$ are weakly acidic.
(c) The H bonded to O is the most acidic hydrogen atom in acetic acid, CH_3CO_2H. The hydrogen atoms on the methyl group α to the carbonyl carbon atom are weakly acidic, and less acidic than the O-H proton.
(d) The methylene hydrogen atoms between the two carbonyl C atoms are the most acidic hydrogen atoms in cyclohexane-1,3-dione because the electron pair on the α carbon atom of the enolate ion can be delocalized onto two adjacent carbonyl oxygen atoms. The other two methylene hydrogen atoms that are α to a single carbonyl group are weaker acids.

Exercise 24.24—Enolate anions

(a)

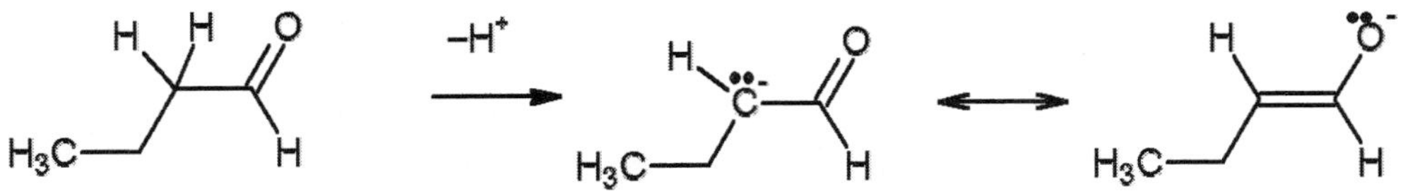

(b)

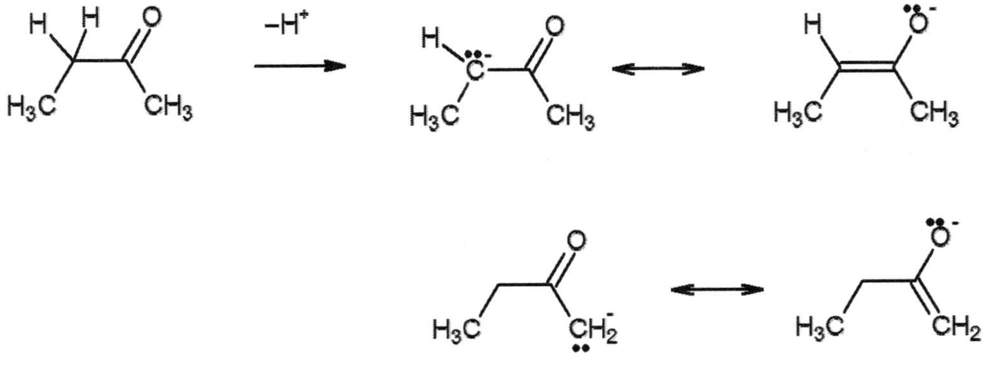

(c)

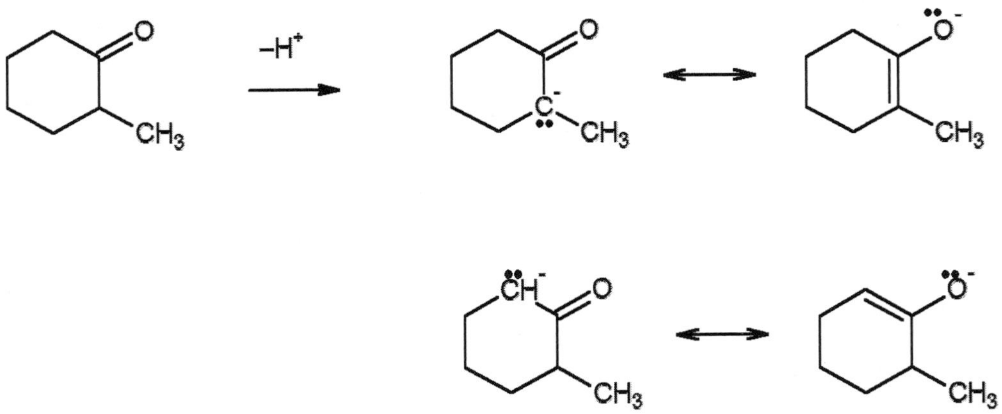

Exercise 24.25—Enolate anions

Exercise 24.26—Malonic ester synthesis

(a)

(b)

(c)

REVIEW QUESTIONS

24.2 Naming carboxylic acids and derivatives

24.27

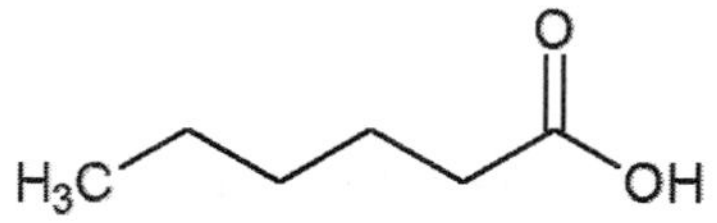

hexanoic acid

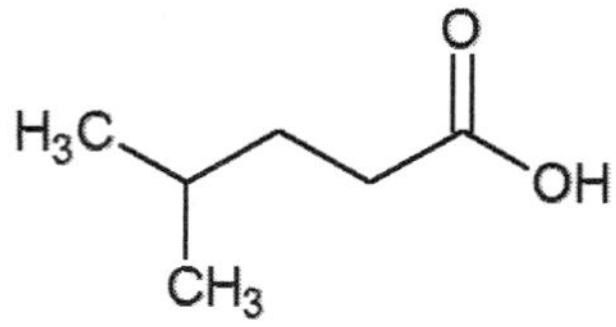

4-methylpentanoic acid

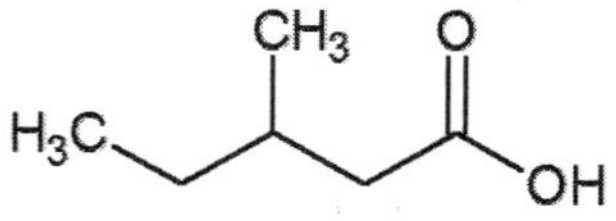

3-methylpentanoic acid

chiral at C3

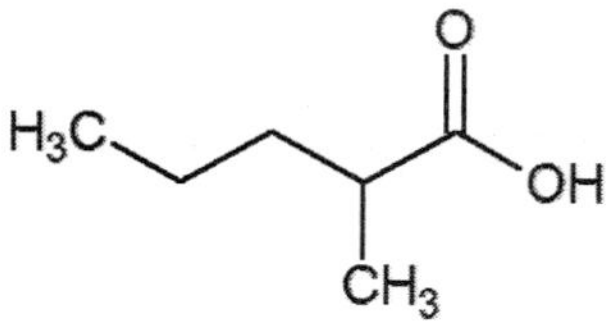

2-methylpentanoic acid

chiral at C2

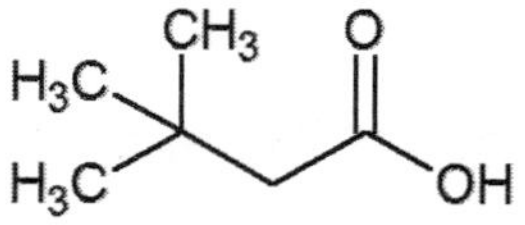

3,3-dimethylbutanoic acid

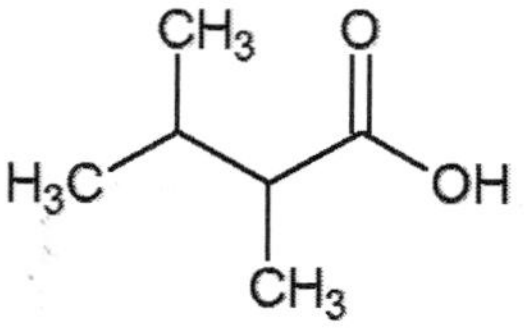

2,3-dimethylbutanoic acid

chiral at C2

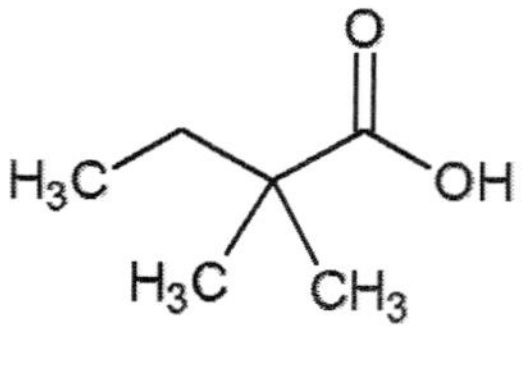

2,2-dimethylbutanoic acid

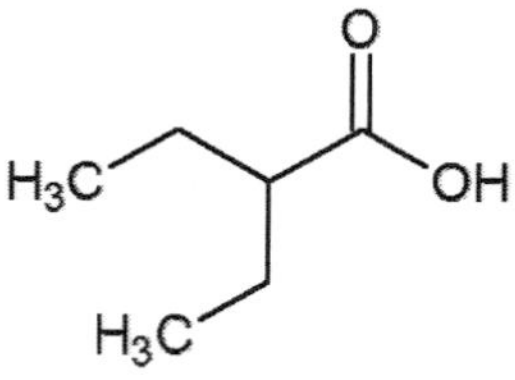

2-ethylbutanoic acid

24.29
(a) *p*-methylbenzamide
(b) 4-ethylhex-2-enenitrile
(c) dimethyl butanedioate
(d) isopropyl 3-phenylpropanoate
(e) phenyl benzoate
(f) N-methyl 3-bromobutanamide
(g) 3,5-dibromobenzoyl chloride
(h) 1-cyanocyclopentene

24.31

(a)

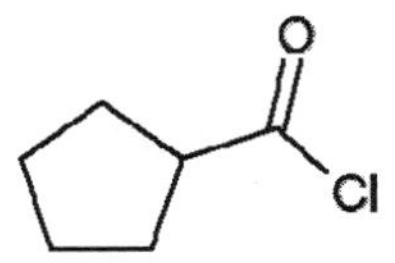

cyclopentanecarbonyl chloride

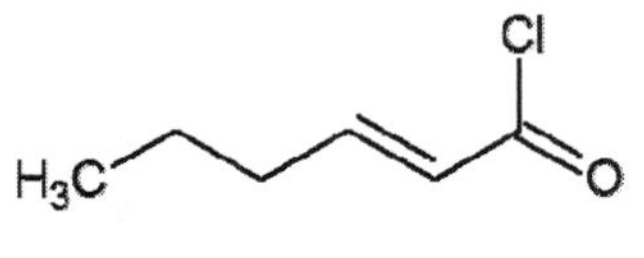

(*E*)-hex-2-enoyl chloride

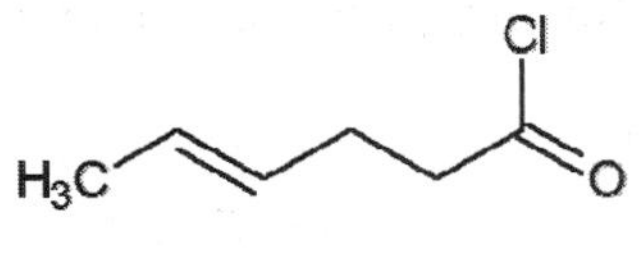

(*E*)-hex-4-enoyl chloride

(b)

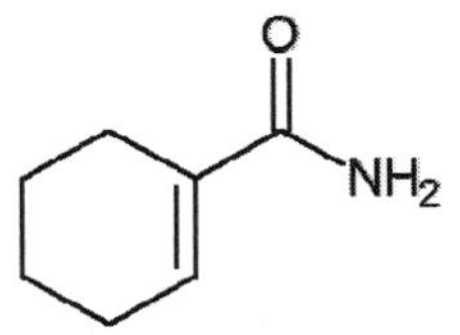

cyclohex-1-enecarboxamide

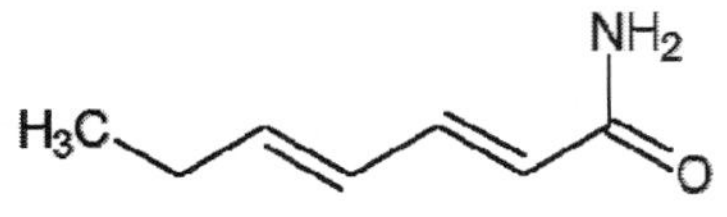

(2*E*,4*E*)-hept-2,4-dienamide

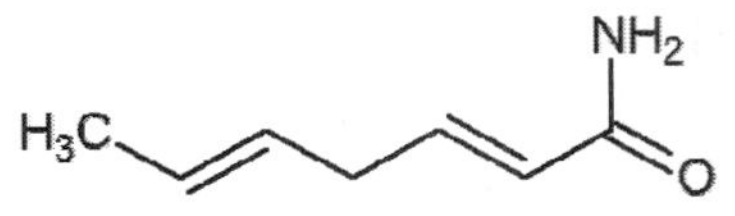

(2*E*,5*E*)-hept-2,5-dienamide

(c)

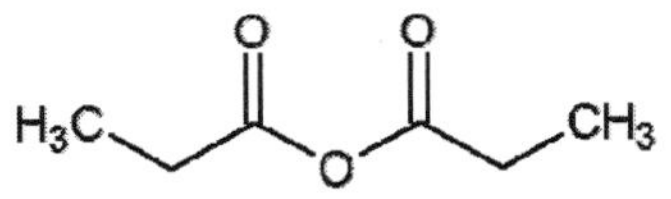

propanoic anhydride

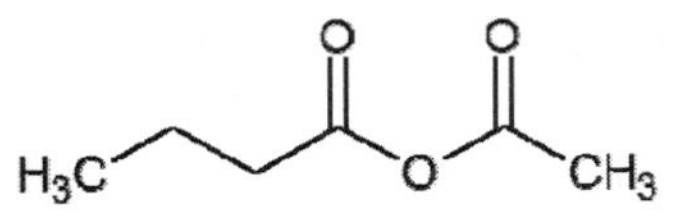

butanoic ethanoic anhydride

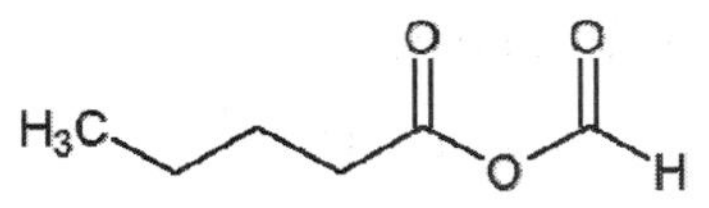

methanoic pentanoic anhydride

(d)

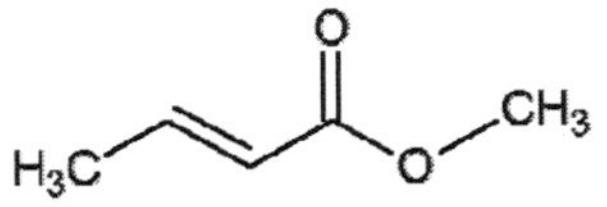

methyl but-2-enoate

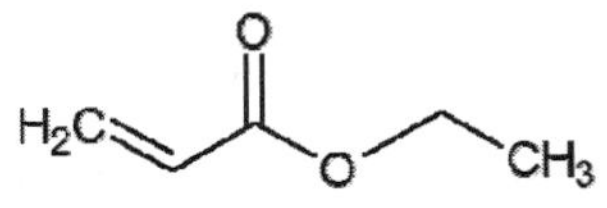

ethyl propenoate

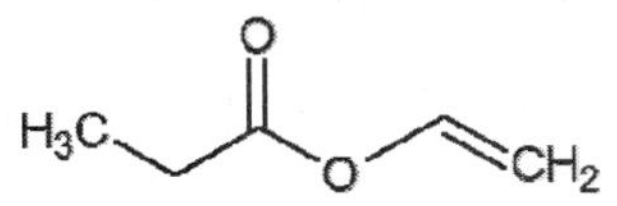

ethenyl propanoate

24.3 Spectroscopy of carboxylic acids and derivatives

24.33
(a) C_4H_7ClO, IR: 1810 cm^{-1}
The IR absorption at 1810 cm^{-1} suggests a C=O stretching vibration of an acid chloride. The ^{1}H NMR spectrum shows three multiplets, a triplet with a relative integration of 2 at 2.9 ppm, a triplet of triplets (possibly) with a relative integration of 2 at 1.8 ppm and a triplet with a relative integration of 3 at 1 ppm. This is consistent with

(b) $C_5H_{10}O_2$, IR: 1735 cm^{-1}
The IR spectrum shows a carbonyl stretch absorption in the right position for a saturated ester, and no O−H stretch. The ^{1}H NMR spectrum shows 3 clusters of peaks at 5, 2, and 1.2 ppm, integrating to 1, 3, and 6 respectively. The downfield peak is a septet split by six equivalent methyl carbons—coupled to the upfield peak that shows a doublet. The peaks at 5 and 1.2 ppm are consistent with an isopropyl group attached to an O atom—the methine H is further downfield than is possible for an H vicinal to only carbon atoms. The peak at 2 ppm with a relative integration of 3 is consistent with a methyl group attached to a carbonyl carbon. The structure is

24.4 Electronic structure and reactivity: carboxylic acids

24.35
According to increasing acidity, we have
(a) acetic acid < chloroacetic acid < trifluoroacetic acid
Halogens are electron withdrawing—this stabilizes the conjugate base anion by spreading out the negative charge. Fluorine is more electronegative than chlorine. Moreover, three fluorine atoms have greater electron-withdrawing strength than one or two fluorine atoms, and they provide more atoms to share the negative charge.
(b) benzoic acid < *p*-bromobenzoic acid < *p*-nitrobenzoic acid
Electron-withdrawing substituents increase the acidity of benzoic acid—especially in the ortho and para positions. Nitro is more electron withdrawing than bromine.
(c) cyclohexanol < phenol < acetic acid
Alcohols are generally weak acids. Phenols are considerably stronger—so much so that they constitute their own class of compounds—they are not just a subset of the alcohols. The stronger acidity of phenols results because of the resonance stabilization of the conjugate base—the negative charge is spread over the phenol ring (specifically, the ortho and para positions).

24.5–24.7 Reactions of carboxylic acids and derivatives

24.37

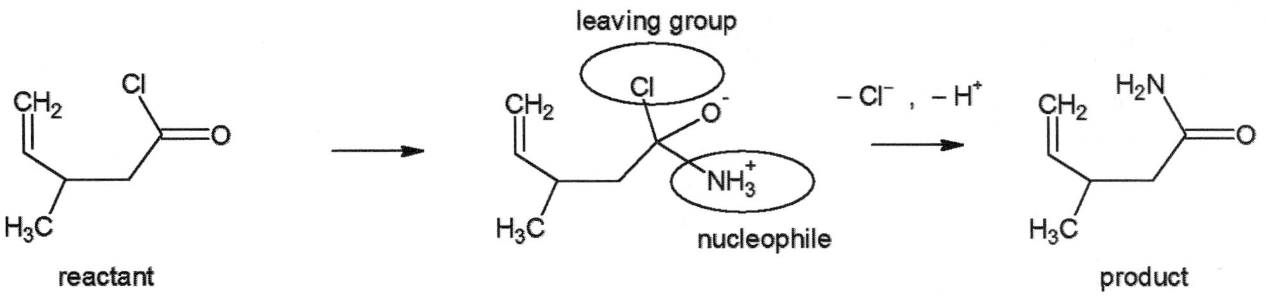

The nucleophile is NH_3, the leaving group is Cl^-, and the reactant and product are shown above.

24.39
According to increasing reactivity with respect to nucleophilic acyl substitution, we have
(c) CH_3CONH_2 < **(a)** $CH_3CO_2CH_3$ < **(d)** $CH_3CO_2COCH_3$ < **(b)** CH_3COCl

24.41

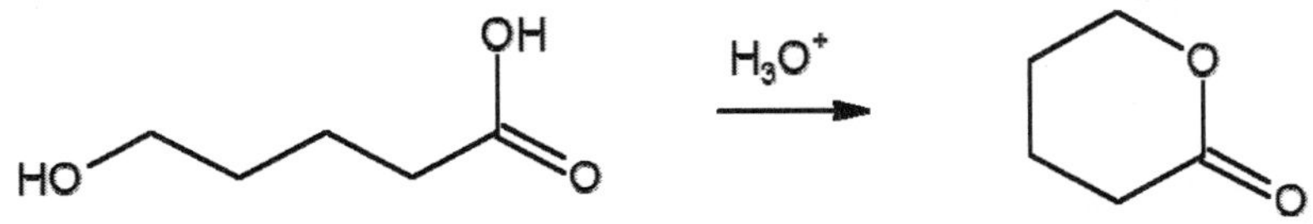

The intramolecular reaction between an alcohol and a carboxylic acid produces a cyclic ester, called a lactone.

24.43

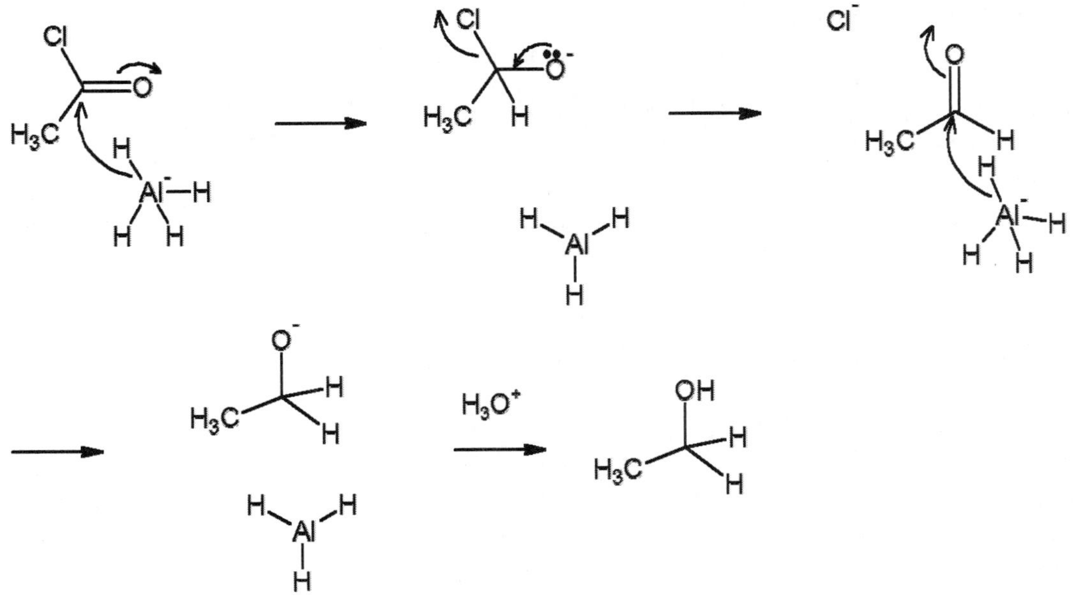

24.45
(a)

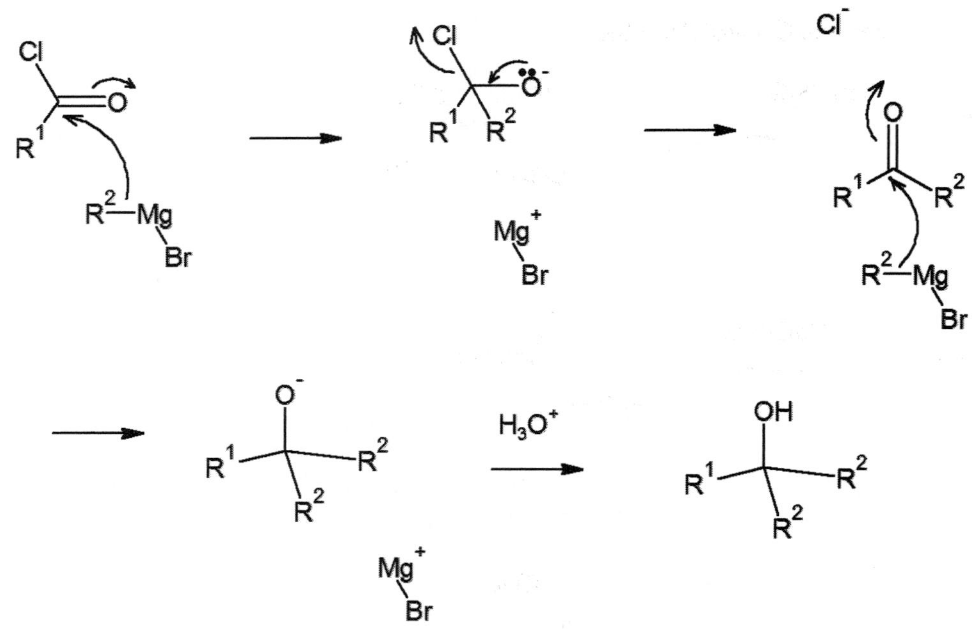
Cl
O
R1
R2
Mg
Br
Cl
O
R1
R2
Mg+
Br
Cl-
O
R1
R2
R2
Mg
Br
O-
R1
R2
R2
Mg+
Br
H3O+
OH
R1
R2
R2

(b)

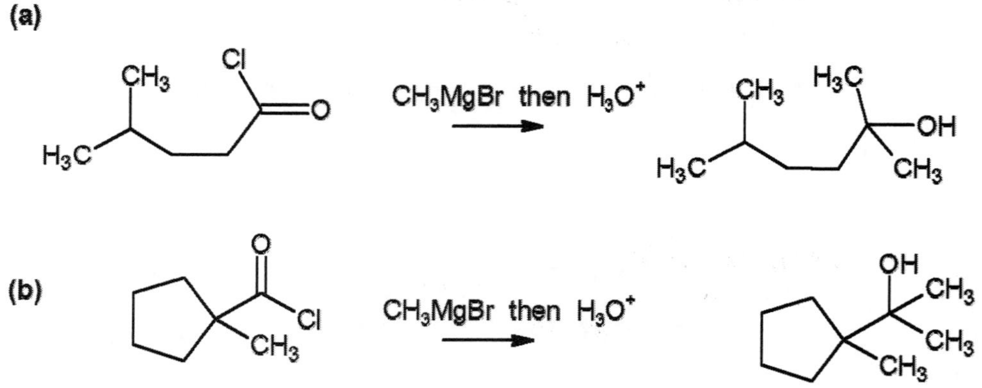
(a)
CH3
Cl
O
H3C
CH3MgBr then H3O+
CH3
H3C
OH
H3C
CH3
(b)
O
Cl
CH3
CH3MgBr then H3O+
OH
CH3
CH3
CH3

24.47

(a)

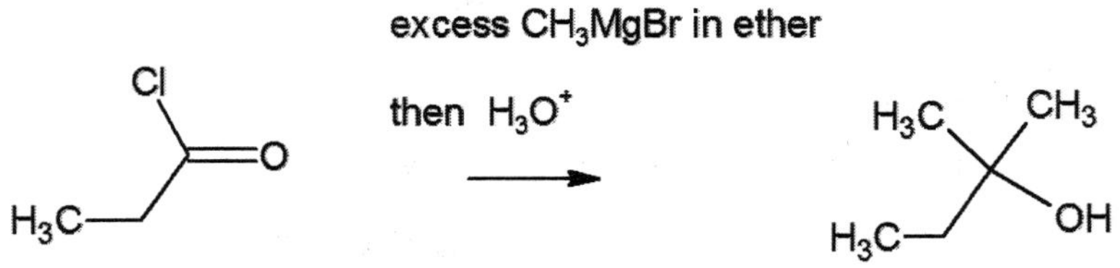
excess CH_3MgBr in ether
then H_3O^+
Cl
O
H_3C
H_3C
CH_3
OH
H_3C

(b)

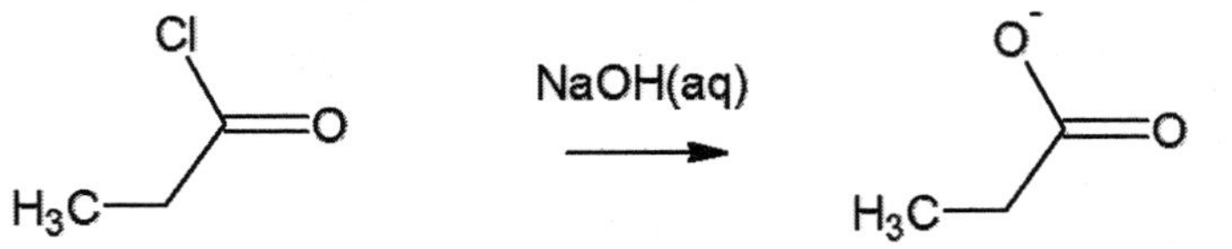
Cl
O
H_3C
NaOH(aq)
O^-
O
H_3C

(c)

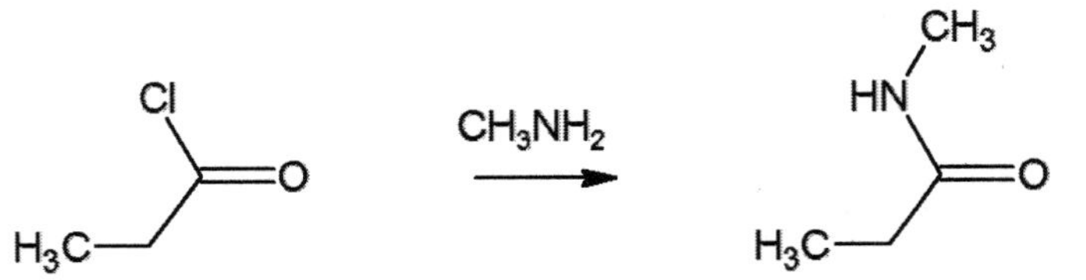
Cl
O
H_3C
CH_3NH_2
CH_3
HN
O
H_3C

(d)

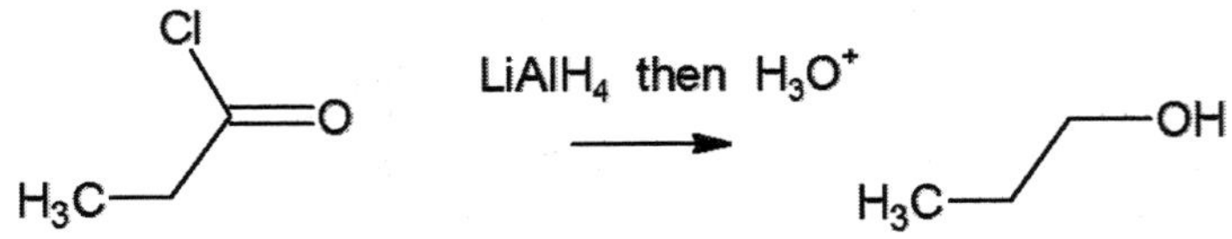
Cl
O
H_3C
$LiAlH_4$ then H_3O^+
OH
H_3C

(e)

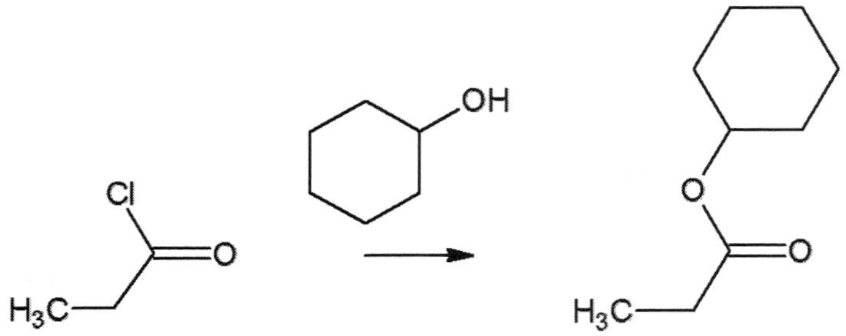

(f)

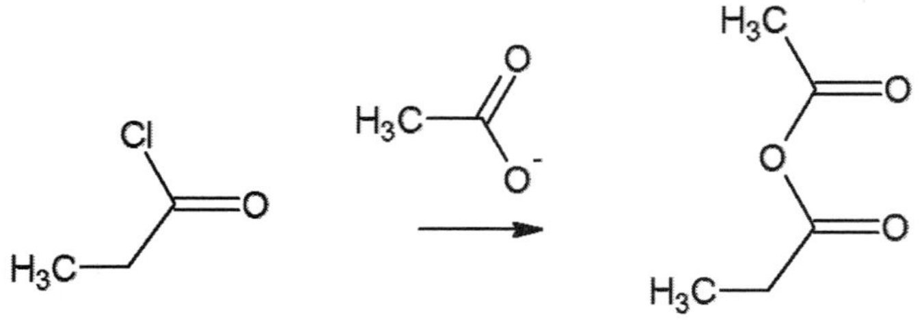

24.49

(a)

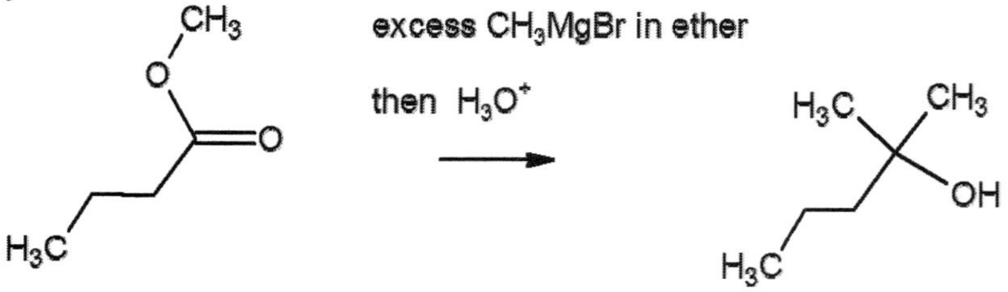

Alternatively, starting with methyl acetate, we could add 1 equivalent of propylmagnesium bromide, then 1 equivalent of methylmagnesium bromide.

(b)

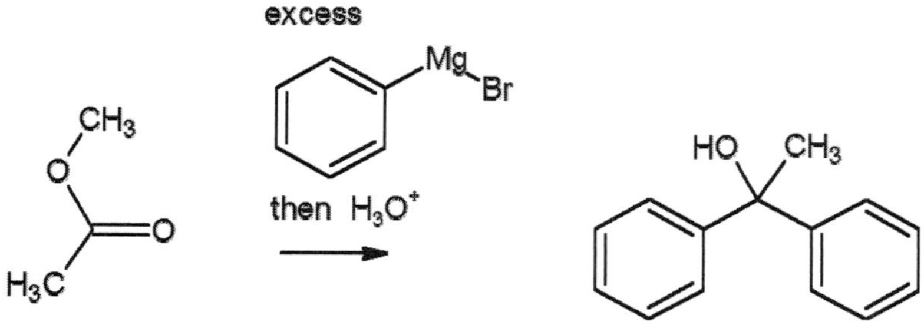

The alternative is to start with methyl benzoate and successively add 1 equivalent of phenylmagnesium bromide and 1 equivalent of methylmagnesium bromide.

24.51

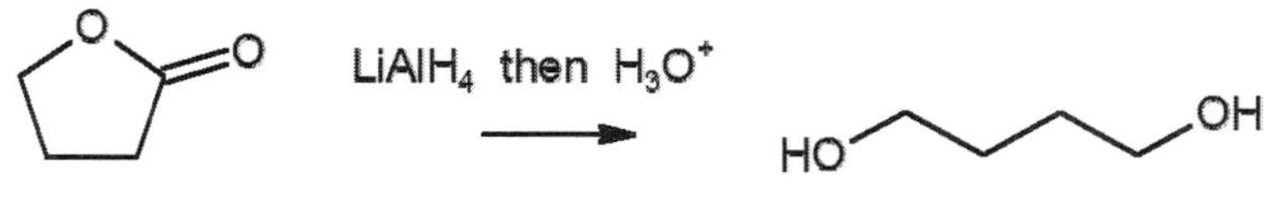

LiAlH₄ then H₃O⁺

24.8 α-Carbon atoms: A second reactive site in carbonyl compounds

24.53

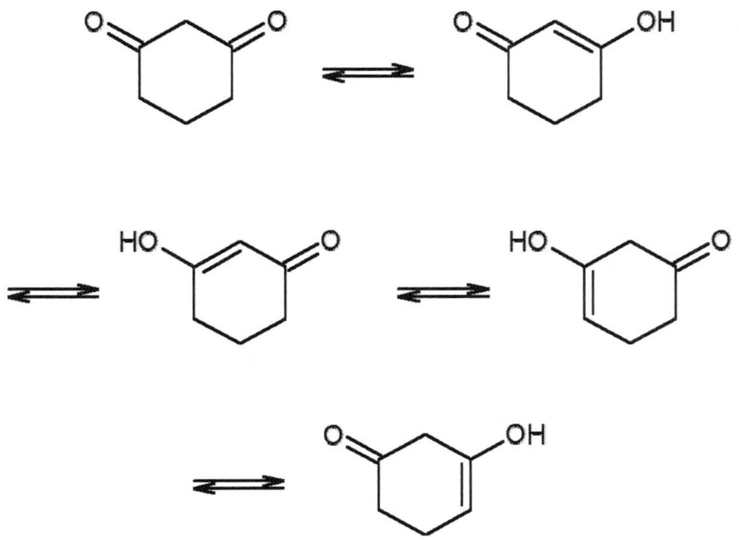

The first two enols (which are equivalent to each other) shown are the most stable because the acidic hydrogen atoms are α to two carbonyl groups.

24.55

(a)

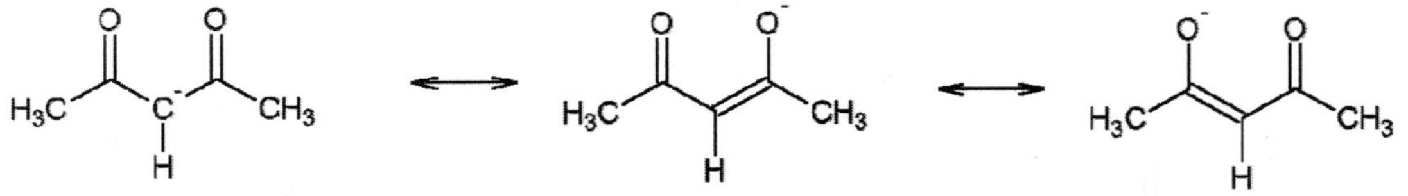

(b)

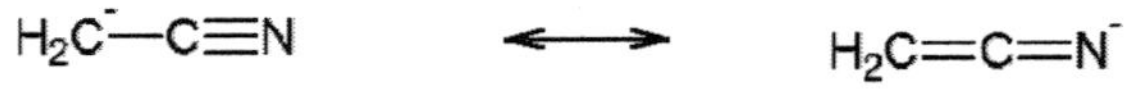

(c)

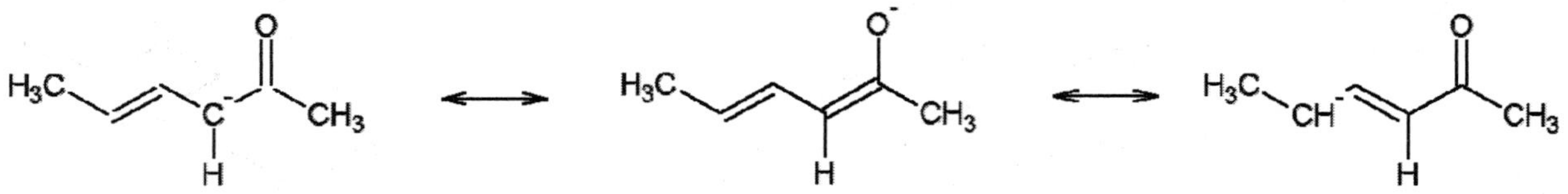

(d)

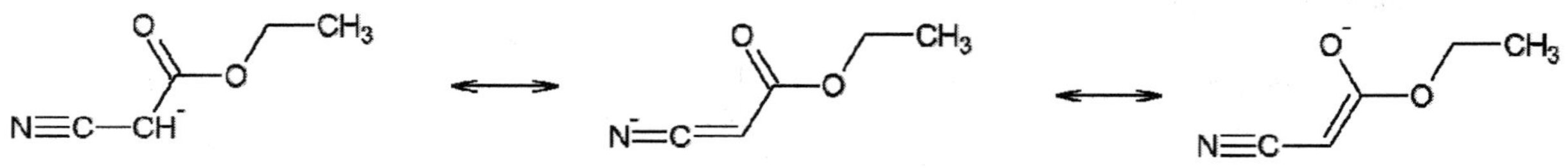

24.57

An enolate is a much stronger base or a nucleophile—it is a stabilized carbanion. It can react with electrophiles by donating a lone pair of electrons through the O atom or through the α carbon atom. The product of such an attack is typically neutral, whereas the same reaction with the enol produces a cationic product.

24.9–24.11 Reactions of enolate ions and condensation reactions

24.59

SUMMARY AND CONCEPTUAL QUESTIONS

24.61

(a)

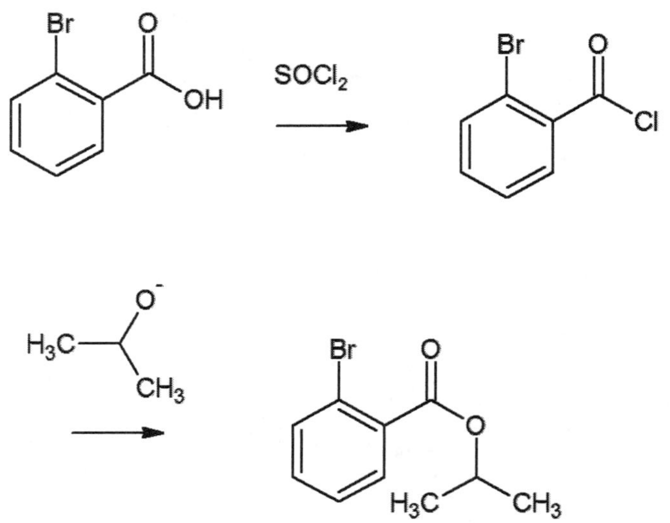

(b)

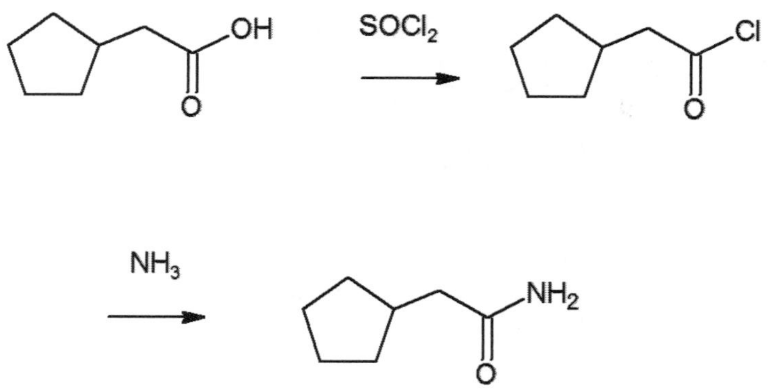

24.63
According to reactivity with hydroxide, we have
methyl acetate > ethyl acetate > isopropyl acetate > *tert*-butyl acetate
This order follows because of the increasing bulk of the alkoxide substituent at the carbonyl carbon.
Hydroxide adds to the carbonyl carbon in the first step of saponification. A large alkoxide
substituent reduces the accessibility of the electrophilic carbonyl carbon to hydroxide.

24.65

(a)

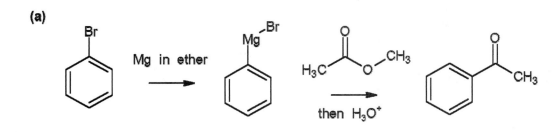

(b)

(c)

24.67

(a)

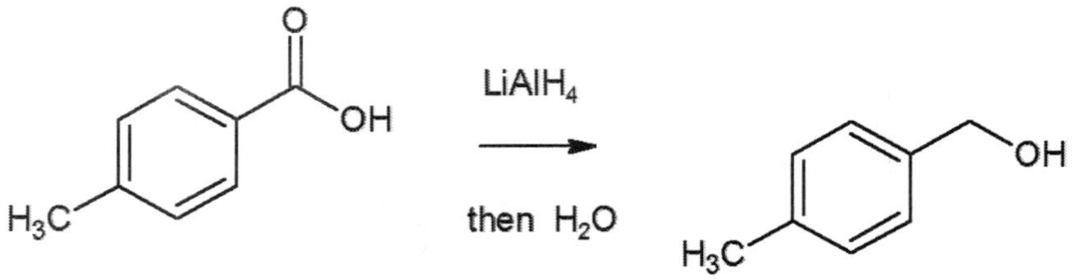

LiAlH₄

then H₂O

(b)

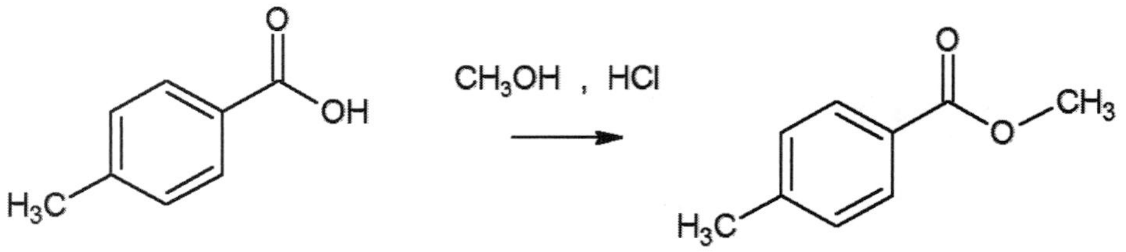

CH₃OH , HCl

(c)

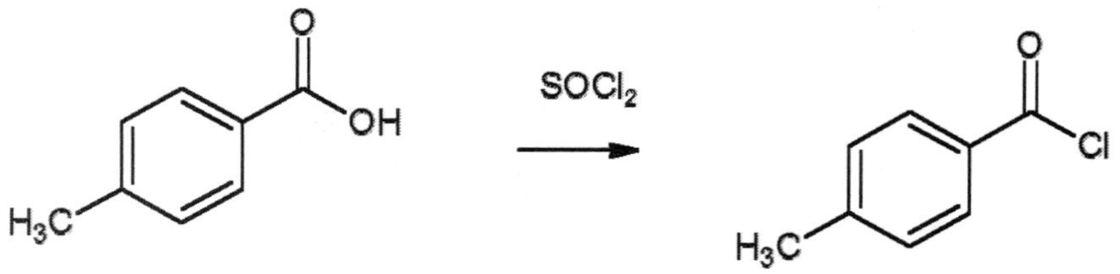

SOCl₂

(d)

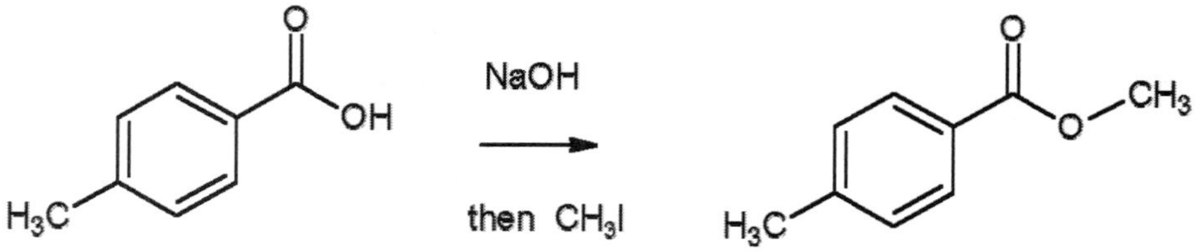

NaOH

then CH₃I

24.69

(a)

H3C
CH3
NaCN
N
H3C
CH3
LiAlH4
then H2O
NH2
H3C
CH3

(b)

Br
Mg in ether
Mg
Br
Mg
Br
H3C
O
CH3
O
then H3O+
H3C
O

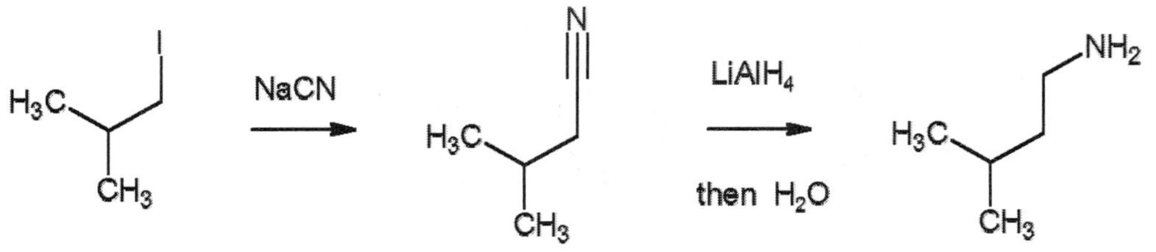

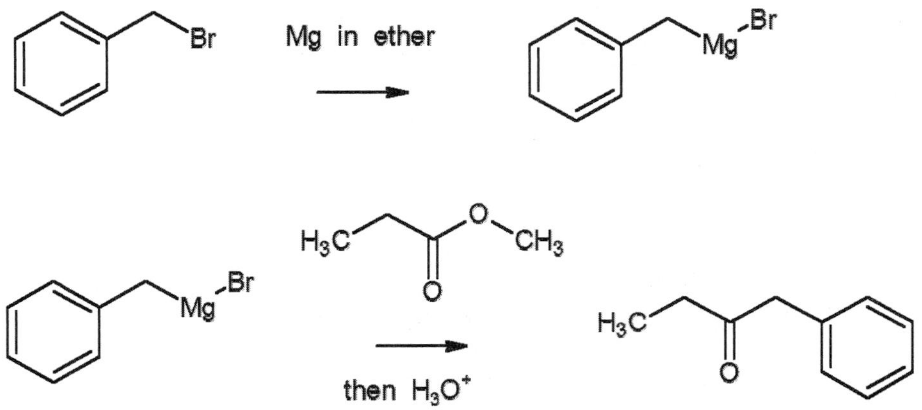

24.71

24.73

24.75
First step:

Polymerization step:

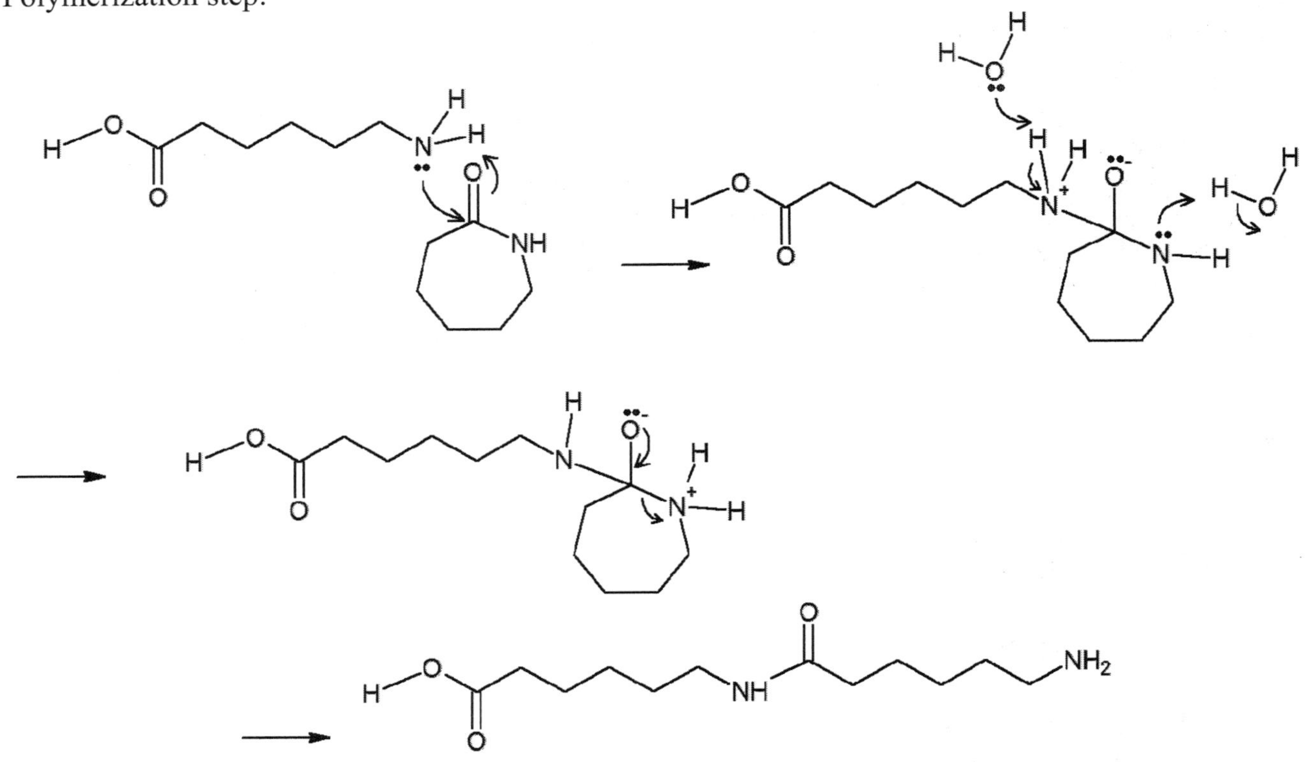

The amine of the amino acid monomer in step one reacts with a second molecule of caprolactam rather than with a second molecule of the intermediate.

The amine group attacks the amide on another caprolactam molecule, extending the chain by another monomer. This process repeats many times to give Nylon 6.

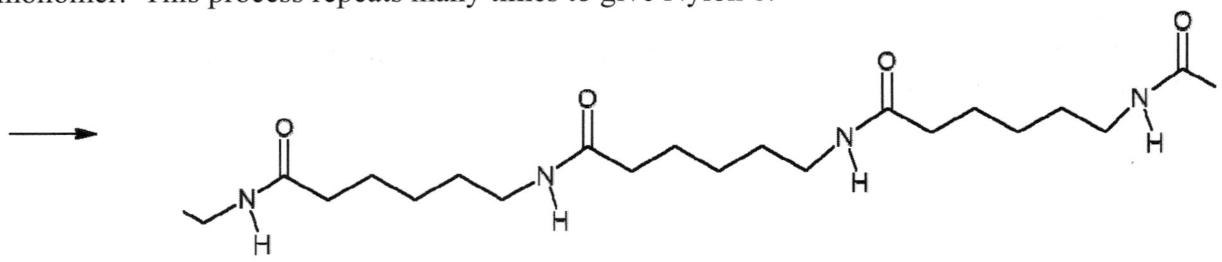

24. 77

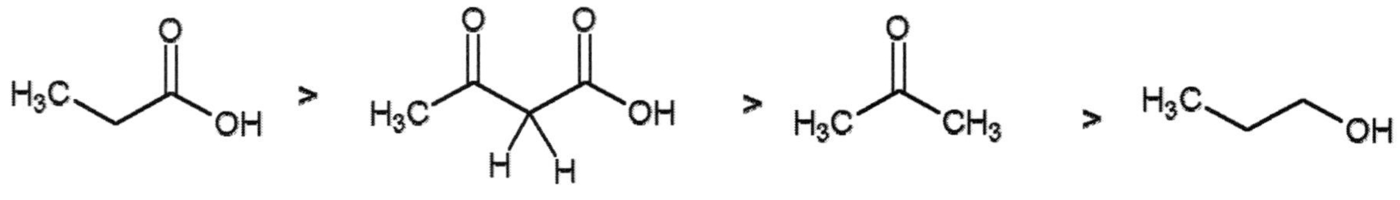

24.79

Hydrolysis of a cocaine molecule yields three molecules:

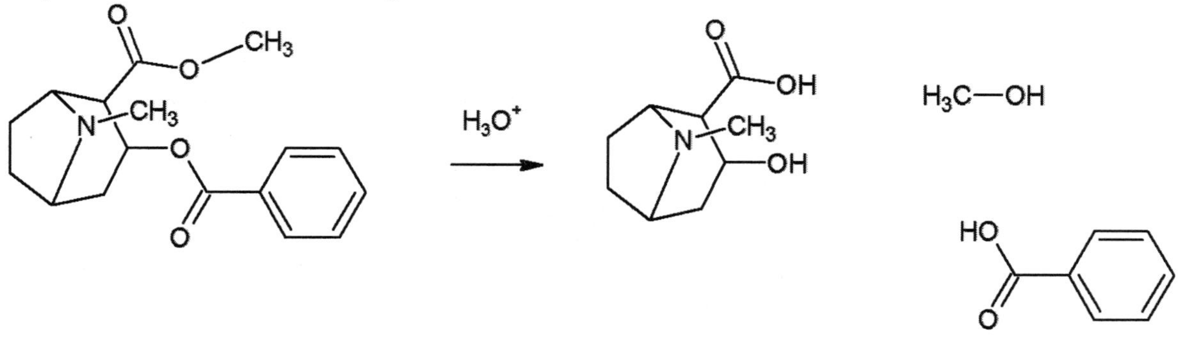

24.81

(a)

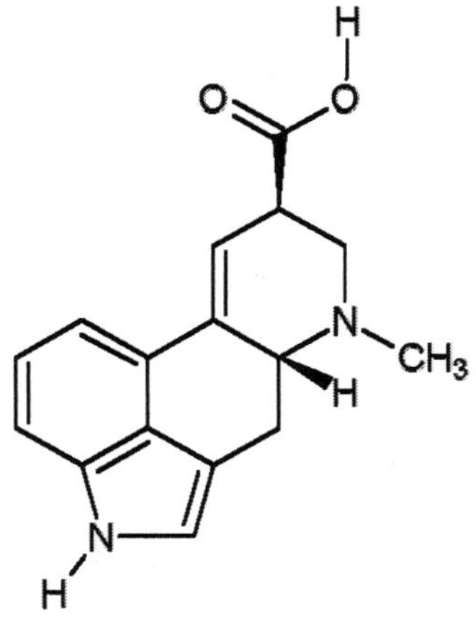

Hydrolysis of an amide produces the carboxylic acid shown, plus an amine corresponding to the portion of the molecule shown in black in your text.

(b)

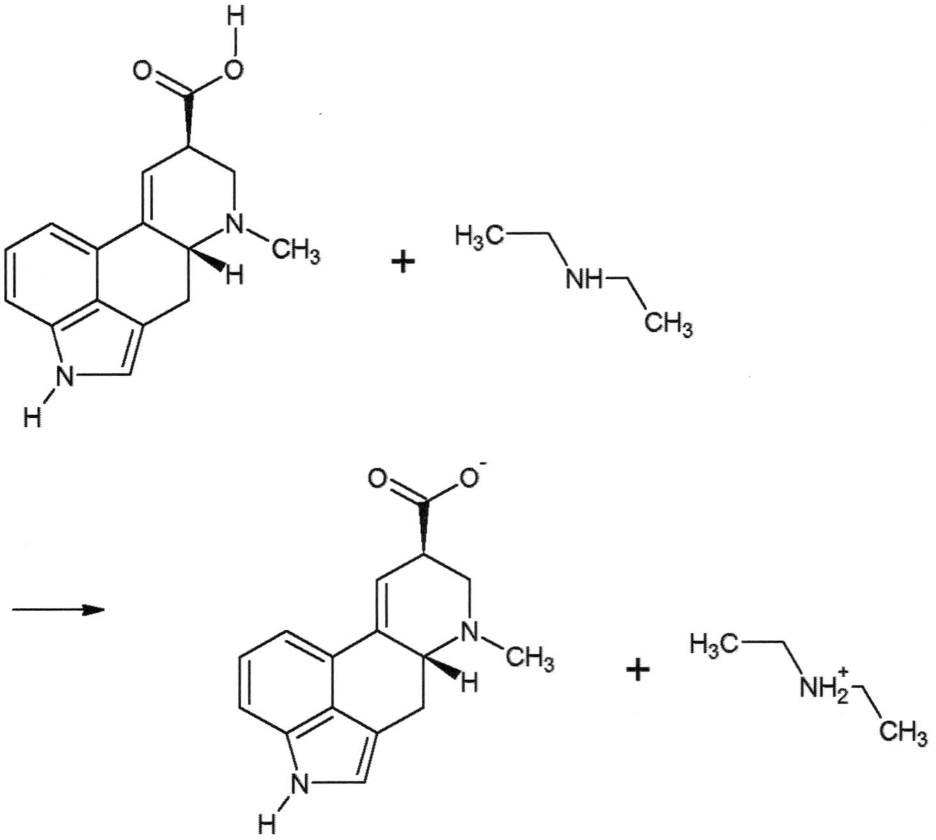

(c)

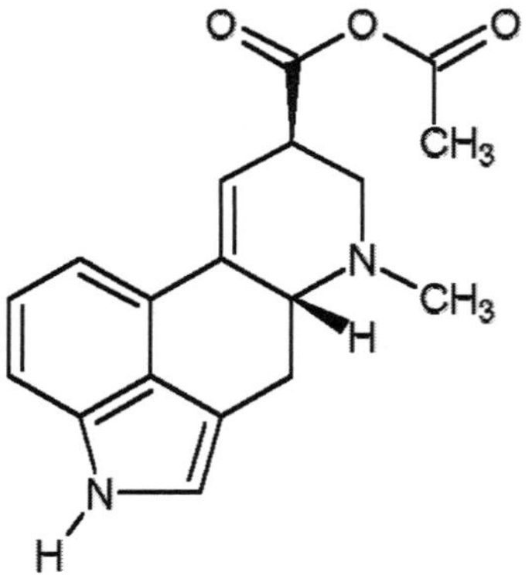

(d) For each mol of the anhydride, one mol of diethylamine is required to form the amide, and the second to react with the H^+ produced when the amide is formed.

24.83
First step:

Hydrolysis of acetylated enzyme:

24.85

As in the malonic ester synthesis, identify the components in the product where C-C bonds will need to form, remembering that the acetoacetic ester synthesis produces a mono- or di-substituted acetone molecule, with the new C-C bonds to the α C atom next to the C=O carbon atom.

(a)

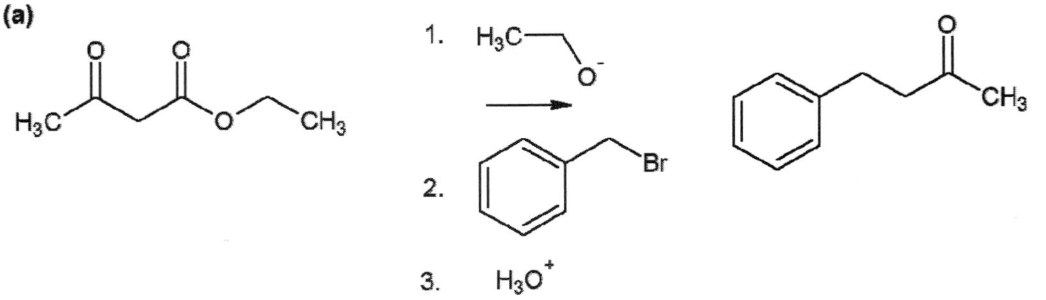

(b)

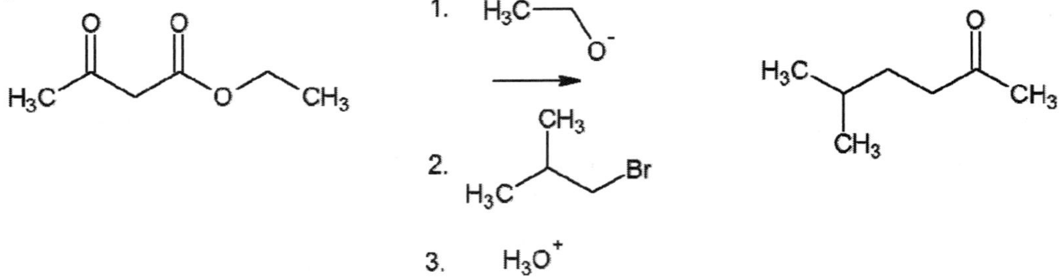

(c)

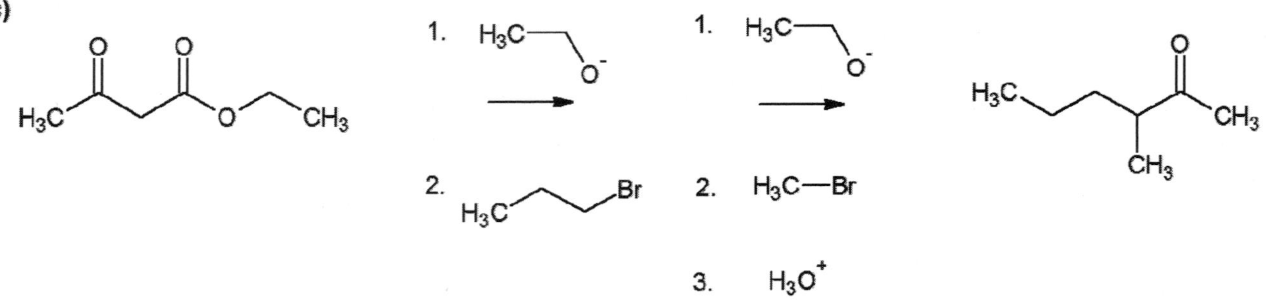

Chapter 25
Amines and Nitrogen Heterocycles

IN-CHAPTER EXERCISES

Exercise 25.2—Nomenclature

(a) primary amine
(b) secondary amine
(c) tertiary amine
(d) quaternary ammonium salt

Exercise 25.3—^{1}H NMR spectra

The NMR spectrum shows five clusters of peaks, consistent with the five non-equivalent sets of protons on *p*-ethoxyaniline.
From high to low chemical shift, we can make sense of them as follows. The two apparent doublets at 6.7 and 6.6 each have a relative integration of 2 and are coupled to each other. Since they come in the chemical shift range for aromatic protons, they are consistent with a para disubstituted benzene ring. The quartet at 3.9 ppm with a relative integration of 2 and triplet at 1.3 ppm with an integration of 3 is characteristic of an ethyl group, with the methylene CH_2 bonded to an electronegative element (an O atom in this case). The two protons on the NH_2 group are not split by neighbouring protons and show up as a broad singlet with an integration of 2 at 3.4 ppm. The broad signal is due to the wide range of extents of H-bonding to neighbouring groups by the N-H protons. Since the extent of H-bonding varies greatly from proton to proton, each proton is in a different magnetic environment around the nucleus and collectively they exhibit a continuum of close chemical shift values. This leads to a broad peak.

Exercise 25.4—^{13}C NMR spectra

There should be 6 peaks in the ^{13}C NMR spectrum. Although there are 8 C atoms in *p*-ethoxyaniline, C-2 and C-6 of the phenyl ring are equivalent, as are C-3 and C-5. The methyl C peak (C bonded to another alkyl C) should appear in the 10-30 ppm range. The methylene C peak (C bonded to O) should appear in the 50-70 ppm. The phenyl C peaks will be in the 110-150 ppm range, with the C's bonded directly to O and N furthest downfield.

Exercise 25.5—Acid-base properties

With respect to basicity we have
(a) $CH_3CH_2NH_2$ > $CH_3CH_2CONH_2$
The nitrogen atom of the amide is much less basic than that of an amine because its electron pair can be delocalized by orbital overlap with the neighbouring π orbital of the C=O group.
(b) NaOH > $C_6H_5NH_2$
Hydroxide is a strong base, whereas amines are weak bases, like ammonia.
(c) CH_3NHCH_3 > $CH_3NHC_6H_5$

Phenyl amines have reduced basicity because the N lone pair is partially delocalized with the phenyl π system—i.e., it is less available to be donated to a proton.

(d) $(CH_3)_3N > CH_3OCH_3$

Just as ammonia is a stronger base than water, amines are stronger bases than ethers.

Exercise 25.6—Acid-base reactions of amines

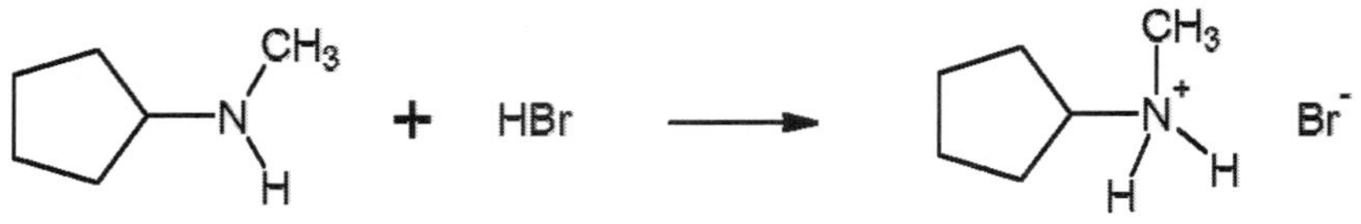

Exercise 25.7—Alkylation reactions

(a)

(b)

In part (a) some quaternary ammonium salt will also be formed as a side reaction.

Exercise 25.8—Alkylation reactions

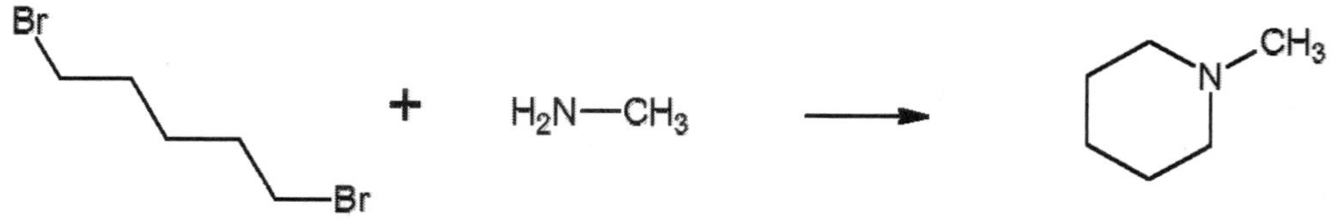

Exercise 25.9—Electrophilic aromatic substitution

(a)

CH₃—N pyrrole →(Br₂)→ CH₃—N pyrrole—Br

(b)

CH₃—N pyrrole →(CH₃Cl & AlCl₃)→ CH₃—N pyrrole—CH₃

(c)

CH₃—N pyrrole →(CH₃COCl & AlCl₃)→ CH₃—N pyrrole—C(=O)CH₃

Exercise 25.10—Electrophilic aromatic substitution

The nitronium ion results from nitric and sulfuric acids, as described in Section 20.6.

Exercise 25.11—Heterocyclic amines

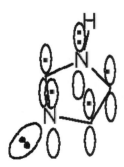

The pyrrole-like N has its lone pair in a p orbital, whereas the pyridine-like N has its lone pair in an sp^2 orbital.

Exercise 25.12 — Heterocyclic amines

The pyridine-like N is more basic—we see the negative charge density appearing as a red patch in the electrostatic potential map. This is because its lone pair is in an sp^2 orbital—it is not delocalized in the π system. The pyrrole-like N lone pair, in contrast, is in a p orbital and is therefore delocalized in the π system, and is not available for donation to a proton.

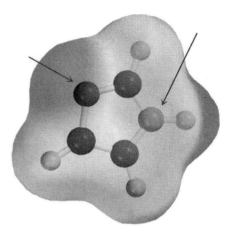

Imidazole

© 2007 Thomson Higher Education

REVIEW QUESTIONS

25.2 Naming amines

25.13
(a) *N*-methylisopropylamine—it is a secondary amine
(b) *trans*-(2-methylcyclopentyl)amine—it is a primary amine
(c) *N*-isopropylaniline—it is a secondary amine

25.15

(a)

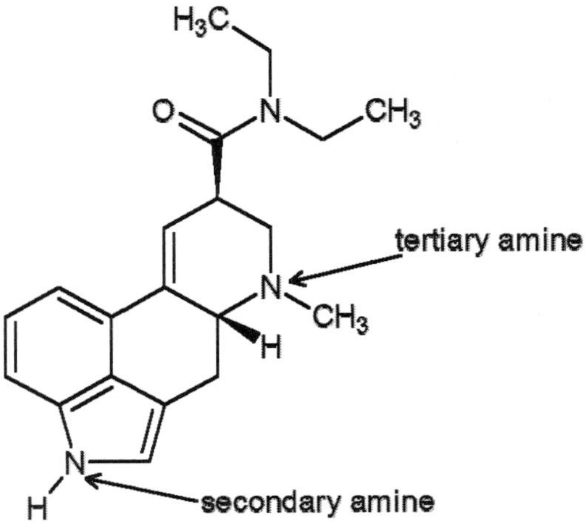

(b)

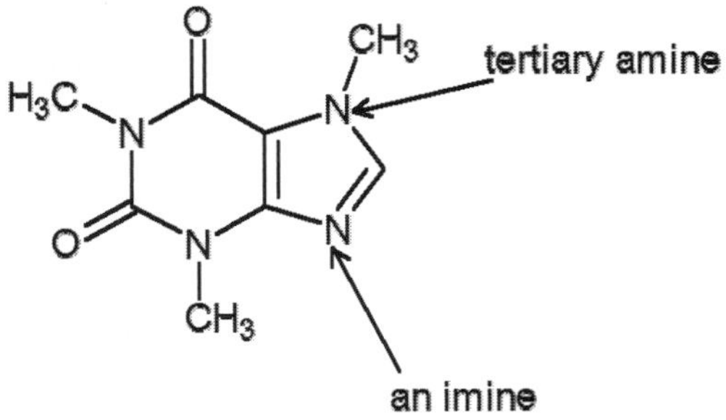

25.17

(a)

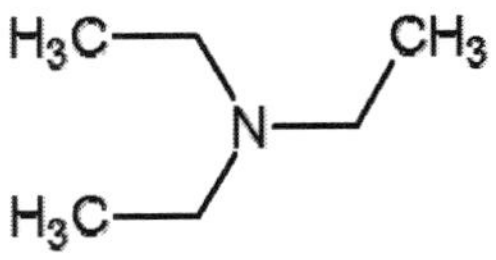

(b)

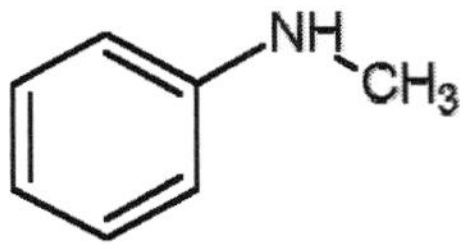

(c)

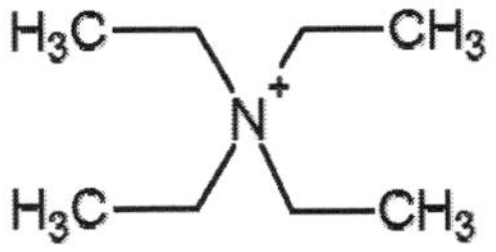

(d)

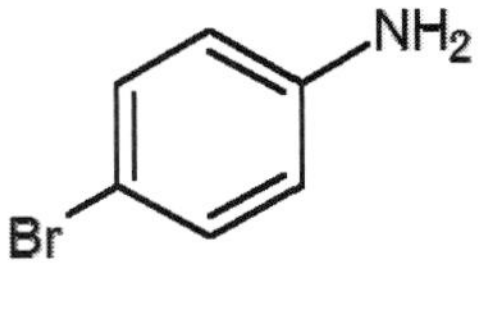

(e)

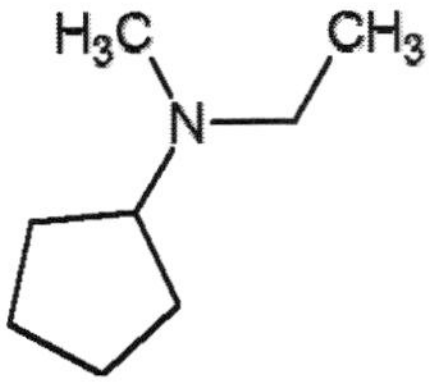

25.19

Mescaline:

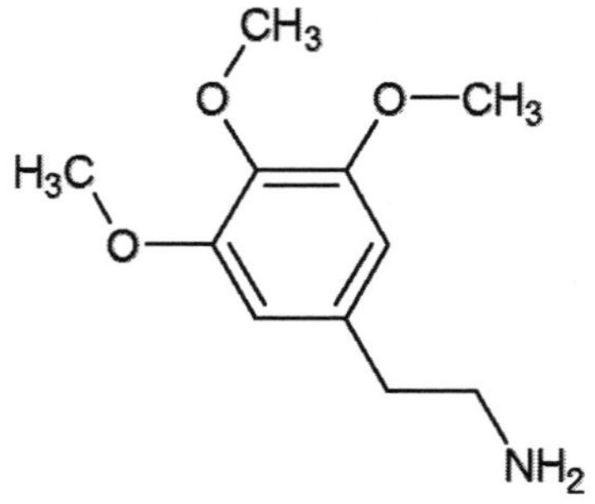

25.21

(a)

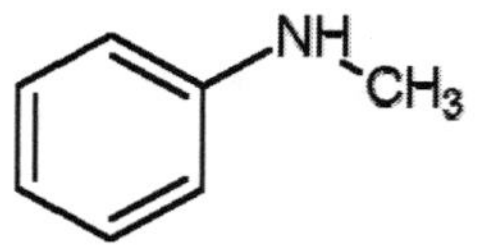

(b)

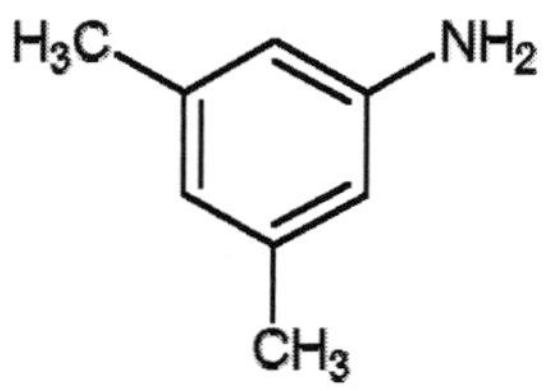

(c)

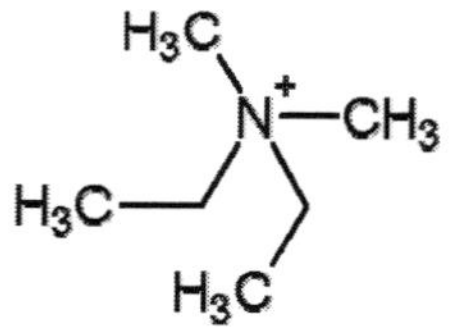

(d)

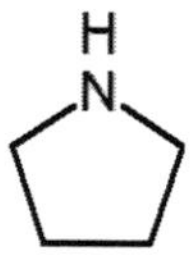

25.4 Electronic structure and reactivity

25.23

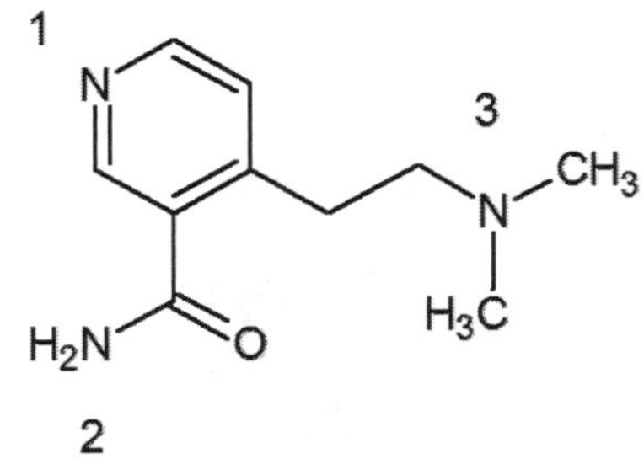

According to basicity, the three N's in this molecule are ranked as follows: 3 > 1 > 2. 3 is an alkyl amine, 1 an aromatic heterocyclic amine, and 2 an amide. Alkyl amines are more basic than aromatic heterocyclic amines such as pyridine, and amides are not very basic at all—because their lone pair of electrons is delocalized over the π system of the adjacent C=O group.

25.25

Trimethylamine has a lower boiling point than dimethylamine even though it has a higher molecular weight because it has no N-H groups resulting in hydrogen bonding as dimethylamine does.

25.5 More about alkaloids—Naturally occurring amines

25.27

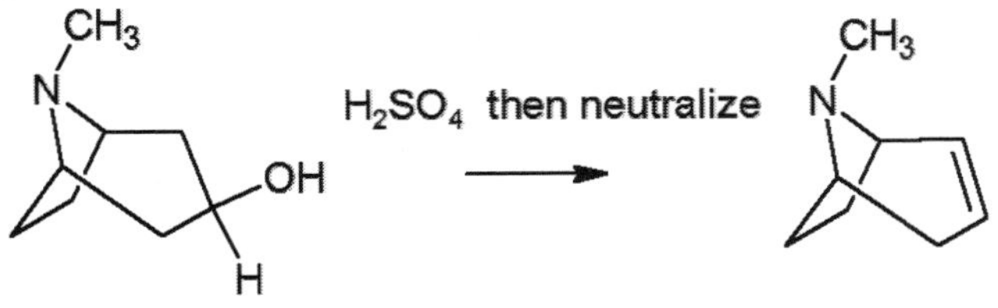

tropine   tropidine

note that the other position for the
hydroxyl corresponds to an
optically active alcohol

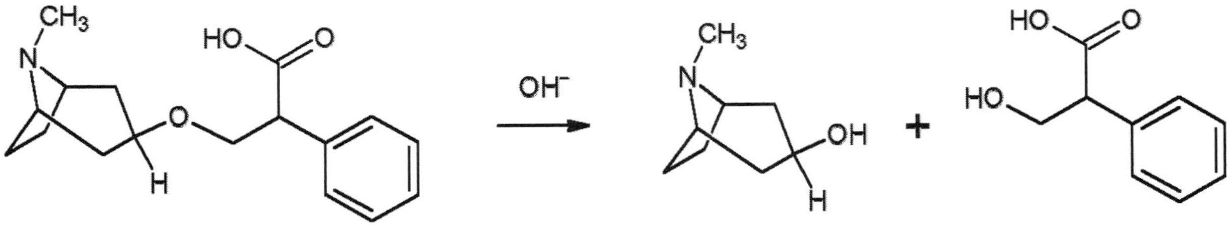

atropine

25.6 Heterocyclic aromatic amines

25.29

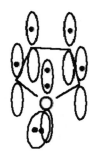

25.31
Nitration at the 2 position:

Nitration at the 3 position:

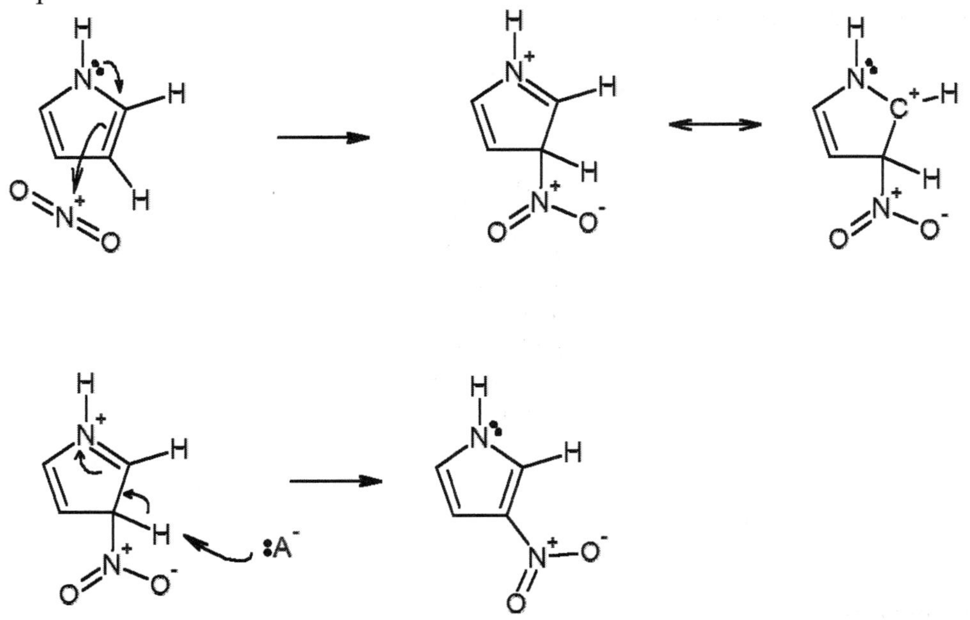

The carbocation intermediate in the case of substitution at the 2 position has three resonance structures, whereas there are only two resonance structures in the case of substitution at the 3 position. The positive charge is more delocalized in the case of the former carbocation intermediate, making that intermediate more stable and the corresponding reaction pathway more favourable. Nitration consequently occurs preferentially at the 2 position.

SUMMARY AND CONCEPTUAL QUESTIONS

25.33

(a)

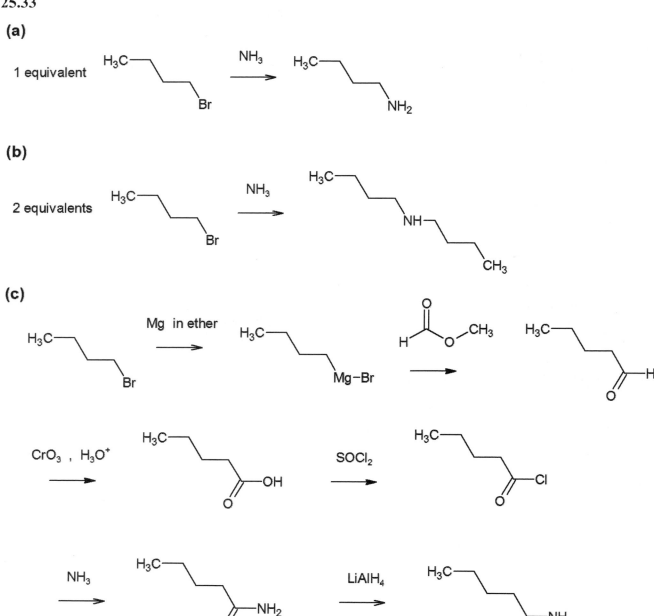

1 equivalent

(b)

2 equivalents

(c)

25.35

$CH_3CH_2NH_2$ is more basic than $CF_3CH_2NH_2$. The fluoro substituents are electron withdrawing—they reduce the availability of electrons at the nitrogen atom.

25.37

Triethylamine is more basic than aniline. The nitrogen lone pair in aniline is partially delocalized into the π system of the benzene ring, making it less available for donation to protons.

This reaction does not go in the direction indicated.

25.39

(a)

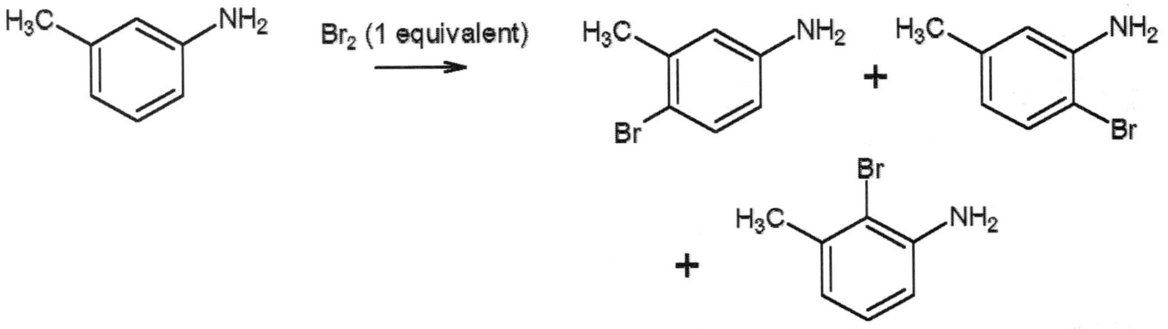

substitution between two other substituents occurs only to a small extent

(b)

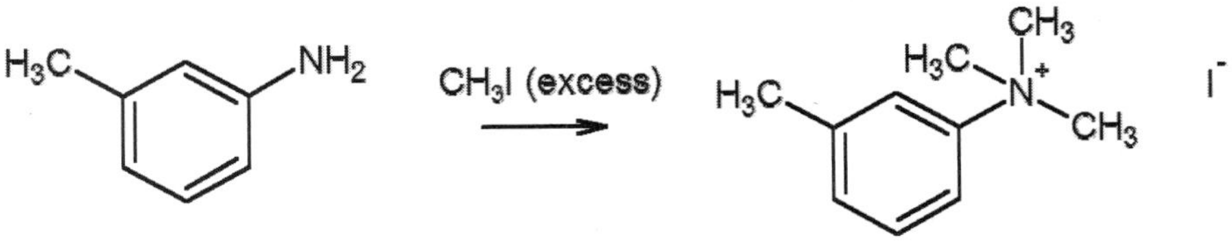

(c)

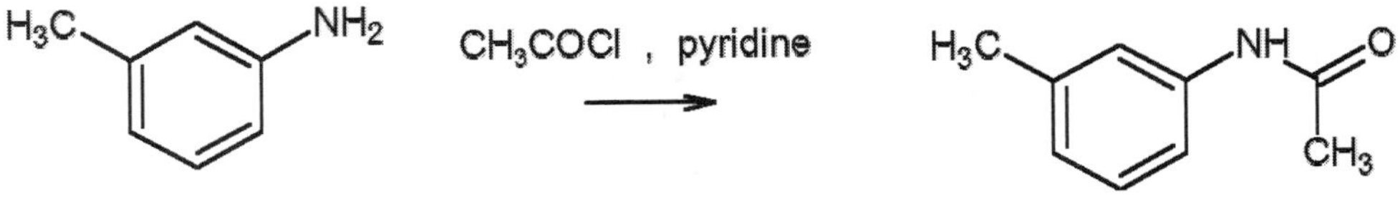

25.41

Diphenylamine is less basic than aniline because the nitrogen lone pair is partially delocalized over two benzene rings, rather than just one.

25.43

The methoxy substituent is electron donating—in the ortho and para positions—and increases the electron density on the amine nitrogen atom.   For example, consider the resonances structures for the para substituted case.

25.45

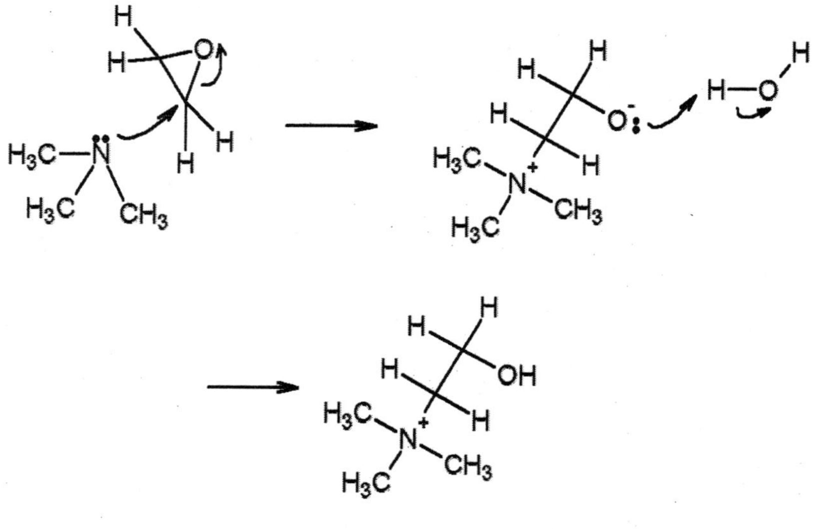

choline

25.47

Stereocentres are marked with arrows below. There are four stereocentres, and as such there are $2^4 =$ 16 stereoisomers.

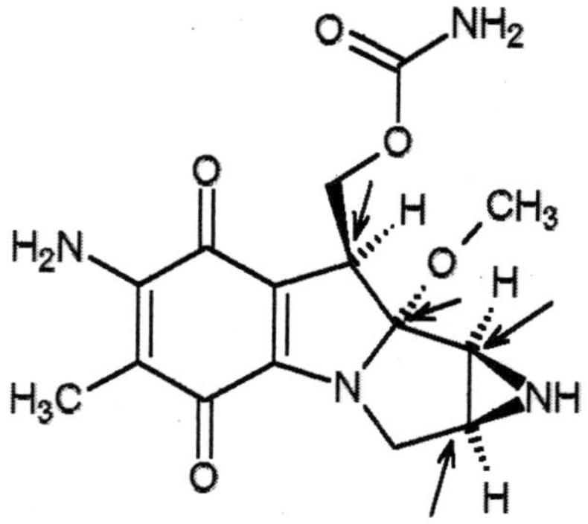

25.49
The aziridine group, like the epoxide group, is very reactive because the bonds are strained by the tight three-membered ring arrangement. The bond angles inside the ring are smaller than what is ideal for tetrahedral carbon and nitrogen (nitrogen is tetrahedral with respect to valence electron pairs).

25.51
DEET is an amide that can be prepared by reaction of an amine, diethyl amine, with a carboxylic acid, *m*-methylbenzoic acid, liberating water in the process. In practice, this reaction is likely best carried out by first preparing the acid chloride from the carboxylic acid.

Chapter 26
Main Group Elements and Their Compounds

IN-CHAPTER EXERCISES

Exercise 26.1—Influence of charge densities of cations on properties

Be^{2+} is more polarizing than Ba^{2+} because it is smaller. It can get much closer to other species, enhancing the effect of its +2 charge.

Exercise 26.2—Influence of charge densities of cations on properties

(a) Chromium(VI) oxide has a very polarizing +6 charged metal cation, producing a chromium oxygen bonds with a strong covalent character. The compound is more like a molecular solid, than a typical metal oxide ionic solid—i.e., like chromium(III) oxide.
(b) Because the chromium(VI) cation is so strongly polarizing, the oxide is a covalent molecular acidic oxide, dissolving in water to produce chromic acid. The more weakly polarizing chromium(III) cation forms the insoluble ionic oxide, Cr_2O_3, which is amphoteric—it dissolves in acid or base and as such can behave as a base or an acid.

Exercise 26.3—Hydrogen and its compounds

$$2\ H_2(g) + O_2(g) \rightarrow 2\ H_2O(l)$$
$$H_2(g) + Cl_2(g) \rightarrow 2\ HCl(g)$$
$$3\ H_2(g) + N_2(g) \rightarrow 2\ NH_3(g)$$

Exercise 26.4—Hydrogen and its compounds

$$2\ K(s) + H_2(g) \rightarrow 2\ KH(s)$$
The product of this reaction is potassium hydride. It is an ionic solid (at room temperature). It is a very strong base—consisting of K^+ and H^- ions. It is so strongly reactive that is generally distributed as a slurry in mineral oil.

Exercise 26.5—The alkali metals, Group 1

Suppose you chose sodium. The reaction with chlorine is

$$2\,Na(s) + Cl_2(g) \rightarrow 2\,NaCl(s)$$

Since all alkali metals have the same valence of 1, the other alkali metals react with chlorine with the same stoichiometry. As sodium is very electropositive and chlorine is very electronegative, the above reaction is exothermic producing positive and negative ions bound electrostatically—i.e., an ionic product. The product is a colourless solid at room temperature—it is made from ions with closed valence shells. Colour is generally associated with open shell systems, which have smaller spacings between electronic energy levels. Ionic materials are generally high melting solids, except in the case of specially constructed materials with very large polyatomic ions (a large spacing between ions reduces the Coulomb attraction). Sodium chloride is soluble in water, as the ions are readily solvated by polar water molecules.

Exercise 26.6—The alkali metals, Group 1

(a) The oxidation half reaction,

$$2\,Cl^-(aq) \rightarrow Cl_2(aq) + 2\,e^-$$

occurs at the anode. Chloride is oxidized to chlorine.
The standard reduction potential for chlorine is 1.36 V. The alternative oxidation process in this aqueous KCl solution is the oxidation of water. The standard reduction potential for oxygen in water is 1.23 V. On the surface, it would appear easier to oxidize water to oxygen than to oxidize chloride (the standard oxidation potential is -1.23 V). However, the standard reduction potential for oxygen applies to a 1 M H^+ solution. Corrected to $[H^+] = 10^{-7}$ M, using the Nernst equation, the reduction potential for water is 1.64 V. Therefore, chloride is more easily oxidized in neutral water than water itself.
The reduction half reaction,

$$2\,H_2O(l) + 2\,e^- \rightarrow 2\,OH^-(aq) + H_2(g)$$

occurs at the cathode. Water is reduced to form hydrogen gas.
The reduction potential for water at pH = 7 is -0.41 V. This is more favourable than the reduction of K^+ to K(s). The standard reduction potential for this alternative process is -2.93 V.
(b) Iodine and hydrogen are the oxidation and reduction products, respectively. The reduction half-reaction is the same as in part (a).

Exercise 26.7—The alkaline earth elements, Group 2

Suppose you chose barium. The reaction with oxygen is

$$2\,Ba(s) + O_2(g) \rightarrow 2\,BaO(s)$$

Since all alkaline earth metals have the same valence of 2, the other alkaline earth metals react with oxygen with the same stoichiometry. As barium is very electropositive and oxygen is very electronegative, the above reaction is exothermic producing positive and negative ions bound electrostatically—i.e., an ionic product. The product is a colourless solid at room temperature—it is made from ions with closed valence shells. Ionic materials are generally high melting solids. In this case, because both ions are doubly charged we expect a quite high melting point. Barium oxide is somewhat soluble in water, as the ions are readily solvated by polar water molecules. However, the lattice energy of the solid is very high because of the doubly charged ions. Had we chosen magnesium for the element, because Mg^{2+} is so much smaller than Ba^{2+}, the lattice energy is so high that MgO is quite insoluble in water—solubility increases as we descend the group since the cations get bigger. Be^{2+} is so small and polarizing that its oxide has significant covalent character, and cannot be considered ionic.

Exercise 26.8—The alkaline earth elements, Group 2

We would not expect to find calcium occurring naturally in the earth's crust as a free element because it is too reactive—it too readily has its valence electrons removed to form compounds with Ca^{2+} ions. It occurs only in the +2 oxidation state.

Exercise 26.9—The Group 13 elements

(a) In HCl solution, we have

$$Ga(OH)_3(s) + 3\,H_3O^+(aq) \rightarrow Ga^{3+}(aq) + 3\,H_2O(l)$$

In NaOH solution,

$$Ga(OH)_3(s) + OH^-(aq) \rightarrow [Ga(OH)_4]^-(aq)$$

(b) $Ga^{3+}(aq)$, with $K_a = 1.2 \times 10^{-3}$, is a stronger acid than $Al^{3+}(aq)$, with $K_a = 1.3 \times 10^{-5}$.
(c) Complete reaction of each mole of $Ga(OH)_3(s)$ with $H_3O^+(aq)$ ions (see part (a) above) requires 3 mol of $H_3O^+(aq)$ ions.

Amount of $Ga(OH)_3(s)$ = 1.25 g / (120.74 g/mol) = 0.01035 mol
required amount of $H_3O^+(aq)$ ions = 3 × 0.0104 mol = 0.03011 mol
volume of HCl solution required = 0.03011 mol / 0.0112 mol L^{-1} = 2.77 L

Exercise 26.10—The Group 13 elements

$$Al(s) + 3\,H_3O^+(aq) \rightarrow Al^{3+}(aq) + \tfrac{3}{2}H_2(g) + 3\,H_2O(l)$$

$$Al(s) + \tfrac{3}{2}Cl_2(g) \rightarrow AlCl_3(s)$$

$$Al(s) + \tfrac{3}{4}O_2(g) \rightarrow \tfrac{1}{2}Al_2O_3(s)$$

Exercise 26.11—The Group 14 elements

Both SiO_2 and CO_2 are covalently bonded, as group 14 elements—with a valence of 4—would be far too polarizing as +4 cations to form ionic compounds. However, whereas the period 2 element carbon forms stable double bonds, period 3 element silicon does not. Consequently, to achieve a valence of 4, silicon forms a network of single bonds with oxygen—i.e., we get a covalently bonded network solid. An entire crystal of SiO_2 is, in a sense, a single molecule. To melt (or break) a crystal of SiO_2 requires the breaking of strong covalent bonds. This requires high temperature (or a strong force). In contrast, carbon achieves its valence of 4 by forming two doubles with oxygen atoms. The small, non-polar, CO_2 molecule is bound to other such molecules by weak dispersion forces that are easily broken. Carbon dioxide is gas at ordinary temperatures.

Exercise 26.12—The Group 14 elements

The strongly polarizing Pb^{4+} cation increases the covalent character of the bonding in $PbCl_4$. Consequently, $PbCl_4$ has considerable covalent character, whereas $PbCl_2$ is best described as a strictly ionic material. This means that $PbCl_4(s)$ can melt (into molecular $PbCl_4(l)$) at lower temperature than $PbCl_2$, which melts to form an ionic liquid.

Exercise 26.13—The Group 15 elements

(a) The Lewis structure of N_2O shown in Table 26.8 of the text (also shown below) is not the only structure possible. Two structures, that are not unreasonable, are shown here. However, the first structure (the one shown in Table 26.8 of the text) is the most important because the negative formal charge is on the more electronegative atom—oxygen as opposed to nitrogen.

(b) $NH_4NO_3(s)$ decomposes explosively to give $N_2O(g)$ and $H_2O(g)$. The reaction is exothermic.

Exercise 26.14—The Group 15 elements

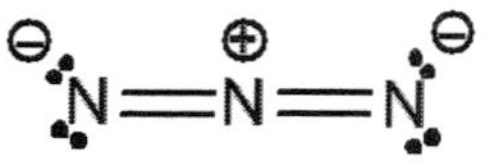

azide ion

Exercise 26.15—The Group 16 elements

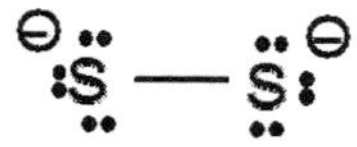

disulfide ion

Exercise 26.16—The Group 17 elements

The order of the halogens according to oxidizing strength is $F_2(aq) > Cl_2(aq) > Br_2(aq) > I_2(aq)$ Therefore, **(a)** $F_2(aq)$ can oxidize $Cl^-(aq)$ ions, $Br^-(aq)$ ions, or $I^-(aq)$ ions (X = F, Y = Cl, Br or I) for example. Other possibilities are X = Cl and Y = Br or I, or X = Br and Y = I.
(b) $Cl_2(aq)$ cannot oxidize $F^-(aq)$ ions (X = Cl, Y = F)—neither can $Br_2(aq)$ or $I_2(aq)$. Neither $Br_2(aq)$ nor $I_2(aq)$ can oxidize $Cl^-(aq)$ ions, and $I_2(aq)$ cannot oxidize $Br^-(aq)$ ions.

Exercise 26.17—Noble gases

(a) Using Slater's rules, we get
$$Z^*(Ar) = +18 - [(7 \times 0.35) + (8 \times 0.85) + (2 \times 1.0)] = +18 - 11.25 = +6.75$$
The valence electrons of Ar are held very tightly. In contrast, the valence electron of K experiences a much smaller effective nuclear charge.
$$Z^*(K) = +19 - [(0 \times 0.35) + (8 \times 0.85) + (10 \times 1.0)] = +19 - 16.8 = +2.2$$
Although the valence electrons of neighbouring Cl are also held tightly, the effective nuclear charge in this case is not quite as large.
$$Z^*(Cl) = +17 - [(6 \times 0.35) + (8 \times 0.85) + (2 \times 1.0)] = +17 - 10.9 = +6.1$$
(b) Using Slater's rules, we get
$$Z^*(Ar^-) = +18 - [(0 \times 0.35) + (8 \times 0.85) + (10 \times 1.0)] = +18 - 16.8 = +1.2$$
The valence electron of Ar^- is held quite weakly. In contrast, the valence electrons of Cl^- experience a much larger effective nuclear charge.
$$Z^*(Cl^-) = +17 - [(7 \times 0.35) + (8 \times 0.85) + (2 \times 1.0)] = +17 - 11.25 = +5.75$$
Even the exotic K^- ion's valence electrons are held more tightly.
$$Z^*(K^-) = +19 - [(1 \times 0.35) + (8 \times 0.85) + (10 \times 1.0)] = +19 - 17.15 = +1.85$$

Exercise 26.18—Noble gases

Assuming ideality (a good assumption for both nitrogen and helium), the density of each gas can be written in terms of its molar mass, M, and the pressure and temperature of the gas.
$$\frac{m}{V} = \frac{nM}{V} = \frac{pM}{RT}$$
Taking the ratio of the densities of nitrogen and helium at the same T and p, gives
$$M_{nitrogen} / M_{helium} = 28.02 \text{ g mol}^{-1} / 4.003 \text{ g mol}^{-1} = 7.00$$

Exercise 26.19—Noble gases
The ionization energies of xenon and krypton are much less than that of neon. So xenon and krypton atoms can enter into reactions by complete removal of electrons to form cations (with highly electronegative elements) or by partial removal to share in covalent bonds. Neon atoms attract their valence electrons too strongly for either of these to happen with common reagents – and yet their electron affinity is too low for them to 'grab' electrons from other species to form anions.

REVIEW QUESTIONS

26.3 Charge density of cations: An explanatory concept

26.21
(a) If $BeCl_2$, BCl_3, $AlCl_3$, $TiCl_4$ and $FeCl_3$ were covalent, the cations would be Be^{2+}, B^{3+}, Al^{3+}, Ti^{4+} and Fe^{3+}. These species are very small and have a high charge. They are consequently very polarizing, and there are no anions that could remain intact next to such cations. In contrast, K^+ ions and Ag^+ ions, which are singly charged and not so small, are not nearly as polarizing. Chloride anions retain their independent anionic form next to these cations—their compounds are ionic.
(b) Although Al^{3+} ions are highly polarizing, giving covalent character to bonds with halides, very small fluoride ions are rather nonpolarizable, so AlF_3 is essentially an ionic compound—the bonding has very little covalent character. Its contrast, chloride and especially bromide are much more polarizable. The compounds, $AlCl_3$ and $AlBr_3$ have significantly more covalent character. These compounds more readily melt to form liquids of mobile $AlCl_3$, and $AlBr_3$ molecules.
(c) Singly charged cations, K^+, Ag^+ and $NH_4{}^+$, are not very polarizing. They do not attract water molecules strongly enough to hold them as coordinating ligands in the ionic crystal lattices of KCl, $AgNO_3$ and NH_4NO_3. In contrast, the polarizing doubly and triply charged cations, Mg^{2+}, Fe^{3+} and Cr^{3+}, retain coordinated water ligands when their salts are precipitated from aqueous solution—i.e., they crystallize as hydrated compounds.
(d) The smaller Li^+ ion is more polarizing than the larger Rb^+ ion. Consequently, Li^+ holds water ligands about it rather tightly, and its mobility in water is that of $[Li(H_2O)_4]^+$ rather than that of free lithium ions. A rubidium ion is insufficiently polarizing to drag water molecules along with it. Its mobility is that of the free cation.

26.23
(a) The molar enthalpy change of hydration of Be^{2+} (2455 kJ mol^{-1}) is greater than that of Ba^{2+} (1275 kJ mol^{-1}) because the former is much smaller. The charges of these ions are the same, but the Coulomb energy of attraction to the negative end of the hydrating water molecules is larger for the smaller beryllium ion.
(b) The molar enthalpy change of hydration of Fe^{3+} (4340 kJ mol^{-1}) is greater than that of Fe^{2+} (1890 kJ mol^{-1}) because the former has a greater charge—it is smaller too since there is one less electron to engage in shielding (electron-electron repulsions).

26.25
The Pb^{4+} cation of lead(IV) chloride is much more polarizing than the Pb^{2+} cation of lead(II) chloride. Pb^{4+} polarizes the chlorides to such an extent that the bonds in lead(IV) chloride have significant covalent character and lead(IV) chloride can exist as individual neutral molecules in a non-polar solvent.

26.4 Hydrogen

26.27

$CH_4(g) + H_2O(g) \rightarrow CO(g) + 3 H_2(g)$

$\Delta H \approx 4\, D(C-H) + 2\, D(O-H) - [D(C\equiv O) + 3\, D(H-H)]$

$\qquad = 4 \times 413 + 2 \times 463 - [1046 + 3 \times 436]$ kJ mol^{-1} = 224 kJ mol^{-1}

The reaction is endothermic because, while the number of product bonds equals the number of reactant bonds (think of the CO bond as 3 bonds), the CO triple bond is less than 3 times either of the reactant single bond energies—i.e., weaker bonds are formed from stronger bonds).

26.29

$\qquad C(s) + H_2O(g) \rightarrow CO(g) + H_2(g)$

$\Delta_r H^\circ = \Delta_f H^\circ[H_2(g)] + \Delta_f H^\circ[CO(g)] - (\Delta_f H^\circ[C(s)] + \Delta_f H^\circ[H_2O(g)])$

$\qquad = \{0 + (-110.53) - (0 + (-241.83))\}$ kJ mol^{-1} = 131.30 kJ mol^{-1}

$\Delta_r S^\circ = S^\circ[H_2(g)] + S^\circ[CO(g)] - (S^\circ[C(s)] + S^\circ[H_2O(g)])$

$\qquad = \{130.7 + 197.7 - (5.6 + 188.84)\}$ J K mol^{-1} = 134.0 J K mol^{-1}

$\Delta_r G^\circ = \Delta_r H^\circ - T\Delta_r S^\circ = 131.30$ kJ $- (298.15$ K$) \times (134.0$ J K mol$^{-1}) \times (1/1000$ kJ/J$) =$
91.3 kJ mol^{-1}

26.5 The alkali metals, Group 1

26.31

$\qquad Na(s) + \frac{1}{2} F_2(g) \rightarrow NaF(s)$

$\qquad Na(s) + \frac{1}{2} Cl_2(g) \rightarrow NaCl(s)$

$\qquad Na(s) + \frac{1}{2} Br_2(l) \rightarrow NaBr(s)$

$\qquad Na(s) + \frac{1}{2} I_2(s) \rightarrow NaI(s)$

All alkali metal halides are ionic solids. They are crystalline, brittle, dissolve in water and are colourless (M^+ and X^- are closed shell species).

26.6 The alkaline earth elements, Group 2

26.33

$\qquad Mg(s) + \frac{1}{2} O_2(g) \rightarrow MgO(s)$

$\qquad 3\, Mg(s) + N_2(g) \rightarrow Mg_3N_2(s)$

26.35

Limestone, $CaCO_3$, is used in agriculture. It is added to fields to neutralize acidic soil, to provide a source of Ca^{2+}, an essential nutrient for plants (and animals). Since there is generally a magnesium impurity, it also provides a source of Mg^{2+}, another essential nutrient. Limestone is also used to make lime, CaO, which is used to make bricks and mortar.

26.37
Start with a balanced chemical reaction,
$$CaO(s) + SO_2(g) \rightarrow CaSO_3(s)$$
Amount of CaO(s), $n = 1.2 \times 10^6$ g / 56.08 g mol^{-1} = 2.1×10^4 mol
Mass of $SO_2(g)$ that can be removed = 2.1×10^4 mol $\times$ 64.07 g mol^{-1} = 1.3×10^6 g = 1.3×10^3 kg

26.39
Because the equilibrium constant for the above reaction is so large, $\sim 10^7$, it can be used to obtain magnesium hydroxide solid (wherein magnesium ions are concentrated in an ionic solid) from sea water with very small concentrations. Magnesium is effectively concentrated by many orders of magnitude in the process.

26.41
Beryllium is a light weight, high strength metal that is stable at high temperatures and has a low thermal expansivity. Consequently, beryllium and beryllium alloys are used in airplane (especially high speed planes), missile, and spacecraft construction. It is also used in high energy physics experiments to make windows and filters. Because of its low density it is relatively transparent to X-rays, for example, yet has the high strength and stiffness of much heavier metals. Beryllium dust generated in manufacturing processes is very toxic. People breathing this dust over a period of time can develop berylliosis—an inflamatory disease of the lungs wherein sufferers experience coughing, shortness of breath, fever, and weight loss.

26.7 Boron, aluminum, and the Group 13 elements

26.43
$$2\,BH_4^- + 2\,I_2 \rightarrow B_2H_6 + 2\,HI + 2\,I^-$$
Iodine gets reduced and the borohydride ion, BH_4^-, gets oxidized (specifically, one of the H atoms goes from oxidation state -1 to $+1$).

26.45

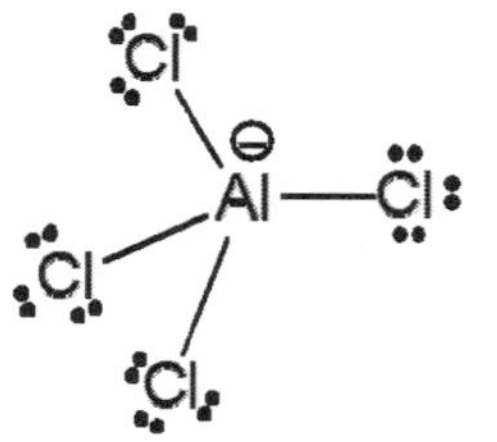

$AlCl_4^-$ is tetrahedral—it has the AX_4 electron domain configuration.

26.47
(a) $BCl_3(g) + 3 H_2O(l) \rightarrow B(OH)_3(s) + 3 H^+(aq) + 3 Cl^-(aq)$
 or $BCl_3(g) + 6 H_2O(l) \rightarrow B(OH)_3(s) + 3 H_3O^+(aq) + 3 Cl^-(aq)$
(b) $\Delta_r H° =$
$\Delta_f H°[B(OH)_3(s)] + 3 \times \Delta_f H°[HCl(aq)] - (\Delta_f H°[BCl_3(g)] + 3 \times \Delta_f H°[H_2O(l)])$
$= \{-1094 + 3(-92.3) - ((-403) + 3(-285.8))\}$ kJ mol^{-1} $= -110.5$ kJ mol^{-1}

26.8 Silicon and the Group 14 elements

26.49
Elemental silicon can be prepared from sand by heating purified silica (i.e., sand) and coke.
 $SiO_2(s) + 2 C(s) \rightarrow Si(l) + 2 CO(g)$
Molten silicon is drawn from the bottom of the furnace. Its purity is not sufficient for electronics applications. Consequently, the elemental silicon is reacted with chlorine to make silicon tetrachloride.
 $Si(s) + 2 Cl_2(g) \rightarrow SiCl_4(l)$
The low boiling silicon tetrachloride (boiling point = 57.6 °C) is purified by distillation, then reacted with extremely pure magnesium (or zinc).
 $SiCl_4(g) + 2 Mg(s) \rightarrow 2 MgCl_2 + Si(s)$
The magnesium chloride product is washed away with water—it is water soluble, while silicon is not. Silicon produced in this way is further purified by zone refining wherein a narrow cylinder of silicon is heated and melted in one segment along the cylinder. The heating slowly traverses the length of the cylinder. Impurities concentrate in the melted segment, leaving higher purity silicon behind. This is because of melting point depression—impure silicon has a lower melting point than pure silicon.

26.51
(a) $Si + 2 CH_3Cl \rightarrow (CH_3)_2SiCl_2$
(b) Amount of Si $= 2.65$ g $/ 28.09$ g mol^{-1} $= 0.0943$ mol
Amount of chloromethane needed $= 2 \times 0.0943$ mol $= 0.1886$ mol
Pressure of chloromethane needed, $p = nRT/V = 0.1886$ mol $\times 8.314$ L kPa K^{-1} mol^{-1} $\times 298.15$ K $/ 5.60$ L $= 83.5$ kPa
(c) Amount of $(CH_3)_2SiCl_2$ produced, $n = 0.0943$ mol
Mass of $(CH_3)_2SiCl_2$ produced, $m = 0.0943$ mol $\times 129.06$ g mol^{-1} $= 12.2$ g

26.9 Group 15 elements and their compounds

26.53
 $Ca(OH)_2(s) + H_3PO_4(aq) \rightarrow CaHPO_4(s) + 2 H_2O(l)$

26.55

$$2\,NO_2(g) \longrightarrow N_2O_4(g)$$

$\Delta_rH° = \Delta_fH°[N_2O_4(g)] - 2 \times \Delta_fH°[NO_2(g)]$
$\quad = (9.66 - 2 \times 33.84)\ kJ\ mol^{-1} = -58.02\ kJ\ mol^{-1}$
$\Delta_rG° = \Delta_fG°[N_2O_4(g)] - 2 \times \Delta_fG°[NO_2(g)]$
$\quad = (98.28 - 2 \times 51.84)\ kJ\ mol^{-1} = -5.40\ kJ\ mol^{-1}$

This reaction is exothermic and at 25 °C and with both substances at 1 bar pressure, the reaction is spontaneous in the direction written.

26.57

Half reactions (it is convenient to use the shorthand, $H^+(aq)$):

$\quad N_2H_5^+(aq) \rightarrow N_2(g) + 5\,H^+(aq) + 4\,e^-$
$\quad IO_3^-(aq) + 6\,H^+(aq) + 5\,e^- \rightarrow \frac{1}{2}I_2(s) + 3\,H_2O(l)$

Balance overall reaction by balancing the electrons (i.e., 5×1st reaction $+ 4 \times 2^{nd}$ reaction):

$\quad 5\,N_2H_5^+(aq) + 4\,IO_3^-(aq) \rightarrow 5\,N_2(g) + H^+(aq) + 2\,I_2(s) + 12\,H_2O(l)$
$E° = E°[IO_3^-(aq)\,|\,I_2(s)] - E°[N_2H_5^+(aq)\,|\,N_2(g)]$
$\quad = 1.195 - (-0.23)\ V = 1.43\ V$

26.59
(a) Lewis structures for N_2O_3

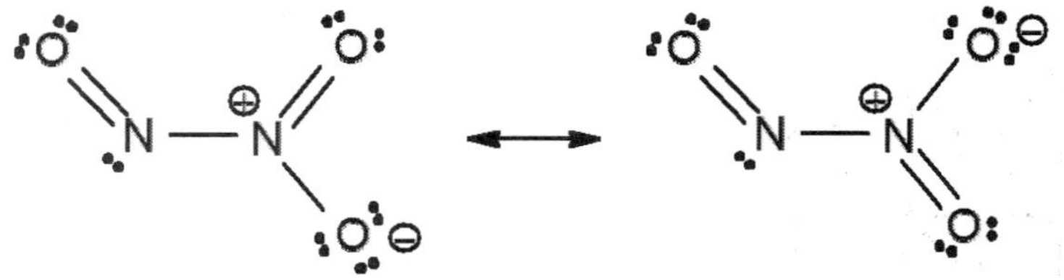

Here, we see that two almost equivalent resonance structures can be drawn, giving the two N–O bonds on the right bond orders of about 1 ½. Resonance structures showing the N–O bond on the left as anything but a double bond are not at all dominant. This latter bond is thus more of a double bond. This is consistent with its observed shorter length.

(b) $\Delta_rS° = (\Delta_rH° - \Delta_rG°)/T =$
$\quad = (40.5 - (-1.59)\ kJ\ mol^{-1}) \times 10^3\ J\ kJ^{-1}/298\ K = 141\ J\ K^{-1}\ mol^{-1}$
$K = \exp(-\Delta_rG°/RT) =$
$\quad = \exp(-(-1.59 \times 10^3\ J\ mol^{-1})/(8.314\ J\ K^{-1}\ mol^{-1} \times 298\ K)) = 141$

(c)

$$N_2O_3(g) \longrightarrow NO(g) + NO_2(g)$$

$\Delta_rH° = \Delta_fH°[NO(g)] + \Delta_fH°[NO_2(g)] - \Delta_fH°[N_2O_3(g)] = 40.5\ kJ\ mol^{-1}$

Therefore,

$\Delta_fH°[N_2O_3(g)] = \Delta_fH°[NO(g)] + \Delta_fH°[NO_2(g)] - 40.5\ kJ\ mol^{-1}$
$\quad = (90.37 + 33.84 - 40.5)\ kJ\ mol^{-1} = 83.71\ kJ\ mol^{-1}$

26.10 The Group 16 elements and their compounds

26.61

$$2\,ZnS(s) + 3\,O_2(g) \longrightarrow 2\,ZnO(s) + 2\,SO_2(g)$$

$\Delta_r H° = 2 \times \Delta_f H°[ZnO(s)] + 2 \times \Delta_f H°[SO_2(g)] - (2 \times \Delta_f H°[ZnS(s)] + 3 \times \Delta_f H°[O_2(g)])$
$\quad\ = \{2 \times (-348.0) + 2 \times (-296.8) - (2 \times (-203) + 3 \times 0)\}\ kJ\ mol^{-1}$
$\quad\ = -884\ kJ\ mol^{-1}$
$\Delta_r G° = 2\,\Delta_f G°[ZnO(s)] + 2\,\Delta_f G°[SO_2(g)] - (2\,\Delta_f G°[ZnS(s)] + 3\,\Delta_f G°[O_2(g)])$
$\quad\ = \{2 \times (-318.2) + 2 \times (-300.2) - (2 \times (-198) + 3 \times 0)\}\ kJ\ mol^{-1}$
$\quad\ = -841\ kJ\ mol^{-1}$

This reaction is product-favoured—$\Delta_r G° < 0$—and it is exothermic which means that it is less product-favoured at higher temperature.

26.63

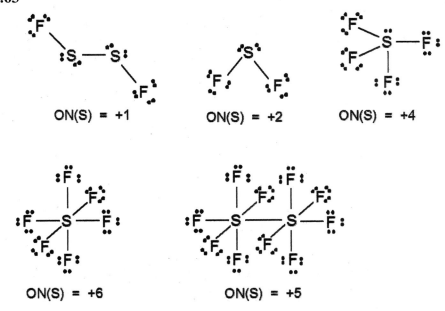

ON(S) = +1   ON(S) = +2   ON(S) = +4

ON(S) = +6   ON(S) = +5

26.11 The halogens, Group 17

26.65
Any species with reduction potential greater than that of chlorine (i.e., > 1.36 V) can be used. For example,

$\quad F_2(g) + 2\,Cl^-(aq) \rightarrow 2\,F^-(aq) + Cl_2(g)$
$\quad H_2O_2(aq) + 2\,H^+(aq) + 2\,Cl^-(aq) \rightarrow 2\,H_2O(l) + Cl_2(g)$
$PbO_2(s) + SO_4^{2-}(aq) + 4\,H^+(aq) + 2\,Cl^-(aq) \rightarrow PbSO_4(s) + 2\,H_2O(l) + Cl_2(g)$
$2\,MnO_4^-(aq) + 16\,H^+(aq) + 10\,Cl^-(aq) \rightarrow 2\,Mn^{2+}(aq) + 8\,H_2O(l) + 5\,Cl_2(g)$

26.67

The reduction potential of $MnO_4^-(aq)$ is 1.51 V under standard conditions. Since this is greater than 1.44 V, it is NOT possible to oxidize $Mn^{2+}(aq)$ with $BrO_3^-(aq)$.

26.69

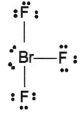

26.71

Amount of CaF_2 required $= 1.0 \times 10^6$ L $\times 2.0 \times 10^{-5}$ mol L^{-1} $= 20.$ mol

Mass of CaF_2 required $= 20.$ mol $\times 78.07$ g mol^{-1} $= 1560$ g $= 1.6$ kg (to two significant figures)

SUMMARY AND CONCEPTUAL QUESTIONS

26.73
$P_4O_{10}(s)$ and $SO_3(g)$ are acidic oxides—potassium and sulfur are non-metals.
$$P_4(s) + 5 O_2(g) \rightarrow P_4O_{10}(s)$$
$$S_8(s) + 12 O_2(g) \rightarrow 8 SO_3(g)$$
The acidic oxides react with water to form acidic solutions:
$$P_4O_{10}(s) + 6 H_2O(l) \rightarrow 4 H_3PO_4(aq)$$ phosphoric acid, which ionizes in aqueous solution to form $H_3O^+(aq)$.
$$SO_3(g) + H_2O(l) \rightarrow HSO_4^-(aq)$$ aquated hydrogensulfate ions, which ionize in aqueous solution to form $H_3O^+(aq)$.

26.75
S^{2-}, Cl^-, K^+ and Ca^{2+} ions are monatomic ions with the same electron configuration as argon.

26.77
In order of increasing basicity, we have
$$CO_2 < SiO_2 < SnO_2$$

26.79
(a) $2 Na(s) + Br_2(l) \rightarrow 2 NaBr(s)$
(b) $2 Mg(s) + O_2(g) \rightarrow 2 MgO(s)$
(c) $2 Al(s) + 3 F_2(g) \rightarrow 2 AlF_3(s)$
(d) $C(s) + O_2(g) \rightarrow CO_2(g)$

26.81
(a) Electrolysis of aqueous sodium chloride:
$$2 Na^+(aq) + 2 Cl(aq) + 2 H_2O(l) \rightarrow 2 Na^+(aq) + 2 OH(aq) + Cl_2(g) + H_2(g)$$
(b) Electrolysis of aqueous sodium chloride is not the only source of sodium hydroxide, or of hydrogen. It is, however, the principle source of chlorine.

26.83
(a) $Li(s)$ and $Be(s)$ are metals
  $B(s)$ is a metalloid
  $C(s)$, $N_2(g)$, $O_2(g)$, $F_2(g)$ and $Ne(g)$ are non-metallic substances.
(b) + (c) $Li(s)$, $Be(s)$ and $B(s)$ are silver-white shiny, with metallic appearance. C is a black solid (graphite) or a lustrous tansparent very hard crystalline material (diamond). $N_2(g)$, $O_2(g)$, $F_2(g)$ and $Ne(g)$ are all gases, colourless except for a yellow tinge in the case of fluorine.

26.85

(a) 2 KClO$_3$(s) + heat → 2 KCl + 3 O$_2$(g)

(b) 2 H$_2$S(g) + 3 O$_2$(g) → 2 H$_2$O(l) + 2 SO$_2$(g)

(c) 2 Na(s) + O$_2$(g) → Na$_2$O$_2$(s)

(d) P$_4$(s) + 3 OH$^-$(aq) + 3 H$_2$O(l) → PH$_3$(g) + 3 H$_2$PO$_4^-$(aq)

(e) 2 NH$_4$NO$_3$(s) + heat → N$_2$O(g) + 2 H$_2$O(g)

(f) 2 In(s) + 3 Br$_2$(l) → 2 InBr$_3$(s)

(g) SnCl$_4$(l) + 4 H$_2$O(l) → Sn(OH)$_4$(aq) + 4 H$^+$(aq)

26.87

Dry, inert powders are used to extinguish sodium fires. A class D fire extinguisher is required. The worst thing you can do is put water on the fire, as the sodium will react violently with water.

26.89

The amount of gas produced by heating 1.00 g A is

$$n(\text{gas}) = \frac{pV}{RT} = \frac{(27.86\text{ kPa})(0.450\text{ L})}{(8.314\text{ L kPa K}^{-1}\text{ mol}^{-1})(298\text{ K})} = 0.00506\text{ mol}$$

Bubbling this gas through Ca(OH)$_2$(aq) gives a white solid, C. This suggests the gas is carbon dioxide. Carbon dioxide reacts in aqueous calcium hydroxide to form insoluble calcium carbonate, CaCO$_3$(s). CO$_2$(g) is also a typical product in the thermal decomposition of carbonates, suggesting A is a carbonate. The other decomposition product, B, would then be an oxide. This is consistent with it forming a basic aqueous solution. If the gas were CO$_2$, then the mass of 0.00506 mol would be

0.00506 mol × 44.01 g mol^{-1} = 0.223 g

This leaves 1.00 − 0.223 = 0.777 g as the mass of the oxide. If we take the decomposition reaction to be (M is replaced by M$_2$ in case of a univalent metal, etc.)

MCO$_3$(s) + heat → MO(s) + CO$_2$(g),

then the molar mass of MO(s) is determined to be

0.777 g / 0.00506 mol = 154 g mol^{-1}

Subtracting the molar mass of O leaves the molar mass of M,

154 g mol^{-1} − 16 g mol^{-1} = 138 g mol^{-1}

This is pretty close to the molar mass of barium. This identification is verified by the green coloured flame, characteristic of barium. Also, the sulfate precipitate is characteristic of barium.

A = BaCO$_3$(s)

B = BaO(s)

C = CaCO$_3$(s)

D = BaCl$_2$(s)

E = BaSO$_4$(s)

Chapter 27
Transition Elements and Their Compounds

IN-CHAPTER EXERCISES

Exercise 27.1—Ligands

(a) methylamine, CH_3NH_2 , **(b)** methyl nitrile, CH_3CN, **(c)** azide ions, N_3^- and **(e)** bromide ions, Br^- are monodentate ligands.
(d) en (ethylenediamine, $H_2NCH_2CH_2NH_2$) and **(f)** phen (phenanthroline, $C_{12}H_8N_2$) are bidentate (special case of polydentate) ligands—i.e., they can form two coordination bonds with metals.

Exercise 27.2—Ligands

NH_3 molecules and NH_2^- ions are ligands. The nitrogen atom has a lone pair (two in the case of the amide ion, NH_2^-) that can form coordination bonds with metals. Ammonium, NH_4^+, has no lone pairs and so cannot be a ligand—it cannot use lone pairs to form covalent bonds.

Exercise 27.3—Formulas of coordination complexes

(a) A complex ion composed of one Co^{3+} ion, three ammonia molecules, and three Cl^- ions, has the formula
$[Co(NH_3)_3Cl_3]$
The ammonia ligands are neutral and there are three chlorides each with a charge of -1. So the net charge coming from the ligands is -3. Since the cobalt ion has a charge of $+3$, the net charge on the complex is zero—it is not an ion.

(b)(i) The net charge on the complex in $K_3[Co(NO_2)_6]$ is -3. We deduce this because each of the three potassium ions has a charge of $+1$. Since each of the nitrite, NO_2^-, ions has a -1 charge, and there are 6 of them in the complex, the net contribution to the complex charge from the ligands is -6. A net charge of -3 for the complex results because the cobalt ion has a charge of $+3$. The coordination number of cobalt is 6: it is bonded to 6 donor atoms.

(ii) The $[Mn(NH_3)_4Cl_2]$ complex is neutral. The contribution to this charge of zero from the ligands is -2, arising solely from the two -1 charged chloride ions (the ammonia ligands are neutral). Thus, the charge on the manganese ion must be $+2$. The coordination number of manganese is 6: it is bonded to 6 donor atoms.

Exercise 27.4—Speciation among complex ions

(a)

$$K_{f1} = \frac{[[Zn(OH_2)_5(CN)]^+]}{[[Zn(OH_2)_6]^{2+}][CN^-]} = 5.0 \times 10^5$$

If $[[Zn(OH_2)_5(CN)]^+] = [[Zn(OH_2)_6]^{2+}]$, then $[CN^-] = 1/(5.0 \times 10^5) = 2.0 \times 10^{-6}$ mol L^{-1}.
(b) From the second equilibrium constant, we get $[[Zn(OH_2)_4(CN)]_2] = [[Zn(OH_2)_5(CN)]^+]$ when $[CN^-] = 1/K_{f2} = 1/(2.5 \times 10^5) = 4.0 \times 10^{-6}$ mol L^{-1}.
(c)

$$\frac{[[Zn(OH_2)_5(CN)]^+]}{[[Zn(OH_2)_6]^{2+}]} = 5.0 \times 10^5 \times [CN^-]$$

$$= 5.0 \times 10^5 \times 0.10 = 5.0 \times 10^4$$

Exercise 27.5—The chelate effect

(i) $\Delta_r G° = \Delta_r H° - T\Delta_r S° = -12.1$ kJ mol^{-1} $-$ 298 K $\times$ 185 J K^{-1} mol^{-1} / 1000 J kJ^{-1} = -67.2 kJ mol^{-1}
The entropy term accounts for the bigger part of this negative change in free energy. The large positive entropy of this process results because 6 ligands are freed with only 3 bidentate ligands taking their place. The final arrangement of atoms is more constrained.
(ii) $K = \exp(-\Delta_r G°/RT) = \exp(67.2 \times 10^3$ J mol^{-1} / (8.314 J K^{-1} mol^{-1} $\times$ 298 K)) = 6.02×10^{11}
This reaction is very product-favoured: at equilibrium, the concentrations of product species will heavily dominate the concentrations of reactants species.

Exercise 27.6—Constitutional isomerism in transition metal complexes

Only 1.0 mol of AgCl(s) will precipitate. The 2 mol of chloride ions within the complex ion are "locked up" and are unavailable to form silver chloride.

Exercise 27.7—Constitutional isomerism in transition metal complexes

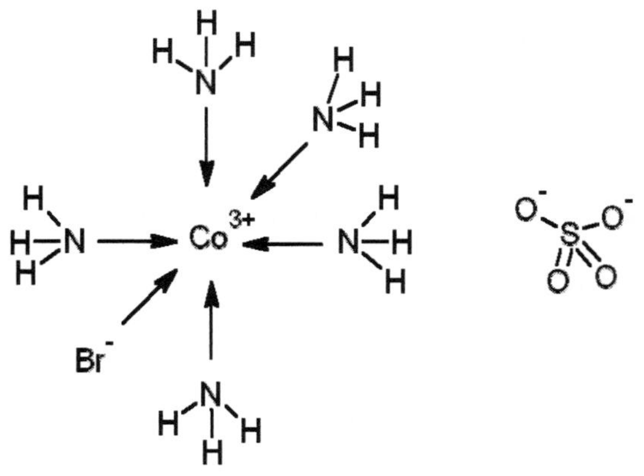

dark violet isomer (A)

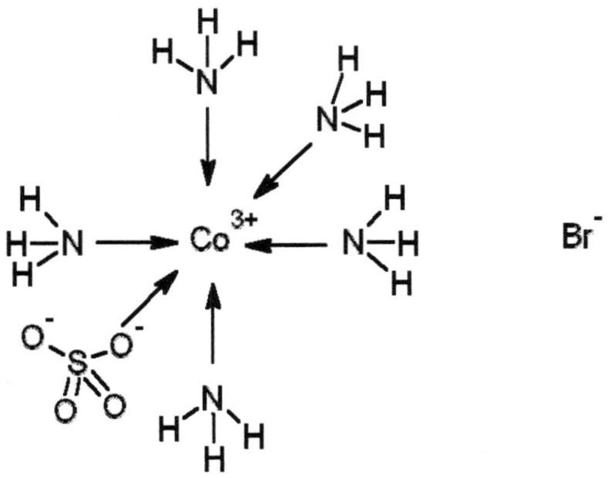

violet-red isomer (B)

(A) = [Co(NH$_3$)$_5$ Br]SO$_4$
(B) = [Co(NH$_3$)$_5$ SO$_4$]Br
Upon dissolution of (A):

$$[Co(NH_3)_5 Br]SO_4(s) \rightarrow [Co(NH_3)_5 Br]^{2+}(aq) + SO_4^{2-}(aq)$$

Upon dissolution of (B):

$$[Co(NH_3)_5 SO_4]Br(s) \rightarrow [Co(NH_3)_5 SO_4]^{+}(aq) + Br^{-}(aq)$$

Adding Ba^{2+}(aq) to the (A) solution gives a barium sulfate precipitate.

$$Ba^{2+}(aq) + SO_4^{2-}(aq) \rightarrow BaSO_4(s)$$

Adding Ba^{2+}(aq) to the (B) solution gives no reaction because the sulfate ions are tied up in the complex and are not available to form the precipitate.

Exercise 27.8—Stereoisomers of coordination complexes

(a) $[Fe(NH_3)_4Cl_2]$ *cis-* and *trans-*isomers

(b) $[Pt(NH_3)_2(SCN)(Br)]$ *cis-* and *trans-*isomers

(c) $[Co(NH_3)_3(NO_2)_3]$ *mer-* and *fac-* isomers

(d) $[Co(en)Cl_4]^-$ No stereoisomerism is possible

Exercise 27.9—Isomerism in coordination complexes

(a) $[Co(NH_3)_2Cl_4]$ has *cis* and *trans* isomers with the ammonia ligands next to each other or on opposite sides of the cobalt ion. Neither of these has enantiomers (neither has chiral molecules).
(b) $[Pt(en)Cl_2]$ does not have *cis* and *trans* isomers: the ethylenediamine ligand must occupy adjacent positions on the metal as, therefore, must the Cl^- ion ligands. There is only one way to do this. The complex has a plane of symmetry, so there are no enantiomers.
(c) $[Co(NH_3)_5Cl]^{2+}$ has no isomers since there is only one chloride ligand—putting it in any of the 6 positions produces the same complex ion. The complex ion has a plane of symmetry, so there are no enantiomers.
(d) There are two enantiomers of $[Ru(phen)_3]^{3+}$ ions. Each phen occupies adjacent positions on the metal ion, limiting the number of possible isomers. If we draw (or build) the structure of this complex ion, it can be seen that it is not superimposable upon its mirror image. It is chiral, and there are enantiomers.

(e) $[MnCl_4]^{2-}$ has no isomers—all the ligands are the same.
(f) $[Co(NH_3)_5NO_2]^{2+}$ There are two linkage isomers with this formula—one with the NO_2^- ion ligand bound to the Co^{3+} ion through a lone pair on the N atom, and one with the NO_2^- ion ligand bound through a lone pair on an O atom.

Exercise 27.10—Isomerism in coordination complexes

(a) Like $[Ru(phen)_3]^{3+}$ in Exercise 27.9(d), $[Fe(en)_3]^{2+}$ has a stereocentre at the Fe^{2+} ion: the complex ion is chiral and there are two enantiomers.
(b) *trans*-$[Co(en)_2Br_2]^+$ ion has a plane of symmetry, so is not chiral. There is not a stereocentre. (The cis isomer has a chiral centre.)
(c) The cobalt ion in *fac*-$[Co(en)(H_2O)Cl_3]$ is not a stereocentre. The complex has a plane of symmetry and is not chiral.
(d) The platinum ion in $[Pt(NH_3)(H_2O)(Cl)(NO_2)]$. There are two stereoisomers of this complex—one with the H_2O ligand *cis* to the Cl ligand, and one with the H_2O ligand *trans* to the Cl ligand. In both cases, the mirror images are superimposable: there is not a stereocentre. (If the complex were tetrahedral, it would be chiral and there would be enantiomers.)

Exercise 27.11—Colours of coordination complexes

According to the size of the crystal field splitting, these species are ordered in terms of energy of photons absorbed as follows:
$[Ti(OH_2)_6]^{3+} < [Ti(NH_3)_6]^{3+} < [Ti(CN)_6]^{3-}$
Here, the strength of the splitting is determined by the ligand, as these species are otherwise the same.

Exercise 27.12—Magnetism of transition metal complexes

(a) In $[Ru(H_2O)_6]^{2+}$, Ru^{2+} is in the +2 oxidation state—it has 6 d electrons. The low and high-spin d electron configurations are depicted as follows:

(b) In $[Ni(NH_3)_6]^{2+}$, Ni^{2+} is in the +2 oxidation state—it has 8 d electrons. This is too many electrons to have high and low spin configurations. Only metals with 4 to 7 d electrons have high and low spin configurations.

$$\frac{\uparrow}{d_{x^2-y^2}} \quad \frac{\uparrow}{d_{z^2}}$$

$$\frac{\uparrow\downarrow}{d_{xy}} \quad \frac{\uparrow\downarrow}{d_{xz}} \quad \frac{\uparrow\downarrow}{d_{yz}}$$

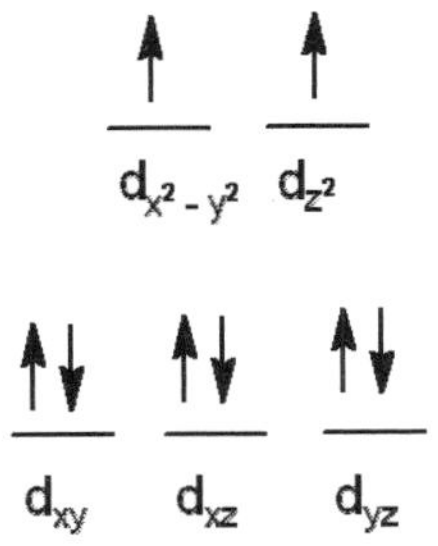

low spin

paramagnetic

Exercise 27.13—Magnetism of transition metal complexes

(a)

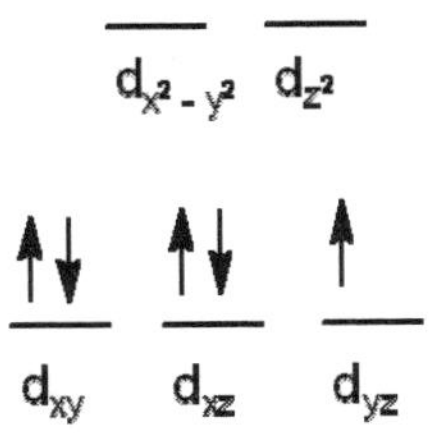

$$\frac{\quad}{d_{x^2-y^2}} \quad \frac{\quad}{d_{z^2}}$$

$$\frac{\uparrow\downarrow}{d_{xy}} \quad \frac{\uparrow\downarrow}{d_{xz}} \quad \frac{\uparrow}{d_{yz}}$$

paramagnetic

$[Mn(CN)_6]^{4-}$ 5 d electrons

(c)

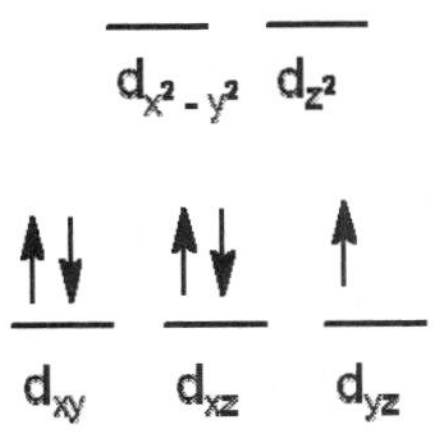

$$\frac{\quad}{d_{x^2-y^2}} \quad \frac{\quad}{d_{z^2}}$$

$$\frac{\uparrow\downarrow}{d_{xy}} \quad \frac{\uparrow\downarrow}{d_{xz}} \quad \frac{\uparrow}{d_{yz}}$$

paramagnetic

$[Fe(H_2O)_6]^{3+}$ 5 d electrons

(b)

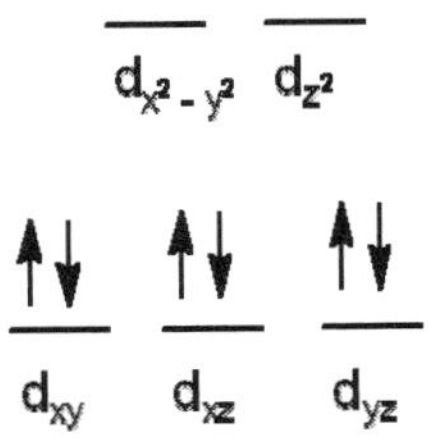

$$\frac{\quad}{d_{x^2-y^2}} \quad \frac{\quad}{d_{z^2}}$$

$$\frac{\uparrow\downarrow}{d_{xy}} \quad \frac{\uparrow\downarrow}{d_{xz}} \quad \frac{\uparrow\downarrow}{d_{yz}}$$

diamagnetic

$[Co(NH_3)_6]^{3+}$ 6 d electrons

(d)

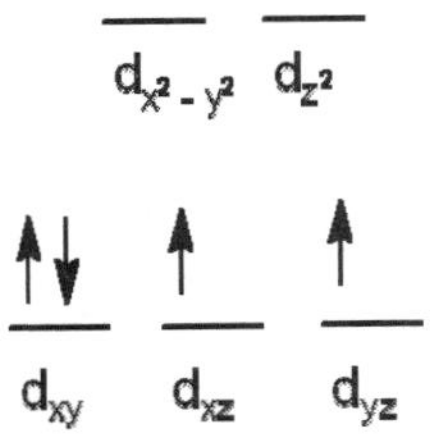

$$\frac{\quad}{d_{x^2-y^2}} \quad \frac{\quad}{d_{z^2}}$$

$$\frac{\uparrow\downarrow}{d_{xy}} \quad \frac{\uparrow}{d_{xz}} \quad \frac{\uparrow}{d_{yz}}$$

paramagnetic

$[Cr(en)_3]^{2+}$ 4 d electrons

REVIEW QUESTIONS

27.2 The *d*-block elements and compounds

27.15
(a) $[Ar] 3d^6$ Fe^{2+} and Co^{3+}
(b) $[Ar] 3d^{10}$ Cu^+ and Zn^{2+}
(c) $[Ar] 3d^5$ Mn^{2+} and Fe^{3+}
(d) $[Ar] 3d^8$ Co^+ and Ni^{2+}

27.3 Metallurgy

27.17
(a) Here, chromium goes from oxidation state +3 to 0, while aluminum goes from 0 to +3. The number of electrons gained by each chromium atom equals the number of electrons lost by each aluminum atom. Thus, it is just a matter of balancing the chromium and aluminum atoms.

$$Cr_2O_3(s) + 2\,Al(s) \longrightarrow Al_2O_3(s) + 2\,Cr(s)$$

(b) Each mol of $TiCl_4$(l) gains 4 mol electrons, while 2 mol electrons are removed from each mol of magnesium atoms. Therefore, there is a 1:2 reaction stoichiometry.

$$TiCl_4(l) + 2\,Mg(s) \longrightarrow Ti(s) + 2\,MgCl_2(s)$$

(c) Ag goes from the +1 to the 0 state, while Zn goes from 0 to +2. Therefore, there is a 2:1 reaction stoichiometry.

$$2\left[Ag(CN)_2\right]^-(aq) + Zn(s) \longrightarrow 2\,Ag(s) + \left[Zn(CN)_4\right]^{2-}(aq)$$

(d) Mn goes from an average oxidation state of 8/3 to 0. Al goes from 0 to +3
Therefore, there is a $3: 8/3 = 9:8$ ratio of amounts of manganese atoms (in Mn_3O_4) to aluminum atoms in Al(s) that react.

$$3\,Mn_3O_4(s) + 8\,Al(s) \longrightarrow 9\,Mn(s) + 4\,Al_2O_3(s)$$

27.4 Coordination compounds

27.19
$[Cr(en)_2(NH_3)_2]^{3+}$
Since both en and NH_3 are neutral, the complex has the charge of the chromium ion, +3.

27.21

First, dissolve a known amount of $CrCl_3$ in a known volume of water. Actually, we would require a volume of water less than desired final volume of solution, then top up the volume with water to account for the change in volume resulting from dissolution. This process is repeated several times, except that sodium chloride is also added to the solution. The resulting solutions all have the same (known) total concentration of Cr^{3+}, and different (known) total concentrations of chloride. Excess $AgNO_3(aq)$ is then added to each of these solutions. There will be a precipitate of silver chloride. After filtering, drying and weighing the precipitate, we can determine how much chloride was available for precipitation in each solution. Since the total amount of chloride is known for each solution, we can determine the amount of chloride that was not available for precipitation in each case. This is the amount of chloride that was part of the complex. Since the total concentration of chromium is known in each case, we now know the ratio of chromium to complex-bound chloride for each solution. These values are then plotted against free chloride concentration—determined by the amount of precipitate. This plot should show whole number plateaus in ranges of chloride concentrations determined by the successive formation constants for the various species. We should be able to identify the ranges that correspond to $[Cr(H_2O)_6]^{3+}$, $[Cr(H_2O)_5Cl]^{2+}$, and $[Cr(H_2O)_4Cl_2]^+$. These correspond to 0, 1, and 2 bound chloride per chromium. According to the information given, we need to add diethylether to obtain $[Cr(H_2O)_3Cl_3]$. Thus, an additional experiment wherein diethylether is added to a solution from the experiment described—before the precipitation reaction —in order to obtain the neutral complex species. The neutral compound will partition into the diethylether layer. It can be separated off, and the neutral complex can be obtained by evaporating the ether. Also, once the solutions with specific charged species are identified, we can evaporate the water and crystallize the distinct ionic solid compounds.

There may be other ways. Discuss as a group.

27.5 Complexation equilibria, stability of complexes

27.23

(a)

$$[Cd(OH_2)_6]^{2+} + en \rightleftharpoons [Cd(OH_2)_4(en)]^{2+} + 2H_2O(l) \qquad K_{f1} = 2.5\times10^5$$

$$[Cd(OH_2)_4(en)]^{2+} + en \rightleftharpoons [Cd(OH_2)_2(en)_2]^{2+} + 2H_2O(l) \qquad K_{f2} = 3.2\times10^4$$

$$\underline{[Cd(OH_2)_2(en)_2]^{2+} + en \rightleftharpoons [Cd(en)_3]^{2+} + 2H_2O(l) \qquad\qquad K_{f3} = 63}$$

$$[Cd(OH_2)_6]^{2+} + 3\,en \rightleftharpoons [Cd(en)_3]^{2+} + 6H_2O(l) \qquad K_f = K_{f1}\times K_{f2}\times K_{f3} = 5.0\times10^{11}$$

(b)

$$\frac{[Cd(OH_2)_4(en)]^{2+}}{[Cd(OH_2)_6]^{2+}[en]} = K_{f1} = 2.5\times10^5$$

$$\frac{[Cd(OH_2)_4(en)]^{2+}}{[Cd(OH_2)_6]^{2+}} = K_{f1}\times[en] = 2.5\times10^5\times1\times10^{-3} = 250$$

(c) $[Cd(OH_2)_4(en)]^{2+} = [Cd(OH_2)_2(en)_2]^{2+}$ when $[en] = 1/K_{f2} = 3.1\times10^{-5}$ M

27.25

Formation constants for the formation of complexes with ammonia.

Complex	Co^{2+}	Ni^{2+}	Cu^{2+}	Zn^{2+}
K_f also known as β	7.7×10^4	5.6×10^8	6.8×10^{12}	2.9×10^9

Here we see the order $Co^{2+} < Ni^{2+} < Cu^{2+} > Zn^{2+}$ for the stability of the ammine complexes. This is consistent with the Irving-Williams series.

27.7 Isomerism in coordination complexes

27.27

The square-planar complex, $[Pt(NH_3)(CN)Cl_2]^-$, has two stereoisomers—*cis* and *trans*.

27.29

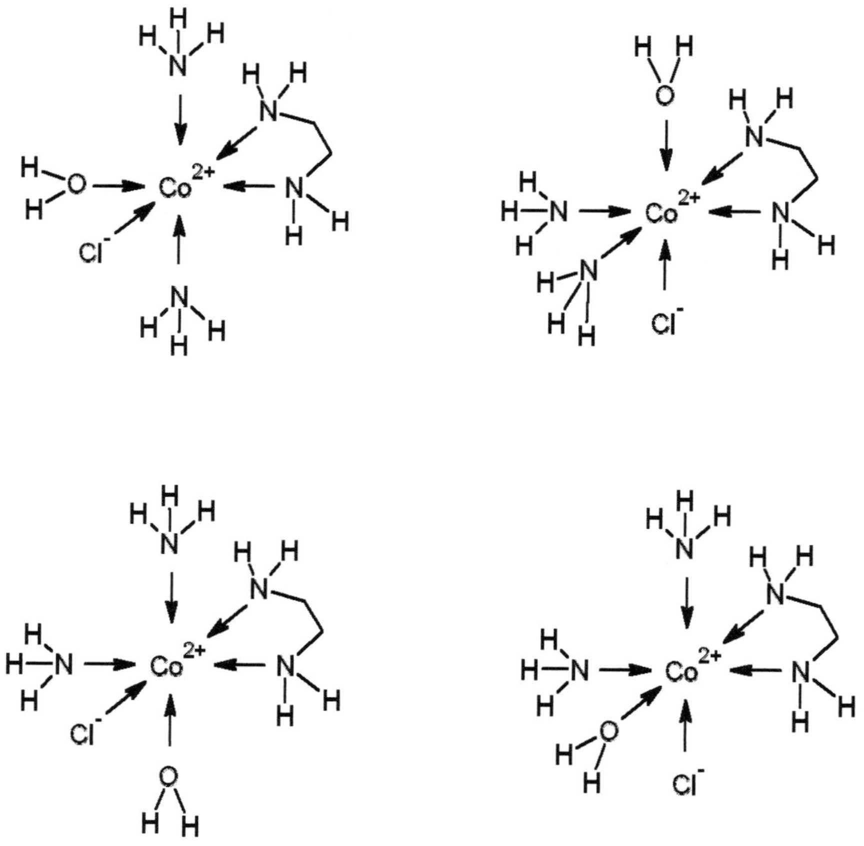

enantiomers

27.31

(a)

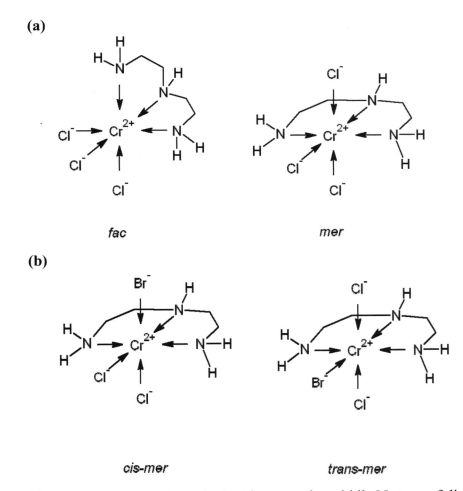

fac   mer

(b)

cis-mer   trans-mer

These complexes are inequivalent because the middle N atom of dien is inequivalent to the outer two N atoms. In the complex on the left, the bromide is *cis* to this N atom—*trans* on the right.

(c)

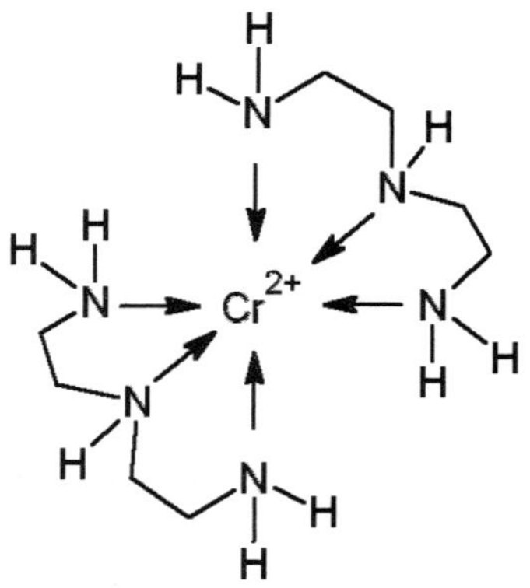

trans-fac 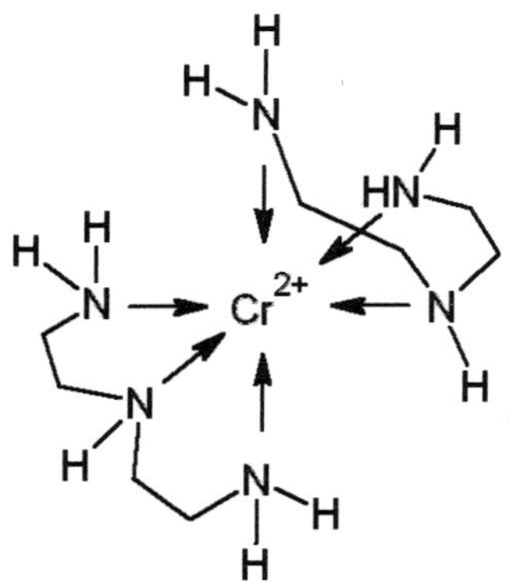   cis-fac

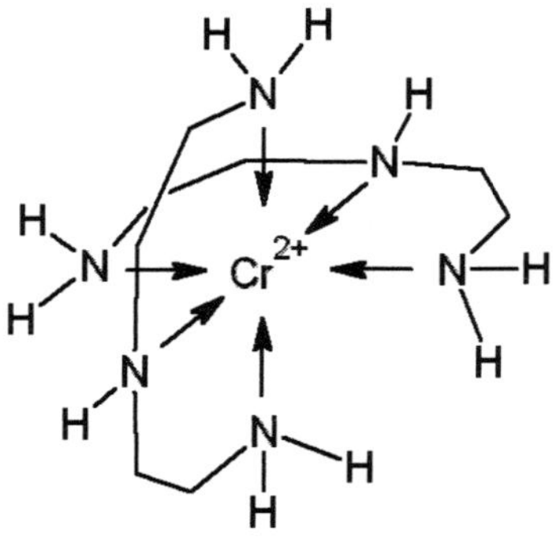

mer

27.33

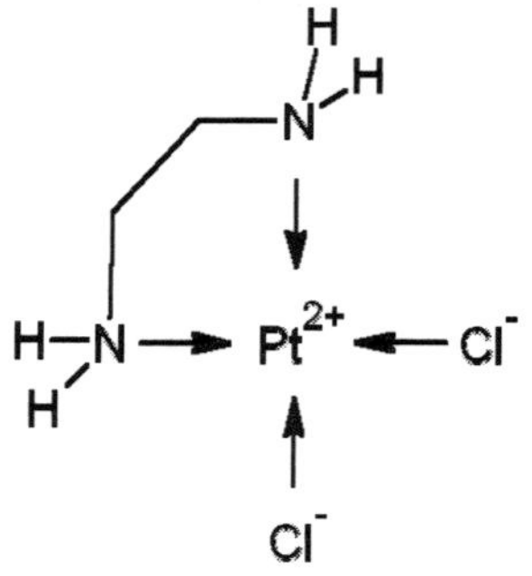

cis

A *trans* isomer would require the N atoms to be on opposite sides of the platinum. The $-CH_2CH_2-$ link between the N atoms is not long enough to allow this.

27.8 Bonding in coordination complexes

27.35

(a) The five *d* orbitals have the same energy. WRONG
A tetrahedral arrangement of ligands breaks the spherical symmetry about the metal atom—it is this spherical symmetry about a free metal atom that gives rise to the equal energies of the 5 *d* orbitals in that case.

(b) The $d_{x^2-y^2}$ and d_{z^2} orbitals are higher in energy than the d_{xz}, d_{yz}, and d_{xy} orbitals. WRONG
This is the order of *d* orbital energies in an octahedral complex.

(c) The d_{xz}, d_{yz}, and d_{xy} orbitals are higher in energy than the $d_{x^2-y^2}$ and d_{z^2} orbitals. CORRECT

(d) The *d* orbitals all have different energies. WRONG
A tetrahedral arrangement of ligands is still quite symmerical. A less symmetrical arrangement of ligands is required to split the *d* orbitals into all different energies.

27.37

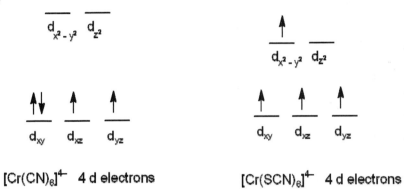

SCN⁻ ions are lower in the spectrochemical series than CN⁻ ions. SCN⁻ ions do not split the *d* orbital energies as much as cyanide ions do.

27.39

(a)

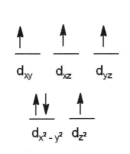

$[FeCl_4]^{2-}$ 6 d electrons

4 unpaired electrons

(b)

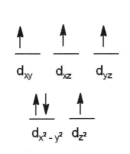

$[CoCl_4]^{2-}$ 7 d electrons

3 unpaired electrons

(c)

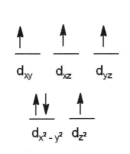

$[MnCl_4]^{2-}$ 5 d electrons

5 unpaired electrons

(d)

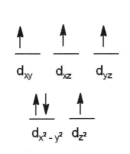

$[ZnCl_4]^{2-}$ 10 d electrons

0 unpaired electrons

27.41

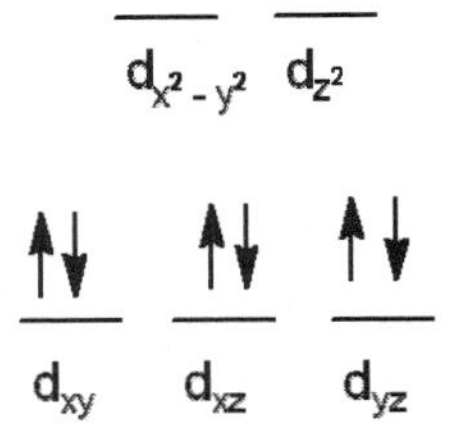

$$\overline{\quad} \quad \overline{\quad}$$
$$d_{x^2-y^2} \quad d_{z^2}$$

$$\underline{\uparrow\downarrow} \quad \underline{\uparrow\downarrow} \quad \underline{\uparrow\downarrow}$$
$$d_{xy} \quad d_{xz} \quad d_{yz}$$

$[Co(en)(NH_3)_2Cl_2]^+$ 6 d electrons

low spin

(a) The coordination number of Co is 6—it has 4 monodentate ligands and 1 bidentate ligand.
(b) The coordination geometry of Co is octahedral—the usual geometry in the case of coordination number 6.
(c) The oxidation state of Co is +3.
(d) There are 0 unpaired electrons.
(e) Because the complex has no unpaired electrons, it is diamagnetic.
(f)

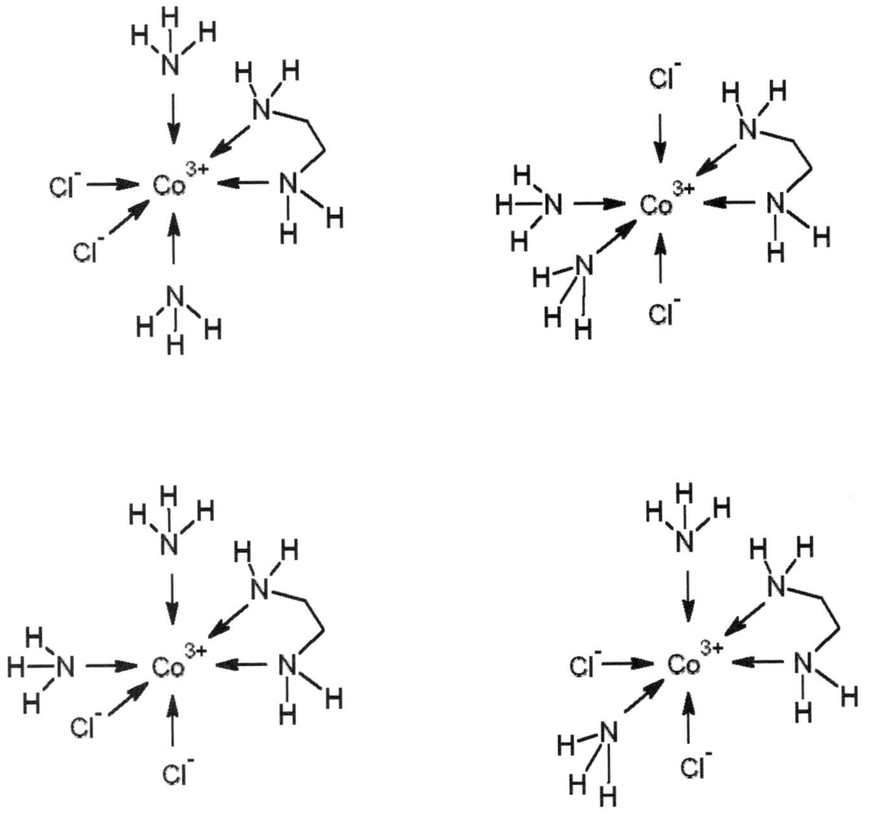

enantiomers

27.43

Hexaaquairon(II) ions, $[Fe(OH_2)_6]^{2+}$, are paramagnetic because the water ligands split the d orbital energies insufficiently to force pairing of the electrons—i.e., it is a high spin complex with 4 unpaired electrons. When NH_3 is added to the solution the ammonia complex ions $[Fe(NH_3)_6]^{2+}$ are formed. This complex is diamagnetic—it is the low spin complex with no unpaired electrons—because the d orbital splitting exceeds the energy required to pair the spins. Ammonia is higher in the spectrochemical series than water.

27.45

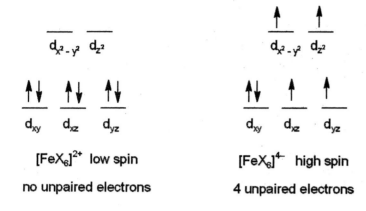

SUMMARY AND CONCEPTUAL QUESTIONS

27.47

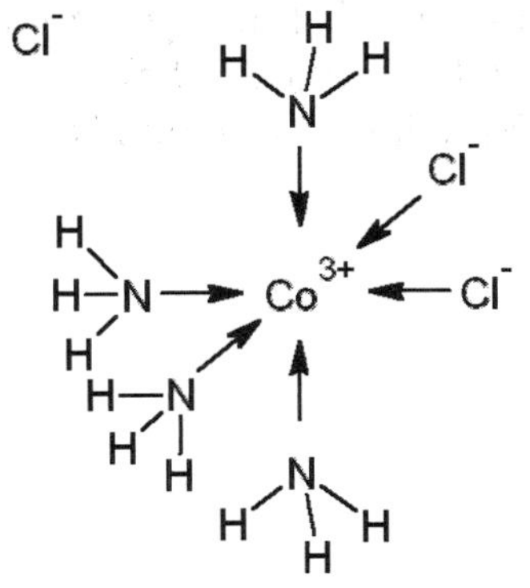

cis-tetraamminedichlorocobalt(III) chloride 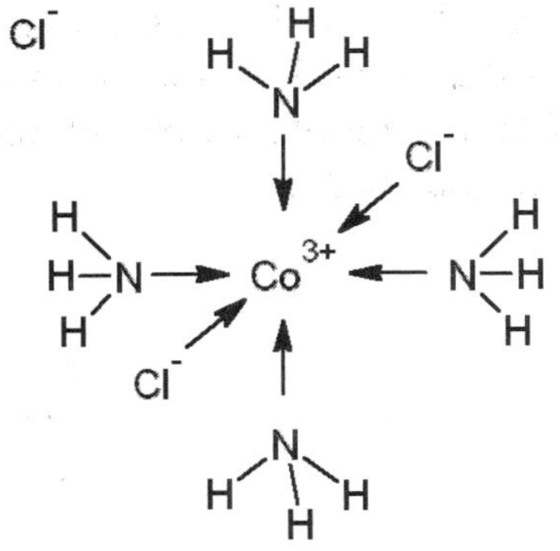   trans-tetraamminedichlorocobalt(III) chloride

27.49

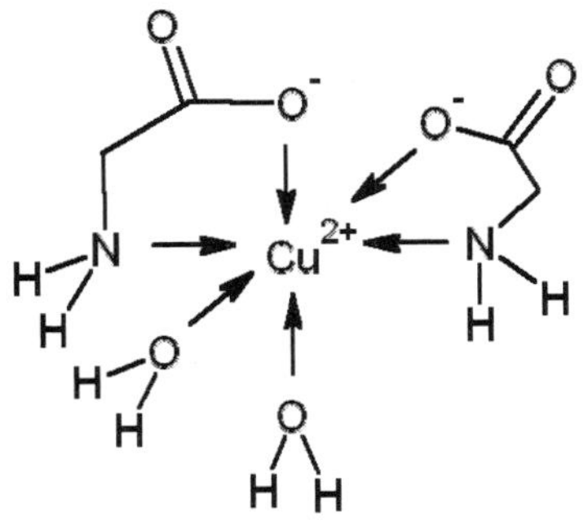

An isomer of $[Cu(H_2NCH_2CO_2)_2(H_2O)_2]$.
(a) The oxidation state of copper is +2.
(b) Copper has coordination number 6.
(c) Copper has 9 d electrons in this complex. Its d electron configuration is shown below.

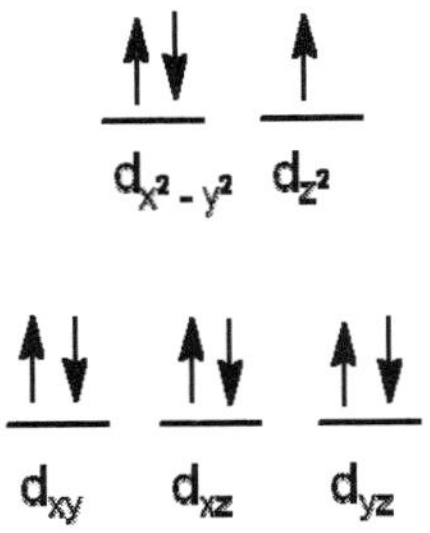

There is one unpaired electron.
(d) The complex is paramagnetic.

461

27.51

(a) $Ni(s) + 4 CO(g) \rightarrow Ni(CO)_4(g)$

$\Delta_r H° = \Delta_f H°[Ni(CO)_4(g)] - (\Delta_f H°[Ni(s)] + 4 \times \Delta_f H°[CO(g)])$
$\quad = \{-602.9 - (0 + 4 \times (-110.525))\} \text{ kJ mol}^{-1} = -160.8 \text{ kJ mol}^{-1}$

$\Delta_r S° = S°[Ni(CO)_4(g)] - (S°[Ni(s)] + 4 S°[CO(g)])$
$\quad = \{410.6 - (29.87 + 4 \times 197.674)\} \text{ J K}^{-1} \text{ mol}^{-1} = -410.0 \text{ J K}^{-1} \text{ mol}^{-1}$

$\Delta_r G° = \Delta_r H° - T \Delta_r S° = -160.8 \text{ kJ mol}^{-1} - 298.15 \text{ K} \times (-410.0 \text{ J K}^{-1} \text{ mol}^{-1}) / (1000 \text{ J/kJ}) = -38.6 \text{ kJ mol}^{-1}$

$\quad K = \exp(-\Delta_r G° / RT) = \exp(38.6 \times 10^3 \text{ J mol}^{-1} / (8.314 \text{ J K}^{-1} \text{ mol}^{-1} \times 298.15 \text{ K})) = 5.792 \times 10^6$

(b) $K \gg 1$. The reaction of $Ni(s)$ and $CO(g)$ is product-favoured.

(c) Because the reaction is exothermic, it is less product-favoured at higher temperatures. Consequently, it is possible to use this reaction to purify nickel metal. Specifically, one would react impure nickel with carbon monoxide at ordinary or even low temperature, where the product is highly favoured. The gaseous product is collected, and then raised in temperature to reverse the reaction and deposit pure nickel. Because the impurities in $Ni(CO)_4(g)$ are other gases, they will not contaminate the solid nickel product. The original solid impurities remain in the solid phase from the first reaction.

Chapter 28
The Chemistry of Modern Materials

REVIEW QUESTIONS

28.2 Metals

28.1
Each Mg atom has 1 $2s$ and 3 $2p$ valence shell orbitals. Thus, there are $4N$ molecular orbitals formed from the valence orbitals of N Mg atoms. Each Mg atom contributes 2 valence electrons, while each molecular orbital can accommodate 2 electrons. Thus, ¼ of the available molecular orbitals are filled with electrons.

28.3
(a) Metal atoms that can form a solid solution in iron, as interstitial atoms, must be smaller than iron atoms. Iron atoms have a radius of 156 pm. Aluminum atoms have a radius of 118 pm. These are considerably smaller than Fe atoms, and might occupy interstitial sites in a solid solution with iron.
(b) Metal atoms that can form a solid solution in iron, as substitutional atoms, must be similar in size to iron atoms, and have similar electronegativity. Neighbouring elements in the same period as iron have similar size and similar electronic properties. Thus, Mn (radius = 161 pm) or Co (radius = 152 pm) might be expected to form such solid solutions.

28.3 Semiconductors

28.5
The band gap of GaAs is 140 kJ mol^{-1}. The energy gap for an individual electron is this energy divided by Avagadro's number—i.e., 140 kJ mol^{-1} / 6.022×10^{23} = 2.32×10^{-22} kJ = 2.32×10^{-19} J (per photon)
The maximum wavelength of light that can excite a transition across the band corresponds to the minimum frequency that can do so. This corresponds to photon energy equal to the band gap.
$$E_g = h\nu = hc/\lambda$$
So,
$$\lambda = hc/E_g = 6.626 \times 10^{-34} \text{ J s} \times 2.998 \times 10^8 \text{ m s}^{-1} / 2.32 \times 10^{-19} \text{ J} = 8.54 \times 10^{-7} \text{ m} =$$
854 nm
This corresponds to light in the near infrared region of the electromagnetic spectrum.

28.4 Ceramics

28.7

Pyrex, or borosilicate glass, is made by adding boron oxide to the melt. The glass that results upon rapid cooling (slow cooling produces opaque ceramic materials) has a lower thermal expansion coefficient (i.e., for a given increase in temperature its degree of expansion is less) than ordinary glass. Thermal expansion (or contraction upon cooling) causes strains in ordinary glass that can cause it to break. This is a problem for glass that is heated in an oven.

28.9

(a) Hydroplasticity is the property of materials that become plastic when water is added to them. Clays have this property. When wet, they are readily deformed into almost any desired shape. They retain this shape after deformation, allowing the wet clay to be heated in an oven—i.e., fired—to make the shape firm and rigid.

(b) A refractory material can withstand very high temperatures without deformation. They also have low thermal conductivities, making them useful for thermal insulation—especially under high temperature conditions. For example, the space shuttle uses refractory ceramic tiles in its heat shield to protect against the extreme temperature conditions of re-entry. The Columbia space shuttle disaster resulted from a missing heat shield tile.

SUMMARY AND CONCEPTUAL QUESTIONS

28.11

To operate a 700 W microwave oven, we would need N cells of the type described in 28.10, where
$$700 \text{ W} = N \times 0.0925 \times 0.25 \text{ W} = N \times 0.0231 \text{ W},$$
accounting for the 25% efficiency, i.e.,
$$N = 700 / 0.0231 = 30300 \text{ cells required.}$$
The total area of the solar panel $= 30300 \text{ cm}^2 = 3.03 \text{ m}^2$.

28.13

In Table 28.1 of the text, the composition of pewter is given as 91% Sn, 7.5% Sb, 1.5% Cu. To get the density of pewter requires knowing the arrangement of the atoms and appropriate values for the atomic radii. Here, we will assume that the atoms are arranged pretty much the way they are in the bulk elements. We will calculate the density of pewter as a weighted average of the densities of the constituent elements. From www.webelements.com we get the bulk element densities:

Sn	7310 kg m^{-3}
Sb	6697 kg m^{-3}
Cu	8920 kg m^{-3}

From these densities, we get
$$0.91 \times 7310 + 0.075 \times 6697 + 0.015 \times 8920 \text{ kg m}^{-3} = 7288 \text{ kg m}^{-3}$$
as the approximate density of pewter.

28.15

While hydrogen bonding is usually stronger than dispersion forces (locally), we cannot make a blanket statement that hydrogen bonding is always stronger. Dispersion forces increase with the size of the molecules involved and, more specifically, with the number of electrons. This is why iodine, a non-polar diatomic molecule, is a solid at normal temperature. Iodine atoms have many more electrons than O atoms and especially H atoms, and they are much further from the nucleus so are much more polarizable. The dispersion forces between iodine molecules are consequently substantial—even, based on the evidence of their phases at ambient temperature, more effective than the hydrogen bonding networks in water.

Chapter 29
Biomolecules

IN-CHAPTER EXERCISES

Exercise 29.1—Nomenclature

(a) Threose is an aldotetrose.
(b) Ribulose is a ketopentose.
(c) Tagatose is a ketohexose.
(d) 2-deoxyribose is an aldopentose.

Exercise 29.2—Fischer projections

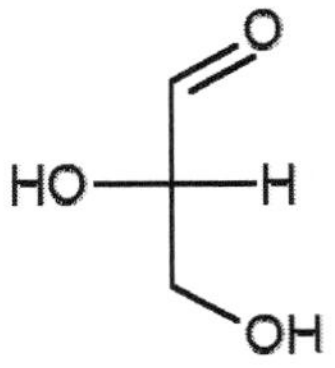

Exercise 29.3—Fischer projections

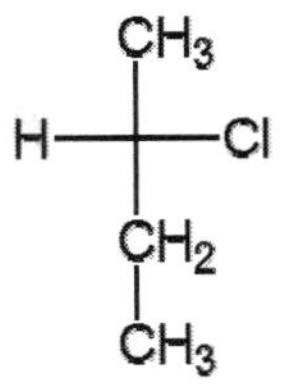

(S)-2-chlorobutane

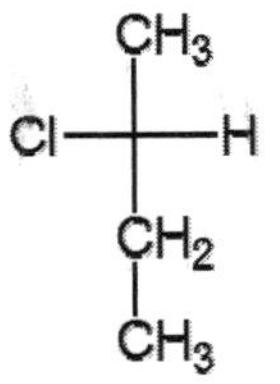

(R)-2-chlorobutane

Exercise 29.4—Stereochemistry

(a)

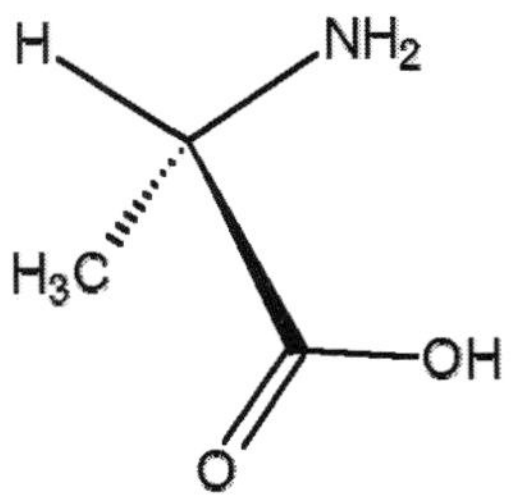

(b)

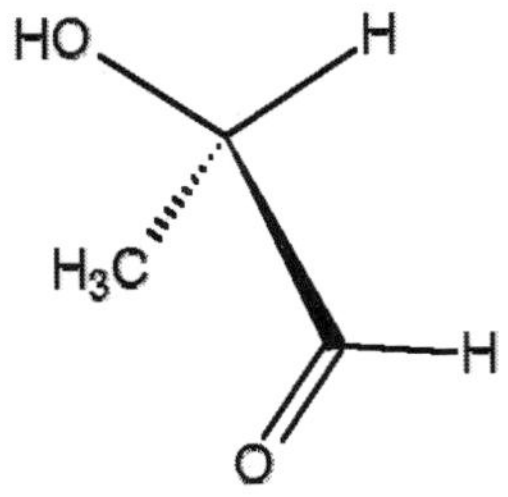

(c)

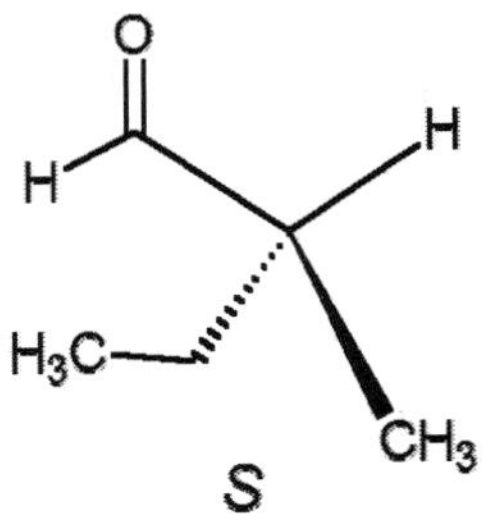

Exercise 29.5—Stereochemistry

(a)

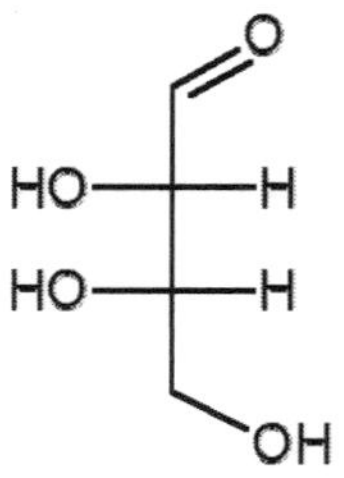

HO——H
HO——H
OH

L-sugar

(b)

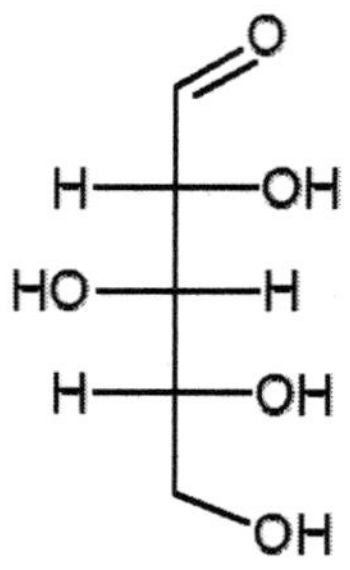

H——OH
HO——H
H——OH
OH

D-sugar

(c)

OH
O
HO——H
H——OH
OH

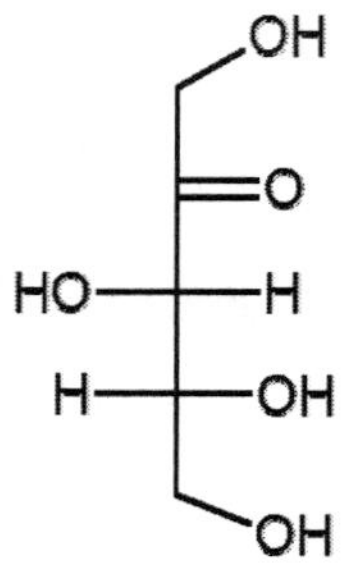

D-sugar

At the bottom stereocentre (furthest from the C=O group), the hydroxyl group is on the right = D-sugar.

Exercise 29.6—Stereochemistry

(a)

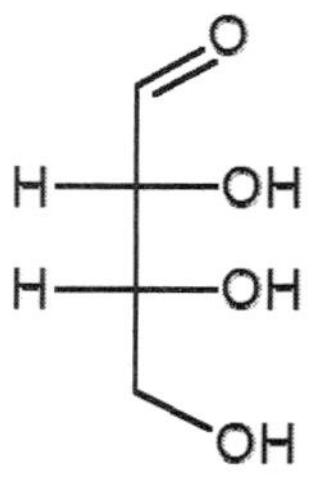

H——OH
H——OH
OH

D-sugar

(b)

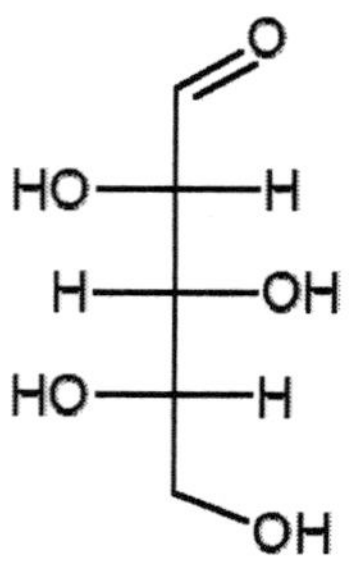

HO——H
H——OH
HO——H
OH

L-sugar

(c)

OH
O
H——OH
HO——H
OH

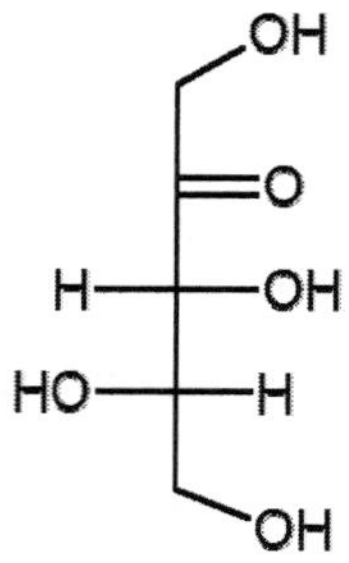

L-sugar

The mirror image of a Fischer projection is obtained by switching the left and right substitutents at each stereocentre.

Exercise 29.7—Carbohydrate structure

α-D-galactopyranose:

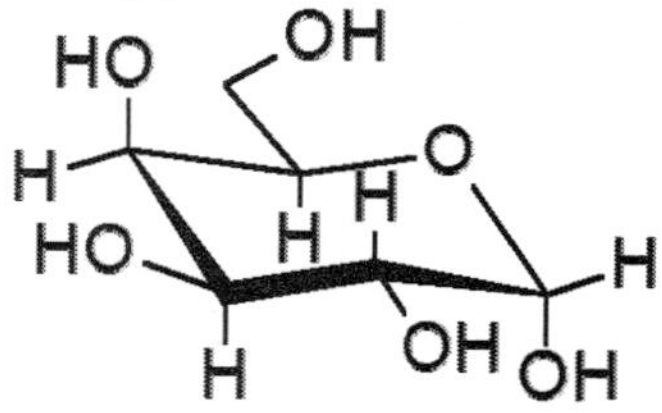

Exercise 29.8—Carbohydrate structure

Furanose forms of ribose:

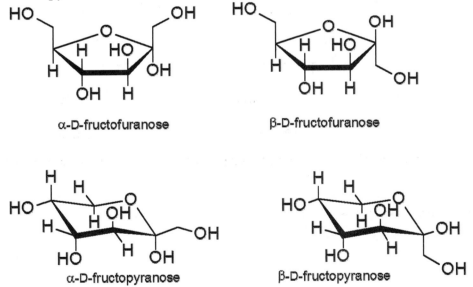

α-D-ribofuranose   β-D-ribofuranose

As seen in the next section, depending on the orientation of reaction of the –OH group with the carbonyl carbon atom, two possible forms of ribofuranose can be formed, with different configuration at the far right hand carbon atom in the structures above, which is called the 'anomeric' carbon atom. The stereochemistry is the same in the two forms at the other three stereocentres.

Exercise 29.9—Carbohydrate structure

Furanose and pyranose forms of D-fructose:

α-D-fructofuranose   β-D-fructofuranose

α-D-fructopyranose   β-D-fructopyranose

Exercise 29.10—Carbohydrate structure

All non-H substituents in β-D-mannopyranose are equatorial except for one axial hydroxyl group.

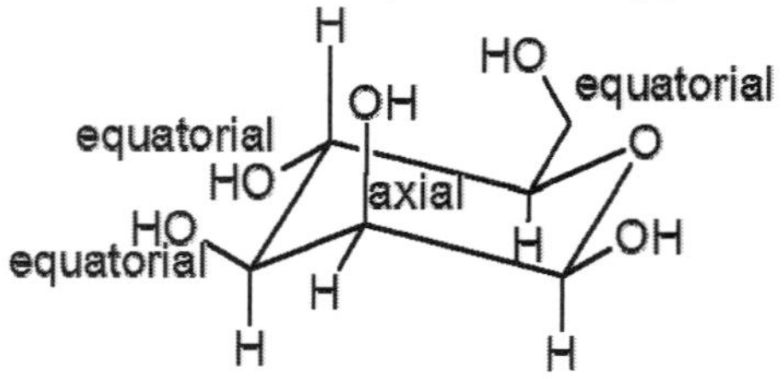

All non-H substituents in β-D-galactopyranose are equatorial except for one axial hydroxyl group, also. β-D-mannopyranose and β-D-galactopyranose should be about equally stable.

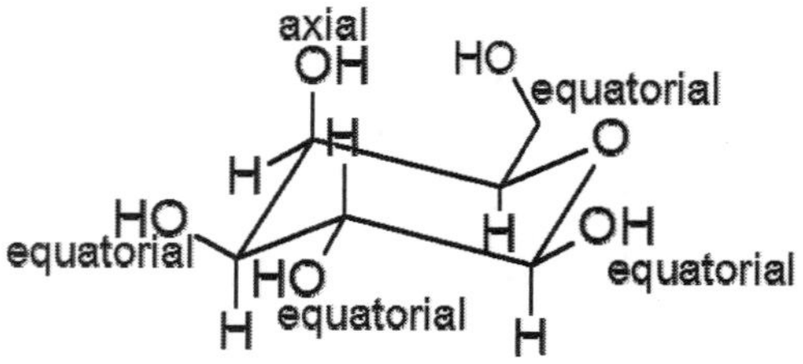

Exercise 29.11—Glycosides

Acetal product of the acid-catalyzed reaction of β-D-galactopyranose with ethanol:

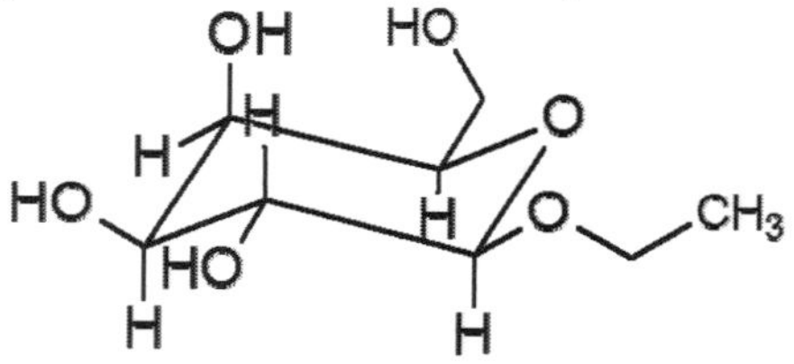

Exercise 29.12—Enzymes

(a) Pyruvate decarboxylase removes CO_2 from substrates and is a lyase. **(b)** Chymotrypsin is used in protein digestion—it cleaves peptides at tyrosine, tryptophan, and phenylalanine. Since this (and other) cleavage reactions are hydrolysis reactions, chymotrypsin is a hydrolase. **(c)** Alcohol dehydrogenase enzymes help with the interconversion of alcohols and aldehydes and ketones and are oxidoreductase enzymes.

Exercise 29.13—Nucleotides

DNA dinucleotide AG:

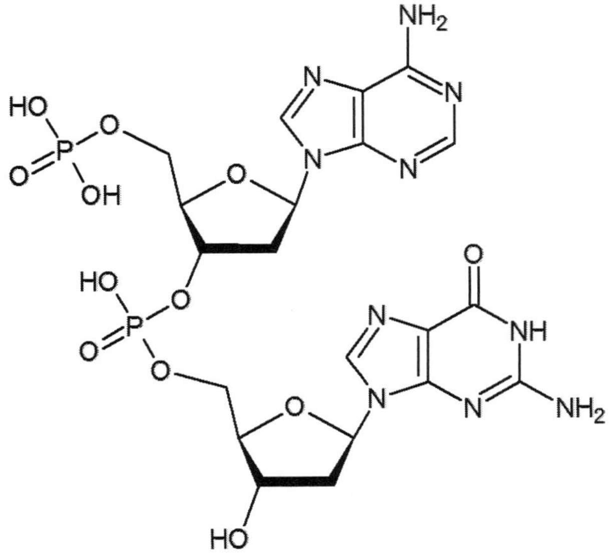

Adenine is at the 5' end.

Exercise 29.14—Nucleotides

RNA dinucleotide UA:

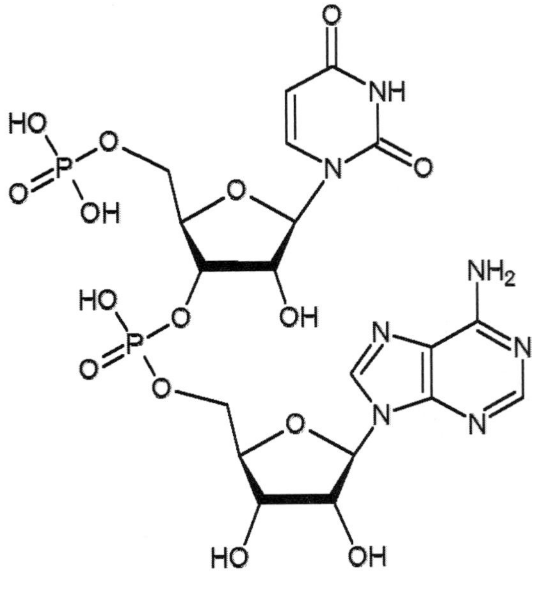

Exercise 29.15—Nucleotides

DNA sequence (3') CCGATTAGGCA(5') is complementary to GGCTAATCCGT (by default this is 5' to 3').

Exercise 29.16—RNA

Hydrogen bonding between uracil and adenine:

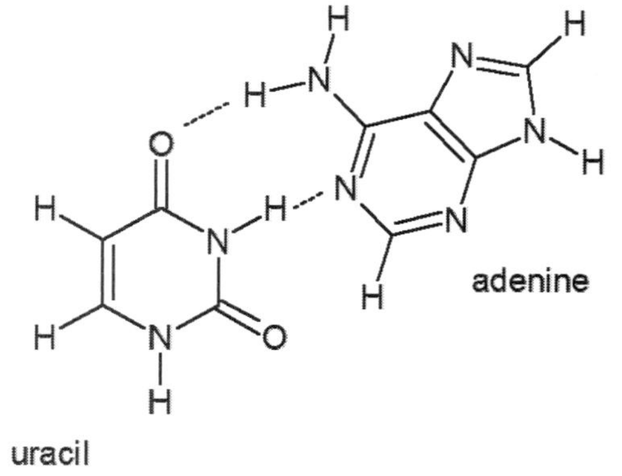

Exercise 29.17—RNA

RNA sequence (5') CUAAUGGCAU (3') is complementary to DNA sequence (3') GATTACCGTA (5').

Exercise 29.18—RNA

RNA sequence (5') UUCGCAGAGU (3') was transcribed from DNA sequence (3') AAGCGTCTCA (5').

Exercise 29.19—RNA

Several different codons can code for each amino acid. The codons are written with the 5' end on the left.
(a) Ala, alanine, is coded by GCU, GCC, GCA, or GCG.
(b) Phe, phenylalanine, is coded by UUU or UUC.
(c) Leu, leucine, is coded by UUA, UUG, CUU, CUC, CUA, or CUG.
(d) Tyr, tyrosine, is coded by UAU or UAC.

Exercise 29.20—RNA

(5')−CUU-AUG-GCU-UGG-CCC-UAA−(3') codes for the polypeptide,
Leu-Met-Ala-Trp-Pro-stop
i.e., leucine, methionine, alanine, tryptophan, and proline are linked together, in this order, and then the chain is terminated at proline.

Exercise 29.21—RNA

mRNA base sequence (5')−CUU-AUG-GCU-UGG-CCC-UAA−(3') gives rise to tRNA sequence, (3')−GAA UAC CGA ACC GGG−(5')

Exercise 29.22—RNA

mRNA base sequence 5' CUU-AUG-GCU-UGG-CCC-UAA 3' arose from transcription of DNA sequence,
(3')−GAA TAC CGA ACC GGG ATT−(5')

REVIEW QUESTIONS

29.2 Carbohydrates

29.23
(a)

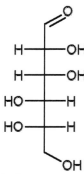

(b) This is an L sugar—so-called because the hydroxyl on the bottom carbon stereocentre is on the left in this Fischer projection. This is L-mannose.
(c) The β anomer of L-mannose in its furanose form:

29.25

29.27
Ascorbic acid has the L configuration because the stereocentre farthest away from the C=O group points to the left.

29.29
(a) A monosaccharide is a simple carbohydrate that can't be hydrolyzed to smaller units.

e.g., D-mannose
(b)

e.g., β- D-mannofuranose—the anomeric centre is indicated by the arrow. It is the carbon atom that becomes the C=O carbon atom in the open chain form. It can be recognized because it is the only C atom in the ring of this molecule that has two O atoms attached to it. In this case it is a hemiacetal carbon atom.
(c) To make a Fischer projection, we arrange the carbon chain vertically such that all C−C bonds point into the paper. To do this, we imagine that the carbon chain that otherwise curls into a spiral is flattened onto the plane of the paper—the vertical lines represent bonds going into the paper and the horizontal lines represent bonds that come out of the plane of the paper.

(d) A glycoside is the acetal of a carbohydrate, formed when an alcohol group on an anomeric carbon atom reacts with an alcohol.

(e) Reducing sugars are sugars that can be reduced by Tollens reagent—a basic aqueous solution of AgNO$_3$. Such sugars must be able to be converted to aldehydes in their open chain form. Acetal linkages do not open up under basic conditions, so glycosides are non-reducing sugars.

(f) Pyranose form is the six-membered ring form of an aldohexose—an aldehyde with 6 carbon atoms.

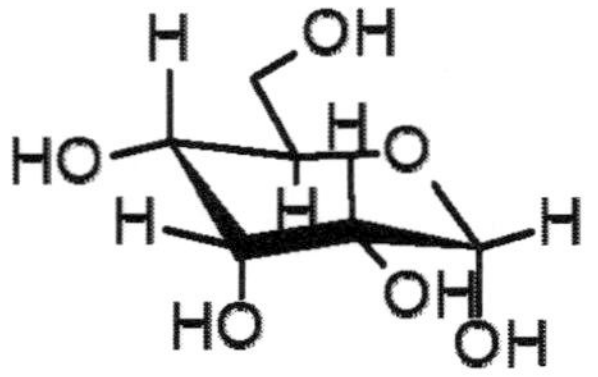

e.g., α- D-allopyranose

(g) A 1,4′ link is a glycosidic linkage between C1(the anomeric carbon atom) of one sugar molecule and the OH group at C4 of a second sugar molecule.

(h) A D-sugar has the –OH group at the stereocentre farthest from the carbonyl group pointing to the right in a Fischer projection.

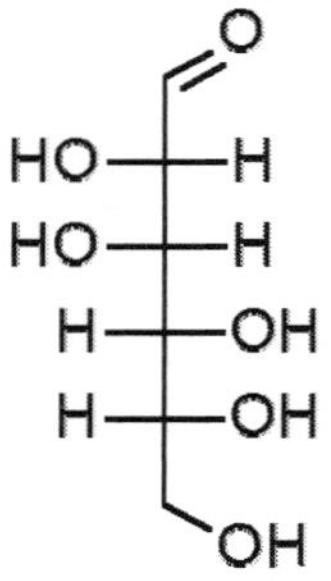

e.g., D-mannose

29.31

D-allose and L-allose are enantiomers. In their pure forms, they have the same values of all physical properties except specific rotation. Thus, they have the same **(a)** melting point, **(b)** solubility in water, and **(d)** density.

(c) The specific rotations are equal in magnitude but opposite in sign.

29.33

(a)

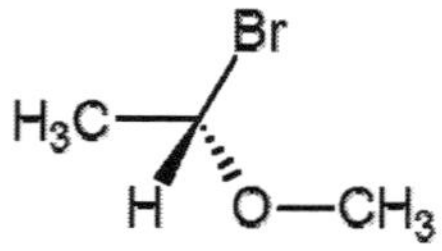

(b)

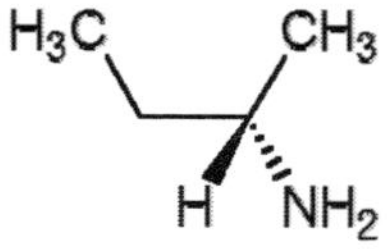

29.35

The specific rotation at equilibrium is a weighted average of the values of the equivalent pure substances. If x is the mole fraction of α-D-galactopyranose at equilibrium, then

$$80.2° = x\,150.7° + (1 - x)\,52.8°$$
$$x = (80.2° - 52.8°)\,/\,(150.7° - 52.8°) = 0.280$$

i.e., the equilibrium mixture is 28% α-D-galactopyranose and 72% β-D-galactopyranose

29.3 Amino acids, peptides, and proteins

29.37

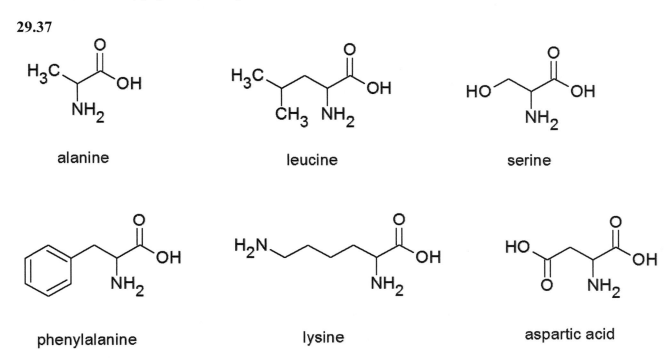

We see that alanine, leucine, and phenylalanine have non-polar R groups. Serine and aspartic acid have polar R groups. The R group of lysine is a non-polar chain with a polar end.

29.39

Tripeptide serine-leucine-valine:

29.41

(a) The amino acid sequence in the protein is its primary structure.

(b) The way different peptide chains in the overall protein are arranged with respect to one another is the quaternary structure.

(c) The way the polypeptide chain is folded, including how amino acids which are far apart in the sequence end up in the overall molecule, is the tertiary structure.

(d) The way the amino acids near one another in the sequence arrange themselves is the secondary structure.

29.4 Nucleic acids and nucleotides

29.43

(a) Sugar ribose:

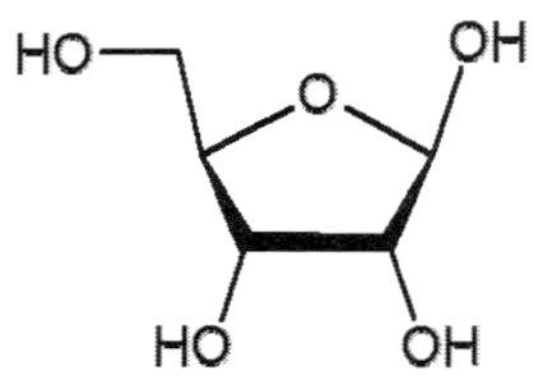

(b) Nucleoside adenosine:

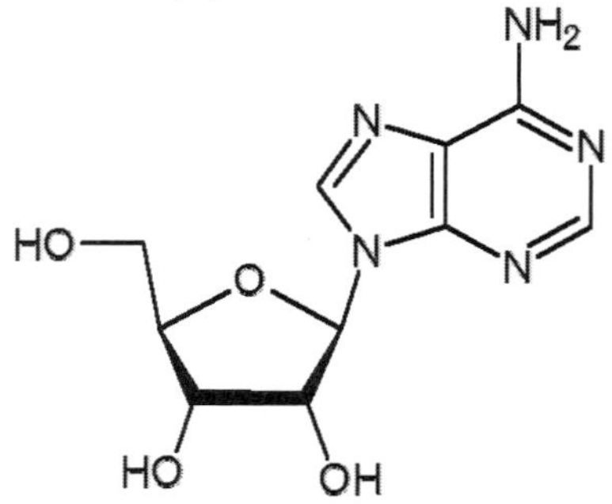

(c) Nucleotide adenosine 5′-monophosphate:

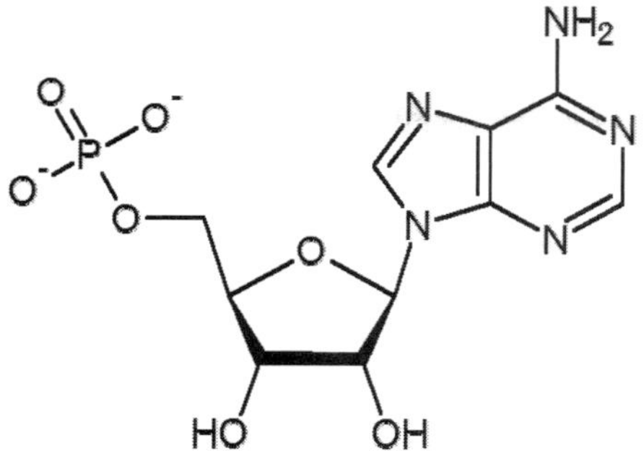

29.45

ATGC and CGTA are not the same molecule. The nucleosides are differently arranged when connected at the 5' and 3' positions to phosphate.

29.47

(a) Transcription is the process wherein mRNA is produced complementary to a strand of DNA. The first step is the unwinding of the DNA double helix. An RNA strand complementary to a DNA strand is synthesized, next to the DNA strand. The RNA strand is then unwound from the DNA. It goes on to protein synthesis. The DNA returns to its double helix form.

(b) Translation is the process wherein the codons of the mRNA are "read" and used to synthesize the specific coded protein. A process of sequentially "reading" the codons is coordinated with the process that makes a peptide linkage to the next amino acid in the specified sequence.

SUMMARY AND CONCEPTUAL QUESTIONS

29.49
Raffinose is not a reducing sugar. The glycosidic linkages cannot be hydrolyzed under basic conditions.

29.51
(a) Glucocerebroside has hydroxyl, acetal, amide, and alkene functional groups.
(b) Hydrolysis of glycosidic linkage:

Hydrolysis of amide:

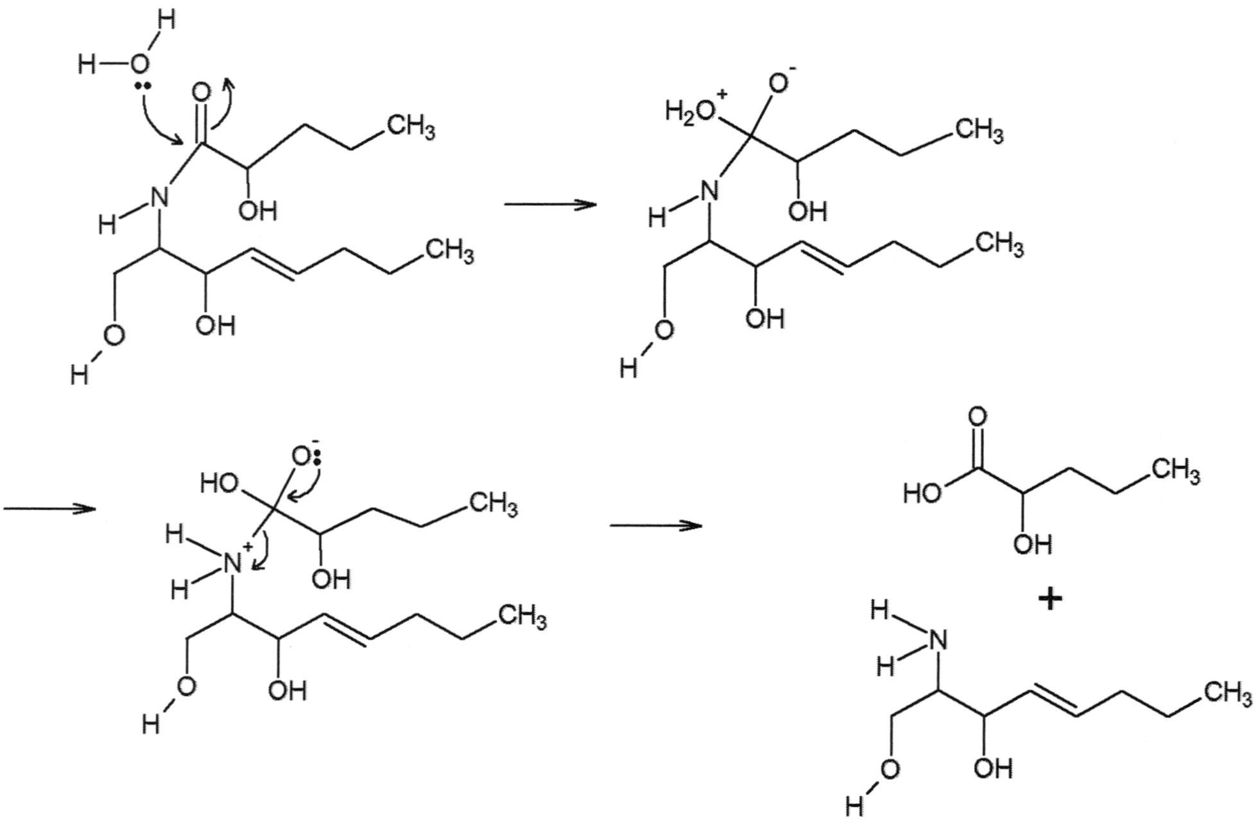

(c)

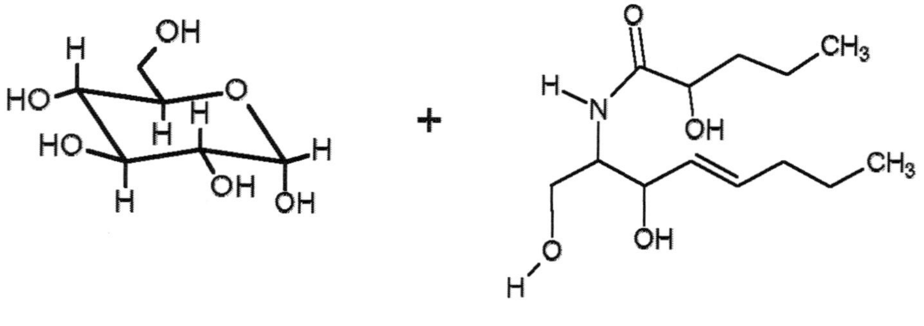

29.53

(a) The complementary strand of DNA is (5′)—GAATCGCGT—(3′)

(b) The complementary strand of mRNA is (5′)—GAAUCGCGU—(3′)

(c) The three anticodons are (5′)—UUC—(3′), (5′)—CGA—(3′), and (5′)—ACG—(3′). They are complementary to the three codons of the mRNA, (5′)—GAA—(3′), (5′)—UCG—(3′), and (5′)—CGU—(3′).

(d) (5′)—GAA—(3′) codes for Glu, glutamic acid, (5′)—UCG—(3′) codes for Ser, serine, and (5′)—CGU—(3′) codes for Arg, arginine.

Chapter 30
Nuclear Chemistry

IN-CHAPTER EXERCISES

In these solutions, the symbols $_{-1}^{0}e$ and $_{-1}^{0}\beta$ can be used interchangeably, as can $_{2}^{4}He$ and $_{2}^{4}\alpha$.

Exercise 30.1—Equations for nuclear reactions

(a)

$$_{86}^{222}Rn \rightarrow {}_{84}^{218}Po + {}_{2}^{4}He$$

The 86 protons of $_{86}^{222}Rn$ (i.e., its atomic number) are divided between the polonium nucleus—84 protons—and the α particle—2 protons. Similarly, the number of neutrons is conserved, as reflected in the atomic mass numbers—222 = 218 + 4.

(b)

$$_{84}^{218}Po \rightarrow {}_{85}^{218}At + {}_{-1}^{0}e$$

The symbols $_{-1}^{0}e$ and $_{-1}^{0}\beta$ can be used interchangeably

Exercise 30.2—Energy of photons

The energy of one photon of γ radiation with a wavelength of 2.0×10^{-12} m is
$$E_{photon} = h\,c\,/\,\lambda = 6.626 \times 10^{-34}\,J\,s \times 2.998 \times 10^{8}\,m\,s^{-1}\,/\,2.0 \times 10^{-12}\,m$$
$$= 9.9 \times 10^{-14}\,J$$
Per mole, the energy is
$$= 6.022 \times 10^{23}\,mol^{-1} \times 9.9 \times 10^{-14}\,J = 5.96 \times 10^{10}\,J\,mol^{-1} = 60\,GJ\,mol^{-1}$$

Exercise 30.3—Radioactive decay

(a)

$$_{90}^{232}Th \rightarrow {}_{82}^{208}Pb + 6\,{}_{2}^{4}He + 4\,{}_{-1}^{0}e$$

(b)

$$_{90}^{232}Th \rightarrow {}_{88}^{228}Ra + {}_{2}^{4}He$$

$$_{88}^{228}Ra \rightarrow {}_{89}^{228}Ac + {}_{-1}^{0}e$$

$$_{89}^{228}Ac \rightarrow {}_{90}^{228}Th + {}_{-1}^{0}e$$

Exercise 30.4—Balanced nuclear reaction equations

(a) $^{13}_{7}N \rightarrow ^{13}_{6}C + ^{0}_{1}\beta$
positron emission

(b) $^{41}_{20}Ca + ^{0}_{-1}\beta \rightarrow ^{41}_{19}K$
electron capture

(c) $^{90}_{38}Sr \rightarrow ^{90}_{39}Y + ^{0}_{-1}\beta$
beta decay

(d) $^{22}_{11}Na \rightarrow ^{22}_{12}Mg + ^{0}_{-1}\beta$
beta decay

Exercise 30.5—Radioactive decay

(a) Silicon-32 has a high neutron to proton ratio. It most likely decays by beta emission to form phosphorus-32.
$^{32}_{14}Si \rightarrow ^{32}_{15}P + ^{0}_{-1}\beta$

(b) Titanium-45 has a low neutron to proton ratio. It most likely decays by positron emission to form
$^{45}_{22}Ti \rightarrow ^{45}_{21}Sc + ^{0}_{1}\beta$

(c) Plutonium-239 is beyond—in mass and atomic number—the band of stability. It most likely decays by alpha emission.
$^{239}_{94}Pu \rightarrow ^{235}_{92}U + ^{4}_{2}He$

(d) Potassium-42 has a high neutron to proton ratio. It most likely decays by beta emission.
$^{42}_{19}K \rightarrow ^{42}_{20}Ca + ^{0}_{-1}\beta$

Exercise 30.6—Binding energy

The binding energy of $^{6}_{3}Li$ is the change in energy for the following process.
$$^{6}_{3}Li \rightarrow 3\,^{1}_{1}H + 3\,^{1}_{0}n$$
The mass change in this process is
$$\Delta m = 3 \times 1.007825 + 3 \times 1.008665 - 6.015125 \text{ g mol}^{-1} = 0.034345 \text{ g mol}^{-1}$$
$$= 3.4345 \times 10^{-5} \text{ kg mol}^{-1}$$
The binding energy is
$$E_b = (\Delta m)c^2 = 3.4345 \times 10^{-5} \text{ kg mol}^{-1} \times (2.998 \times 10^8 \text{ m s}^{-1})^2$$
$$= 3.087 \times 10^{12} \text{ J mol}^{-1} = 3.087 \times 10^9 \text{ kJ mol}^{-1}$$
Per nucleon, we have
$$E_b / n = 3.087 \times 10^9 / 6 \text{ kJ mol}^{-1} \text{ nucleon}$$
$$= 5.145 \times 10^8 \text{ kJ mol}^{-1} \text{ nucleon}$$

Exercise 30.7—Half-life

(a) Starting with 1.5 mg of this isotope, after 49.2 years we have
$$1.5 \text{ mg} \times (1/2)^{49.2/12.3} = 1.5 \text{ mg} \times (1/2)^4 = 1.5 \text{ mg} \times (1/16) = 0.094 \text{ mg}$$
(b) A sample of tritium decays to one eighth of its activity in 3 half-lives—i.e., in 36.9 years.
(c) A sample decays to 1.00% of its original activity in n half-lives.
i.e., $\quad 0.0100 = (1/2)^n$
Therefore, $\quad \ln(0.01) = n \ln(1/2) = n(-\ln 2)$
and $\quad n = 6.64$
This corresponds to 6.64×12.3 years $= 81.7$ years.

Exercise 30.8—Kinetics of nuclear decay

The activity decreases by the factor $53.0/55.8 = 0.950 = (1/2)^n$ in 2 days.
$$n = \ln(0.950)/(-\ln 2) = 0.0743 = 2 \text{ days}/t_{1/2}$$
$$t_{1/2} = 2 \text{ days}/0.0743 = 26.9 \text{ days} \text{ (assuming the 2 days is known to 3 digits)}$$

Exercise 30.9—Kinetics of nuclear decay
The activity decreases by the factor $5.00 \times 10^1/1.08 \times 10^{11} = 4.63 \times 10^{-11}$ in n half-lives.
$$n = \ln(4.63 \times 10^{-11})/(-\ln 2) = 31.0 = t/200 \text{ years}$$
It will take $t = 31.0 \times 200$ years $= 6200$ years.

Exercise 30.10—Carbon dating
Carbon-14 is a β emitter with a half-life of 5730 years.
The activity decreased by the factor $0.155/0.223 = 0.695$ in n half-lives.
$$n = \ln(0.695)/(-\ln 2) = 0.525 = t/5730 \text{ years}$$
The sample is $t = 0.525 \times 5730$ years $= 3010$ years old
The ring data give the age of the tree to be $1874 + 979 \pm 52 = 2853 \pm 52$ years. 3010 years is close, and the two values are likely within uncertainties if the uncertainty in A_0 were taken into account.

Exercise 30.11—Isotope dilution
The isolated sample consists of the fraction, $20/50 = 0.40$, of the original threonine. Since the isolated sample weighs 60.0 mg, there must have been $60.0 \text{mg}/0.40 = 150.0$ mg of threonine in the original sample.

REVIEW QUESTIONS

30.2–30.3 Radioactivity and nuclear reaction equations

30.13
(a) $^{54}_{26}\text{Fe} + ^4_2\text{He} \longrightarrow 2^1_0\text{n} + ^{56}_{28}\text{Ni}$
(b) $^{27}_{13}\text{Al} + ^4_2\text{He} \longrightarrow ^{30}_{15}\text{P} + ^1_0\text{n}$
(c) $^{32}_{16}\text{S} + ^1_0\text{n} \longrightarrow ^1_1\text{H} + ^{32}_{15}\text{P}$
(d) $^{96}_{42}\text{Mo} + ^2_1\text{H} \longrightarrow ^1_0\text{n} + ^{97}_{43}\text{Tc}$

(e) $^{98}_{42}\text{Mo} + ^{1}_{0}\text{n} \longrightarrow ^{99}_{43}\text{Tc} + ^{0}_{-1}\beta$

(f) $^{18}_{9}\text{F} \longrightarrow ^{18}_{8}\text{O} + ^{0}_{1}\beta$

30.15

(a) $^{111}_{47}\text{Ag} \longrightarrow ^{111}_{48}\text{Cd} + ^{0}_{-1}\beta$

(b) $^{87}_{36}\text{Kr} \longrightarrow ^{0}_{-1}\beta + ^{87}_{37}\text{Rb}$

(c) $^{231}_{91}\text{Pa} \longrightarrow ^{227}_{89}\text{Ac} + ^{4}_{2}\text{He}$

(d) $^{230}_{90}\text{Th} \longrightarrow ^{4}_{2}\text{He} + ^{226}_{88}\text{Ra}$

(e) $^{82}_{35}\text{Br} \longrightarrow ^{82}_{36}\text{Kr} + ^{0}_{-1}\beta$

(f) $^{24}_{11}\text{Na} \longrightarrow ^{24}_{12}\text{Mg} + ^{0}_{-1}\beta$

30.17

$$^{235}_{92}\text{U} \longrightarrow ^{231}_{90}\text{Th} + ^{4}_{2}\text{He}$$

$$^{231}_{90}\text{Th} \longrightarrow ^{231}_{91}\text{Pa} + ^{0}_{-1}\beta$$

$$^{231}_{91}\text{Pa} \longrightarrow ^{227}_{89}\text{Ac} + ^{4}_{2}\text{He}$$

$$^{227}_{89}\text{Ac} \longrightarrow ^{227}_{90}\text{Th} + ^{0}_{-1}\beta$$

$$^{227}_{90}\text{Th} \longrightarrow ^{223}_{88}\text{Ra} + ^{4}_{2}\text{He}$$

$$^{223}_{88}\text{Ra} \longrightarrow ^{219}_{86}\text{Rn} + ^{4}_{2}\text{He}$$

$$^{219}_{86}\text{Rn} \longrightarrow ^{215}_{84}\text{Po} + ^{4}_{2}\text{He}$$

$$^{215}_{84}\text{Po} \longrightarrow ^{211}_{82}\text{Pb} + ^{4}_{2}\text{He}$$

$$^{211}_{82}\text{Pb} \longrightarrow ^{211}_{83}\text{Bi} + ^{0}_{-1}\beta$$

$$^{211}_{83}\text{Bi} \longrightarrow ^{211}_{84}\text{Po} + ^{0}_{-1}\beta$$

$$^{211}_{84}\text{Po} \longrightarrow ^{207}_{82}\text{Pb} + ^{4}_{2}\text{He}$$

30.4 Stability of atomic nuclei

30.19

(a) Gold-198 decays to mercury-198.

$^{198}_{79}\text{Au} \longrightarrow ^{198}_{80}\text{Hg} + ^{0}_{-1}\beta$

The atomic number increases by 1, while the mass number does not change. This is beta decay.

(b) Radon-222 decays to polonium-218.

$^{222}_{86}\text{Rn} \longrightarrow ^{218}_{84}\text{Po} + ^{4}_{2}\text{He}$

The atomic number decreases by 2, while the mass number decreases by 4. This is alpha decay.

(c) Cesium-137 decays to barium-137.

$$^{137}_{55}\text{Cs} \longrightarrow {}^{137}_{56}\text{Ba} + {}^{0}_{-1}\beta$$

This is beta decay.

(d) Indium-110 decays to cadmium-110.

$$^{110}_{49}\text{In} \longrightarrow {}^{110}_{48}\text{Cd} + {}^{0}_{1}\beta$$

This is positron emission.

30.21

(a) Bromine-80m is a metastable state of bromine. It most likely decays by gamma emission to form bromine-80.

$$^{80m}_{35}\text{Br} \rightarrow {}^{80}_{35}\text{Br} + n\gamma$$

n is the number of gamma photons emitted.

(b) Californium-240 is beyond—in mass and atomic number—the band of stability. However, its neutron to proton ratio is very low. It most likely decays by positron emission.

$$^{240}_{98}\text{Cf} \rightarrow {}^{240}_{97}\text{Bk} + {}^{0}_{1}\beta$$

Alternatively: $^{240}_{98}\text{Cf} \rightarrow {}^{236}_{96}\text{Cm} + {}^{4}_{2}\text{He}$

(c) Cobalt-61 has a high neutron to proton ratio. It most likely decays by beta emission.

$$^{61}_{27}\text{Co} \rightarrow {}^{61}_{28}\text{Ni} + {}^{0}_{-1}\beta$$

(d) Carbon-11 has a low neutron to proton ratio. It most likely decays by positron emission.

$$^{11}_{6}\text{C} \rightarrow {}^{11}_{5}\text{B} + {}^{0}_{1}\beta$$

30.23

(a) Of the nuclei ^{3}H ^{16}O ^{20}F and ^{13}N

^{3}H and ^{20}F have high neutron to proton ratios and are likely to decay by beta emission.

(b) Of the nuclei ^{238}U ^{19}F ^{22}Na and ^{24}Na

^{22}Na has a low neutron to proton ratio and is likely to decay by positron emission.

30.25

For ^{10}B,

$\Delta m = 5 \times 1.00783 + 5 \times 1.00867 - 10.01294 \text{ g mol}^{-1} = 0.069510 \text{ g mol}^{-1} = 6.9510 \times 10^{-5} \text{ kg mol}^{-1}$

$E_b = \Delta m\, c^2$

$= 6.9510 \times 10^{-5} \text{ kg mol}^{-1} \times (2.998 \times 10^8 \text{ m s}^{-1})^2$

$= 6.248 \times 10^{12} \text{ J mol}^{-1} = 6.248 \times 10^9 \text{ kJ mol}^{-1}$

Per nucleon,

$E_b / n = 6.248 \times 10^9 \text{ kJ mol}^{-1} / 10 = 6.248 \times 10^8 \text{ kJ mol}^{-1}$ nucleons

For ^{11}B,

$\Delta m = 5 \times 1.00783 + 6 \times 1.00867 - 11.00931 \text{ g mol}^{-1} = 0.081805 \text{ g mol}^{-1} = 8.1805 \times 10^{-5} \text{ kg mol}^{-1}$

$E_b = \Delta m\, c^2$

$= 8.1805 \times 10^{-5} \text{ kg mol}^{-1} \times (2.998 \times 10^8 \text{ m s}^{-1})^2$

$= 7.353 \times 10^{12} \text{ J mol}^{-1} = 7.353 \times 10^9 \text{ kJ mol}^{-1}$

Per nucleon,
$$E_b / n = 7.353 \times 10^9 \text{ kJ mol}^{-1} / 11 = 6.684 \times 10^8 \text{ kJ mol}^{-1} \text{ nucleons}$$
We see that ^{11}B is the more stable nucleus—it has a higher binding energy per nucleon.

30.27
For ^{40}Ca,
$$\Delta m = 20 \times 1.00783 + 20 \times 1.00867 - 39.96259 \text{ g mol}^{-1} = 0.36721 \text{ g mol}^{-1} =$$
$3.6721 \times 10^{-4} \text{ kg mol}^{-1}$
$$\begin{aligned} E_b &= \Delta m\, c^2 \\ &= 3.6721 \times 10^{-4} \text{ kg mol}^{-1} \times (2.998 \times 10^8 \text{ m s}^{-1})^2 \\ &= 3.300 \times 10^{13} \text{ J mol}^{-1} = 3.300 \times 10^{10} \text{ kJ mol}^{-1} \end{aligned}$$
Per nucleon,
$$E_b / n = 3.300 \times 10^{10} \text{ kJ mol}^{-1} / 40 = 8.252 \times 10^8 \text{ kJ mol}^{-1} \text{ nucleons}$$
This result is consistent with Figure 30.6.

30.29
For ^{16}O,
$$\Delta m = 8 \times 1.00783 + 8 \times 1.00867 - 15.99492 \text{ g mol}^{-1} = 0.13700 \text{ g mol}^{-1} =$$
$1.3700 \times 10^{-4} \text{ kg mol}^{-1}$

$$\begin{aligned} E_b &= \Delta m\, c^2 \\ &= 1.3700 \times 10^{-4} \text{ kg mol}^{-1} \times (2.998 \times 10^8 \text{ m s}^{-1})^2 \\ &= 1.231 \times 10^{13} \text{ J mol}^{-1} = 1.231 \times 10^{10} \text{ kJ mol}^{-1} \end{aligned}$$
Per nucleon,
$$E_b / n = 1.231 \times 10^{10} \text{ kJ mol}^{-1} / 16 = 7.696 \times 10^8 \text{ kJ mol}^{-1} \text{ nucleons}$$

30.5 Rates of nuclear decay

30.31
64 h corresponds to 64 h / 12.7 h = 5.04 half-lives. At this time, the mass of ^{64}Cu is
$$25.0 \text{ μg} \times (1/2)^{5.04} = 0.760 \text{ μg}$$

30.33
(a) $^{131}_{53}\text{I} \rightarrow {}^{131}_{54}\text{Xe} + {}^{0}_{-1}\beta$
(b) 40.2 days corresponds to 40.2 days / 8.04 days = 5.00 half-lives. At this time, the mass of ^{198}Au is
$$2.4 \text{ μg} \times (1/2)^5 = 0.075 \text{ μg}$$

30.35
13 days corresponds to 13 days / (78.25 / 24) days = 4.0 half-lives. At this time, the mass of ^{67}Ga is
$$0.015 \text{ mg} \times (1/2)^{4.0} = 9.4 \times 10^{-4} \text{ mg}$$

30.37

(a) $^{222}_{86}Rn \rightarrow {}^{218}_{84}Po + {}^{4}_{2}He$

(b) 20.0% of ^{222}Rn remains after n half-lives.
$\qquad 0.200 = (1/2)^n$
$\qquad n = \ln(0.200) / \ln(1/2) = 2.32$
$\qquad$ 2.32 half-lives corresponds to 2.32 × 3.82 days = 8.87 days

30.39

$\qquad$ 0.72 of the ^{14}C remains after n half-lives.
$\qquad 0.72 = (1/2)^n$
$\qquad n = \ln(0.72) / \ln(1/2) = 0.47$
$\qquad$ 0.47 half-lives corresponds to $0.17 \times 5.73 \times 10^3$ years = 2700 years
$\qquad$ This is the age of the bone fragment.

30.41

(a) A cobalt-60 source will drop to 1/8 of its original activity after n half-lives.
$\qquad 0.125 = (1/2)^n$
$\qquad n = \ln(0.125) / \ln(1/2) = 3$
$\qquad$ 3 half-lives corresponds to 3 × 5.27 years = 15.81 years

(b) 1 year corresponds to 1 / 5.27 = 0.190 half-lives. At this time, the fraction of ^{60}Co remaining is
$\qquad (1/2)^{0.190} = 0.877$

30.43

(a)

$$^{23}_{11}Na + {}^{1}_{0}n \rightarrow {}^{24}_{11}Na$$

$$^{24}_{11}Na \rightarrow {}^{24}_{12}Mg + {}^{0}_{-1}\beta$$

(b)

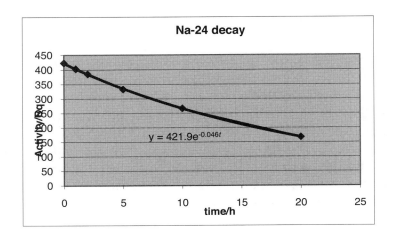

The data is clearly well fit by a simple exponential decay, $421.9\, e^{-0.046\, t}$. The parameters of this fit are well approximated using just the first and last data. $k = 0.046$ corresponds to
$\qquad t_{1/2} = \ln 2 / k = \ln 2 / 0.046\ h = 15\ h$

30.6 Artificial nuclear reactions

30.45

$$_{94}^{239}\text{Pu} + 2\,_{0}^{1}\text{n} \rightarrow \,_{94}^{241}\text{Pu}$$

$$_{94}^{241}\text{Pu} \rightarrow \,_{95}^{241}\text{Am} + \,_{-1}^{0}\beta$$

30.47

$$_{92}^{238}\text{U} + \,_{6}^{12}\text{C} \rightarrow \,_{98}^{246}\text{Cf} + 4\,_{0}^{1}\text{n}$$

It must be carbon-12.

30.49

(a) $_{48}^{114}\text{Cd} + \,_{1}^{2}\text{H} \longrightarrow \,_{48}^{115}\text{Cd} + \,_{1}^{1}\text{H}$

(b) $_{3}^{6}\text{Li} + \,_{1}^{2}\text{H} \longrightarrow \,_{4}^{7}\text{Be} + \,_{0}^{1}\text{n}$

(c) $_{20}^{40}\text{Ca} + \,_{1}^{2}\text{H} \longrightarrow \,_{19}^{38}\text{K} + \,_{2}^{4}\text{He}$

(d) $_{29}^{63}\text{Cu} + \,_{1}^{2}\text{H} \longrightarrow \,_{30}^{65}\text{Zn} + \gamma$

30.51

$$_{5}^{10}\text{B} + \,_{0}^{1}\text{n} \longrightarrow \,_{3}^{7}\text{Li} + \,_{2}^{4}\text{He}$$

30.7 Nuclear fission

30.53

$$_{3}^{6}\text{Li} + \,_{0}^{1}\text{n} \rightarrow \,_{1}^{3}\text{H} + \,_{2}^{4}\text{He}$$

SUMMARY AND CONCEPTUAL QUESTIONS

30.55

Carbon-14 levels in the atmosphere are in steady state—determined by the flux of cosmic rays that produce it in the upper atmosphere. Plants take up carbon-14 (via carbon dioxide) in proportion to its concentration in the atmosphere. When they die, no more carbon-14 is absorbed. The relative amount of carbon-14 (to carbon-12) decays exponentially with the known half-life of carbon-14. Measuring this isotope ratio allows the age of the sample to be determined—via the logarithm of the isotope ratio divided by its initial value (the atmospheric level). The method, as described here, assumes the atmospheric level has remained constant. This level is known to have varied by up to 10% in the past. The method can be corrected for this variation, however. Nevertheless, the method is limited to objects between 100 and 40,000 years old. There has not been enough carbon-14 decay in objects less than 100 years old, while the level of carbon-14 is too small to give an accurate age measurement beyond 40,000 years.

30.57
Radiation can cause transmutation of elements within cells, or can simply ionize or otherwise disrupt cellular molecules. The resulting chemical reactions can significantly affect cellular function and even damage DNA molecules. This can lead to cell death, if the cell is unable to repair the damage at the rate it occurs. Thus, exposure of humans to radiation can cause radiation sickness and even death, at sufficiently exposure levels. The killing of tissue by radiation can be used to treat diseases such as cancer, however. By focusing radiation on a tumour, we can kill the cancer tissue and possibly cure the patient.

30.59
The amount of ^{87}Rb is decreased by the factor,
$$0.951 = (1/2)^n$$
$n = \ln(0.951) / \ln(1/2) = 0.072$ half-lives
This corresponds to $0.072 \times 4.8 \times 10^{10}$ years $= 3.5 \times 10^9$ years.

30.61
The amount of ^{235}U decreases by the factor
$$0.72 / 3.0 = 0.24 = (1/2)^n$$
$n = \ln(0.75) / \ln(1/2) = 2.06$ half-lives
This corresponds to $2.06 \times 7.04 \times 10^8$ years $= 1.45 \times 10^9$ years.

30.63
Amount of ^{235}U $= 1000$ g $/ 235.0439$ g mol^{-1} $= 4.255$ mol
Amount of energy released by fission of uranium 235 $= 4.255$ mol $\times 2.1 \times 10^{10}$ kJ mol^{-1} $= 8.9 \times 10^{10}$ kJ
To obtain the same amount of energy from coal requires
$$8.9 \times 10^{10} \text{ kJ} / 2.9 \times 10^7 \text{ kJ t}^{-1} = 3100 \text{ t of coal}$$

30.65
The time between the sample injection and the taking of the blood sample is negligible compared to the half-life of tritium. This is a sample dilution problem. A 1.0 mL sample was injected, and a 1.0 mL blood sample was taken. The activities measured are proportional to the sample concentrations. The ratio of the activities equals the ratio of the blood volume to 1.0 mL. The blood volume is just
$$2.0 \times 10^6 \text{ Bq} / 1.5 \times 10^4 \text{ Bq} \times 1.0 \text{ mL} = 130 \text{ mL}$$

30.67
We could label the methanol with the radioactive isotope, ^{15}O, carry out the reaction, and then sample the water and look for ^{15}O that could only have come from methanol. The same experiment could be carried out with labelled acetic acid to see if acid ^{15}O ends up in the water product.

30.69
The number of atoms in 1.0 mg of ^{238}U is

6.022×10^{23} mol $\times 1.0 \times 10^{-3}$ g / 238.050782 g mol^{-1} = 2.530×10^{18}

From $\Delta N / \Delta t = -kN$

we get (note that 1 Bq = 1 s^{-1})

$k = (\Delta N / \Delta t) / N$

$= 12$ s^{-1} / 2.530×10^{18} = 4.74×10^{-18} s^{-1}

$= 4.74 \times 10^{-18}$ s$^{-1} \times (60 \times 60 \times 24 \times 365)$ s y^{-1} = 1.50×10^{-10} y^{-1}

$t_{1/2} = \ln(2) / k = 4.6 \times 10^{9}$ y

This is close to the literature value, and consistent with the uncertainty expected with a measurement of only 12 events.